GIS Fundamentals:

A First Text on Geographic Information Systems

4th Edition

Paul Bolstad
University of Minnesota - St. Paul

Front cover: A view of the Iberian Peninsula and northern Africa from space, developed by Tom Patterson of the U.S. National Park Service, Harper's Ferry National Monument. As ably described by Mr. Patterson at his website www.shadedrelief.com, this map uses a combination of techniques to convey elevation, vegetation, clouds, and the heavens, using a variety of data. Data and images for download, a description of techniques, and links to techniques may be found at his website.

Errata and other helpful information about this book may be found at http://www.paulbolstad.net/gisbook.html

GIS Fundamentals: A first text on geographic information systems, 4th edition.
copyright (c) 2012 by Paul Bolstad

Second printing August, 2015

XanEdu
Change the course.

530 Great Road
Acton, MA 01720
800-562-2147

Instructor resources available at www.paulbolstad.net, or at www.paulbolstad.org

ISBN 978-0-9717647-3-6

Acknowledgements

I must thank many people for their contributions to this book. My aim in the first three editions was to provide a readable, thorough, and affordable introductory text on GIS. Although gratified by the adoption of the book at more than 350 colleges and universities, there was ample room for improvement, and the theory and practice of GIS has evolved. Students and faculty offered corrections, helpful suggestions, and enthusiastic encouragement, and for these I give thanks. They have led to substantial improvements in this fourth edition, improvements that include new text, modifications of most sections in most chapters, improved clarity in a few of the more difficult passages, and more worked examples and homework problems with answers.

Many friends and colleagues deserve specific mention. Tom Lillesand pointed me on this path and inspired by word and deed. Lynn Usery helped clarify my thinking on many concepts in geography, Paul Wolf on surveying and mapping, and Harold Burkhart was a great a mentor as one could hope to encounter early on in a career. Several colleagues and students caught both glaring and subtle errors. David Doyle provided great help on datums. Andy Jenks, Lynn Usery, Esther Brown, Margaret Bolstad, and Ryan Kirk spent uncounted hours reviewing draft manuscripts, honing the form and content of this book. Colleagues too numerous to mention have graciously shared their work, as have a number of businesses and public organizations.

Finally, this project would not have been possible save for the encouragement and forbearance of Holly, Sam, and Sheryl, and the support of Margaret.

While many helped in the 4th edition of this book and I've read it too many times to count, I'm sure I've left many opportunities for improvement. If you have comments to share or improvements to suggest, please send them to Eider Press, 2303 4th Street, White Bear Lake, MN 55110, or pbolstad@gmail.com.

Paul Bolstad

Companion Resources

There are a number of resources available to help instructors use this book. These may be found at the website:

www.paulbolstad.org/gisbook.html

Perhaps the most useful are the book figures, made available in presentation-friendly formats. Most of the figures used in the book are organized by chapter, and may be downloaded and easily incorporated into common slide presentation packages. A few graphics are not present because I do not hold copyright, or could not obtain permission to distribute them.

Sample chapters are available for download, although figure detail has been downsampled to reduce file sizes. These are helpful for those considering adoption of the textbook, or when a bookstore doesn't order enough copies. Copies sometimes fall short, and early chapters may help during the first few weeks of class.

Lecture and laboratory materials are available for the introductory GIS course I teach at the University of Minnesota, via a link to a campus website. Lectures are available in presentation and note format, as well as laboratory exercises, homeworks, and past exams.

A updated list of errors is also provided, with corrections. Errors are listed by printing, since errors are corrected at each subsequent print run. Errors that change meaning are noted, as these are perhaps more serious than the distracting errors in grammar, spelling, or punctuation.

Chapter 1: An Introduction to GIS.......................1

Introduction 1
What is a GIS? 1
    GIS: A Ubiquitous Tool 3
    Why Do We Need GIS? 3
    GIS in Action 8
    Geographic Information Science 11
GIS Components 13
    Hardware for GIS 13
    GIS Software 13
    ArcGIS 14
    GeoMedia 15
    MapInfo 15
    Idrisi 16
    Manifold 16
    AUTOCAD MAP 16
    QGIS 17
    GRASS 17
    MicroImages 17
    ERDAS 17
    Microstation 1..8
    Smallworld 18
GIS in Organizations 19
    Summary 20
    The Structure of This Book 20

Chapter 2: Data Models.................................1

Introduction 25
    Coordinate Data 27
    Attribute Data and Types 31
Common Spatial Data Models 33
Vector Data Models 34
    Polygon Inclusions and Boundary Generalization 37
    Vector Topology 38
    Vector Features and Attribute Tables 42

Raster Data Models **44**
   Models and Cells 44
   Raster Features and Attribute Tables 48
   A Comparison of Raster and Vector Data Models 48
   Conversion Between Raster and Vector Models 51
Triangulated Irregular Networks **53**
   Multiple Models 54
   Object Data Models 55
Data and File Structures **58**
   Binary and ASCII Numbers 58
   Pointers and Indexes 59
   Data Compression 61
   Raster Pyramids 62
   Common File Formats 63
   Summary 63

Chapter 3: Geodesy, Datums, Map Projections, and Coordinate Systems 71

Introduction **71**
   Early Measurements 73
   Specifying the Ellipsoid 75
   The Geoid 78
   Geographic Coordinates, Latitude, and Longitude 81
   Horizontal Datums 83
   Datum Adjustment 85
   Commonly Used Datums 88
   Datum Transformations 91
   Vertical Datums 96
   Dynamic Heights 98
   Control Accuracy Specification 100
Map Projections and Coordinate Systems **101**
   Common Map Projections in GIS 107
   The State Plane Coordinate System 110
   UTM Coordinate System 113
   Continental and Global Projections 115
   Conversion Among Coordinate Systems 118
   The Public Land Survey System 120
   Summary 123

Chapter 4: Maps, Data Entry, Editing, and Output 131

Building a GIS Database **131**
   Introduction 131
   Map Types 134
   Map Scale 135
   Map Generalization 137
   Map Boundaries and Spatial Data 138
Digitizing: Coordinate Capture **140**
   On-screen Digitizing 140
   Hardcopy Map Digitization 142
   Characteristics of Manual Digitizing 143
   The Digitizing Process 144
   Digitizing Errors, Node and Line Snapping 145
   Reshaping: Line Smoothing and Thinning 147
   Scan Digitizing 149
   Editing Geographic Data 150
   Features Common to Several Layers 152
Coordinate Transformation **153**
   Control Points 153
   The Affine Transformation 154
   Other Coordinate Transformations 156
   A Caution When Evaluating Transformations 156
   Control Point Sources: Surveying 159
   Control Points from Maps and Digital Data 160
   GNSS Control Points 161
   Raster Geometry and Resampling 162
   Map Projection vs. Transformation 163
Output: Maps, Digital Data, and Metadata **164**
   Cartography and Map Design 164
   Digital Data Output 171
   Metadata: Data Documentation 171
   Summary 174

Chapter 5: Global Navigation Satellite Systems and Coordinate Surveying 183

Introduction **183**
  GNSS Basics 184
  GNSS Broadcast Signals 186
  Range Distances 188
  Positional Uncertainty 190
  Sources of Range Error 190
  Satellite Geometry and Dilution of Precision 191
Differential Correction **193**
  Real-Time Differential Positioning 197
  WAAS and Satellite-based Corrections 199
  RTK and Virtual Reference Stations 199
  A Caution on Datums 199
Optical and Laser Coordinate Surveying **201**
GNSS Applications **206**
  Field Digitization 206
  Field Digitizing Accuracy and Efficiency 209
  Rangefinder Integration 213
  GNSS Tracking 213
  Summary 216

Chapter 6: Aerial and Satellite Images 223

Introduction **223**
Basic Principles **225**
Aerial Images **228**
  Camera Aircraft, Formats and Systems 228
  Digital Aerial Cameras 230
  Film and Film Cameras 232
  Geometric Quality of Aerial Images 234
  Terrain and Tilt Distortion in Aerial Images 236
  Media, Lens, and Camera Distortion 239
  Stereo Photographic Coverage 241
  Geometric Correction of Aerial Images 243
  Photointerpretation 247

Satellite Images **250**
    Basic Principles of Satellite Image Scanners 250
    Sub-meter Satellite Systems 252
    SPOT 254
    Landsat 255
    Resourcesat-1 257
    MODIS and VEGETATION 257
    MERIS 258
    LiDAR 260
    Other Systems 262
    Satellite Images in GIS 262
    Aerial or Satellite Images in GIS: Which to Use? 262
    Image Sources 264
    Summary 264

Chapter 7: Digital Data **271**

Introduction **271**
    Map Services vs. Locally Storable Data 272
National and Global Digital Data **273**
Digital Data for the United States **276**
    National Spatial Data Infrastructure 276
    The National Atlas 276
    The U.S. National Map 276
    Digital Elevation Models 278
    Hydrologic Data 282
    Digital Images 287
    Legacy Aerial Photographs: SCS, DOQ, NHAP,
       and NAPP Photos 288
    NAIP Digital Images 289
    National Land Cover Data 290
    NASS CDL 293
    National Wetlands Inventory 294
    Digital Soils Data 295
    Digital Floodplain Data 298
    Climate, Geology, and Other Environmental Data 300
    Digital Census Data 301
    Summary 303

Chapter 8: Attribute Data and Tables 307

Introduction **307**
   Database Components and Characteristics 310
   Physical, Logical, and Conceptual Structures 313
   Relational Databases 313
   Primary Operators 316
   Hybrid Database Designs in GIS 319
Selection Based on Attributes **321**
   The Restrict Operator: Table Queries 321
Joining (or Relating) Tables **326**
   Concatenated Keys 330
   Multi-table Joins 331
Normal Forms in Relational Databases **332**
   Keys and Functional Dependencies 332
   The First and Second Normal Forms 335
   The Third Normal Form 338
   Trends in Spatial DBMS 339
   Summary 339

Chapter 9: Basic Spatial Analysis....................347

Introduction **347**
   Input, Operations, and Output 348
   Scope 349
Selection and Classification **352**
   Set Algebra 352
   Boolean Algebra 354
   Spatial Selection Operations 356
   Classification 359
   The Modifiable Areal Unit Problem 366
Dissolve **368**
Proximity Functions and Buffering **370**
   Buffers 371
   Raster Buffers 372
   Vector Buffers 372
Overlay 377
   Raster Overlay 378
   Vector Overlay 379

Clip, Intersect, and Union:...384
A Problem in Vector Overlay 385
An Example Spatial Analysis 388
Network Analysis **390**
Geocoding 395
Summary 397

Chapter 10: Topics in Raster Analysis 407

Introduction **407**
Map Algebra **408**
Local Functions 412..
Mathematical Functions 412 ....
Logical Operations 413
Reclassification 415
Nested Functions 4 17
Overlay 418
Neighborhood, Zonal, and Global Functions **424**
Zonal Functions 433
Cost Surfaces 434
Summary 437

Chapter 11: Terrain Analysis 443

Introduction **443**
Slope and Aspect 445
Hydrologic Functions 452
Viewsheds 458
Profile Plots 460
Contour Lines 460
Shaded Relief Maps 463
Terrain Analysis Software 464
Summary 465

Spatial Estimation: Interpolation, Prediction, and Core Area Delineation 473

Introduction **473**
Sampling **475**
    Sampling Patterns 475
Spatial Interpolation Methods **478**
    Nearest Neighbor Interpolation 480
    Fixed Radius – Local Averaging 481
    Inverse Distance Weighted Interpolation 483
    Splines 486
Spatial Prediction **487**
    Spatial Regression 491
    Trend Surface and Simple Spatial Regression 491
    Kriging and Co-Kriging 493
    Interpolation Accuracy 497
Core Area Mapping **499**
    Mean Center and Mean Circle 499
    Convex Hulls 500
    Characteristic Hull Polygons 502
    Kernel Mapping 503
    Time-Geographic Density Estimation 509
    Summary 513

Spatial Models and Modelling 521

Introduction **521**
Cartographic Modeling **525**
    Designing a Cartographic Model 527
    Weightings and Rankings 527
    Rankings Within Criteria 528
    Weighting Among Criteria 530
    Cartographic Models: A Detailed Example 533
    Simple Spatial Models 542
    Spatio-temporal Models 545
    Cell-Based Models 547
    Agent-based Modeling 548
    Example 1: Process-based Hydrologic Models 549

Example 2: LANDIS, a Stochastic Model of Forest Change 552
LANDIS Design Elements 553
Summary 556

Data Standards and Quality 561
Introduction 561
The Geospatial Competency Model 563
Spatial Data Standards 564
Data Accuracy 565
Documenting Spatial Data Accuracy 566
Positional Accuracy 568
A Standard Method for Measuring
    Positional Accuracy 571
Accuracy Calculations 573
Errors in Linear or Area Features 576
Attribute Accuracy 577
Error Propagation in Spatial Analysis 579
Summary 580

New Developments in GIS 585
Introduction 585
GNSS 586
Fixed and Mobile Three-Dimensional Mapping 588
Datum Modernization 591
National Adjustment of 2011 (NA2011) 592
Improved Remote Sensing 593
Cloud-Based GIS 597
Open GIS 598
Open Standards for GIS 598
Open Source GIS 599
A Hybrid Model 599
Summary 600

Appendix A: Glossary **603**
Appendix B: Sources **617**
Appendix C: Useful Relationships................................**623**
    Length 623
    Area 623
    Angles 623
    Scale 623
    State Plane Zones 624
    Trigonometric Relationships 625

Appendix D: Answers to Selected Questions **635**
    Chapter 1 635
    Chapter 2 636
    Chapter 3 638
    Chapter 4 640
    Chapter 5 643
    Chapter 6 645
    Chapter 7 647
    Chapter 8 648
    Chapter 9 650
    Chapter 10 65
    Chapter 11 656
    Chapter 12 661
    Chapter 13 663
    Chapter 14 664

Index **665**

1 An Introduction to GIS

Introduction

Geography has always been important to humans. Stone-age hunters anticipated the location of their quarry, early explorers lived or died by their knowledge of geography, and current societies work and play based on their understanding of who belongs where. Applied geography, in the form of maps and spatial information, has served discovery, planning, cooperation, and conflict for at least the past 3000 years (Figure 1-1). Maps are among the most beautiful and useful documents of human civilization.

Most often our geographic knowledge is applied to routine tasks, such as puzzling a route in an unfamiliar town or searching for the nearest metro station. Spatial information has a greater impact on our lives than we realize, by helping us produce the food we eat, the energy we burn, the clothes we wear, and the diversions we enjoy.

Because spatial information is so important, we have developed tools called geographic information systems (GIS) to help us with our geographic knowledge. A GIS helps us gather and use spatial data (we will use the abbreviation GIS to refer to both singular, system, and plural, systems). Some GIS components are purely technological; these include space-age data collectors, advanced communications networks, and sophisticated computing. Other GIS components are very simple, for example, a pencil and paper used to field-verify a map.

As with many aspects of life in the last five decades, how we gather and use spatial data has been profoundly altered by modern electronics, and GIS software and hardware are primary examples of these technological developments. The capture and treatment of spatial data has accelerated over the past three decades, and continues to evolve.

Key to all definitions of a GIS are "where" and "what." GIS and spatial analyses are concerned with the absolute and relative location of features (the "where"), as well as the properties and attributes of those features (the "what"). The locations of important spatial objects such as rivers and streams may be recorded, and also their size, flow rate, or water quality. Indeed, these attributes often depend on the spatial arrangement of "important" features, such as land use above or adjacent to streams. A GIS aids in the analysis and display of these spatial relationships.

What is a GIS?

A GIS is a tool for making and using spatial information. Among the many definitions of GIS, we choose:

A GIS is a computer-based system to aid in the collection, maintenance, storage, analysis, output, and distribution of spatial data and information.

When used wisely, GIS can help us live healthier, wealthier, and safer lives.

Figure 1-1: A map of southern South America, published 1562 by Diego Gutierrez. The Straits of Magellan are near the lower center, and the Parana River near top-center. Early maps were key to the European exploration of new worlds.

GIS and spatial analyses are concerned with the quantitative location of important features, as well as properties and attributes of those features. Mount Everest is in Asia, Pierre is in South Dakota, and the cruise ship Titanic is at the bottom of the Atlantic Ocean. A GIS quantifies these locations by recording their *coordinates*, numbers that describe the position of these features on Earth. The GIS may also be used to record the height of Mount Everest, the population of Pierre, or the depth of the Titanic, as well as any other defining characteristics of each spatial feature.

Each GIS user may decide what features are important, and what is important about them. For example, forests are important to us. They protect our water supplies, yield wood, harbor wildlife, and provide space to recreate (Figure 1-2). We are concerned about the level of harvest, the adjacent land use, pollution from nearby industries, or when and where forests burn. Informed management of our forests requires at a minimum knowledge of all these related factors, and, perhaps above all, the spatial arrangement of these factors. Buffer strips near rivers may protect water supplies, clearings may prevent the spread of fire, and polluters downwind may not harm our forests while polluters upwind might. A GIS aids immensely in analyzing these spatial relationships and interactions. A GIS is also particularly useful at displaying spatial data and reporting the results of spatial analysis. In many instances GIS is the only way to solve spatially-related problems.

GIS: A Ubiquitous Tool

GIS are essential tools in business, government, education, and non-profit organizations, and GIS use has become mandatory in many settings. GIS have been used to fight crime, protect endangered species, reduce pollution, cope with natural disasters, treat the AIDS epidemic, and to improve public health; in short, GIS have been instrumental in addressing some of our most pressing societal problems.

GIS tools in aggregate save billions of dollars annually in the delivery of governmental and commercial goods and services. GIS regularly help in the day-to-day management of many natural and man-made resources, including sewer, water, power, and transportation networks. GIS are at the heart of one of the most important processes in U.S. democracy, the constitutionally mandated reshaping of U.S. congressional districts, and hence the distribution of tax dollars and other government resources.

Why Do We Need GIS?

GIS are needed in part because human populations and consumption have reached levels such that many resources, including air and land, are placing substantial limits on human action (Figure 1-3). Human populations have doubled in the last 50 years, surpassing 6 billion, and we will likely add

Figure 1-2: GIS allow us to analyze the relative spatial location of important geographic features. The satellite image at the center shows a forested area in western Oregon, USA, with a patchwork of lakes (dark area, upper left and mid-right), forest and clearings (middle), and snow-covered mountains (right). Spatial analyses in a GIS may aid in ensuring sustainable recreation, timber harvest, environmental protection, and other benefits from this and other globally important regions (courtesy NASA).

another 4 billion humans in the next 50 years. The first 100,000 years of human existence caused scant impacts on the world's resources, but in the past 300 years humans have permanently altered most of the Earth's surface. The atmosphere and oceans exhibit a decreasing ability to benignly absorb carbon dioxide and nitrogen, two primary waste products of humanity. Silt chokes many rivers and there are abundant examples of smoke, ozone, or other noxious pollutants substantially harming public health (Figure 1-4). By the end of the 20th century most lands south of the boreal region has been farmed, grazed, cut, built over, drained, flooded, or otherwise altered by humans (Figure 1-5).

GIS help us identify and address environmental problems by providing crucial information on where problems occur and who are affected by them. GIS help us identify the source, location, and extent of adverse environmental impacts, and may

help us devise practical plans for monitoring, managing, and mitigating environmental damage.

Human impacts on the environment have spurred a strong societal push for the adoption of GIS. Conflicts in resource use, concerns about pollution, and precautions to protect public health have led to legislative mandates that explicitly or implicitly require the consideration of geography. The U.S. Endangered Species Act of 1973 (ESA) is an example of the importance of geography in resource management. The ESA requires adequate protection of rare and threatened organisms. Effective protection entails mapping the available habitat and analyzing species range and migration patterns. The location of viable remnant plant and animal populations relative to current and future human land uses must be analyzed, and action taken to ensure species survival. GIS have proven to be useful tools in all of these tasks. GIS is mandated in other endeavors,

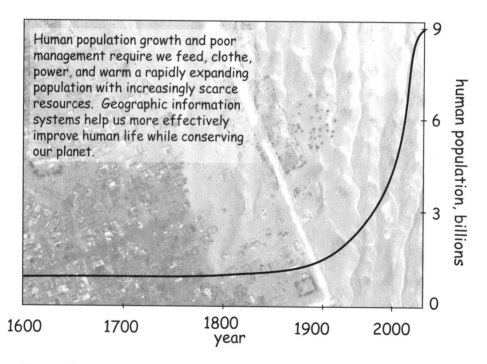

Figure 1-3: Human population growth during the past 400 years has increased the need for efficient resource use (courtesy United Nations and Ikonos).

Figure 1-4: The long-dormant Chaiten volcano, in south-central Chile, erupted in May, 2008. Advanced satellite imaging allows us to track the eruptions and plumes, new space-based surveying aids in planning evacuation and mapping damage, and repeated observation allows us to overlay observations, measure impacts, and plan for recovery (courtesy NASA).

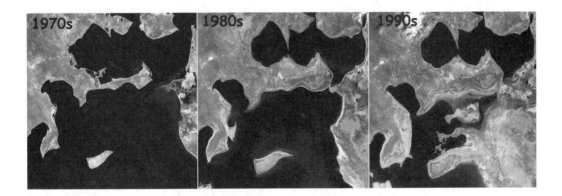

Figure 1-5: The environmental impacts wrought by humans have accelerated in many parts of the world during the past century. These satellite images from the 1970s (left) to 1990s (right) show a shrunken Aral Sea due to the overuse of water. Diversion for irrigation has destroyed a rich fishery, the economic base for many seaside villages. GIS may be used to document change, mitigate damage, and to effectively manage our natural resources (courtesy NASA).

including emergency services, flood protection, disaster assessment and management (Figure 1-6), and infrastructure development.

Public organizations have adopted GIS because of legislative mandates, and because GIS aid in governmental functions. For example, emergency service vehicles are regularly dispatched and routed using GIS. Callers to emergency response dispatchers are automatically identified by telephone number, and their address recalled. The GIS software matches this address to the nearest fire, police, or ambulance station. A map or route description is immediately generated by the software, based on information on location and the street network, and sent to the appropriate station with a dispatch alarm.

Many businesses have adopted GIS because they provide increased efficiency in the delivery of goods and services. Retail businesses locate stores based on a number of spatially-related factors. Where are the

potential customers? What is the spatial distribution of competing businesses? Where are potential new store locations? What is traffic flow near current stores, and how easy is it to park near and access these stores? Spatial analyses are used every day to answer these questions. GIS are also used in hundreds of other business applications, such as to route delivery vehicles, guide advertising, design buildings, plan construction, and sell real estate.

The societal push to adopt GIS has been complemented by a technological pull in the development and application of GIS. Thousands of lives and untold wealth have been lost because ship captains could not answer the simple question, "Where am I?" Robust nautical navigation methods emerged in the 18th century, and have continually improved to today, when anyone can quickly locate their outdoor position to within a few meters. A remarkable positioning technology, generically known as Global Naviga-

Figure 1-6: GIS may aid in disaster assessment and recovery. These satellite images from Banda Aceh, Indonesia, illustrate tsunami-caused damage to a shoreline community. Emergency response and longer-term rebuilding efforts may be improved by spatial data collection and analysis (courtesy DigitalGlobe).

Figure 1-7: Portable computing is one example of the technological pull driving GIS adoption. These hand-held devices substantially improve the speed and accuracy of spatial data collection (courtesy Cogent3D, www.GISRoam.com.).

tion Satellite Systems (GNSS), is now incorporated in cars, planes, boats, and trucks. GNSS are indispensable tools in commerce, planning, and safety.

The technological pull has developed on several fronts. Spatial analysis in particular

has been helped by faster computers with more storage. Most real-world spatial problems were beyond the scope of all but the largest government and business organizations until the 1990s. GIS computing expenses are becoming an afterthought, as computing resources often cost less than a few weeks salary for a qualified GIS professional. Costs decrease and performance increases at dizzying rates, with predicted plateaus pushed back each year. Powerful field computers are lighter, faster, more capable, and less expensive each year, so spatial data display and analysis capabilities may always be at hand (Figure 1-7). GIS on rugged, field-portable computers has been particularly useful in field data entry and editing.

In addition to the computing improvements and the development of GNSS, current "cameras" deliver amazingly detailed aerial and satellite images. Initially, advances in image collection and interpretation were spurred by World War II and then the Cold War because accurate maps were required, but unavailable. Turned toward peacetime endeavors, imaging technologies now help us map food and fodder, houses and highways, and most other natural and human-built objects. Images may be rapidly converted to accurate spatial information over broad areas (Figure 1-8). A broad range

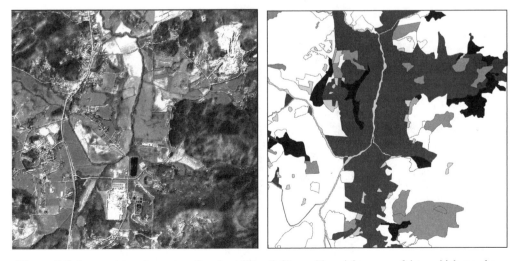

Figure 1-8: Images taken from aircraft and satellites (left) provide a rich source of data, which may be interpreted and converted to information about the Earth surface (right).

of techniques have been developed for extracting information from image data, and ensuring this information faithfully represents the location, shape, and characteristics of features on the ground. Visible light, laser, thermal, and radar scanners are currently being developed to further increase the speed and accuracy with which we map our world. Thus, advances in these three key technologies -- imaging, GNSS, and computing -- have substantially aided the development of GIS.

GIS in Action

Spatial data organization, analyses, and delivery are widely applied to improve life. Here we describe three examples that demonstrate how GIS are in use.

Marvin Matsumota is alive today because of GIS. The 60 year-old hiker became lost in Joshua Tree National Park, a 300,000 hectare desert landscape famous for its distinct and rugged terrain. Between six and eight hikers become lost there in a typical year, sometimes fatally so. Because of the danger of hypothermia, dehydration, and death, the U.S. National Park Service (NPS) organizes search and rescue operations that include foot patrols, horseback, vehicle, and helicopter searches (Figure 1-9).

The search and rescue operation for Mr. Matsumota was organized and guided using GIS. Search and rescue teams carried field positioning devices that recorded team location and progress. Position data were downloaded from the field devices to a field GIS center, and frequently updated maps were produced. On-site incident managers used these maps to evaluate areas that had been searched, and to plan subsequent efforts in real time. Accurate maps showed exactly what portions of the park had been searched and by what method. Appropriate teams

Figure 1-9: Search and rescue operations, such as the one for Marvin Matsumoto (upper left, inset) are spatial activities. Searchers must combine information on where the lost person was last seen, likely routes of travel, maps of the areas already searched, time last searched, and available resources to effectively mount a search campaign (courtesy Tom Patterson, US-NPS).

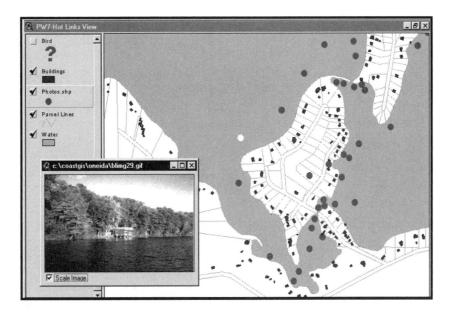

Figure 1-10: Parcel information entered in a GIS may substantially improve government services. Here, images of the shoreline taken from lake vantage points are combined with digital
maps of the shoreline, buildings, and parcel boundaries. The image in the lower left was obtained
from the location shown as a light dot near the center of the figure (courtesy Wisconsin Sea Grant
and LICGF).

were tasked to unvisited areas. Ground
crews could be assigned to areas that had
been searched by helicopters, but contained
vegetation or terrain that limited visibility
from above. Marvin was found on the fifth
day, alive but dehydrated and with an injured
skull and back from a fall. The search team
was able to radio its precise location to a rescue helicopter. Another day in the field and
Marvin likely would have died, a day saved
by the effective use of GIS. After a week in
the hospital and some months convalescing
at home, Marvin made a full recovery.

GIS are also widely used in planning
and environmental protection. Oneida
County is located in northern Wisconsin, a
forested area characterized by exceptional
scenic beauty. The County is in a region with
among the highest concentrations of fresh-
water lakes in the world, a region that is also
undergoing a rapid expansion in the permanent and seasonal human population. Retir-
ees, urban exiles, and vacationers are
increasingly drawn to the scenic and recre-

ational amenities available in Oneida
County. Permanent county population grew
by nearly 30% from 1990 to 2010, and the
seasonal influx almost doubles the total
county population each summer.

Population growth has caused a boom
in construction and threatened the lakes that
draw people to the County. A growing number of building permits are for near-shore
houses, hotels, or businesses. Seepage from
septic systems, runoff from fertilized lawns,
or erosion and sediment from construction
all decrease lake water quality. Increases in
lake nutrients or sediment may lead to turbid
waters, reducing the beauty and value of the
lakes and nearby properties.

In response to this problem, Oneida
County, the Sea Grant Institute of the University of Wisconsin, and the Land Information and Computer Graphics Facility of the
University of Wisconsin have developed a
Shoreland Management GIS Project. This
project helps protect valuable nearshore and
lake resources, and provides an example of

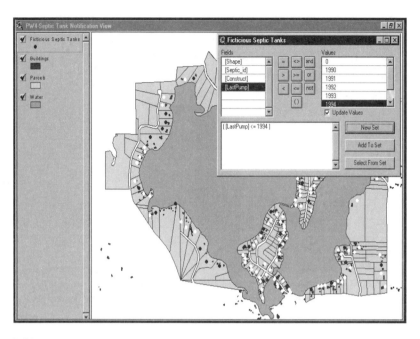

Figure 1-11: GIS may be used to streamline government function. Here, septic systems not compliant with pollution prevention ordinances are identified by white circles (courtesy Wisconsin Sea Grant Institute and LICGF).

how GIS tools are used for water resource management (Figure 1-10).

Oneida County has revised zoning and other ordinances to protect shoreline and lake quality and to ensure compliance without undue burden on landowners. The County uses GIS technology in the maintenance of property records. Property records include information on the owner, tax value, and any special zoning considerations. The county uses these digital records when creating parcel maps, processing sale, subdivision, or other parcel transactions, and when integrating new data such as aerial or boat-based images to help detect property changes and zoning violations.

GIS may also be used to administer shoreline zoning ordinances, or to notify landowners of routine tasks, such as septic system maintenance. Northern lakes are particularly susceptible to nutrient pollution from near-shore septic systems (Figure 1-11). Timely maintenance of each septic system must be verified. The GIS can automati-

cally identify owners out of compliance and generate an appropriate notification.

Our third example illustrates how GIS helps save endangered species. The black-footed ferret is a small carnivore of western North America, and is one of the most endangered mammals on the continent (Figure 1-12). The ferret lives in close association with prairie dogs, communally-living rodents once found over much of North America. Ferrets feed on prairie dogs and live in their burrows, and prairie dog colonies provide refuge from coyotes and other larger carnivores that prey on the ferret. The blackfooted ferret has become endangered because of declines in the range and number of prairie dog colonies, coupled with ferret sensitivity to canine distemper and other diseases.

The U.S. Fish and Wildlife Service (USFWS) has been charged with preventing the extinction of the blackfooted ferret. This entails establishing the number and location of surviving animals, identifying the habitat requirements for a sustainable population, and analyzing what factors are responsible for the decline in ferret numbers, so that a recovery plan may be devised.

Because blackfooted ferrets are nocturnal animals that spend much of their time underground, and because ferrets have always been rare, relatively little was known about their life history, habitat requirements, and the causes of mortality. For example, young ferrets often disperse from their natal prairie dog colonies in search of their own territories. Dispersal is good when it leads to an expansion of the species. However, there are limits on how far a ferret may successfully travel. If the nearest suitable colony is too far away, the dispersing young ferret may likely die of starvation or be eaten by a larger predator. The dispersing ferret may reach a prairie dog colony that is too small to support it. Ferret recovery has been ham-

pered because we don't know when prairie dog colonies are too far apart, or if a colony is too small to support a breeding pair of ferrets. Because of this lack of spatial knowledge, wildlife managers have had difficulty selecting among a number of activities to enhance ferret survival. These activities include the establishment of new prairie dog colonies, fencing colonies to prevent the entry of larger predators, removing predators, captive breeding, and the capture and transport of young or dispersing animals.

GIS have been used to provide data necessary to save the blackfooted ferret (Figure 1-12). Individual ferrets are tracked in nighttime spotlighting surveys, often in combination with radiotracking devices. Ferret location and movement are combined with detailed data on prairie dog colony boundaries, burrow locations, surrounding vegetation, and other spatial data (Figure 1-13). Individual ferrets can be identified and vital characteristics monitored, including home range size, typical distance travelled, number of offspring, and survival. These data are combined and analyzed in a GIS to improve the likelihood of species recovery.

Geographic Information Science

Although we have defined GIS as geographic information *systems*, there is another GIS: geographic information *science*, sometimes abbreviated as GISci. The distinction is important, because the future development of GIS depends on progress in GIScience.

GIScience is the theoretical foundation on which GIS are based. GIS research is typically concerned with technical aspects of GIS implementation or application. GIScience includes these technical aspects, but also seeks to redefine concepts in geography and geographic information in the context of the digital age. GIScience is concerned with how we conceptualize geography and how we collect, represent, store, visualize, analyze, use, and present these geographic concepts. The work draws from many fields, including traditional geography, geodesy, remote sensing, surveying, com-

Figure 1-12: Specialized equipment is used to collect spatial data. Here a burrow location is recorded using a GPS receiver, as an interested black footed ferret looks on (courtesy Randy Matchett, USFWS).

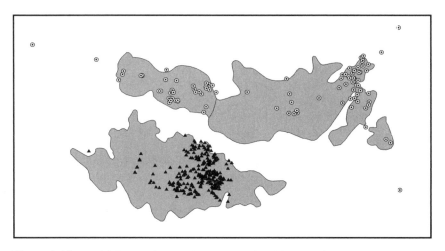

Figure 1-13: Spatial data, such as the boundaries of prairie dog colonies (gray polygons) and individual blackfooted ferret positions (triangle and circle symbols) may be combined to help understand how best to save the blackfooted ferret (courtesy Randy Matchett, USFWS).

puter science, cartography, mathematics, statistics, cognitive science, and linguistics. GIScience investigates not only technical questions of interest to applied geographers, business-people, planners, public safety officers, and others, but GIScience is also directed at more basic questions. How do we perceive space? How might we best represent spatial concepts, given the new array of possibilities provided by our advancing technologies? How does human psychology help or hinder effective spatial reasoning?

Science has been described as a hand-maiden of technology in the applied world. A more apt analogy is perhaps that science is a parent of technology. GIS, narrowly defined, is more about technology than science. But since GIS is the tool with which we solve problems, we are mistaken if we consider it as the starting and ending point in geographic reasoning. An understanding of GIScience is crucial to the further development of GIS, and in many cases, crucial to the effective application of GIS. This book focuses primarily on GIS, but provides relevant information related to GIScience as appropriate for an introductory course.

GIS Components

A GIS is comprised of hardware, software, data, humans, and a set of organizational protocols. These components must be well integrated for effective use of GIS, and the development and integration of these components is an iterative, ongoing process. The selection and purchase of hardware and software is often the easiest and quickest step in the development of a GIS. Data collection and organization, personnel development, and the establishment of protocols for GIS use are often more difficult and time-consuming endeavors.

Hardware for GIS

A fast computer, large data storage capacities, and a high-quality, large display form the hardware foundation of most GIS (Figure 1-14). A fast computer is required because spatial analyses are often applied over large areas and/or at high spatial resolutions. Calculations often have to be repeated over tens of millions of times, corresponding to each space we are analyzing in our geographical analysis. Even simple operations may take substantial time on general-purpose computers when run over large areas,

and complex operations can be unbearably long-running. While advances in computing technology during the 1990s have substantially reduced the time required for most spatial analyses, computation times are still unacceptably long for a few applications.

While most computers and other hardware used in GIS are general purpose and adaptable for a wide range of tasks, there are also specialized hardware components that are specifically designed for use with spatial data. While many non-GIS endeavors require the entry of large data volumes, GIS is unique in the volume of data that must be entered to define the shape and location of geographic features, such as roads, rivers, and parcels. Specialized equipment, described in Chapters 4 and 5, has been developed to aid in these data entry tasks.

GIS Software

GIS software provides the tools to manage, analyze, and effectively display and disseminate spatial information (Figure 1-15). GIS by necessity involves the collection and manipulation of coordinates. We also must collect qualitative or quantitative informa-

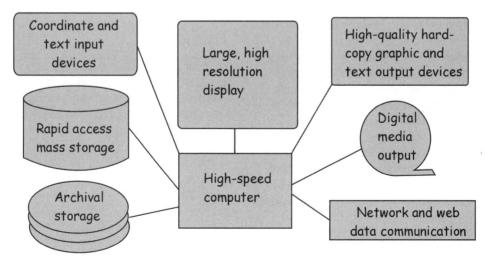

Figure 1-14: GIS are typically used with a number of general-purpose and specialized hardware components.

Data entry
- manual coordinate capture
- attribute capture
- digital coordinate capture
- data import

Editing
- manual point, line and area
  feature editing
- manual attribute editing
- automated error detection
  and editing

Data management
- copy, subset, merge data
- versioning
- data registration and projection
- summarization, data reduction
- documentation
- compression
- indexing

Analysis
- spatial query
- attribute query
- interpolation
- connectivity
- proximity and adjacency
- buffering
- terrain analyses
- boundary dissolve
- spatial data overlay
- moving window analyses
- map algebra

Output
- map design and layout
- hardcopy map printing
- digital graphic production
- export format generation
- metadata output
- digital map serving

Figure 1-15: Functions commonly provided by GIS software.

tion on the non-spatial attributes of geographic features. We need tools to view and edit these data, manipulate them to generate and extract the information we require, and produce the materials to communicate the information we have developed. GIS software provides the specific tools for some or all of these tasks.

There are many public domain and commercially available GIS software packages, and many of these packages originated at academic or government-funded research laboratories. The Environmental Systems Research Institute (ESRI) line of products, including ArcGIS, is a good example. Much of the foundation for early ESRI software was developed during the 1960s and 1970s at Harvard University in the Laboratory of Computer Graphics and Spatial Analysis. Alumni from Harvard carried these concepts with them to Redlands, California when forming ESRI, and included them in their commercial products.

Our description below, while including most of the major or widely used software packages, is not meant to be all-inclusive.

There are many additional software tools and packages available, particularly for specialized tasks or subject areas. Appendix B lists sources that may be helpful in identifying the range of software available, and for obtaining detailed descriptions of specific GIS software characteristics and capabilities.

ArcGIS

ArcGIS and its predecessors, ArcView and Arc/Info, are the most popular GIS software suites at the time of this writing. The Arc suite of software has a larger user base and higher annual unit sales than any other competing product. ESRI, the developer of ArcGIS, has a world-wide presence. ESRI has been producing GIS software since the early 1980s, and ArcGIS is its most recent and well-developed integrated GIS package. In addition to software, ESRI also provides substantial training, support, and fee-consultancy services at regional and international offices.

ArcGIS is designed to provide a large set of geoprocessing procedures, from data entry through analysis to most forms of data output. As such, ArcGIS is a large, complex, sophisticated product. It supports multiple data formats, many data types and structures, and literally thousands of possible operations that may be applied to spatial data. It is not surprising that substantial training is required to master the full capabilities of ArcGIS.

ArcGIS provides wide flexibility in how we conceptualize and model geographic features. Geographers and other GIS-related scientists have conceived of many ways to think about, structure, and store information about spatial objects. ArcGIS provides for the broadest available selection of these representations. For example, elevation data may be stored in at least four major formats, each with attendant advantages and disadvantages. There is equal flexibility in the methods for spatial data processing. This broad array of choices, while responsible for the large investment in time required for mastery of ArcGIS, provides concomitantly substantial analytical power.

GeoMedia

GeoMedia and the related MGE products are a popular GIS suite. GIS and related products have been developed and supported by Intergraph, Inc. of Huntsville, Alabama for over 30 years. GeoMedia offers a complete set of data entry, analysis, and output tools. A comprehensive set of editing tools may be purchased, including those for automated data entry and error detection, data development, data fusion, complex analyses, and sophisticated data display and map composition. Scripting languages are available, as are programming tools that allow specific features to be embedded in custom programs, and programing libraries to allow the modification of GeoMedia algorithms for special-purpose software.

GeoMedia is particularly adept at integrating data from divergent sources, formats,

and platforms. Intergraph appears to have dedicated substantial effort toward the OpenGIS initiative, a set of standards to facilitate cross-platform and cross-software data sharing. Data in any of the common commercial databases may be integrated with spatial data from many formats. Image, coordinate, and text data may be combined.

GeoMedia also provides a comprehensive set of tools for GIS analyses. Complex spatial analyses may be performed, including queries, for example, to find features in the database that match a set of conditions, and spatial analyses such as proximity or overlap between features. Worldwide web and mobile phone-based applications and application development are well supported.

MapInfo

MapInfo is a comprehensive set of GIS products developed and sold by the MapInfo Corporation, of Troy, New York. MapInfo products are used in a broad array of endeavors, although use seems to be concentrated in many business and municipal applications. This may be due to the ease with which MapInfo components are incorporated into other applications. Data analysis and display components are supported through a range of higher language functions, allowing them to be easily embedded in other programs. In addition, MapInfo provides a flexible, stand-alone GIS product that may be used to solve many spatial analysis problems.

Specific products have been designed for the integration of mapping into various classes of applications. For example, MapInfo products have been developed for embedding maps and spatial data into wireless handheld devices such as telephones, data loggers, or other portable devices. Products have been developed to support internet mapping applications, and serve spatial data in worldwide web-based environments. Extensions to specific database products such as Oracle are provided.

Idrisi

Idrisi is a GIS system developed by the Graduate School of Geography of Clark University, in Massachusetts. Idrisi differs from the previously discussed GIS software packages in that it provides both image processing and GIS functions. Image data are useful as a source of information in GIS. There are many specialized software packages designed specifically to focus on image data collection, manipulation, and output. Idrisi offers much of this functionality while also providing a large suite of spatial data analysis and display functions.

Idrisi has been developed and maintained at an educational and research institution, and was initially used primarily as a teaching and research tool. Idrisi has adopted a number of very simple data structures, a characteristic that makes the software easy to modify. Some of these structures, while slow and more space-demanding, are easy to understand and manipulate for the beginning programmer. The space and speed limitation have become less relevant with improved computers. File formats are well documented and data easy to access. The developers of Idrisi have expressly encouraged researchers, students, and users to create new functions for Idrisi. The Idrisi project has then incorporated user-developed enhancements into the software package. Idrisi is an ideal package for teaching students both to use GIS and to develop their own spatial analysis functions.

Idrisi is relatively low cost, perhaps because of its affiliation with an academic institution, and is therefore widely used in education. Low costs are an important factor in many developing countries, where Idrisi has also been widely adopted.

Manifold

Manifold is a relatively inexpensive GIS package with a surprising number of capabilities. Manifold combines GIS and some remote sensing capabilities. Basic spatial data entry and editing support are provided, as well as projections, basic vector and raster analysis, image display and editing, and output. The program is extensible through a series of software modules. Modules are available for surface analysis, business applications, internet map development and serving, database support, and advanced analyses.

Manifold GIS differs from other packages in providing sophisticated image editing capabilities in a spatially-referenced framework. Portions of images and maps may be cut and pasted into other maps while maintaining proper geographic alignment. Transparency, color-based selection, and other capabilities common to image editing programs are included in Manifold GIS.

AUTOCAD MAP

AUTOCAD is the world's largest-selling computer drafting and design package. Produced by Autodesk, Inc., of San Rafael, California, AUTOCAD began as an engineering drawing and printing tool. A broad range of engineering disciplines are supported, including surveying and civil engineering. Surveyors have traditionally developed and maintained the coordinates for property boundaries, and these are among the most important and often-used spatial data. AUTOCAD MAP adds substantial analytical capability to the already complete set of data input, coordinate manipulation, and data output tools provided by AUTOCAD.

The latest version, AUTOCAD MAP 3D, provides a substantial set of spatial data analysis capability. Data may be entered, verified, and output. Data may also be searched for features with particular conditions or characteristics. More sophisticated spatial analysis may be performed, including path finding or data combination. AUTOCAD MAP 3D incorporates many of the specialized analysis capabilities of other, older GIS packages, and is a good example of the convergence of GIS software from a number of disciplines.

QGIS

QGIS is an open-source software project, an initiative under the Open Source Geospatial Foundation. The software is a collaborative development by a community of developers and users. QGIS is free, stable, changes smoothly through time, with the source code available so that it can be extended as needed for specific tasks. It provides a graphical user interface, support of a wide variety of data types and formats, and runs on Unix, MacOSX, and Microsoft Windows operating systems. As with most open source software, the original offering had limited capabilities, but with an average of approximately two updates a year since 2002, QGIS provides a large number of basic GIS display and analysis functions. An interface has been developed with GRASS, another open source GIS that is analytically richer, but that lacks a point-and-click, graphical user interface.

GRASS

GRASS, the Geographic Resource Analysis Support System, is a free, open source GIS that runs on many platforms. The system was originally developed by the U.S. Army Construction Engineering Laboratory (CERL), starting in the early 1980s, when much GIS software was limited in access and applications. CERL followed an open approach to development and distribution, leading to substantial contributions by a number of university and other government labs. Development was discontinued in the military, and taken up by an open source "GRASS Development Team," a self-identified group of people donating their time to maintain and enhance GRASS. The software provides a broad array of raster and vector operations, and is used in both research and applications worldwide. Detailed information and the downloadable software are available at http://grass.itc.it/index.php.

MicroImages

MicroImages produces TNTmips, an integrated remote sensing, GIS, and CAD software package. MicroImages also produces and supports a range of other related products, including software to edit and view spatial data, software to create digital atlases, and software to publish and serve data on the internet.

TNTmips is notable both for its breadth of tools and the range of hardware platforms supported in a uniform manner. MicroImages recompiles a basic set of code for each platform so that the look, feel, and functionality is nearly identical irrespective of the hardware platform used. Image processing, spatial data analysis, and image, map, and data output are supported uniformly across this range.

TNTmips provides an impressive array of spatial data development and analysis tools. Common image processing tools are available, including ingest of a broad number of formats, image registration and mosaics, reprojection, error removal, subsetting, combination, and image classification. Vector analyses are supported, including support for point, line, and area features, multi-layer combination, viewshed, proximity, and network analyses. Extensive online documentation is available, and the software is supported by an international network of dealers.

ERDAS

ERDAS (Earth Resources Data Analysis System), now owned and developed by Leica Geosystems, began as an image processing system. The original purpose of the software was to enter and analyze satellite image data. ERDAS led a wave of commercial products for analyzing spatial data collected over large areas. Product development was spurred by the successful launch of the U.S. Landsat satellite in the 1970s. For the first time, digital images of the entire Earth surface were available to the public.

The ERDAS image processing software evolved to include other types of imagery, and to include a comprehensive set of tools for cell-based data analysis. Image data are supplied in a cell-based format. Cell-based analysis is a major focus of sections in three chapters of this book, so there will be much more discussion in later pages. For now, it is important to note that the "checkerboard" format used for image data may also be used to store and manipulate other spatial data. It is relatively easy and quite useful to develop cell-based spatial analysis tools to complement the image processing tools.

ERDAS and most other image processing packages provide data output formats that are compatible with most common GIS packages. Many image processing software systems are purchased explicitly to provide data for a GIS. The support of ESRI data formats is particularly thorough in ERDAS. ERDAS GIS components can be used to analyze these spatial data.

Microstation

Microstation, by Bentley Systems, is a software system for flexible, integrated infrastructure design and development. Although its origins are as a computer assisted drafting and design program, Microstation has evolved into a general set of tools, including field data collection, photogrammetry, sophisticated map composition, database management, analysis, and reporting.

Microstation products are particularly focused on infrastructure and development activities. Tools include a comprehensive suite for property records, including surveying parcel data management, terrain analysis and calculations for excavation and earthworks, rainfall runoff analysis and drainage design, street and utility layout, and 3D viewing of design alternatives. Microstation also supports more specialized spatial design and construction tools, including transportation and utility networks, mining, and power generation systems and networks.

Smallworld

Smallworld, currently owned by General Electric Energy, is a product suite focused primarily on power and other utility management, and other network systems. Primarily targeted for large organizations, the suite supports common spatial data formats, field data entry, complex, topological network models, integration with corporate databases, component and network design, and graphics and output in support of static and dynamic network characteristics.

This review of spatial data collection and analysis software is in no way exhaustive. There are many other software tools available, many of which provide unique, novel, or particularly clever combinations of geoprocessing functions. ILWIS, MapWindow, PCI, and ENVI are just a few of the available software packages with spatial data development or analysis capabilities. In addition, there are thousands of add-ons, special purpose tools, or specific modules that complement these products. Websites listed in Appendix B at the end of this book will provide more information on these and other GIS software products.

GIS in Organizations

Although new users often focus on GIS hardware and software components, we must recognize that GIS exist in an institutional context. Effective use of GIS requires an organization to support various GIS activities. Most GIS also require trained people to use them, and a set of protocols guiding how the GIS will be used. The institutional context determines what spatial data are important, how these data will be collected and used, and ensures that the results of GIS analyses are properly interpreted and applied. GIS share a common characteristic of many powerful technologies. If not properly used, the technology may lead to a significant waste of resources, and may do more harm than good. The proper institutional resources are required for GIS to provide all its potential benefits.

GIS are often employed as decision support tools (Figure 1-16). Data are collected, entered, and organized into a spatial database, and analyses performed to help make specific decisions. The results of spatial analyses in a GIS often uncover the need for more data, and there are often several iterations through the collection, organization, analysis, output, and assessment steps before a final decision is reached. It is important to recognize the organizational structure within which the GIS will operate, and how GIS

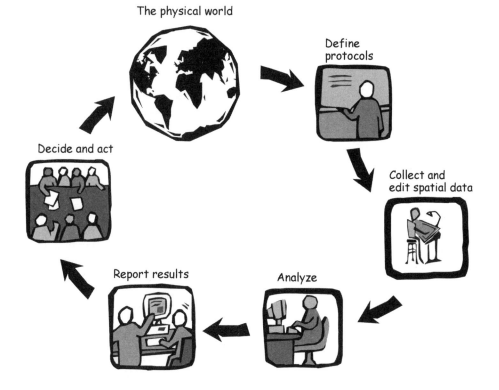

Figure 1-16: GIS exist in an institutional context. Effective use of GIS depends on a set of protocols and an integration into the data collection, analysis, decision, and action loop of an organization.

will be integrated into the decision-making processes of the organization.

One important question that must be answered early is "what problem(s) are we to solve with the GIS?" GIS add significant analytical power through the ability to measure distances and areas, identify vicinity, analyze networks, and through the overlay and combination of different information. Unfortunately, spatial data development is often expensive, and effective GIS use requires specialized knowledge or training, so there is often considerable expense in constructing and operating a GIS. Before spending this time and money there must be a clear identification of the new questions that may be answered, or the process, product, or service that will be improved, made more efficient, or less expensive through the use of GIS. Once the ends are identified, an organization may determine the level of investment in GIS that is warranted.

Summary

GIS are computer-based systems that aid in the development and use of spatial data. There are many reasons we use GIS, but most are based on a societal push, our need to more effectively and efficiently use our resources, and a technological pull, our interest in applying new tools to previously insoluble problems. GIS as a technology is based on geographic information science, and is supported by the disciplines of geography, surveying, engineering, space science, computer science, cartography, statistics, and a number of others.

GIS are comprised of both hardware and software components. Because of the large volumes of spatial data and the need to input coordinate values, GIS hardware often have large storage capacities, fast computing speed, and ability to capture coordinates. Software for GIS are unique in their ability to manipulate coordinates and associated attribute data. A number of software tools and packages are available to help us develop GIS.

While GIS are defined as tools for use with spatial data, we must stress the importance of the institutional context in which GIS fit. Because GIS are most often used as decision-support tools, the effective use of GIS requires more than the purchase of hardware and software. Trained personnel and protocols for use are required if GIS are to be properly applied. GIS may then be incorporated in the question-collect-analyze-decide loop when solving problems.

The Structure of This Book

This book is designed to serve a semester-long, 15-week course in GIS at the university level. We seek to provide the relevant information to create a strong basic foundation on which to build an understanding of GIS. Because of the breadth and number of topics covered, students may be helped by knowledge of how this book is organized. Chapter 1 (this chapter), sets the stage, providing some motivation and a background for GIS. Chapter 2 describes basic data representations. It treats the main ways we use computers to represent perceptions of geography, common data structures, and how these structures are organized. Chapter 3 provides a basic description of coordinates and coordinate systems, how coordinates are defined and measured on the surface of the Earth, and conventions for converting these measurements to coordinates we use in a GIS.

Chapters 4 through 7 treat spatial data collection and entry. Data collection is often a substantial task and comprises one of the main activities of most GIS organizations. General data collection methods and equipment are described in Chapter 4. Chapter 5 describes Global Navigation Satellite Systems (GNSS), a common technology for coordinate data collection. Chapter 6 describes aerial and space-based images as a source of spatial data. Most historical and contemporary maps depend in some way on image data, and this chapter provides a background on how these data are collected and used to create spatial data. Chapter 7 provides a brief description of common digital data sources available in the United States, their formats, and uses.

Chapters 8 through 13 treat the analysis of spatial data. Chapter 8 focuses on attribute data, attribute tables, database design, and analyses using attribute data. Attributes are half our spatial data, and a clear understanding of how we structure and use them is key to effective spatial reasoning. Chapters 9, 10, 11, and 12 describe basic spatial analyses, including adjacency, inclusion, overlay, and data combination for the main data models used in GIS. They also describe more complex spatio-temporal models. Chapter 13 describes various methods for spatial prediction and interpolation. We typically find it impractical or inefficient to collect "wall-to-wall" spatial and attribute data. Spatial prediction allows us to extend

our sampling and provide information for unsampled locations. Chapter 14 describes how we assess and document spatial data quality, while Chapter 15 provides some musings on current conditions and future trends.

We give preference to the International System of Units (SI) throughout this book. The SI system is adopted by most of the world, and is used to specify distances and locations in the most common global coordinate systems and by most spatial data collection devices. However, some English units are culturally embedded, for example, the survey foot, or 640 acres to a Public Land Survey Section, and so these are not converted. Because a large portion of the target audience for this book is in the United States, English units of measure often supplement SI units.

Suggested Reading

Amdahl, G. (2001). *Disaster Response: GIS for Public Safety*, ESRI Press: Redlands, California.

Burrough, P.A.& Frank, A.U. (1995). Concepts and paradigms in spatial information: Are current geographical information systems truly generic?, *International Journal of Geographical Information Systems*, 9:101-116.

Burrough, P. A. & McDonnell, R. A. (1992). *Principles of Geographical Information Systems*, Oxford University Press: New York.

Campbell, H. J. & Masser, I. (1992), GIS in local government: some findings from Great Britain, *International Journal of Geographical Information Systems*, 6:529-546.

Commission on Geoscience (1997). *Rediscovering Geography: New Relevance for Science and Society*, National Academy Press: Washington D.C.

Goodchild, M. F. (1992). Geographical information science, *International Journal of Geographical Information Systems*, 6:31-45.

Grimshaw, D. (2000). *Bringing Geographical Information Systems Into Business*, 2nd Edition. Wiley: New York.

Haklay, M. (2010). *Interacting with Geospatial Technologies*. Wiley: New York

Haining, R. (1990). *Spatial Data Analysis in the Social and Environmental Sciences*, Cambridge University Press: Cambridge.

Johnston, C. (1998). *Geographic Information Systems in Ecology*, Blackwell Scientific: Boston.

Kouyoumijian, V. (2011). *GIS in the Cloud: The New Age of Cloud Computing and Geographic Information Systems*. ESRI Press: Redlands.

MaGuire, D.J., Goodchild, M.F., & Rhind, D.W., (Ed.). (1991) *Geographic Information Systems*, Longman Scientific: New York.

Martin, D. (1996). *Geographical Information Systems: Socio-economic Applications* (2nd ed.). Routledge: London.

McHarg, I. (1995) *Design with Nature*, Wiley: New York.

National Research Council of the National Academies (2006). *Beyond Mapping: Meeting National Needs through Enhanced Geographic Information Science*. National Academies Press: Washington D.C.

Peuquet, D. J. & Marble, D. F. (Eds.). (1990). *Introductory Readings in Geographic Information Systems*, Taylor and Francis: Washington D.C.

Pickles, J. (Ed.). (1995). *Ground Truth: The Social Implications of Geographic Information Systems*, Guilford: New York.

de Smith, M.G., Goodchild, M.F. & Longley, P.A. (2007). *Geospatial Analysis: A Comprehensive Guide to Principles, Techniques, and Software Tools.* Winchelsea Press: Leicester.

Smith, D. A. & Tomlinson, R. F. (1992). Assessing costs and benefits of geographical information systems: methodological and implementation issues, *International Journal of Geographical Information Systems*, 6:247-256.

Theobald, D. M. (2003). *GIS Concepts and ArcGIS Methods*, Conservation Planning Technologies: Fort Collins.

Tillman Lyle, J. (1999). Design for Human Ecosystems: Landscape, Land Use, and Natural Resources, Island Press: Washington.

Tomlinson, R. (1987). Current and potential uses of geographical information systems. The North American experience, *International Journal of Geographical Information Systems*, 1:203-218.

Exercises

1.1 - Why are we more interested in spatial data today than 100 years ago?

1.2 - You have probably collected, analyzed, or communicated spatial data in one way or another during the past month. Describe each of these steps for a specific application you have used or observed.

1.3 - How are GIS software different from most other software?

1.4 - Describe the ways in which GIS are hardware different from other computer hardware?

1.5 - What are the limitations of using a GIS? Under what conditions might the technology hinder problem solving, rather than help?

1.6 - Are paper maps and paper data sheets a GIS? Why or why not?

2 Data Models

Introduction

Data in a GIS represent a simplified view of physical *entities*, the roads, mountains, accident locations, or other features we wish to identify. Data include information on the spatial location and extent of the entities, and information on their non-spatial properties.

Each entity is represented by a *spatial feature* or *cartographic object* in the GIS, and so there is an entity-object correspondence. Because every computer system has limits, only a subset of essential characteristics are recorded. As illustrated in Figure 2-1, we may represent lakes in a region by a set of

polygons. These polygons are associated with a set of essential characteristics that define each lake, perhaps average lake depth, the name, or some measure of water quality.

Essential characteristics are subjectively chosen by the spatial data developer. The essential characteristics of a forest would be different in the eyes of a logger than those of a conservation officer, a hunter, or a hiker. Objects are abstract representations of reality that we store in a spatial database, and they are imperfect representations because we can only record a subset of the characteristics of

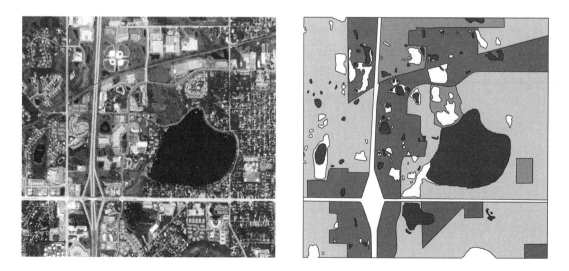

Figure 2-1: A physical entity is represented by a spatial object in a GIS. Here, lakes (dark areas) and other land cover types are represented by polygons.

any entity. No one abstraction is universally better than any other, and the goal of the GIS user is to define objects that support the intended use at the desired level of detail and accuracy.

A *spatial data model* (Figure 2-2) may be defined as the objects in a spatial database plus the relationships among them. The term "model" is fraught with ambiguity because it is used in many disciplines to describe many things. Here the purpose of a spatial data model is to provide a formal means of representing and manipulating spatially-referenced information. In Figure 2-1, our data model consists of two parts. The first part is a set of polygons (closed areas) recording the edges of distinct land uses, and the sec-

ond part is a set of numbers or letters associated with each polygon. The data model may be considered the most recognizable level in our computer abstraction of the real world. Data structures and binary machine code are successively less recognizable, but more computer-compatible forms of the spatial data.

Coordinates are used to define the spatial location and extent of geographic objects. A coordinate most often consists of a pair of numbers that specify location in relation to a point of origin. The coordinates quantify the distance from the origin when measured along a standard direction. Single or groups of coordinates are organized to represent the shapes and boundaries that

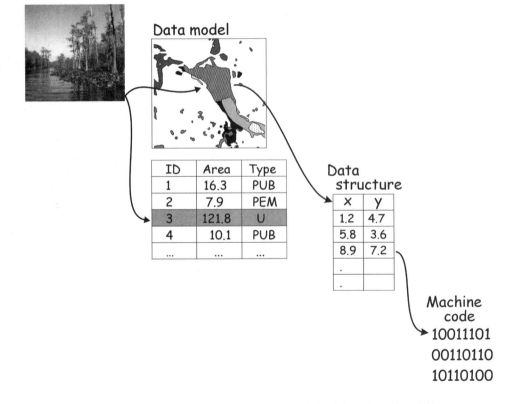

Figure 2-2: Levels of abstraction in the representation of spatial entities. The real world is represented in successively more machine-compatible but humanly obscure forms.

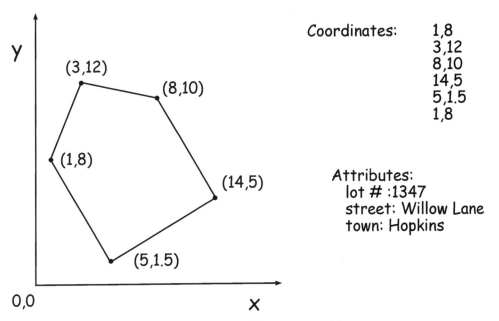

Figure 2-3: Coordinate and attribute data are used to represent entities.

define the objects. Coordinate information is an important part of the data model, and models differ in how they represent these coordinates. Coordinates are usually expressed in one of many standard coordinate systems. The coordinate systems are usually based upon standardized map projections (discussed in Chapter 3) that unambiguously define the coordinate values for every point in an area.

Typically, attribute data complement coordinate data to define cartographic objects (Figure 2-3). Attribute data are collected and referenced to each object. These attribute data record the non-spatial components of an object, such as a name, color, pH, or cash value. Keys, labels, or other indices are used so that the coordinate and attribute data may be viewed, related, and manipulated together.

Most conceptualizations view the world as a set of layers (Figure 2-4). Each layer organizes the spatial and attribute data for a given set of cartographic objects in the region of interest. These are often referred to as *thematic layers*. As an example consider a

GIS database that includes a soils data layer, a population data layer, an elevation data layer, and a roads data layer. The roads layer contains only roads data, including the location and properties of roads in the analysis area (Figure 2-4). There are no data regarding the location and properties of any other geographic entities in the roads layer. Information on soils, population, and elevation are contained in their respective data layers. Through analyses we may combine data to create a new data layer, for example, we may identify areas that have high elevation and join this information with the soils data. This combination may create a new data layer with a new, composite soils/elevation variable.

Coordinate Data

Coordinates define location in two or three-dimensional space. Coordinate pairs, x and y, or coordinate triples, x, y, and z, are used to define the shape and location of each spatial object or phenomenon.

Spatial data in a GIS most often use a *Cartesian* coordinate system, so named after Rene Descartes, the system's originator. Cartesian systems define two or three *orthogonal* (right-angle) axes. Two-dimensional Cartesian systems define x and y axes in a plane (Figure 2-5, left). Three-dimensional Cartesian systems define a z axis, orthogonal to both the x and y axes. An ori-

gin is defined with zero values at the intersection of the orthogonal axes (Figure 2-5, right). Cartesian coordinates are usually specified as decimal numbers that, by convention, increase from bottom to top and from left to right.

Two-dimensional Cartesian coordinate systems have historically been the logical

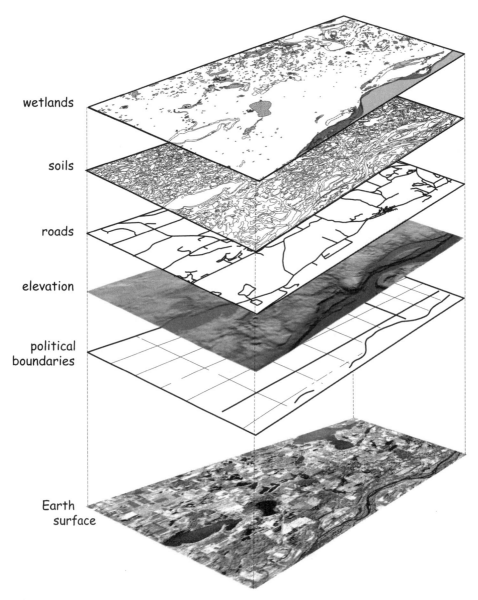

wetlands

soils

roads

elevation

political
boundaries

Earth
surface

Figure 2-4: Spatial data are often stored as separate thematic layers, with objects grouped based on a set of properties, e.g., water, roads, or land cover, or some other agreed-upon set.

2 - Dimensional

3 - Dimensional

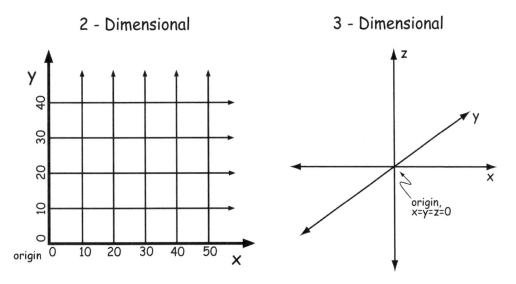

Figure 2-5: Two dimensional (left) and three dimensional (right) Cartesian coordinate systems.

and most common choice for mapping small areas. Small is a relative term, but here we mean maps of farm fields, land and property, cities, and counties. We typically introduce acceptably small errors for most applications when we ignore the Earth's curvature over these small areas. When we map over larger areas, or need the highest precision and accuracy, we usually must choose a three-dimensional system.

Coordinate data may also be specified in a *spherical coordinate system*. Hipparchus, a Greek mathematician of the 2nd century B.C., was among the first to specify locations on the Earth using angular measurements on a sphere. The most common spherical system uses two angles of rotation and a radius distance, r, to specify locations on a modeled Earth surface (Figure 2-6). The first angle of rotation is along the equator. The equator is the imaginary line halfway between the north and south poles. The equator splits the globe into northern and southern hemispheres. The longitude (λ) measures an angle in an east-west direction along the equator. A second angle of rotation, measured in a north-south direction, is used to define a latitude (ϕ).

Spherical coordinates

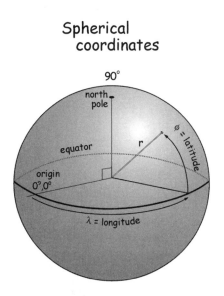

Figure 2-6: Three-dimensional, spherical coordinates may define location by two angles of rotation, λ and ϕ, and a radius vector, r, to a point on a sphere.

There are two primary conventions used for describing latitude and longitude values (Figure 2-7). The first uses a leading letter, N, S, E, or W to indicate direction, followed by a number to indicate location. Latitudes increase from 0 at the equator to 90 degrees at the poles. Northern latitudes are preceded by an N and southern latitudes by an S, for example, N90°, S10°. Longitude values are preceded by an E or W, respectively, east or west of a longitude of origin, for example W110°. Longitudes range from 0 to 180 degrees east or west. Note that the east and west longitudes meet at 180 degrees, so that E180° equals W180°.

Signed coordinates is the second common convention for specifying latitude and longitude in a spherical system. Northern latitudes are positive and southern latitudes are negative, and eastern longitudes positive and western longitudes negative. Latitudes vary from -90 degrees to 90 degrees, and longitudes vary from -180 degrees to 180 degrees. By this convention the longitudes "meet" at the maximum and minimum values, so -180° equals 180°.

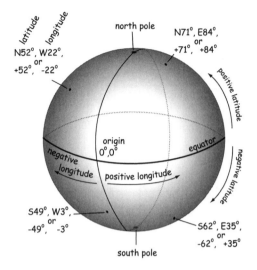

Figure 2-7: Spherical coordinates of latitude and longitude are most often expressed as directional (N/S, E/W), or as signed numbers. Latitudes are positive north, negative south, longitudes are positive east, negative west.

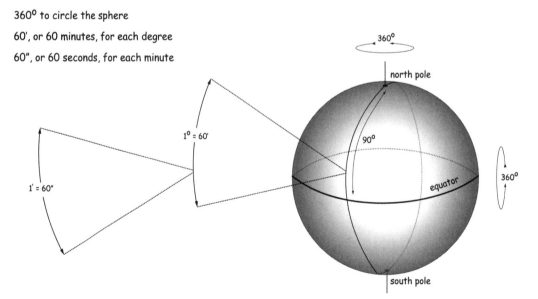

Figure 2-8: There are 360 degrees in a complete circle, with each degree composed of 60 minutes, and each minute composed of 60 seconds.

Coordinates may easily be converted between these two conventions. North latitudes and east longitudes are converted by removing the leading N or E, respectively. South latitudes and west longitudes by removing the leading S or W, respectively, and then changing the sign of the remaining number from a positive to a negative value.

Spherical coordinates are most often recorded in a degrees-minutes-seconds (DMS) notation, for example N43° 35′ 20″, signifying 43 degrees, 35 minutes, and 20 seconds of latitude. In DMS, each degree is made up of 60 minutes of arc, and each minute is in turn divided into 60 seconds of arc (Figure 2-8). This yields 60 times 60 or 3600 seconds for each degree of latitude or longitude. Note that the ancient Babylonians established these splits, of 360 degrees for a complete circle, with degrees and minutes subsequently divided into 60 units, and we've carried this convention down to today.

Spherical coordinates may also be expressed as decimal degrees (DD). When using DD the degrees take their normal -180 to 180 (longitude) and -90 to 90 (latitude) ranges, but minutes and seconds are reported as a decimal portion of a degree (from 0 to 0.99999...). In our previous example, N43° 35′ 20″ would be reported as 43.5888. DMS may be converted to DD by:

$$DD = DEG + MIN/60 + SEC/3600 \quad (2.1)$$

Examples of the forward and reverse conversion between decimal degrees and degrees-minutes-seconds units are shown in Figure 2-9.

Attribute Data and Types

Attribute data are used to record the non-spatial characteristics of an entity. Attributes, also called *items* or *variables*, may be envisioned as a list of characteristics that describe the features we represent in a GIS. Color, depth, weight, owner, vegetation type, or landuse are examples of variables

```
DD from DMS
DD = D + M/60 + S/3600
e.g.
DMS = 32° 45′ 28″

DD = 32 + 45/60 + 28/3600
   = 32 + 0.75 + 0.0077778
   = 32.7577778

DMS from DD
D = integer part
M = integer of decimal part x 60
S = 2nd decimal x 60
e.g.
DD = 24.93547
D = 24
M = integer of 0.93547 x 60
   = integer of 56.1282
   = 56
S = 2nd decimal x 60
   = 0.1282 * 60 = 7.692
so DMS is
   24° 56′ 7.692″
```

Figure 2-9 Examples for converting between DMS and DD expressions of spherical coordinates.

that may appear as attributes. Attributes have values, for example, a fire hydrant may be colored red, yellow, or orange, have 1 to 4 flanges, and a rating of high, medium, or low. Attributes are often presented in tables and arranged in rows and columns (Figure 2-10). Each row corresponds to an individual spatial object, and each column corresponds to an attribute. Tables are often organized and managed using a specialized computer program called a database management system (DBMS, described more fully in Chapter 8).

Attributes may be in many forms, but all attributes can be categorized as nominal, ordinal, or interval/ratio attributes.

Nominal attributes are variables that provide descriptive information about an object. The color is recorded for each hydrant in Figure 2-10. Other examples of nominal data are vegetation type, a city name, the owner of a parcel, or soil series. There is no implied order, size, or quantitative information contained in nominal attributes.

Nominal attributes may also be images, film clips, audio recordings, or other descriptive information. Just as the color or type attributes provide nominal information for an entity, an image may also provide descriptive information. GIS for real estate often have images of the buildings or surroundings as part of the database. Images provide information not easily conveyed any other way. These image or sound attributes are sometimes referred to as "BLOBs" for *binary large objects*.

Ordinal attributes imply a ranking or order by their values. An ordinal attribute may be descriptive, such as high, mid, or low (the rating column in Figure 2-10), or it may be numeric; for example, an erosion class may be given a value from 1 to 10. The order reflects only rank, and does not specify

the form of the scale. An ordinal attribute value of four has a higher rank than a value of two, but we can't infer that the attribute value is twice as large, because we can't assume the scale is linear.

Interval/ratio attributes are used for numeric items where both rank order and absolute difference in magnitudes are reflected in the numbers, for example, the number of flanges in the second column of Figure 2-10. These data are often recorded as real numbers, most often on a linear scale. Area, length, weight, height, or depth are a few examples of attributes that are represented by interval/ratio variables.

Items have a *domain*, a range of values they may take. Colors might be restricted to red, yellow, and green, cardinal direction to north, south, east, or west, and size to all positive real numbers.

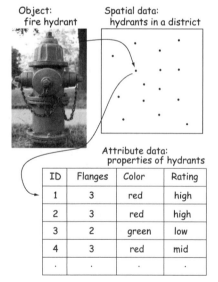

Object:
fire hydrant

Spatial data:
hydrants in a district

Attribute data:
properties of hydrants

ID	Flanges	Color	Rating
1	3	red	high
2	3	red	high
3	2	green	low
4	3	red	mid
.	.	.	.

Figure 2-10: Attributes are typically envisioned as arranged by columns and rows, with objects arranged in rows, and attributes aligned in columns.

Common Spatial Data Models

Spatial data models begin with a conceptualization, a view of real world phenomena or entities. Consider a road map suitable for use at a statewide or provincial level. This map is based on a conceptualization that defines roads as lines. These lines connect cities and towns that are shown as discrete points or polygons on the map. Road properties may include only the road type, e.g., a limited access interstate, state highway, county road, or some other type of road. The roads have a width represented by the drawing symbol on the map, however this width, when scaled, may not represent the true road width. This conceptualization identifies each road as a linear feature that fits into a small number of categories. All state highways are represented by the same type of line, even though the state highways may vary. Some may be paved with concrete, others with asphalt. Some may have wide shoulders, others not, or dividing barriers of concrete, versus a broad vegetated median. We realize these differences can exist within this conceptualization.

There are two main conceptualizations used for digital spatial data. The first conceptualization defines discrete objects using a *vector data model*. Vector data models use discrete elements such as points, lines, and polygons to represent the geometry of real-world entities (Figure 2-11).

A farm field, a road, a wetland, cities, and census tracts are examples of discrete entities that may be represented by discrete objects. Points are often used to define the locations of "small" objects such as wells, buildings, or ponds. Lines may be used to represent linear objects, e.g., rivers or roads, or to identify the boundary between what is a part of the object and what is not a part of the object. We may map landcover for a region of interest, and we categorize discrete areas as a uniform landcover type. A forest may share an edge with a pasture, and this boundary is represented by lines. The boundaries between two polygons may not be discrete on the ground, for example, a forest edge may grade into a mix of trees and grass, then to pasture; however in the vector conceptualization, a line between two landcover types will be drawn to indicate a discrete, abrupt transition between the two types. Lines and points have coordinate locations, but points have no dimension, and lines have no dimension perpendicular to their direction. Area features may be defined by a closed, connected set of lines.

Vector

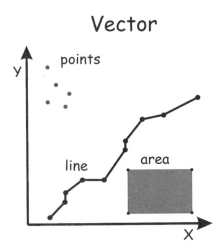

Raster

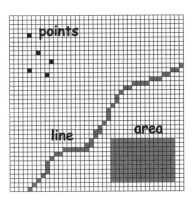

Figure 2-11: Vector and raster data models.

The second common conceptualization identifies and represents grid cells for a given region of interest. This conceptualization employs a *raster* data model (Figure 2-11). Raster cells are arrayed in a row and column pattern to provide "wall-to-wall" coverage of a study region. Cell values are used to represent the type or quality of mapped variables. The raster model is used most commonly with variables that may change continuously across a region. Elevation, mean temperature, slope, average rainfall, cumulative ozone exposure, or soil moisture are examples of phenomena that are often represented as continuous fields. Raster representations are also commonly used to represent discrete features, for example, class maps of vegetation or political units.

Data models are at times interchangeable in that many phenomena may be represented with either the vector or raster conceptual approach. For example, elevation may be represented as a surface (continuous field) or as a series of lines representing contours of equal elevation (discrete objects). Data may be converted from one conceptual view to another, for example, the location of contour lines (lines of equal elevation) may be determined by evaluating the raster surface, or a raster data layer may be derived from a set of contour lines. These conversions entail some costs both computationally and perhaps in data accuracy.

The decision to use either a raster or vector conceptualization often depends on the type of operations to be performed. For example, slope is more easily determined when elevation is represented as a continuous field in a raster data set. However, discrete contours are often the preferred format for printed maps, so the discrete conceptualization of a vector data model may be preferred for this application. The best data model for a given organization or application depends on the most common operations, the experiences and views of the GIS users, the form of available data, and the influence of the data model on data quality.

In addition to these two main data models, there are other data models that may be described as variants, hybrids, or special forms by some GIS users, and as different families of data models by others. A triangulated irregular network (TIN) is an example of such a data model. This model is most often used to represent surfaces, such as elevations, through a combination of point, line, and area features. Many consider this a special type of vector data model. Variants or other representations related to raster data models also exist. We will introduce and discuss variants later in this and other chapters.

Vector Data Models

A vector data model uses sets of coordinates and associated attribute data to define discrete objects. Groups of coordinates define the location and boundaries of discrete objects, and these coordinate data plus their associated attributes are used to create vector objects representing the real-world entities (Figure 2-12).

There are three basic types of vector objects: points, lines, and polygons (Figure 2-13). A point uses a single coordinate pair to represent the location of an entity that is considered to have no dimension. Gas wells, light poles, accident location, and survey points are examples of entities often represented as point objects. Some of these have real physical dimension, but for the purposes of the GIS users they may be represented as points. In effect, this means the size or dimension of the entity is not important, only its location.

Attribute data are attached to each point, and these attribute data record the important non-spatial characteristics of the point entities. When using a point to represent a light pole, important attribute information might

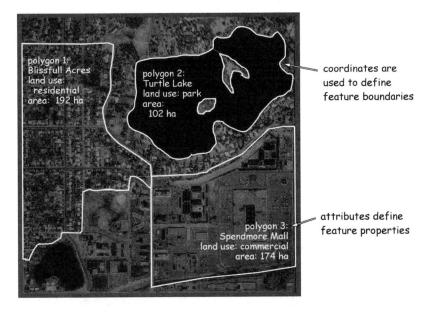

Figure 2-12: Coordinates define spatial location and shape. Attributes record the important non-spatial characteristics of features in a vector data model.

be the height of the pole, the type of light and power source, and the last date the pole was serviced.

Linear features, often referred to as lines or *arcs*, are represented as lines when using vector data models. Lines are most often represented as an ordered set of coordinate pairs. Each line is made up of line segments that run between adjacent coordinates in the ordered set (Figure 2-13). A long, straight line may be represented by two coordinate pairs, one at the start and one at the end of the line. Curved linear entities are most often represented as a collection of short, straight, line segments, although curved lines are at times represented by a mathematical equation describing a geometric shape. Lines typically have a starting point, an ending point, and intermediate points to represent the shape of the linear entity. Starting points and ending points for a line are sometimes referred to as *nodes*, while intermediate points in a line are referred to as *vertices* (Figure 2-13). Attributes may be attached to the whole line, line segments, or to nodes and vertices along the lines.

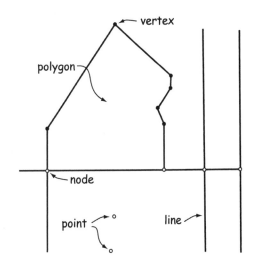

Figure 2-13: Points, nodes and vertices define points, line, and polygon features in a vector data model.

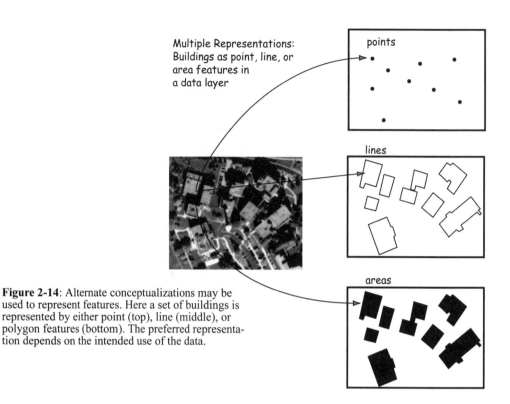

Multiple Representations:
Buildings as point, line, or
area features in
a data layer

Figure 2-14: Alternate conceptualizations may be
used to represent features. Here a set of buildings is
represented by either point (top), line (middle), or
polygon features (bottom). The preferred representation depends on the intended use of the data.

Area entities are most often represented
by closed polygons. These polygons are
formed by a set of connected lines, either
one line with an ending point that connects
back to the starting point, or as a set of lines
connected start-to-end. Polygons have an
interior region and may entirely enclose
other polygons in this region. Polygons may
be adjacent to other polygons and thus share
"bordering" or "edge" lines with other polygons. Attribute data such as area, perimeter,
landcover type, or county name may be
linked to each polygon.

Note that there is no uniformly superior
way to represent features. Some feature
types may appear to be more "naturally" represented one way: manhole covers as points,
roads as lines, and parks as polygons. However, in a very detailed data set, the manhole
covers may be represented as circles, and
both edges of the roads may be drawn and
the roads represented as polygons. The representation depends as much on the detail,

accuracy, and intended use of the data set as
our common conception or general shape of
the objects.

A single set of features may be represented differently, depending on the interests
and purposes of the GIS users (Figure 2-14).
A point layer may be chosen when general
feature position is needed, e.g., general
building location (Figure 2-14, top). Other
users may be interested in the outline of the
feature and so require representation by
lines, while polygon representations may be
preferred for other applications (Figure 2-14
mid and lower, respectively). Our intended
use often determines our conceptual model
and hence vector type used to represent a
feature.

Polygon Inclusions and Boundary Generalization

Vector data often contain two characteristics, polygon inclusions and boundary generalization, that must be noted. These characteristics are oft-ignored, but may affect the use of vector data, occasionally with dire consequences. These concepts must be understood, their presence evaluated, and effects weighed in the use of vector data sets.

Polygon inclusions are areas in a polygon that are different from the rest of the polygon, but still part of the polygon. Inclusions occur because we typically assume an area represented by a polygon is homogeneous, but this assumption may be wrong, as

illustrated in Figure 2-15. The figure shows a vector polygon layer representing raised landscaping beds (a). The general attributes for the polygon may be coded, for example, the surface type may be recorded as cedar mulch. The area noted in Figure 2-15b shows a walkway that is an inclusion in a raised bed. This walkway has a concrete surface. Hence, this walkway is an unresolved inclusion within the polygon.

One solution to the problem of inclusions is to create a polygon for each inclusion. This often is not done because it may take too much effort to identify and collect the boundary location of each inclusion, and there typically is some lower limit, or minimum mapping unit, on the size of objects we care to record in our data. Inclusions are

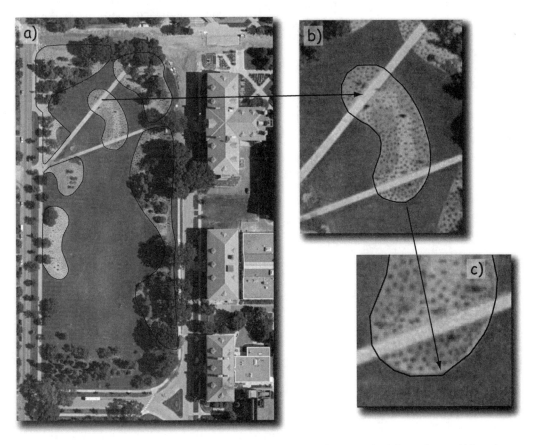

Figure 2-15: Examples of polygon inclusions (sidewalk inclusion in flower bed shown in a and b), and boundary generalization (c) in a vector data model. These approximations typically occur as a consequence of adopting a vector representation, and their impacts must be considered when using vector data.

present in some form in nearly all polygon data layers.

Boundary generalization is the incomplete representation of boundary locations. This problem stems from the typical way we represent linear and area features in vector data sets. As shown in Figure 2-15c, polygon boundaries are typically represented as a set of connected, short, straight-line segments. The segments are a means to trace the position of a line feature or the boundary separating two area features. For curved lines, these straight line segments may be viewed as a sampling of the true curve, and there is typically some deviation of the line segment from the "true" curved boundary. The amount of generalization depends on many factors, and generally should be so small as to be unimportant for any intended use of the spatial data. However, since many data sets may have unforeseen uses, or may be obtained from a third party, the boundary generalization should be recognized and evaluated relative to the specific requirements of any given spatial analysis. There are additional forms of generalization in spatial data, and these are described more thoroughly in Chapter 4.

Vector Topology

Vector data often contain *vector topology*, enforcing strict connectivity and recording adjacency and planarity. Early systems employed a spaghetti data model (Figure 2-16a), in which lines may not intersect when they should, and may overlap without connecting. The spaghetti model severely limits spatial data analysis and is little used except for very basic data entry or translation. Topological models create an intersection and place a node at each line crossing, record connectivity and adjacency, and maintain information on the relationships between and among points, lines, and polygons in spatial data. This greatly improves the speed, accuracy, and utility of many spatial data operations.

Topological properties are conserved when converting vector data among common coordinate systems, a common practice in GIS analysis (described in Chapter 3). Polygon adjacency is an example of a topologically invariant property, because the list of neighbors for any given polygon does not change during geometric stretching or bending (Figure 2-16, b and c). These relationships may be recorded separately from the coordinate data.

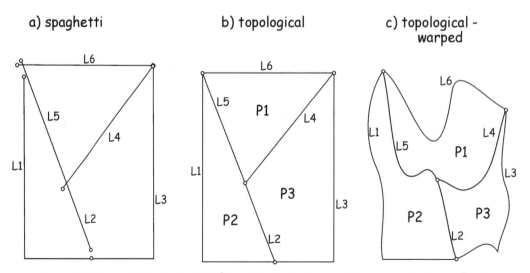

Figure 2-16: Spaghetti (a), topological (b), and topological-warped (c) vector data. Figures b and c are topologically identical because they have the same connectivity and adjacency.

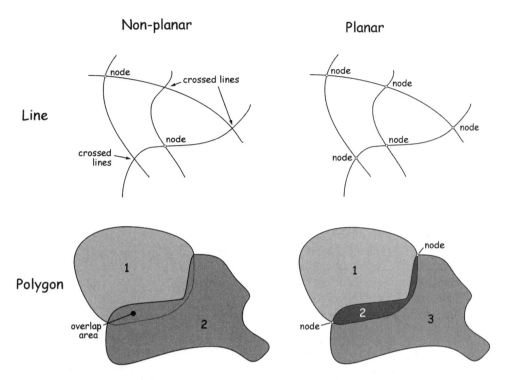

Figure 2-17: Non-planar and planar topology in lines and polygons.

Topological vector models may vary, and enforce particular types of topological relationships. *Planar topology* requires that all features occur on a two-dimensional surface. There can be no overlaps among lines or polygons in the same layer (Figure 2-17). When planar topology is enforced, lines may not cross over or under other lines. At each line crossing there must be an intersection.

The left side of Figure 2-17 shows non-planar graphs. In the top left figure, four line segments coincide. At some locations the lines intersect at a node, shown as white-filled circles, but at some locations a line passes over or under another line segment. These lines are non-planar. The top right of Figure 2-17 shows planar topology enforced for these same four line segments. Nodes are found at each line crossing.

Polygons can also be non-planar, as shown at the bottom left of Figure 2-17. Two polygons overlap slightly at an edge. This may be due to an error, for example, the two

polygons share a boundary but have been recorded with an overlap, or there may be two areas that overlap in some way. If topological planarity is enforced, these two polygons must be resolved into three separate, non-overlapping polygons. Nodes are placed at the intersections of the polygon boundaries (lower right, Figure 2-17).

There are additional topological constructs besides planarity that may be specified. For example, polygons may be exhaustive, in that there are no gaps, holes or "islands" allowed. Line direction may be recorded, so that a "from" and "to" node are identified in each line. Directionality aids the representation of river or street networks, where there may be a natural flow direction.

There is no single, uniform set of topological relationships that are included in all topological data models. Different vendors have incorporated different topological information in their data structures. Planar topology is often included, as are representa-

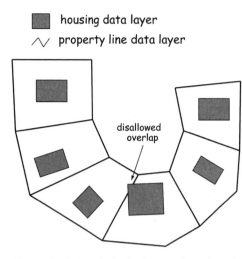

housing data layer

property line data layer

disallowed overlap

Figure 2-18: Topological rules may be enforced across data layers. Here, rules may be specified to avoid overlap between objects in different layers.

tions of *adjacency* (which polygons are next to which) and *connectivity* (which lines connect to which).

Some GIS software create and maintain detailed topological relationships in their data. This results in more complex and perhaps larger data structures, but access is often faster, and topology provides more consistent, "cleaner" data. Other systems maintain little topological information in the data structures, but compute and act upon topology as needed during specific processing.

Topology may also be specified between layers, because we may wish to enforce spatial relationships between entities that are stored separately. As an example, consider a data layer that stores property lines (cadastral data), and a housing data layer that stores building footprints (Figure 2-18). Rules may be specified that prevent polygons in the housing data layer from crossing property lines in the cadastral data layer. This would indicate a building that crosses a property line. Most such instances occur as a result of small errors in data entry or misalignment among data layers. Topological restrictions between two data layers avoid these inconsistencies. Exceptions may be

granted in those few cases when a building truly does cross property lines.

There are many other types of topological constraints that may be enforced, both within and between layers. *Dangles*, lines that do not connect to other lines, may be proscribed, or limited to be greater or less than some threshold length. Lines and points may be required to coincide, for example, water pumps as points in one data layer and water pipes as lines in another, or lines in separate layers may be required to intersect or be coincident. While these topological rules add complexity to vector data sets, they may also improve the logical consistency and value of these data.

Topological vector models often use codes and tables to record topology. As described above, nodes are the starting and ending points of lines. Each node and line is given a unique identifier. Sequences of nodes and lines are recorded as a list of identifiers, and point, line, and polygon topology recorded in a set of tables. The vector features and tables in Figure 2-19 illustrate one form of this topological coding.

Many GIS software systems are written such that the topological coding is not visible to users, nor directly accessible by them. Tools are provided to ensure the topology is created and maintained, that is, there may be directives that require that polygons in two layers do not overlap, or to ensure planarity for all line crossings. However, the topological tables these commands build are often quite large, complex, and linked in an obscure way, and therefore hidden from users.

Point topology is often quite simple. Points are typically independent of each other, so they may be recorded as individual identifiers, perhaps with coordinates included, and in no particular order (Figure 2-19, top).

Line topology typically includes substantial structure, and identifies at a minimum the beginning and ending points of each line (Figure 2-19, middle). Variables record the topology and may be organized in a table. These variables may include a line identifier, the starting node, and the ending node for each line. In addition, lines may be assigned a direction, and the polygons to the left and right of the lines recorded. In most cases left and right are defined in relation to the direction of travel from the starting node to the ending node.

Polygon topology may also be defined by tables (Figure 2-19, bottom). The tables may record the polygon identifiers and the list of connected lines that define the polygon. Edge lines are often recorded in sequential order. The lines for a polygon form a closed loop, and thus, the starting node of the first line in the list also serves as the ending node for the last line in the list. Note that there may be a "background" polygon defined by the outside area. This background polygon is not a closed polygon like all the rest, however it may be defined for

consistency and to provide entries in the topology tables.

Topological vector models greatly enhance many vector data operations. Adjacency analyses are reduced to a "table lookup", a quick and easy operation in most software systems. For example, an analyst may want to identify all polygons adjacent to a city. Assume the city is represented as a single polygon. Adjacency analysis reduces to 1) scanning the polygon topology table to find the polygon labeled "city" and reading the list of lines that bound the polygon, and 2) scanning this list of lines for the city polygon, accumulating a list of all left and right polygons. Polygons adjacent to the city may be identified from this list. List searches on topological tables are typically much faster than searches involving coordinate data.

Topological vector models also enhance many other spatial data operations. Network

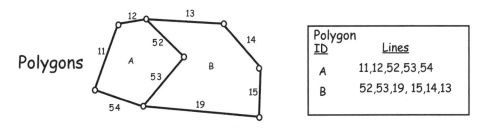

Figure 2-19: An example of vector features and corresponding topology tables. Information on the adjacency, connectivity, and other spatial relationships may be stored in topology tables, and joined to features by indices, here represented by values in the ID columns.

and other connectivity analyses are concerned with the flow of resources through defined pathways. Topological vector models explicitly record the connections of a set of pathways and so facilitate network analyses. Overlay operations are also enhanced when using topological vector models. The mechanics of overlay operations are discussed in greater detail in Chapter 9, however we will state here that they involve identifying line adjacency, intersection, and resultant polygon formation. The interior and exterior regions of existing and new polygons must be determined, and these regions depend on polygon topology. Hence, topological data are useful in many spatial analyses.

Topological data models often have an advantage of smaller file sizes, largely because coordinate data are recorded once. For example, a non-topological approach often stores polygon boundaries twice. Lines 52 and 53 at the bottom of Figure 2-19 will be recorded for both polygon A and polygon B. Long, complex boundaries in polygon datasets may double their size. This increases both storage requirements and processing.

There are limitations and disadvantages to topological vector models. First, there are computational costs in defining the topological structure of a vector data layer. Software must determine the connectivity and adjacency information, assign codes, and build the topological tables. Computational costs are typically quite modest with current computer technologies.

Second, the data must be very "clean", in that all lines must begin and end with a node, all lines must connect correctly, and all polygons must be closed. Unconnected lines or unclosed polygons will cause errors during analyses. Significant human effort may be required to ensure clean vector data because each line and polygon must be checked. Software may help by flagging or fixing "dangling" nodes that do not connect to other nodes, and by automatically identifying all polygons. Each dangling node and

polygon may then be checked, and edited as needed to correct errors.

Limitations and the extra editing are far outweighed by the gains in efficiency and analytical capabilities provided by topological vector models. Many current vector GIS packages use topological vector models in some form.

Vector Features and Attribute Tables

Topological vector models are used to define spatial features in a data layer. As we described earlier in this chapter, geographic features are associated with non-spatial attributes. Typically a table is used to organize the attributes, and there is a linkage between rows in the attribute table and the spatial components of a data layer (Figure 2-20, top). In most GIS software, we can most easily view the tables and a graphic representation of the spatial data as a linked table and digital map (Figure 2-20, top).

There is commonly a *one-to-one linkage* between each entry in the attribute table and each feature in the data layer. This means for each feature in the data layer there is one and only one entry in the attribute table. Occasionally there may be layers with a many-to-one relationship between attribute table entries and multiple features in a data layer.

The attribute table typically has an "identifier" (ID) column. This ID column has a unique value for each unique spatial feature in the data layer. The ID item is often used to distinguish each unique feature in the data layer when editing, in analysis, or when combining data across several layers. Additional attributes are organized in respective columns, with values appropriate for the corresponding spatial feature. For example, the area of a polygon may be stored in an area column, and associated with the polygon through the unique ID (Figure 2-20).

Most GIS employ underlying file structures to organize components of the spatial data. An example organization is shown in the bottom half of Figure 2-20, where the

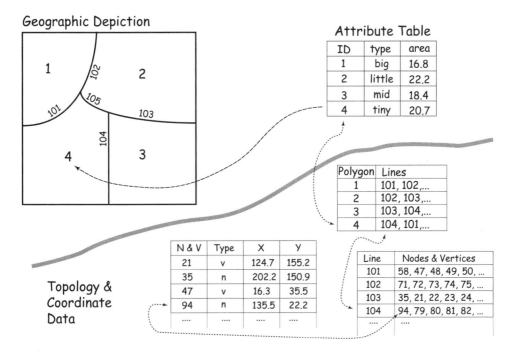

Figure 2-20: Features in a topological data layer typically have a one-to-one relationship with entries in an associated attribute table. The attribute table typically contains a column with a unique identifier, or ID, for each feature. Topology and coordinate data are often hidden from the user, but linked to the attribute and geographic features through pointers and index variables, described in the Data and File Structures section, later in this chapter.

topological elements are recorded in a linked set of tables, here, one each for the polygons, lines, and nodes and vertices. Most GIS maintain the spatial and topological data as a single or cluster of linked files. This internal file structure is often insulated from direct manipulation by the GIS user, but underlies nearly all spatial data manipulations. A user may directly edit or otherwise manipulate table values, usually with the exception of the ID, and the underlying topology and coordinate data are accessed via requests to display, change, or analyze the spatial data components. Data layers may also include additional information (not shown) on the origin, region covered, date of creation, edit history, coordinate system, or other characteristics of a data set.

Note that not all GIS store coordinate and topological data in non-tabular file structures. Coordinates, points, lines, polygons, and other composite features may be stored in tables similar to attribute tables. It is premature to discuss the details of these *spatially-enabled* databases, because they are based on something called a *relational data model*, described in detail in Chapter 8. Faster computers support this generally more flexible approach, allowing simpler and more transparent access across different types of GIS software.

Raster Data Models

Models and Cells

Raster data models define the world as a regular set of cells in a grid pattern (Figure 2-21). Typically these cells are square and evenly spaced in the x and y directions. The phenomena or entities of interest are represented by attribute values associated with each cell location.

Raster data models are the natural means to represent "continuous" spatial features or phenomena. Elevation, precipitation, slope, and pollutant concentration are examples of continuous spatial variables. These variables characteristically show significant changes in value over broad areas. The gradients can be quite steep (e.g., at cliffs), gentle (long, sloping ridges), or quite variable (rolling hills). Raster data models depict these gradients by changes in the values associated with each cell.

Raster data sets have a *cell dimension*, defining the size of the cell (Figure 2-21). The cell dimension specifies the length and width of the cell in surface units. For example, the cell dimension may be specified as a square 30 meters on each side. The cells are usually oriented parallel to the x and y direc-

tions, and the coordinates of a corner location are specified.

When the cells are square and aligned with the coordinate axes, the calculation of a cell location is a simple process of counting and multiplication. A cell location may be calculated from the cell size, known corner coordinates, and cell row and column number. For example, if we know the lower-left cell coordinate, all other cell coordinates may be determined by the formulas:

$$N_{cell} = N_{lower-left} + row * cell\ size \quad (2.2)$$

$$E_{cell} = E_{lower-left} + column * cell\ size \quad (2.3)$$

where N is the coordinate in the north direction (y), E is the coordinate in the east direction (x), and the **row** and **column** are counted starting with zero from the lower left cell.

There is often a trade-off between spatial detail and data volume in raster data sets. The number of cells needed to cover a given area increases four times when the cell size is cut in half (Figure 2-22). Smaller cells provide greater spatial detail, but at the cost of larger data sets.

The cell dimension also affects the spatial precision of the data set, and hence positional accuracy. The cell coordinate is usually defined at a point in the center of the cell. The coordinate applies to the entire area covered by the cell. Positional accuracy is typically expected to be no better than approximately one-half the cell size. No matter the true location of a feature, coordinates are truncated or rounded up to the nearest cell center coordinate. Thus, the cell size should be no more than twice the desired accuracy and precision for the data layer represented in the raster, and often it is specified to be smaller.

Each raster cell represents a given area on the ground and is assigned a value that may be considered to apply to the entire cell.

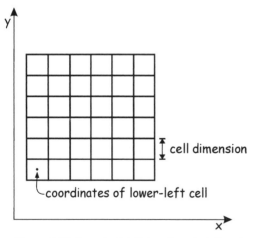

Figure 2-21: Important defining characteristics of a raster data model.

If the variable is uniform across the raster cell, the value will be correct over the cell. However, under most conditions there is within-cell variation, and the raster cell value represents the average, central, or most common value found in the cell. Consider a raster data set representing annual weekly income with a cell dimension that is 300 meters (980 feet) on a side. Further suppose that there is a raster cell with a value of 710. The entire 300 by 300 meters area is considered to have this value of 710 pesos per week. There may be many households within the raster cell that do not earn exactly 710 pesos per week. However the 710 pesos may be the average, the highest point, or some other representative value for the area covered by the cell. While raster cells often represent the average or the value measured at the center of the cell, they may also represent the median, maximum, or another statistic for the cell area.

An alternative interpretation of the raster cell applies the value to the central point of the cell. Consider a raster grid containing elevation values. Cells may be specified as 200 meters square, and an elevation value assigned to each square. A cell with a value of 8000 meters (26,200 feet) may be assumed to have that value at the center of the cell, but this value will not be assumed to apply to the entire cell.

A raster data model may also be used to represent discrete data (Figure 2-23), for

a	a	a	a	r	f	f	a	a	a	a	a
a	a	a	a	r	f	f	a	a	a	a	a
a	a	a	f	r	f	f	a	a	a	a	a
a	a	a	r	r	f	f	a	a	a	a	a
a	a	a	r	f	f	f	a	a	a	a	a
a	f	f	r	f	f	f	a	a	a	a	a
a	f	f	r	f	u	f	a	a	a	a	a
h	h	h	h	h	h	h	h	h	h	h	h
f	f	r	u	u	u	u	a	a	a	a	a
f	f	r	f	u	u	a	a	a	a	a	a
f	f	f	r	f	f	a	a	a	a	a	a
f	f	f	f	r	f	a	a	a	a	a	a

a = agriculture   u = developed
f = forest   r = river
h = highways

Figure 2-23: Discrete or categorical data may be represented by codes in a raster data layer.

example, to represent landcover in an area. Raster cells typically hold numeric or single-letter alphabetic characters. A coding scheme defines what land cover type the discrete values signify. Each code may be found at many raster cells.

Raster cell values may be assigned and interpreted in at least seven different ways (Table 2-1). We have described three: a raster cell as a point physical value (elevation), as a statistical value (average income), and as discrete data (landcover). Raster values may also be used to represent points and

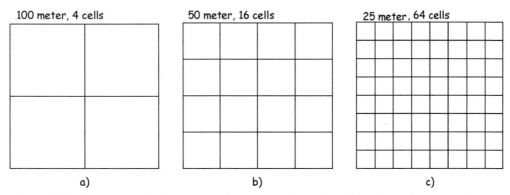

100 meter, 4 cells   50 meter, 16 cells   25 meter, 64 cells

a)   b)   c)

Figure 2-22: The number of cells in a raster data set depends on the cell size. For a given area, a linear decrease in cell size causes an exponential increase in cell number, e.g., halving the cell size causes a four-fold increase in cell number.

Table 2-1: Types of data represented by raster cell values (from L. Usery, pers. comm.).

Data Type	Description	Example
point ID	alpha-numeric ID of closest point	hospital
line ID	alpha-numeric ID of closest line	nearest road
contiguous region ID	alpha-numeric ID for dominant region	state
class code	alpha-numeric code for general class	vegetation type
table ID	numeric position in a table	row
physical analog	numeric value representing surface value	elevation
statistical value	numeric value from a statistical function	population density

lines, as the IDs of lines or points that occur closest to the cell center.

Point and line assignment to raster cells may be complicated when there are multiple features within a single cell. For example, when light poles are represented in a raster data layer, cell value assignment is straight-forward when there is only one light in a cell (Figure 2-24, near **A**). When there are multiple poles in a single cell there is some ambiguity, or generalization in the assignment (Figure 2-24, near **B**). One common solution represents one feature from the group, and retains information on the attributes and characteristics of that feature. This entails some data loss. Another solution is to reduce the raster cell size so that there are no multiple features in a cell. This may result in impractically large data sets. More complex schemes may record multiple instances of features in a cell, but these then may slow access or otherwise decrease the utility that comes from the simple raster structure.

Similar problems may occur when there are multiple line segments within a raster cell, for example, when linear features such as roads are represented in a raster data set.

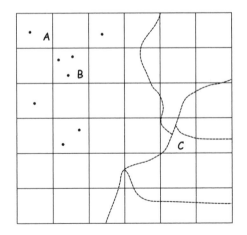

Figure 2-24: Raster cell assignment requires decisions when multiple objects occur in the same cell.

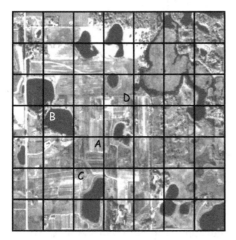

Figure 2-25: Raster cell assignment with mixed landscapes. Upland areas are lighter greys, water the darkest greys.

When two or more roads meet, they will do so within a raster cell, and some set of attributes must be assigned (Figure 2-24C). Since attributes are assigned by cells, some precedence must be established, with one line given priority over others.

Raster cell assignment also may be complicated when representing what we typically think of as discrete, uniform areas. Consider the area in Figure 2-25. We wish to represent this area with a raster data layer, with cells assigned to one of two class codes, one each for land or water. Water bodies appear as darker areas in the image, and the raster grid is shown overlain. Cells may contain substantial areas of both land and water, and the proportion of each class may span from zero to 100 percent. Some cells are purely one class and the assignment is unambiguous, for example, the cell labelled A in the Figure 2-25 contains only land. Others are unambiguous, such as cell B (water) or D (land). Some are nearly equal in their proportion of land and water, as in cell C.

One common method to assign classes for mixed cells is called "winner-take-all". The cell is assigned the class of the largest area type. Cells A, C, and D would be assigned the land class, cell B the water class. Another option applies preference in cell assignment. If any of an "important" type is found, then the cell is assigned that value, regardless of the proportion. If we specify a preference for water, then cells B, C, and D in Figure 2-25 would be assigned the water type, and cell A the land type.

Regardless of the assignment method used, Figure 2-25 illustrates two phenomena when discrete objects are represented using a raster data model. First, some areas that are not the assigned class are included in some rasters cells. These "inclusions" are inevitable because cells must be assigned to a discrete class, the cell boundaries are rigidly assigned, and the class boundaries on the ground rarely line up with the cell boundaries. Some mixed cells occur in nearly all raster layers. The GIS user must acknowledge these inclusions, and consider their impact on the intended spatial analyses.

Second, differences in class assignment rules may substantially alter the data layer, as shown in our simple example. In more complex landscapes, there will be more potential cell types, which may increase the assignment sensitivity. Decreasing the raster cell size reduces the significance of classes in the assignment rule, but at the cost of increased data volumes.

The occurrence of more than one line or point within a raster cell can result in similar assignment problems. If two points occur, then which point ID is assigned? If two lines occur, then which line ID should be assigned? Some rule must be developed, for example, the point that falls nearest the center may be assigned, or the line with the longest segment within the raster cell. Similar to when area features are assigned to rasters, "inclusions" and dependence on the class assignment rules affect the output.

Raster Features and Attribute Tables

Raster layers may also have associated attribute tables. This is most common when nominal data are represented, but may also be used with ordinal or interval/ratio data. Just as with topological vector data, features in the raster layer may be linked to rows in an attribute table, and these rows may describe the essential non-spatial character-istics of the features.

Figure 2-26a and b show data repre-sented in both vector (a) and raster (b) data models. Vector data are shown in Figure 2-26a with a one-to-one correspondence between polygon features and rows in the attribute table. The IDorg column identifies each polygon, and an attributes class and area are assigned for each. Figure 2-26b shows a raster data set that represents the same data and maintains a one-to-one rela-tionship in the data table. An additional col-umn, cell-ID, must be added to uniquely identify each raster location. The corre-sponding attributes IDorg, class, and area are repeated for each cell. Note that the area values are the same for all cells and thus all rows in the table. This one-to-one correspon-dence is rarely used with raster data sets, for reasons described below.

The nature of the raster data model often affects the characteristics of associated attribute tables and may require adjustments in how attribute and spatial data are repre-sented. Note that maintaining a one-to-one correspondence between raster cells and rows in the attribute table comes at some cost - the large size of the attribute table. While the vector representation of these data requires an attribute table with five rows, representing these same features using a ras-ter data model results in 100 rows. As the cells are smaller, the number of raster cells increases to cover a given dimension, and the data volume grows exponentially.

We often use raster data sets with bil-lions of cells. If we insist on a one-to-one cell/attribute relationship, the table may become too large. Even simple processes such as sorting, searching, or subsetting records become prohibitively time consum-ing. Display and redraw rates become low, reducing the utility of these data, and decreasing the likelihood that GIS will be effectively applied to the problem.

To avoid these problems, a many-to-one relationship is usually allowed between the raster cells and the attribute table (Figure 2-26c). Many raster cells may refer to a single row in the attribute column. This substan-tially reduces the size of the attribute table for most data sets although it does so at the cost of some spatial ambiguity. There may be multiple, non-contiguous patches for a specific type. For example, the upper left and lower right portion of the raster data set in Figure 2-26c are both of class 10. Both are recognized as distinct features in the vector and one-to-one raster representation, but are represented by the same attribute entry in the many-to-one raster representation. This reduces the size of the attribute table, but at the cost of reducing the flexibility of the attribute table. Many-to-one relationships effectively create multipart areas. The data for the represented variable may be summa-rized by class; however, these classes may or may not be spatially contiguous.

An alternative is to maintain the one-to-one relationship, but to index all the raster cells in a contiguous group, thereby reducing the number of rows in the attribute table. This requires software to develop and main-tain the indices, and to create them and reconstitute the indexing after spatial opera-tions. These indexing schemes add overhead and increase data model complexity, thereby removing one of the advantages of raster data sets over vector data sets.

A Comparison of Raster and Vec-tor Data Models

The question often arises, "Which are better, raster or vector data models?" The answer is neither and both. Neither of the two classes of data models is better in all conditions or for all data. Both have advan-tages and disadvantages relative to each

a) Vector, one-to-one

attribute table

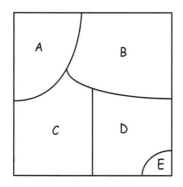

IDorg	class	area
A	10	16.8
B	11	22.2
C	15	18.4
D	21	16.4
E	10	3.8

b) Raster, one-to-one

A	A	A	A	B	B	B	B	B	B
A	A	A	A	B	B	B	B	B	B
A	A	A	A	B	B	B	B	B	B
A	A	A	B	B	B	B	B	B	B
A	A	A	C	C	B	B	B	B	B
C	C	C	C	C	D	D	D	D	D
C	C	C	C	C	D	D	D	D	D
C	C	C	C	C	D	D	D	D	D
C	C	C	C	C	D	D	D	E	E
C	C	C	C	C	D	D	E	E	E

attribute table
(cell 1 is upper-left corner)

cell-ID	IDorg	class	area
1	A	10	0.8
2	A	10	0.8
3	A	10	0.8
4	A	10	0.8
5	B	11	0.8
6	B	11	0.8
7	B	11	0.8
.	.	.	.
.	.	.	.
.	.	.	.
100	E	10	0.8

c) Raster, many-to-one

10	10	10	10	11	11	11	11	11	11
10	10	10	10	11	11	11	11	11	11
10	10	10	10	11	11	11	11	11	11
10	10	10	11	11	11	11	11	11	11
10	10	10	15	15	11	11	11	11	11
15	15	15	15	15	21	21	21	21	21
15	15	15	15	15	21	21	21	21	21
15	15	15	15	15	21	21	21	21	21
15	15	15	15	15	21	21	21	10	10
15	15	15	15	15	21	21	10	10	10

attribute table

class	area
10	18.4
11	24.0
15	21.6
21	13.6

Figure 2-26: Raster data models may combine spatial and attribute data in manners similar to or different from vector data. A typical vector representation of area features is shown in a, above, that maintains a one-to-one relationship between polygons and table rows. Raster data may also maintain this one-to-one relationship (b). A many-to-one relationship between cells and table rows may be adopted (c) because raster data sets often contain a prohibitively large number of cells.

Table 2-2: A comparison of raster and vector data models.

Characteristic	Raster	Vector
data structure	usually simple	usually complex
storage require-ments	larger for most data sets without compression	smaller for most data sets
coordinate conver-sion	may be slow due to data volumes, and require resampling	simple
analysis	easy for continuous data, simple for many layer combinations	preferred for network analyses, many other spatial operations more complex
spatial precision	floor set by cell size	limited only by positional measurements
accessibility	easy to modify or program, due to simple data structure	often complex
display and output	good for images, but discrete features may show "stairstep" edges	map-like, with continuous curves, poor for images

other and to additional, more complex data models. In some instances it is preferable to maintain data in a raster model, and in others in a vector model. Most data may be represented in both, and may be converted among data models. As an example, land cover may be represented as a set of polygons in a vector data model or as a set of identifiers in each cell in a raster grid. The choice often depends on a number of factors, including the predominant type of data (discrete or continuous), the expected types of analyses, available storage, the main sources of input data, and the expertise of the human operators.

Raster data models exhibit several advantages relative to vector data models. First, raster data models are particularly suitable for representing themes or phenomena that change frequently in space. Each raster cell may contain a value different than its neighbors. Thus trends as well as more rapid variability may be represented.

Raster data structures are generally simpler than vector data models, particularly when a fixed cell size is used. Most raster models store cells as sets of rows, with cells organized from left to right, and rows stored from top to bottom. This organization is quite easy to code in an array structure in most computer languages.

Raster data models also facilitate easy overlays, at least relative to vector models. Each raster cell in a layer occupies a given position corresponding to a given location on the Earth's surface. Data in different layers align cell-to-cell over this position. Thus, overlay involves locating the desired grid cell in each data layer and comparing the values found for the given cell location. This cell look-up is quite rapid in most raster data structures, and thus layer overlay is quite simple and rapid when using a raster data model.

Finally, raster data structures are the most practical method for storing, displaying, and manipulating digital image data, such as aerial photographs and satellite imagery. Digital image data are an important source of information when building, viewing, and analyzing spatial databases. Image display and analysis are based on raster operations to sharpen details on the image, specify the brightness, contrast, and colors for display, and to aid in the extraction of information.

Vector data models provide some advantages relative to raster data models. First, vector models often lead to more compact data storage, particularly for discrete objects. Large homogenous regions are recorded by the coordinate boundaries in a vector data model. These regions are recorded as a set of cells in a raster data model. The perimeter grows more slowly than the area for most feature shapes, so the amount of data required to represent an area increases much more rapidly with a raster data model. Vector data are much more compact than raster data for most themes and levels of spatial detail.

Vector data are a more natural means for representing networks and other connected linear features. Vector data by their nature store information on intersections (nodes) and the linkages between them (lines). Traffic volume, speed, timing, and other factors may be associated with lines and intersections to model many kinds of networks.

Vector data models are easily presented in a preferred map format. Humans are familiar with continuous line and rounded curve representations in hand- or machine-drawn maps, and vector-based maps show these curves. Raster data often show a "stair-step" edge for curved boundaries, particularly when the cell resolution is large relative to the resolution at which the raster is displayed. Vector data may be plotted with more visually appealing continuous lines and rounded edges.

Vector data models facilitate the calculation and storage of topological information. Topological information aids in performing adjacency, connectivity, and other analyses in an efficient manner. Topological information also allows some forms of automated error and ambiguity detection, leading to improved data quality.

Conversion Between Raster and Vector Models

Spatial data may be converted between raster and vector data models. Vector-to-raster conversion involves assigning a cell value for each position occupied by vector features. Vector point features are typically assumed to have no dimension. Points in a raster data set must be represented by a value in a raster cell, so points have at least the dimension of the raster cell after conversion from vector-to-raster models. Points are usually assigned to the cell containing the point coordinate. The cell in which the point resides is given a number or other code identifying the point feature occurring at the cell location. If the cell size is too large, two or more vector points may fall in the same cell, and either an ambiguous cell identifier assigned, or a more complex numbering and assignment scheme implemented. Typically a cell size is chosen such that the diagonal cell dimension is smaller than the distance between the two closest point features.

Vector line features in a data layer may also be converted to a raster data model. Raster cells may be coded using different criteria. One simple method assigns a value to a cell if a vector line intersects with any part of the cell (Figure 2-27a, left). This ensures the maintenance of connected lines in the raster form of the data. This assignment rule often leads to wider than appropriate lines because several adjacent cells may be assigned as part of the line, particularly when the line meanders near cell edges. Other assignment rules may be applied, for example, assigning a cell as occupied by a line only when the cell center is near a vector line segment (Figure 2-27a, right). "Near" may be defined as some sub-cell distance, for instance, 1/3 the cell width. Lines passing through the corner of a cell will not be recorded as in the cell. This may lead to thinner linear features in the raster data set, but often at the cost of line discontinuities.

The output from vector-to-raster conversion depends on the algorithm used, even though you use the same input. This brings

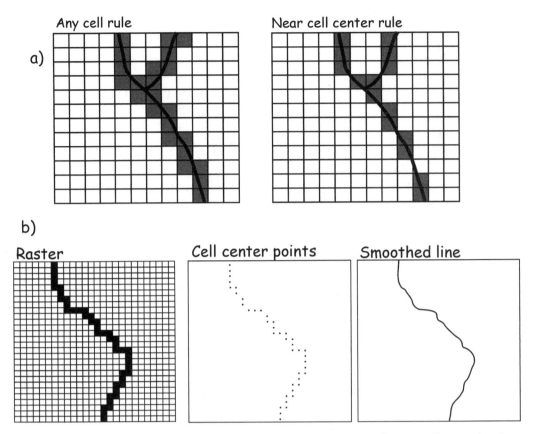

Figure 2-27: Vector-to-raster conversion (a) and raster-to-vector conversion (b). In a, cells are assigned in a raster if they intersect with a converted vector. The left and right panels show how two assignment rules result in different raster coding near lines. Panels in b show how raster data may be converted to vector formats, and may involve line smoothing or other operations to remove the "stair-step" effect.

up an important point to remember when applying any spatial operation. The output often depends in subtle ways on the spatial operation. What appear to be quite small differences in the algorithm or key defining parameters may lead to quite different results. The ease of spatial manipulation in a GIS provides a powerful and often easy-to-use set of tools. The GIS user should bear in mind that these tools can be more efficient at producing errors as well as more efficient at providing correct results. Until sufficient experience is obtained with a suite of algorithms, in this case vector-to-raster conversion, small, controlled tests should be performed to verify the accuracy of a given method or set of constraining parameters.

Up to this point we have covered vector-to-raster data conversion. Data may also be converted in the opposite direction, from raster to vector data. Point, line, or area features represented by raster cells are converted to corresponding vector data coordinates and structures. Point features are represented as single raster cells. Each vector point feature is usually assigned the coordinate of the corresponding cell center.

Linear features represented in a raster environment may be converted to vector lines. Conversion to vector lines typically involves identifying the continuous connected set of grid cells that form the line. Cell centers are typically taken as the locations of vertices along the line (Figure 2-27b). Lines may then be "smoothed" using a mathematical algorithm to remove the "stair-step" effect.

Triangulated Irregular Networks

A *triangulated irregular network* (TIN) is a data model commonly used to represent terrain heights. Typically the x, y, and z locations for measured points are entered into the TIN data model. These points are distributed in space, and the points may be connected such that the smallest triangle possible spans any three adjacent points. The TIN forms a connected network of triangles (Figure 2-28). *Delaunay triangles* are created such that the lines from one triangle do not cross the lines of another. Line crossings are avoided by identifying the *convergent circle* for a set of three points (Figure 2-28).The convergent circle is defined as the circle passing through all three points. A triangle is drawn only if the corresponding convergent circle contains no other sampling points. Each triangle defines a terrain surface, or facet, assumed to be of uniform slope and aspect over the triangle.

The TIN model typically uses some form of indexing to connect neighboring points. Each edge of a triangle connects to two points, which in turn each connect to other edges. These connections continue recursively until the entire network is spanned. Thus, the TIN is a rather more complicated data model than the simple raster grid.

While the TIN model may be more complex than simple raster models, it may also be much more appropriate and efficient when storing terrain data in areas with variable relief. Relatively few points are required to represent large, flat, or smoothly continuous areas. Many more points are desirable when representing variable, discontinuous terrain. Surveyors often collect more samples per unit area where the terrain is highly variable. A TIN easily accommodates these differences in sampling density, resulting in more, smaller triangles in the densely sampled area. Rather than imposing a uniform cell size and having multiple measurements for some cells, one measurement for others, and no measurements for most cells, the TIN preserves each measurement point at each location.

Figure 2-28: A TIN data model defines a set of adjacent triangles over a sample space (left). Sample points, facets, and edges are components of TIN data models. Triangles are placed by convergent circles. These intersect the vertices of a triangle and contain no other possible vertices (below).

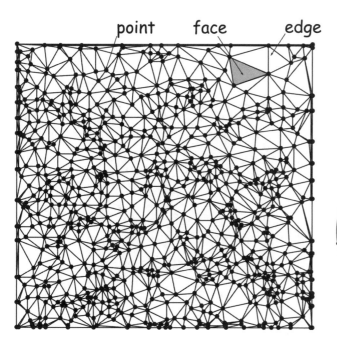

point   face   edge

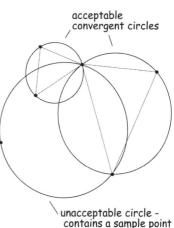

acceptable convergent circles

unacceptable circle - contains a sample point

Multiple Models

Digital data may often be represented using any one of several data models. The analyst must choose which representation to use. Digital elevation data are perhaps the best example of the use of multiple data models to represent the same theme (Figure 2-29). Digital representations of terrain height have a long history and widespread use in GIS. Elevation data and derived surfaces such as slope and aspect are important in hydrology, transportation, ecology, urban and regional planning, utility routing, and a number of other activities that are analyzed using GIS. Because of this widespread importance, digital elevation data are commonly represented in a number of data models.

Raster grids, triangulated irregular networks (TINs), and vector contours are the most common data structures used to organize and store digital elevation data. Raster and TIN data are often called *digital elevation models* (DEMs) or *digital terrain models* (DTMs) and are commonly used in terrain analysis. Contour lines are most often used as a form of input, or as a familiar form of output. Historically, hypsography (terrain heights) was depicted on maps as contour lines (Figure 2-29). Contours represent lines of equal elevation, typically spaced at fixed elevation intervals across the mapped areas. Because many important analyses are more difficult using contour lines, most digital elevation data are stored using raster or TIN models.

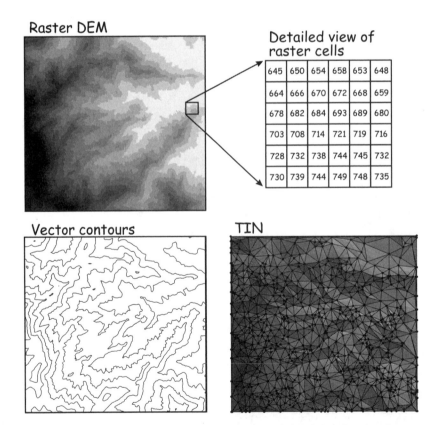

Figure 2-29: Data may often be represented in several data models. Digital elevation data are commonly represented in raster (DEM), vector (contours), and TIN data models.

Object Data Models

The *object data model* is a relatively recent alternative for structuring spatial data. The object data model incorporates much of the philosophy of object-oriented programming into a spatial data model. A main goal is to raise the level of abstraction so that the data objects may be conceptualized and addressed in a more natural way. Object models attempt to encapsulate the information and operations (often called "methods") into discrete objects. These objects could be geographic features, such as a city. Spatial and attribute data associated with a given city would be incorporated in a single city object. This object may include not only information on the city boundary, but also streets, building locations, waterways, or other features that might be in separate data structures in a layered topological vector model. The topology could be included, but would likely be incorporated within the single object. Topological relationships to exterior objects such as adjacent cities or counties may also be represented.

Object model approaches have been adopted by at least one major vendor of GIS software and are gaining acceptance and finding application in a number of fields. We typically conceptualize a small set of features, or types of real-world objects, that we wish to represent in our GIS database. These features themselves might be composed of other features, and these features by yet other features. Buildings might contain floors, and floors contain rooms. We easily conceptualize a building as a nesting of these sometimes complex objects or features.

As another example, we might want to represent an electrical power distribution system using a GIS. This would include a number of different types of features, including power plants, transportation networks to bring fuel to the plants, plant buildings, power lines, and customer buildings. We might represent these as vector points, lines, and polygons. However, these might be difficult to conceptualize and cumbersome to manage if we separate the features into a set

of vector layers. Complex objects, for example, a substation, might be composed of transformers, regulators, interconnected networks, and powerlines. We might have difficulty representing this with simple vector layers. We might need multiple point layers or types to represent transformers vs. regulators, and one or more vector layers to represent the interconnections and the power lines. We might also need to enforce rules about specific kinds of connections or relationships, for example, there is always a regulator between a power source and a transformer.

Object models for spatial data often follow a *logical model*, a user's view of the real objects we portray with a GIS (Figure 2-30). This model includes all the "things" of interest, and the relationships among them. Things, or objects, might include power poles, transformers, powerlines, meters, and customer buildings, and relationships among them would include a transformer on a pole, lines between poles, and meters at points

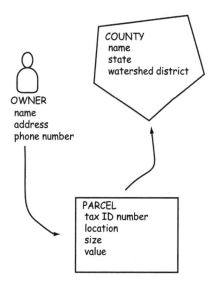

Figure 2-30: Objects in a GIS database may be conceptualized in a diagram, or logical model of how they are related. Here, three types of objects are represented, with owners associated with parcels, and parcels associated with counties.

along the lines. The logical model is often represented as a box-and-line diagram.

Most object models define the properties of each object, and the relationships among the same and different types of objects. Pipes may have a diameter, material type, and date of last inspection, and be connected to valves and tanks, and run through fields or under buildings. The pipes may be represented by lines and the valves by points, but these vector elements are enhanced in the object model because the specific pipe and pipe properties may be linked to the specific valve attached to a given location. The properties of the pipe are encapsulated within the pipe object. Valves may be required to be coincident with pipes, to match pipe properties, and pipelines required to begin and end with a valve.

Because relationships are stored in the database, we can transparently represent the topology and change the topological rules as needed. Relationships are an explicit part of the object-oriented design, so rules on object properties, such as "pipe materials may be only clay, steel, PVC, or copper," or "service lines must connect to main lines" may be explicitly embedded in the database. In other data models these relationships are often embedded in specific, hard-to-modify computer code, a disadvantage that adds cost and time to modifying a database.

Object models can be specified to automatically transfer properties within classes of objects, a capability called *inheritance*. We may create a generic valve object, with a maximum pressure rating, cost, and material type. Within this general class we may create a number of valve subclasses, for example, emergency cut-off valves, primary control valves, or shunt valves. These subclasses will inherit all the property variables from a generic valve in that each has a cost, maximum pressure, and material, and is required to exist on a pipe, but each subclass may also have additional, unique properties.

Figure 2-31 shows an example of an object data model for hydrologic basins and related stream networks and features. The top frame shows graphics of the features, in this example basins, sub-basins, a stream network, and features on the stream network such as measurement or sampling stations. The bottom panel shows the feature types, attributes, and properties in the object model. Note that there are both object properties and topological relationships represented, and that multiple feature types may be represented in the object model. Main basins, sub-basins, and stations are all types of water features. Basins may be made up of multiple kinds of area features. Sub-basin objects may be composed of point, line, and area features, all constrained topologically to be within basins or coincident with stream segments, as appropriate. Stream networks are defined as segments and junctions, and stations of several types may be present, all constrained to be located on stream segments.

The object data model has both advantages and disadvantages when compared to traditional topological vector and raster data models. Some geographic entities may be naturally and easily identified as discrete units for particular problems, and so may be naturally amenable to an object-oriented approach. Some proponents claim object models are more easily implemented across a wider range of database software, particularly for complex models. However, object-data models are less useful for representing continuously varying features, such as elevation. In addition, for many problems object definition and indexing may be quite complex. Software developers have had difficulty developing generic tools that may quickly implement object models, so there is an added level of specialized training required. Finally, we note that there is no widely-accepted, formal definition of what constitutes an object data model.

Objects

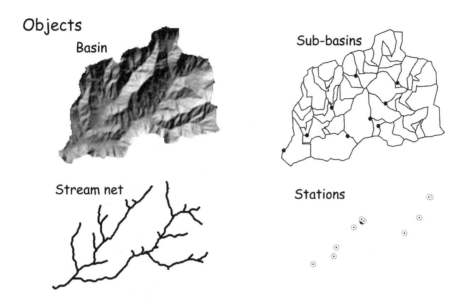

Schematic Diagram

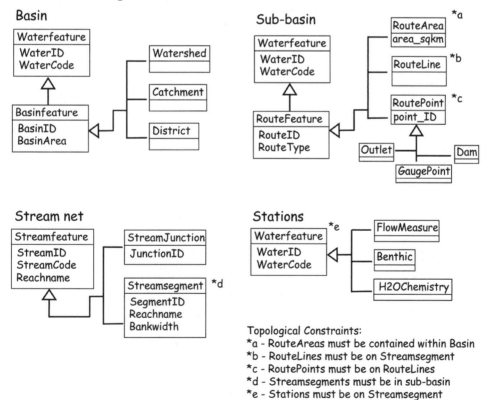

Topological Constraints:
*a - RouteAreas must be contained within Basin
*b - RouteLines must be on Streamsegment
*c - RoutePoints must be on RouteLines
*d - Streamsegments must be in sub-basin
*e - Stations must be on Streamsegment

Figure 2-31: Object-oriented data models allow us to encapsulate complex objects that may be a combination of many different features and feature types, while explicitly identifying the embedded complexity in a standard way. Constraints such as topological relationships across objects may also be represented.

Data and File Structures

Binary and ASCII Numbers

No matter which spatial data model is used, the concepts must be translated into a set of numbers stored on a computer. All information stored on a computer in a digital format may be represented as a series of 0's and 1's. These data are said to be stored in a *binary* format, because each digit may contain one of two values, 0 or 1. Binary numbers are in a base of 2, so each successive column of a number represents a power of two.

We use a similar column convention in our familiar ten-based (decimal) numbering system. As an example, consider the number 47 which we represent using two columns. The seven in the first column indicates there are seven units of one. The four in the tens column indicates there are four units of ten.

Each higher column represents a higher power of ten. The first column represents one (10^0=1), the next column represents tens (10^1=10), the next column hundreds (10^2=100), and upward for successive powers of ten. We add up the values represented in the columns to decipher the number.

Binary numbers are also formed by representing values in columns. In a binary system each column represents a successively higher power of two (Figure 2-32). The first (right-most) column represents 1 ($2^0 = 1$), the second column (from right) represents twos ($2^1 = 2$), the third (from right) represents fours ($2^2 = 4$), then eight ($2^3 = 8$), sixteen ($2^4 = 16$), and upward for successive powers of two. Thus, the binary number 1001 represents the decimal number 9: a one from the rightmost column, and eight from the fourth column (Figure 2-32).

Binary columns

Equivalent numbers

binary	decimal
00000001	1
00000010	2
00000011	3
00000100	4
00000101	5
00000110	6
00000111	7
00001000	8
00001001	9
00001010	10
....	....

Figure 2-32: Binary representation of decimal numbers.

Each digit or column in a binary number is called a *bit*, and eight columns, or bits, are called a *byte*. A byte is a common unit for defining data types and numbers, for example, a data file may be referred to as containing 4-byte integer numbers. This means each number is represented by 4 bytes of binary data (or 8 x 4 = 32 bits).

Several bytes are required when representing larger numbers. For example, one byte may be used to represent 256 different values. When a byte is used for non-negative integer numbers, then only values from 0 to 255 may be recorded. This will work when all values are below 255, but consider an elevation data layer with values greater than 255. If the data are not rescaled, then more than one byte of storage is required for each value. Two bytes will store up to 65,536 different numbers. Terrestrial elevations measured in feet or meters are all below this value, so two bytes of data are often used to store elevation data. Real numbers such as 12.19 or 865.3 typically require more bytes, and are effectively split, that is, two bytes for the whole part of the real number, and four bytes for the fractional portion.

Binary numbers are often used to represent codes. Spatial and attribute data may then be represented as text or as standard codes. This is particularly common when raster or vector data are converted for export or import among different GIS software systems. For example, ArcGIS, a widely used GIS, produces several export formats that are in text or binary formats. Idrisi, another popular GIS, supports binary and alphanumeric raster formats.

One of the most common number coding schemes uses ASCII designators. ASCII stands for the American Standard Code for Information Interchange. ASCII is a standardized, widespread data format that uses seven bits, or the numbers 0 through 126, to represent text and other characters. An extended ASCII, or ANSI (American National Standards Institute) scheme, uses these same codes, plus an extra binary bit to represent numbers between 127 and 255. These codes are then used in many pro-grams, including GIS, particularly for data export or exchange.

ASCII codes allow us to easily and uniformly represent alphanumeric characters such as letters, punctuation, other characters, and numbers. ASCII converts binary numbers to alphanumeric characters through an index. Each alphanumeric character corresponds to a specific number between 0 and 255, which allows any sequence of characters to be represented by a number. One byte is required to represent each character in extended ASCII coding, so ASCII data sets are typically much larger than binary data sets. Geographic data in a GIS may use a combination of binary and ASCII data stored in files. Binary data are typically used for coordinate information, and ASCII or other codes may be used for attribute data.

Pointers and Indexes

Data files may be linked by file *pointers*, *indexes*, or other structures. A pointer is an address or index that connects one file location to another. Pointers are a common way to organize information within and across multiple files. Figure 2-33 depicts an example of the use of pointers to organize spatial data. In Figure 2-33, the polygon is composed of a set of lines. Pointers are used to link the set of lines that form each polygon. There is a pointer from each line to the next line, forming a chain that defines the polygon boundary.

Pointers help by organizing data in such a way as to improve access speed. Unorganized data would require time-consuming searches each time a polygon boundary was to be identified. Pointers also allow efficient use of storage space. In our example, each line segment is stored only once. Several polygons may point to the line segment as it is typically much more space-efficient to add pointers than to duplicate the line segment.

Shapefiles are a common vector spatial data format that uses an index to link files. Shapefiles were originally developed by ESRI, inc., as a way to store point, line, and polygon features, although they have since been adopted as a common format for data interchange and analysis. Shapefiles are supported by Autocad, QGIS, MapWindow, Manifold, and most other GIS softwares that process vector data.

Shapefiles represent layers with a cluster of files. Each file has the same base name but a different filename extension, indicated by a suffix, e.g. the ".shp" in the filename "boundary.shp." A transportation data layer stored in shapefile format might have the base name of roads, with different suffixes for different files:

roads.shp

roads.shx

roads.dbf

roads.prj

etc....

The first three files above are all required to represent a vector data layer using shapefiles. These files are connected using indices, numbers that identify connections and groupings for various components. The .shp files contain the coordinates that represent each road, organized by line segments. There is general information for each segment, and then a list of coordinates and other data for the segment. This is followed

by general information for the next segment, and another list. Since road lengths vary, so will each record (string of numbers) for each road. Note that adjacent road segments are often near each other in the file, but don't have to be. When multiple segments connect at a junction, for example, at a crossroad, not all connections can be sequentially ordered in the list.

Segments in the roads.shp file are indexed by pointers in the roads.shx file. Part of the information stored for a segment is the identifiers of connecting segments. The roads.shx file contains indices that point to the segment records in the .shp files, based on these identifiers. This speeds access, because without indexing, the software would have to search the .shp file each time it needed to find adjacent segments in a road.

The roads.dbf file also uses an index to point to the combined roads in the .shp and .shx files. A group of segments may be used to form a line, and associated with a set of attributes stored in a .dbf file, for example, attributes on road name, surface type, or speed limit. By appropriate use of pointers and indices, largely hidden to the user, this group of three shapefiles implements our vector data model.

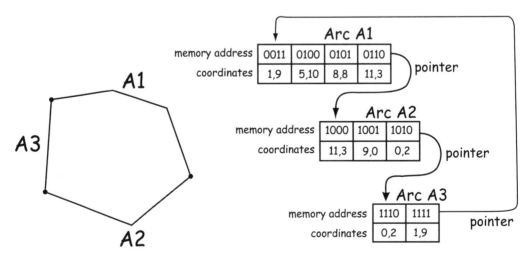

Figure 2-33: Pointers are used to organize vector data. Pointers reduce redundant storage and increase speed of access.

Because pointers and indices are key elements in organizing the spatial data, altering them directly will usually cause problems. Typically these indices are created by the software during processing, and updated as needed when data are added, modified, or analyzed. Pointers may be visible, for example the OID columns in the .dbf tables used with shapefiles, but manually changing the values will often ruin the data layer. You should know the identity and use of pointers in your data sets, so that you don't change them inadvertently.

Pointers, indexing, and multi-file layers are not limited to vector data. Many raster formats store a majority of the cell data in one file, and additional, linked information in an associated file. You must be careful when transferring a data layer to include all the associated files. For example, copying the roads.shp and roads.dbf files to a new location does not copy a usable data layer. The software expects a .shx file; an incomplete file set is often useless.

Data Compression

We often compress spatial data files because they are large. Data compression reduces file size while maintaining the information contained in the file. Compression algorithms may be "lossless," in that all information is maintained during compression, or "lossy," in that some information is lost. A lossless compression algorithm will produce an exact copy of the original when it is applied and then the appropriate decompression algorithm applied. A lossy algorithm will alter the data when it is applied and the appropriate decompression algorithm applied. Lossy algorithms are most often used with image data and are uncommonly applied to thematic spatial data.

Data compression is most often applied to discrete raster data, for example, when representing polygon or area information in a raster GIS. There are redundant data elements in raster representations of large homogenous areas. Each raster cell within a homogenous area will have the same code as most or all of the adjacent cells. Data compression algorithms remove much of this redundancy.

Run-length coding is a common data compression method. This compression technique is based on recording sequential runs of raster cell values. Each run is recorded as the value found in the set of adjacent cells and the run-length, or number of cells with the same value. Seven sequential cells of type A might be listed as A7 instead of AAAAAAA. Thus, seven cells would be represented by two characters. Consider the data recorded in Figure 2-34, where each line of raster cells is represented by a set of run-length codes. In general run-length coding reduces data volume, as shown for the top three rows in Figure 2-34. Note that in some instances run-length coding increases the data volume, most often when there are no long runs. This occurs in

Raster								Run-length codes
9	9	6	6	6	6	6	7	2:9, 5:6, 1:7
6	6	6	6	6	6	6	6	8:6
9	9	6	6	6	6	7	7	2:9, 4:6, 2:7
9	8	9	6	6	7	7	5	1:9, 1:8, 1:9, 2:6, 2:7, 1:5

Figure 2-34: Run-length coding is a common and relatively simple method for compressing raster data. The left number in the run-length pair is the number of cells in the run, and the right is the cell value. Thus, the 2:9 listed at the start of the first line indicates a run of length two for the cell value 9.

the last line of Figure 2-34, where frequent changes in adjacent cell values result in many short runs. However, for most thematic data sets containing area information, run-length coding substantially reduces the size of raster data sets.

There is also some data access cost in run-length coding. Standard raster data access involves simply counting the number of cells across a row to locate a given cell. To locate a cell in run-length coding we must sum along the run-length codes to identify a cell position. This is typically a minor additional cost, but in some applications the trade-off between speed and data volume may be objectionable.

Quad tree representations are another raster compression method. Quad trees are similar to run-length codings in that they are most often used to compress raster data sets when representing area features. Quad trees may be thought of as a raster data structure with a variable spatial resolution. Raster cell sizes are combined and adjusted within the data layer to fit into each specific area feature (Figure 2-35). Large raster cells that fit entirely into one uniform area are assigned the value corresponding to that area, for example, the three largest cells in Figure 2-35 are all assigned the value a. Successively smaller cells are then fit, halving the cell

dimension at each iteration, again fitting the largest cell that will fit in each uniform area. This is illustrated in the top-left corner of Figure 2-35. Successively smaller cells are defined by splitting "mixed cells" into four quadrants, and assigning the values a or b to uniform areas. This is repeated down to the smallest cell size that is needed to represent uniform areas at the required detail.

The varying cell size in a quad tree representation requires more sophisticated indexing than simple raster data sets. Pointers are used to link data elements in a tree-like structure, hence the name quad trees. There are many ways to structure the data pointers, from large to small, or by dividing quandrants, and these methods are beyond the scope of an introductory text. Further information on the structure of quad trees may be found in the references at the end of this chapter.

There are many other data compression methods that are commonly applied. JPEG and wavelet compression algorithms are often applied to reduce the size of spatial data, particularly image or other data. Generic bit and byte-level compression methods may be applied to any files for compression or communications. There is usually some cost in time to the compression and decompression.

Raster Pyramids

We sometimes intentionally increase the size of our raster datasets without increasing the resolution, in a process known as *pyramiding*. We create pyramids to increase display speeds when viewed at small scales ("zoomed out"). Long redraw times is often an issue when displaying large data sets, particularly when panning frequently. When displayed at very small scales, the cell size of a data set may be smaller than the resolution of the computer screen. A raster data set 1,000,000 pixels across has 1000 times the horizontal pixels that can be displayed on a monitor with 1000 pixel horizontal resolution. However, display software must wade through all 1,000,000 data elements in a row

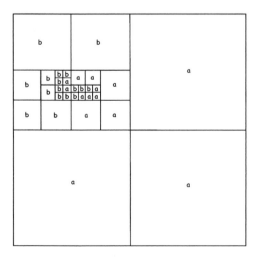

Figure 2-35: Quad tree compression.

to pick the 1 cell in 1000 to display. While clever software can help, there are limits to how much we can speed up the redraws.

Pyramiding in effect saves sub-sampled copies of the cells at various resolutions. In our example above, pyramids may do the equivalent of saving every two, every four, every 10, every 30, and every 100 cells, all within the same raster data set. The software then compares the display scale to the dimensions of the data set, and chooses the most appropriate cell resolution to display. Redraws are much faster, and transparent to the user.

Note that we say pyramids "in effect" save copies of cells at various resolutions. This is the simplest method, but often not the most efficient for space or speed of access. Sophisticated indexing may be used to point to the cells at the appropriate resolutions.

Note that pyramiding comes at a cost, both in the size and complexity of the raster data set. Indexing schemes complicate the simple raster data structure, and the software must be able to navigate the indexing scheme. Already large raster data sets may be inflated from a few percent to several times, although in practice it is typically less than a doubling.

Common File Formats

A small number of file formats are commonly used to store and transfer spatial data. Some of these file structures arose from distribution formats adopted by governmental departments or agencies. Others are based on formats specified by software vendors, and some have been devised by standards-making bodies. Some knowledge of the types, properties, and common naming conventions of these file formats is helpful to the GIS practitioner.

Common geographic data formats may be placed into three large classes: raster, vector, and attribute. Raster formats may be further split into single-band and multi-band file types. Multi-band raster data sets are most often used to store and distribute image data, while single-band raster data sets are used to store both single-band images and non-image spatial data. Table 2-3 summarizes some of the most common spatial data formats.

Summary

In this chapter we have described the main ways of conceptualizing spatial entities, and of representing these entities as spatial features in a computer. We commonly employ two conceptualizations, also called spatial data models: a raster data model and a vector data model. Both models use a combination of coordinates, defined in a Cartesian or spherical system, and attributes, to represent our spatial features. Features are usually segregated by thematic type in layers.

Vector data models describe the world as a set of point, line, and area features. Attributes may be associated with each feature. A vector data model splits that world into discrete features, and often supports topological relationships. Vector models are most often used to represent features that are considered discrete, and are compatible with vector maps, a common output form.

Raster data models are based on grid cells and represent the world as a "checkerboard," with uniform values within each cell. A raster data model is a natural choice for representing features that vary continuously across space, such as temperature or precipitation. Data may be converted between raster and vector data models.

We use data structures and computer codes to represent our conceptualizations in more abstract, but computer-compatible forms. These structures may be optimized to reduce storage space and increase access speed, or to enhance processing based on the nature of our spatial data.

Table 2-3: Common formats for spatial data.

Type and source	Extension or other naming convention	Characteristics (R=Raster, V=Vector, A=Attribute, I=Image)
DLG, USGS	.opt, .ddf, .dbf	Digital Line Graph data (V, A).
DXF, AutoDesk	.dxf	Drawing exchange file, an ASCII or binary file for exchanging spatial data (V).
DWG, Autodesk	.dwg	Native binary file used by AutoDesk to store geographic data and drawings in AutoCAD (V).
Interchange, ESRI	.e00	ASCII text file for vector and identifying attribute data (V).
shapefile, ESRI	.shp, .shx, .dbf, .prj, and others	Three or more binary files that include the vector coordinate, attribute, and other information (V).
VPF, US Dept. of Defense.	various	Vector Products Format, Defense Mapping Agency distribution specification.
TIGER, U.S. Census	tgrxxyyyy, stfzz	Set of files by U.S. census areas, xx is a state code, yyy an area code, zz numbers for various file types
MIF/MID, MapInfo	.mif, .mid	Map Interchange File, vector and raster data transport from MapInfo (V,R).
SDTS, U.S. Government	none	Spatial Data Transfer Standard, specifies the spatial objects, attributes, reference system, data dictionary, and other information (R,V, A).
GeoTIFF, Work Group	.TIF, .TFF	An extension for geo-referencing Aldus-Adobe public domain TIFF format (R).
DEM, U.S.G.S	.dem	ASCII text format used to distribute elevation information (R).
Imagine, ERDAS	.img	Multiband-capable image format (R)
NetCDF, Open Geospatiol Consortium	.cdf	machine-independent data formats for scientific data arrays, common for storing two- and three-dimensional rasters

Suggested Reading

Batcheller, J.K., Gittings, B.M., & Dowers, S. (2007). The performance of vector oriented data storage in ESRI's ArcGIS. *Transactions in GIS*, 11:47-65.

Batty, M. & Xie, Y. (1991). Model structures, exploratory spatial data analysis, and aggregation. *International Journal of Geographical Information Systems*, 8:291-307.

Bhalla, N. (1991). Object-oriented data models: a perspective and comparative review. *Journal of Information Science*, 17:145-160.

Bregt, A.K., Denneboom, J, Gesink, H.J., & van Randen, Y. (1991). Determination of rasterizing error: a case study with the soil map of The Netherlands. *International Journal of Geographical Information Systems*, 5:361-367.

Carrara, A., Bitelli, G., & Carla, R. (1997). Comparison of techniques for generating digital terrain models from contour lines. *International Journal of Geographical Information Systems*, 11:451-473.

Congalton, RG. (1997). Exploring and evaluating the consequences of vector-to-raster and raster-to-vector conversion. *Photogrammetric Engineering and Remote Sensing*, 63:425-434.

Downs, R.M. (1998). The geographic eye: seeing through GIS. *Transactions in GIS,* 2:111-121.

Holroyd, F., & Bell, S. B. M. (1992). Raster GIS: Models of raster encoding. *Computers and Geosciences*, 18:419-426.

Joao, E.M. (1998). *Causes and Consequences of Map Generalization.* Taylor and Francis: London.

Kumler, M.P. (1994). An intensive comparison of triangulated irregular networks (TINs) and digital elevation models. *Cartographica*, 31:1-99.

Langram, G. (1992). *Time in Geographical Information Systems.* Taylor and Francis: London.

Laurini, R. & Thompson, D. (1992). *Fundamentals of Spatial Information Systems,* Academic Press: London.

Lee, J. (1991). Comparison of existing methods for building triangular irregular network models of terrain from grid digital elevation models. *International Journal of Geographical Information Systems*, 5:267-285.

Maquire, D.J., Goodchild, M.F., & Rhind, D. (Eds.) (1991). *Geographical Information Systems: Principles and Applications*, Longman Scientific: Harlow.

Masser, I. (2005). *GIS Worlds: Creating Spatial Data Infrastructures*. ESRI Press: Redlands

Nagy, G. & Wagle, S.G. (1979). Approximation of polygonal maps by cellular maps. *Communications of the Association of Computational Machinery*, 22:518-525.

Peuquet, D.J. (1984). A conceptual framework and comparison of spatial data models. *Cartographica*, 21:66-113.

Peuquet, D.J. (1981). An examination of techniques for reformatting digital cartographic data. Part II: the raster to vector process. *Cartographica*, 18:375-394.

Piwowar, J.M., LeDrew, E.F., & Dudycha, D.J. (1990). Integration of spatial data in vector and raster formats in geographical information systems. *International Journal of Geographical Information Systems*, 4:429-444.

Peuker, T. K. & Chrisman, N. (1975). Cartographic Data Structures. *The American Cartographer*, 2:55-69.

Rana, S. (2004). *Topological Data Structures for Surfaces: An Introduction to Geographical Information Science*. Wiley: New York.

Rigaux, P., Scholl, M., & Voisard, A. (Eds.) (2002). *Spatial databases: with application to GIS*. Elsevier: New York

Rossiter, D.G. (1996). A theoretical framework for land evaluation, *Geoderma*, 72:165-190.

Shaffer, C.A., Samet, H., & Nelson R C. (1990). QUILT: a geographic information system based on quadtrees, *International Journal of Geographical Information Systems*, 4:103-132.

Slocum, T.A., McMaster, R.B, Kessler, F.C. & Howard, H.H. (2005). *Thematic Cartography and Geographic Visualization*, 2nd Ed., Prentice-Hall: New York.

Tomlinson, R.F. (1988). The impact of the transition from analogue to digital cartographic representation. *The American Cartographer,* 15:249-262.

Wedhe, M. (1992). Grid cell size in relation to errors in maps and inventories produced by computerized map processes. *Photogrammetric Engineering and Remote Sensing*, 48:1289-1298.

Wise, S. (2002). *GIS Basics*, Taylor & Francis: New York.

Worboys, M.F. & Duckham, M. (2004). *GIS: A Computing Perspective* (2nd ed.). CRC Press: Boca Raton.

Zeiler, M. (1999). *Modeling Our World: The ESRI Guide to Geodatabase Design*. ESRI Press: Redlands.

Zhi-Jun L. & D.E. Weller (2007). A stream network model for integrated watershed modeling. *Environmental Modeling and Assessment*, DOI:10.1007/s10666-007-9083-9.

Study Questions

2.1 - How is an entity different from a cartographic object?

2.2 - Describe the successive levels of abstraction when representing real-world spatial phenomena on a computer. Why are there multiple levels, instead of just one level in a spatial data representation?

2.3 - Define a data model and describe three primary differences between the two most commonly used data models.

2.4 - Characterize the following lists as nominal, ordinal, or interval/ratio:

a) 1.1, 5.7, -23.2, 0.4, 6.67

b) green, red, blue, yellow, sepia

c) white, light grey, dark grey, black

d) extra small, small, medium, large, extra large

e) forest, woodland, grassland, bare soil

f) 1, 2, 3, 4, 5, 6, 7.

2.5 - Complete the following coordinate conversion table, converting the listed points from degrees-minutes-seconds (DMS) to decimal degrees (DD), or from DD to DMS. See Figure 2-9 for the conversion formula.

Point	DMS	Decimal Degrees
1	36°45'12"	36.75333
2	114°58'2"	
3	85°19'7"	
4		14.00917
5		275.00001
6		0.99528
7	183°19'22"	

2.6 - What is topology, and why is it important? What is planar topology, and when might non-planar be more useful than planar topology?

2.7 - What are the respective advantages and disadvantages of vector data models vs. raster data models?

2.8 - Under what conditions are mixed cells a problem in raster data models? In what ways may the problem of mixed cells be addressed?

2.9 - Indicate which of the following are allowable geographic coordinates:

      a) N45 45' 45"   b) longitude -127.34795   c) S96 12' 33"

      d) E 66 15' 60"   e) W -12 23' 55"   f) N 56.9999

2.10 - Indicate which of the following are allowable geographic coordinates:

      a) W145 45'12"   b) latitude -62.34795   c) E110 52' 43"

      d) S 49 15' 59"   e) N 89 59' 60"   e) S 46.6000

2.11 - Express following base 10 numbers in binary notation:

    a) 2   b) 8   c) 9   d) 17

    e) 0   f) 128   g) 22   h) 19

2.12 - Express following base 10 numbers in binary notation:

    a) 1   b) 23   c) 256   d) 4

    e) 11   f) 10   g) 3   h) 20

2.13 - Express the following binary numbers in base 10 notation:

    a) 0101   b) 0001   c) 1111   d) 00101101

    e) 1101   f) 1011   g) 10000001   h) 11111111

2.14 - Express the following binary numbers in base 10 notation:

    a) 1110   b) 1001   c) 0011   d) 10000101

    e) 1000   f) 1010   g) 10010001   h) 11110000

2.15 - The following figure shows change in raster resolution, combining four small cells on the left to create and output for each corresponding larger cell on the right. Fill in the two rasters on the right, for the interval/ratio data (top), and the nominal data (bottom).

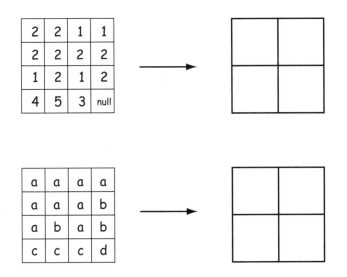

2.16 - What is a triangulated irregular network?

2.17 - What are binary and ASCII numbers? What are the binary equivalents of the following decimal numbers to a binary form: 8, 12, 244?

2.18 - Why do we need to compress data? Which are most commonly compressed, raster data or vector data? Why?

2.19 - What are pointers when used in the context of spatial data, and how are they helpful in organizing spatial data?

2.20 - What are the main concepts behind object data models, and how do they differ from other data models?

2.21 - Write the run length coding for each of the rows in this raster:

b	b	a	a	a	c	a	a	a
c	c	b	b	d	d	d	a	a
b	b	b	b	b	b	b	b	b
e	c	f	b	a	d	f	b	a
a	s	a	f	f	f	b	b	a

2.22 - Write the run length coding for each of the rows in this raster:

c	c	c	c	a	a	a	a	a
a	a	b	b	d	d	d	a	a
e	e	e	f	f	f	f	f	e
a	a	a	a	a	a	a	a	a
c	c	a	a	a	b	f	d	e

3 Geodesy, Datums, Map Projections, and Coordinate Systems

Introduction

Geographic information systems are different from other information systems because they contain spatial data. These spatial data include coordinates that define the location, shape, and extent of geographic objects. To effectively use GIS, we must develop a clear understanding of how coordinate systems are established for the Earth, how these coordinates are measured on the Earth's curving surface, and how these coordinates are transferred to flat maps. This chapter introduces *geodesy*, the science of measuring the shape of the Earth, and *map projections*, the transformation of coordinate locations from the Earth's curved surface onto flat maps.

Defining coordinates for the Earth's surface is complicated by three main factors. First, most people best understand geography in a Cartesian coordinate system on a flat surface. Humans naturally perceive the Earth's surface as flat, because at human scales the Earth's curvature is barely perceptible. Humans have been using flat maps for more than 40 centuries, and although globes are quite useful for perception and visualization at extremely small scales, they are not practical for most purposes.

A flat map must distort geometry in some way because the Earth is curved. When we plot latitude and longitude coordinates on a Cartesian system, "straight" lines will appear bent, and polygons will be distorted. This distortion may be difficult to detect on detailed maps that cover a small area, but the distortion is quite apparent on large-area maps. Because measurements on maps are affected by the distortion, we must somehow reconcile the portrayal of the Earth's truly curved surface onto a flat surface.

The second main problem in defining a coordinate system results from the irregular shape of the Earth. We learn early on that the Earth is shaped as a sphere. This is a valid approximation for many uses, however, it is only an approximation. Past and present natural forces yield an irregularly shaped Earth. These deformations affect how we best map the surface of the Earth, and how we define Cartesian coordinate systems for mapping and GIS.

Thirdly, our measurements are rarely perfect, and this applies when measuring both the shape of the Earth, and the exact position of features on it. All locations depend on measurements that contain some error, and on analyses that must make some assumptions. Our measurements improve through time, and so does the sophistication of our models, so our positional estimates improve; this evolution means our estimates of positions change through time.

Because of these three factors, we often have several different sets of coordinates to define the same location on the surface of the Earth. Remember, coordinates

Coordinates for a Point Location

From Surveyor Data:

	Latitude (N)	Longitude (W)	
NAD83(2007)	44 57 23.23074	093 05 58.28007	
NAD83(1986)	44 57 23.22405	093 05 58.27471	
NAD83(1996)	44 57 23.23047	093 05 58.27944	
	X	Y	
SPC MNS	317,778.887	871,048.844	MT
SPC MNS	1,042,579.57	2,857,766.08	sFT
UTM15	4,978,117.714	492,150.186	MT

From Data Layers:

	X	Y	
MN-Ramsey	573,475.592	160,414.122	sFT
MN-Ramsey	174,195.315	48,893.966	MT
SPC MNC	890,795.838	95,819.779	MT
SPC MNC	2,922,552.206	314,365.207	sFT
LCC	542,153.586	18,266.334	MT

Figure 3-1: An example of different coordinate values for the same point. We may look up the coordinates for a well-surveyed point, and we may also obtain the coordinates for the same point from a number of different data layers. We often find multiple latitude/longitude values (surveyor data, top), or x and y values for the same point (surveyor data, or from data layers, bottom).

are sets of numbers that unambiguously define locations. They are usually x and y values, or perhaps x, y, and z values, or latitude and longitude values unique to a location. But these values are only "unique" to the location for a specified set of measurements and time. The coordinates depend on how we translate points from a curved Earth to a flat map surface (first factor, above), the estimate we use for the real shape of the Earth (second factor), and what set of measurements we reference our coordinates to (the third factor). We may, and often do, address these three factors in a number of different ways, and the coordinates for the same point will be different for these different choices.

An example will help clarify this concept. Figure 3-1 shows the location of a U.S. bench mark, a precisely surveyed and monumented point. Coordinates for this point are maintained by Federal and State government surveyors, and resulting coordinates shown at the top right of the figure. Note that there are three different versions of the latitude/longitude location for this point. In this case, the three versions differ primarily due to differences in the measurements used to establish the point's location, and how measurement errors were adjusted (the third factor, discussed above). The GIS practitioner may well ask, which latitude/longitude pair should I use? This chapter contains the information that should allow you to choose wisely.

Note that there are also several versions of the x and y coordinates for the point in Figure 3-1. The difference in the coordinate values are too great to be due solely to measurement errors. They are due primarily to how we choose to project from the curved Earth to a flat map (the first factor), and in part to the Earth shape we adopt and the measurement system we use (the second two factors).

We first must define a specific coordinate system, meaning we choose a specific way to address the three main factors of projection distortion, an irregularly shaped Earth, and measurement imprecision. Thereafter the coordinates for a given point are fixed, as are the spatial relationships to other measured points. But it is crucial to realize that different ways of addressing 1) the Earth's curvature, 2) the Earth's deviation from our idealized shape, and 3) inevitable inaccuracies in measurement, will result in different coordinate systems, and these differences are the root of much confusion and many errors in spatial analysis. As a rule, you should understand the coordinate system used for all of your data, and convert all data to the same coordinate system prior to analysis. The remainder of this chapter describes how we define, measure, and convert among coordinate systems.

Early Measurements

In specifying a coordinate system, we must first define the size and shape of the Earth. Humans have long speculated on this. Babylonians believed the Earth was a flat disk floating in an endless ocean, a notion adopted by Homer, one of the more widely known Greek writers. The Greeks were early champions of geometry, and they had many competing views of the shape of the Earth. One early Greek, Anaximenes, believed the Earth was a rectangular box, while Pythagoras and later Aristotle reasoned that the Earth must be a sphere. He observed that ships disappeared over the horizon, the moon appeared to be a sphere, that the stars moved in circular patterns, and that constellations shift when viewed from different ends of the Mediterranean Sea. These observations were all consistent with a spherical Earth.

The Greeks next turned toward estimating the size of the sphere. The early Greeks measured locations on the Earth's surface relative to the Sun or stars, reasoning they provided a stable reference frame. This assumption underlies most geodetic observations taken over the past 2000 years, and still applies today, with suitable refinements.

Eratosthenes, a Greek scholar in Egypt, performed one of the earliest well-founded measurements of the Earth's circumference. He noticed that on the summer solstice the Sun at noon shone to the bottom of a deep well in Syene. He believed that the well was located on the Tropic of Cancer, so that the Sun would be exactly overhead during the summer solstice. He also observed that 805 km north in Alexandria, at exactly the same date and time, a vertical post cast a shadow. The shadow/post combination defined an angle which was about $7^{\circ}12'$, or about 1/50th of a circle (Figure 3-2).

Eratosthenes deduced that the Earth must be 805 multiplied by 50, or about 40,250 kilometers in circumference. His calculations were all in stadia, the unit of measure of the time, and have been converted

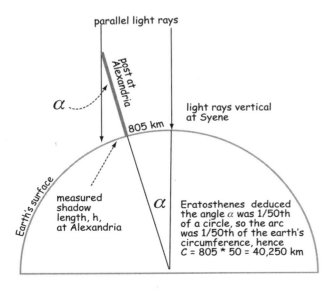

Figure 3-2: Measurements made by Eratosthenes to determine the circumference of the Earth.

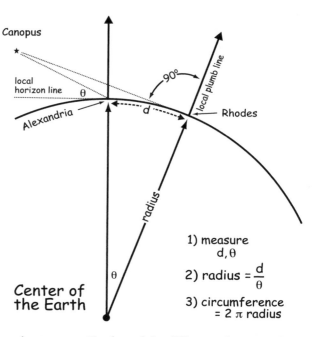

Figure 3-3: Posidonius approximated the Earth's radius by simultaneous measurement of zenith angles at two points. Two points are separated by an arc distance d measured on the Earth surface. These points also span an angle θ defined at the Earth center. The Earth radius is related to d and θ. Once the radius is calculated, the Earth circumference may be determined. Note this is an approximation, not an exact estimate, but was appropriate for the measurements available at the time (adapted from Smith, 1997).

here to the metric equivalent, using our best idea of the length of a stadia. Eratosthenes' estimate differs from our modern measurements of the Earth's circumference by less than 4%.

The accuracy of Eratosthenes' estimate is quite remarkable, given the equipment for measuring distance and angles at that time, and because a number of his assumptions were incorrect. The well at Syene was located about 60 kilometers off the Tropic of Cancer, so the Sun was not directly overhead. The true distance between the well location and Alexandria was about 729 kilometers, not 805, and the well was 3°3' east of the meridian of Alexandria, and not due north. However these errors either compensated for or were offset by measurement errors to end up with an amazingly accurate estimate.

Posidonius, another Greek scholar, made an independent estimate of the size of the Earth by measuring angles from local vertical (plumb) lines to a star near the horizon (Figure 3-3). Stars visible in the night sky define a uniform reference. The angle between a plumb line and a star location is called a *zenith angle.* The zenith angle can be measured simultaneously at two locations

on Earth, and the difference between the two zenith angles can be used to calculate the circumference of the Earth. Figure 3-3 illustrates the observation by Posidonius at Rhodes. The star named Canopus was on the horizon at Rhodes, meaning the zenith angle at Rhodes was 90 degrees. He also noticed Canopus was above the horizon at Alexandria, meaning the zenith angle was less than 90 degrees. The surface distance between these two locations was also measured, and the measurements combined with an approximate geometric relationships to calculate the Earth's circumference. Posidonius calculated the difference in the zenith angles at Canopus as about 1/48th of a circle between Rhodes and Alexandria. By estimating these two towns to be about 800 kilometers apart, he calculated the circumference of the Earth to be 38,600 kilometers. Again there were compensating errors, resulting in an accurate value. Another Greek scientist determined the circumference to be 28,960 kilometers, and unfortunately this shorter measurement was adopted by Ptomely for his world maps. This estimate was widely accepted until the 1500s, when Gerardus Mercator revised the figure upward.

During the 17th and 18th centuries two developments led to intense activity directed at measuring the size and shape of the Earth. Sir Isaac Newton and others reasoned the Earth must be flattened somewhat due to rotational forces. They argued that centrifugal forces cause the equatorial regions of the Earth to bulge as it spins on its axis. They proposed the Earth would be better modeled by an *ellipsoid*, a sphere that was slightly flattened at the poles. Measurements by their French contemporaries taken north and south of Paris suggested the Earth was flattened in an equatorial direction and not in a polar direction. The controversy persisted until expeditions by the French Royal Academy of Sciences between 1730 and 1745 measured the shape of the Earth near the equator in South America and in the high northern latitudes of Europe. Complex, repeated, and highly accurate measurements established that the curvature of the Earth was greater at the equator than the poles, and that an ellipsoid flattened at the poles was indeed the best geometric model of the Earth's surface.

Note that the words spheroid and ellipsoid are often used interchangeably. For example, the Clarke 1880 ellipsoid is often referred to as the Clarke 1880 spheroid, even though Clarke provided parameters for an ellipsoidal model of the Earth's shape. GIS software often prompts the user for a spheroid when defining a coordinate projection, and then lists a set of ellipsoids for choices.

An ellipsoid is sometimes referred to as a special class of spheroid known as an "oblate" spheroid. Thus, it is less precise but still correct to refer to an ellipsoid more generally as a spheroid. It would perhaps cause less confusion if the terms were used more consistently, but the usage is widespread.

Specifying the Ellipsoid

Once the general shape of the Earth was determined, geodesists focused on precisely measuring the size of the ellipsoid. The ellipsoid has two characteristic dimensions (Figure 3-4). These are the *semi-major axis*, the radius a in the equatorial direction, and the *semi-minor axis*, the radius b in the polar direction. The equatorial radius is always greater than the polar radius for the Earth ellipsoid. This difference in polar and equatorial radii can also be described by the flattening factor, as shown in Figure 3-4.

Earth radii have been determined since the 18th century using a number of methods. The most common methods until recently have involved astronomical observations similar to the those performed by Posidonius. These astronomical observations, also called celestial observations, are combined with long-distance surveys over large areas (Figure 3-5). The distance and associated angles are measured in polar and equatorial directions, and used to estimate radii along the arcs. Several measurements were often combined to estimate semi-major and semi-minor axes.

Star and sun locations have been observed and cataloged for centuries, and combined with accurate clocks, the positions of these celestial bodies may be measured to precisely establish the latitudes and longitudes of points on the surface of the Earth. Measurements during the 18th, 19th and early 20th centuries used optical instruments for celestial observations (Figure 3-6).

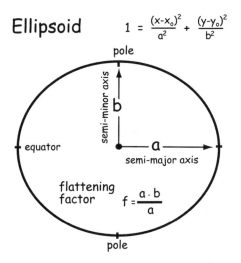

Figure 3-4: An ellipsoidal model of the Earth's shape.

An ellipsoid is defined in part
by two radii, a and b

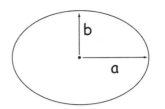

We may use the relationship
$r = d \cdot \theta$ to estimate radii:

$$a = \frac{d_1}{\theta_1}$$

$$b = \frac{d_2}{\theta_2}$$

Generally, the measurements are
not at the poles and equator, and
the math is more complicated, but
the principle is the same.

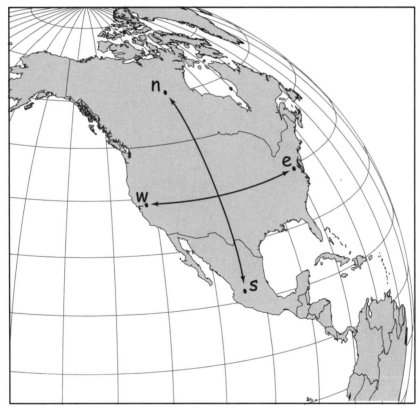

Figure 3-5: Two arcs illustrate the surface measurements and calculations used to estimate the semi-major and semi-minor axes, here for North America. The arc lengths may be measured by surface surveys, and the angles from astronomical observations, as illustrated in Figure 3-2 and Figure 3-3.

Figure 3-6: An instrument used in the early 1900s for measuring the position of celestial bodies.

Measurement efforts through the 19th and 20th centuries led to the establishment of a set of official ellipsoids (Table 3-1). Why not use the same ellipsoid everywhere on Earth, instead of the different ellipsoids listed in Table 3-1? Different ellipsoids were adopted in various parts of the world primarily because there were different sets of measurements used in each region or continent, and these measurements often could not be combined in a unified analysis.

Historically, geodetic surveys were isolated by large water bodies. For example, surveys in Australia did not span the Pacific Ocean to reach Asia. Geodetic surveys relied primarily on optical instruments prior to the early 20th century. These instruments were essentially precise telescopes, and sighting distances were limited by the Earth's curvature. Individual survey legs greater than 50 kilometers (30 miles) were rare, so during

Table 3-1: Official ellipsoids. Radii may be specified more precisely than the 0.1 meter shown here (from Snyder, 1987 and other sources).

Name	Year	Equatorial Radius, a meters	Polar Radius, b meters	Flattening Factor	Users
Airy	1830	6,377,563.4	6,356,256.9	1/299.32	Great Britain
Bessel	1841	6,377,397.2	6,356,079.0	1/299.15	Central Europe, Chile, Indonesia, U.S.
Clarke	1866	6,378,206.4	6,356,583.8	1/294.98	North America; Philippines
Clarke	1880	6,378,249.1	6,356,514.9	1/293.46	Most of Africa; France
International	1924	6,378,388.0	6,356,911.9	1/297.00	Much of the world
Australian	1965	6,378,160.0	6,356,774.7	1/298.25	Australia
WGS72	1972	6,378,135.0	6,356,750.5	1/298.26	NASA, US Def. Dept.
GRS80	1980	6,378,137.0	6,356,752.3	1/298.26	Worldwide
WGS84	1987 - current	6,378,137.0	6,356,752.3	1/298.26	US DOD, Worldwide

this period there were no good ways to connect surveys between continents.

Because continental surveys were isolated, ellipsoidal parameters were fit for each country, continent, or comparably large survey area. These ellipsoids represented continental measurements and conditions. Because of measurement errors, differences in methods for ellipsoidal calculation, and because the Earth's shape is not a perfect ellipsoid (described in the next section), different ellipsoids around the world usually had slightly different origins, axis orientations, and radii. These differences, while small, often result in quite different estimates for coordinate location at any given point, depending on the ellipsoid used.

More recently, data derived from satellites, lasers, and broadcast timing signals have been used for extremely precise measurements of relative positions across continents and oceans. Global measurements and faster computers allow us to estimate globally-applicable ellipsoids. These ellipsoids provide a "best" overall fit ellipsoid to observed measurements across the globe. Global ellipsoids such as the GRS80 or WGS84 are now preferred and most widely used.

The Geoid

As noted in the previous section, the true shape of the Earth varies slightly from the mathematically smooth surface of an ellipsoid. Differences in the density of the Earth cause variation in the strength of the gravitational pull, in turn causing regions to dip or bulge above or below a reference ellipsoid (Figure 3-7). This undulating shape is called a *geoid*.

Geodesists have defined the geoid as the three-dimensional surface along which the pull of gravity is a specified constant. The geoidal surface may be thought of as an imaginary sea that covers the entire Earth and is not affected by wind, waves, the Moon, or forces other than Earth's gravity. The surface of the geoid extends across the

Earth, approximately at mean sea level across the oceans, and continuing under continents at a level set by gravity. The surface is always at right angles to the direction of local gravity, and this surface is the reference against which heights are measured.

Figure 3-8 shows how differences in the Earth's shape due to geoidal deviations will produce different best local ellipsoids. Surveys of one portion of the Earth that best fit the surveyed points will produce different best estimates of the ellipsoid origin, axis orientation, and of r_1 and r_2 than surveys of other parts of the Earth. Measurements based on Australian surveys yielded a different "best" ellipsoid than those in Europe. Likewise, Europe's best ellipsoidal estimate was different from Asia's, and from South America's, North America's, or those of other regions. One ellipsoid could not be fit to all the world's survey data because during the 18th and 19th centuries there was no clear way to combine a global set of measurements.

We must emphasize that a geoidal surface differs from mean sea level. Mean sea

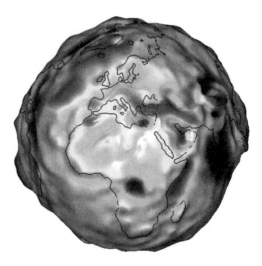

Figure 3-7: Depictions of the Earth's gravity field, as estimated from satellite measurements. These show the undulations, greatly exaggerated, in the Earth's gravity, and hence the geoid (courtesy University of Texas Center for Space Research, and NASA).

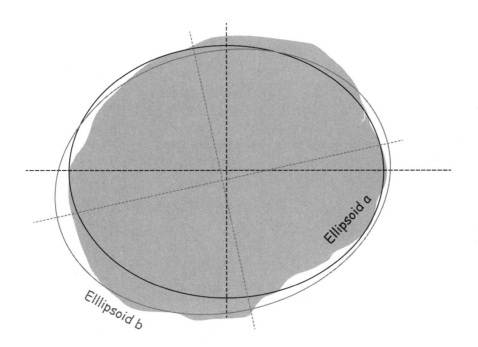

Figure 3-8: Different ellipsoids were estimated due to local irregularities in the Earth's shape. Local best-fit ellipsoids varied from the global best fit, but until the 1970s, there were few good ways to combine global geodetic measurements.

level may be higher or lower than a geoidal surface because ocean currents, temperature, salinity, and wind variations can cause persistent high or low areas in the ocean. These differences are measurable, in places over a meter (3 feet), perhaps small on global scale, but large in local or regional analysis. We historically referenced heights to mean sea level, and many believe we still do, but this is no longer true for most spatial data analyses.

Because we have two reference surfaces, a geiod and an ellipsoid, we also have two bases from which to measure height. Elevation is typically defined as the distance above a geoid. This height above a geoid is also called the *orthometric height* (Figure 3-9). Heights above the ellipsoid are often referred to as *ellipsoidal height*. These are illustrated in Figure 3-9, with the ellipsoidal height labeled h, and orthometric height labeled H. The difference between the ellip-

soidal height and geoidal height at any location, shown in Figure 3-9 as N, has various names, including *geoidal height* and *geoidal separation*.

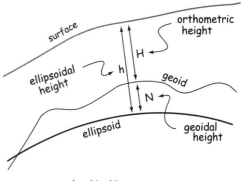

$$h = H + N$$

ellipsoidal height = orthometric height + geoidal height

Figure 3-9: Ellipsoidal, orthometric, and geoidal height are interrelated. Note that values for N are highly exaggerated in this figure - values for N are typically much less than H.

Geoidal Height

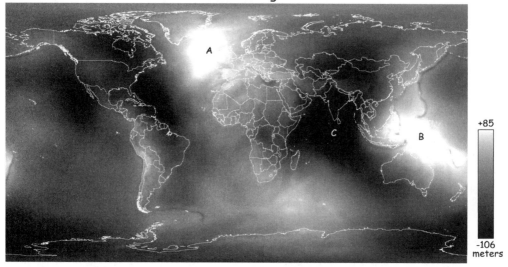

Figure 3-10: Geoidal heights vary across the globe. This figure depicts positive geoidal heights in lighter tones (geoid above the ellipsoid) and negative geoidal heights in darker tones. Note that geoidal heights are positive for large areas near Iceland and the Philippines (**A** and **B**, respectively), while large negative values are found south of India (**C**). Continental and country borders are shown in white.

The absolute value of the geoidal height is less than 100 meters over most of the Earth (Figure 3-10), Although it may at first seem difficult to believe, the "average" ocean surface near Iceland is more than 150 meters "higher" than the ocean surface northeast of Jamaica. This height difference is measured relative to the ellipsoid. Since gravity pulls in a direction that is perpendicular to the geoidal surface, the force is at a right angle to the surface of the ocean, resulting in permanent bulges and dips in the mean ocean surface due to variations in the gravitational pull. Variation in ocean heights due to swells and wind-driven waves are more apparent at local scales, but are much smaller than the long-distance geoidal undulations.

The geoidal height is quite small relative to the polar and equatorial radii. As noted in Table 3-1, the Earth's equatorial radius is about 6,780,000 meters, or about 32,000 times the range of the highest to lowest geoidal heights. This small geoidal height is imperceptible in an object at human scales. For example, the largest geoidal height is less than the relative thickness of a coat of

paint on a ball three meters (10 feet) in diameter. However, while relatively small, the geoidal variations in shape must still be considered for accurate vertical and horizontal mapping over continental or global distances.

The geoid is a measured and interpolated surface, and not a mathematically defined surface. The geoid's surface is measured using a number of methods, initially by a combination of *plumb bob*, a weight suspended by a string that indicates the direction of gravity, and horizontal and vertical distance measurements, and later with various types of *gravimeters*, devices that measure the gravitational force.

Satellite-based measurements in the late 20th century substantially improved the global coverage, quality, and density of geoidal height measurements. The GRACE experiment, initiated with the launch of twin satellites in 2002, is an example of such improvements. Distances between a pair of satellites are constantly measured as they orbit the Earth. The satellites are pulled closer or drift farther from the Earth due to variation in the gravity field. Because the orbital path changes slightly each day, we eventually have nearly complete Earth coverage of the strength of gravity, and hence the location of the reference gravitational surface. The ESA GOCE satellite, launched in 2009, uses precision accelerometers to measure gravity-induced velocity change. GRACE and GOCE observations have substantially improved our estimates of the gravitational field and geoidal shape.

Satellite and other observations are used by geodesists to develop geoidal models. These support a series of geoid estimates, e.g., by the U.S. NGS with GEOID90 in 1990, with succeeding geoid estimates in 1993, 1996, 1999, 2003, 2009, and one planned for 2012. These are called models because we measured geoidal heights at points or along lines at various parts of the globe, but we need geoidal heights everywhere. Equations are statistically fit that relate the measured geoidal heights to geographic coordinates. Given any set of geographic coordinates, we may then predict the geoidal height. These models provide an accurate estimation of the geoidal heights for the entire globe.

Geographic Coordinates, Latitude, and Longitude

Once a size and shape of the reference ellipsoid has been determined, the Earth poles and equator are also defined. The poles are defined by the axis of revolution of the ellipsoid, and the equator is defined as the circle mid-way between the two poles, at a right angle to the polar axis, and spanning the widest dimension of the ellipsoid. We estimate these locations from precise surface and astronomical measurements. Once the locations of the polar axis and equator have been estimated, we can define a set of geographic coordinates. This creates a reference system by which we may specify the position of features on the ellipsoidal surface.

As noted in Chapter 2, geographic coordinate systems consist of latitude, which varies from north to south, and longitude, which varies from east to west (Fig-

ure 3-11). Lines of constant longitude are called meridians, and lines of constant latitude are called parallels. Parallels run parallel to each other in an east-west direction around the Earth. The meridians are geographic north/south lines that converge at the poles.

By convention, the equator is taken as zero degrees latitude, and latitudes increase in absolute value to the north and south. Latitudes are thus designated by their magnitude and direction, for example 35°N or 72°S. When signed values are required, northern latitudes are designated positive and southern latitudes designated negative. An international meeting in 1884 established a longitudinal origin intersecting the Royal Greenwich Observatory in England. Known as the *prime* or *Greenwich meridian*, this north-to-south line was the origin, or zero value, for longitudes. East or west longitudes are specified as angles of rotation away from the Prime Meridian. When required, west is considered negative and east positive.

Improvements in measurements, crustal movements, and changes in conventions have resulted in the present zero longitude about 102 meters (335 feet) east of the Greenwich observatory.

There is often confusion between magnetic north and geographic north. Magnetic north and the geographic north do not coin-

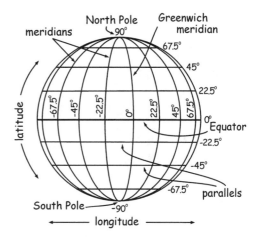

Figure 3-11: Nomenclature of geographic latitudes and longitudes.

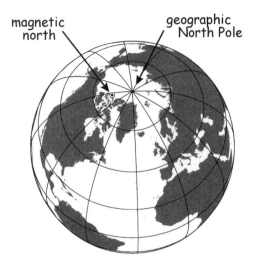

Figure 3-12: Magnetic north and the geographic North Pole.

The compass will usually point east or west of geographic north, defining an angular difference in direction to the poles. This angular difference is called the magnetic *declination* and varies across the globe. The specification of map projections and coordinate systems is always in reference to the geographic North Pole, not magnetic north.

Geographic coordinates do not form a Cartesian system (Figure 3-13). A Cartesian system defines lines of equal value in a right-angle grid. Geographic coordinates occur on a curved surface, and the longitudinal lines converge at the poles. This convergence means the distance spanned by a degree of longitude varies from south to north. A degree of longitude spans approximately 111.3 kilometers at the equator, but 0 kilometers at the poles. In contrast, the ground distance for a degree of latitude varies only slightly, from 110.6 kilometers at the equator to 111.7 kilometers at the poles.

cide (Figure 3-12). Magnetic north is the location towards which a compass points. The geographic North Pole is the northern pole of the Earth's axis of rotation. If you were standing on the geographic North Pole with a compass, it would point approximately in the direction of northern Canada, towards magnetic north some 600 kilometers away.

Because magnetic north and the geographic North Pole are not in the same place, a compass does not point at geographic north when observed from most places on Earth.

Convergence causes regular geometric figures specified in geographic coordinates to appear distorted when drawn on a globe. For example, "circles" with a fixed radius in geographic units, such as 5°, are not circles on the surface of the globe, although they may appear as circles when the Earth surface is "unrolled" and plotted with distortion on a flat map; note the erroneous size and shape of Antarctica at the bottom of Figure 3-13.

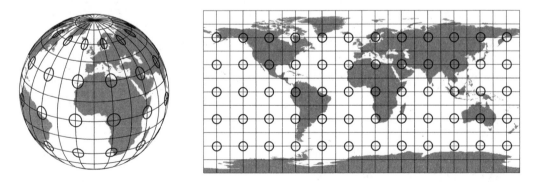

Figure 3-13: Geographic coordinates on a spherical (left) and Cartesian (right) representation. Notice the circles with a 5 degree radius appear distorted on the spherical representation, illustrating the change in surface distance represented by a degree of longitude from the equator to near the poles.

Horizontal Datums

The geographic coordinate system we have just described provides for specifying locations on the Earth. However, this gives us the exact longitude of only one arc, the zero line of longitude. We must estimate the longitudes and latitudes of all other locations through surveying measurements, until quite recently by observing stars and by measuring distances and directions between points. These surveying methods have since been replaced by modern, satellite-based positioning, but even these new methods are ultimately dependent on astronomical observations. Through these methods we establish a set of points on Earth for which the horizontal and vertical positions have been accurately determined.

These well-surveyed points allow us to specify a *reference frame*, usually an origin or starting point. If we are using a spherical reference frame, we must also specify the orientation and scale of our ellipsoid. If we are using a three-dimensional Cartesian reference frame, we must specify the X, Y, and Z axes, including their origin and orientation. All other coordinate locations we use are measured with reference to this set of precisely surveyed points, including the coordinates we enter in our GIS to represent spatial features.

Many countries have a government body charged with making precise geodetic surveys. For example, most surveys in the United States are related back to high accuracy points maintained by the National Geodetic Survey (NGS). The NGS establishes geodetic latitudes and longitudes of known points most of which are monumented with a bronze disk, concrete posts, or other durable markers. These points, taken together, underpin *geodetic datums*, upon which most subsequent surveys and positional measurements are based.

A *datum* is a reference surface. A geodetic datum consists of two major components. The first component is an ellipsoid with a spherical or three-dimensional Cartesian coordinate system and an origin. Eight parameters are needed to specify the ellipsoid: a and b to define the size/shape of the ellipsoid, the X, Y, and Z values of the origin, and an orientation angle for each of the three axes.

The second part of a useful datum consists of a set of points and lines that have been painstakingly surveyed using the best methods and equipment, and an estimate of the coordinate location of each point in the datum, e.g., the NGS points described in the previous paragraph. Some authors define the datum as a specified reference surface, and a *realization of a datum* as that surface plus a physical network of precisely measured points. In this nomenclature, the measured points describe a *Terrestrial Reference Frame*. This clearly separates the theoretical surface, the reference system or datum, from the terrestrial reference frame, a specific set of measurement points that help fix the datum. While this more precise language may avoid some confusion, datum will con-

Figure 3-14: Astronomical observations were used in early geodetic surveys to measure datum locations (courtesy NMSI)

tinue to refer to both the defined surface and the various realizations of each datum.

Different datums are specified through time because our realizations, or estimates of the datum, change through time. New points are added and survey methods improve. We periodically update our datum when a sufficiently large number of new survey points has been measured. We do this by re-estimating the coordinates of our datum points after including these newer measurements, thereby improving our estimate of the position of each point.

Historically, the relative positions of a set of datum points were determined using celestial measurements in combination with high-accuracy ground measurements. Most early measurements involved precise field surveys with optical instruments (Figure 3-14). These methods have been replaced in recent years by sophisticated electronic and satellite-based surveying systems.

Precisely surveyed points are also known as *bench marks*. Bench marks usually consist of a brass disk embedded in rock or concrete (Figure 3-15), although they also may consist of marks chiseled in rocks, embedded iron posts, or other long-term marks. Due to the considerable effort and cost of establishing the coordinates for each

Figure 3-16: Signs are often placed near control points to warn of their presence and aid in their location.

bench mark, they are often redundantly monumented, and their distance and direction from specific local features are recorded. Control survey points are often identified with a number of nearby signs to aid in recovery (Figure 3-16).

Geodetic surveys in the 18th and 19th centuries combined horizontal measurements with repeated, excruciatingly precise astronomical observations to determine latitude and longitude of a small set of points. Only a few datum points were determined using astronomical observations. Astronomical observations were typically used at the starting point, a few intermediate points, and near the end of geodetic surveys. This is because star positions required repeated measurements over several nights. Clouds, haze, or a full moon often lengthened the measurement times. In addition, celestial measurements required correction for atmospheric refraction, a process which bends light and changes the apparent position of stars. Refraction depends on how high the star is in the sky at the time of measurement, as well as temperature, atmospheric humidity, and other factors.

Horizontal measurements were as precise and much faster than astronomical measurements when surveys originated at known locations. These horizontal surface measure-

Figure 3-15: A brass disk used to monument a survey bench mark.

ments were then used to connect these astronomically surveyed points and thereby create an expanded, well-distributed set of known datum points. Figure 3-17 shows an example survey, where open circles signify points established by astronomical measurements and filled circles denote points established by surface measurements.

Figure 3-17 shows a *triangulation survey*, until the mid 1980s (and the advent of GPS) the method commonly used to establish datum points via horizontal surface measurements. Triangulation surveys utilize a network of interlocking triangles to determine positions at survey stations. Triangulation surveys were adopted because we can create them through angle measurement, with few surface distance measurements, an advantage in the late 18th century when many datums were first developed. Triangulation also improves accuracy; because there are multiple measurements to each survey station, the location at each station may be computed by various paths. The survey

accuracy can be field-checked, because large differences in a calculated station location via different paths indicate a survey error. There are always some differences in the measured locations when traversing different paths. An acceptable error limit was often set, usually as a proportion of the distance surveyed. In one common standard, differences in the measured location of more than 1 part in 100,000 would be considered unacceptable. When unacceptable errors were found, survey lines were re-measured.

Triangulation networks spanned long distances, from countries to continents (Figure 3-18). Individual measurements of these triangulation surveys were rarely longer than a few kilometers, however triangulations were nested, in that triangulation legs were combined to form larger triangles spanning hundreds of kilometers. These are demonstrated in Figure 3-18 where the sides of each large triangle are made up themselves of smaller triangulation traverses.

Datum Adjustment

Once a sufficiently large set of points have been surveyed, the survey measurements must be harmonized into a consistent set of coordinates. Small inconsistencies are inevitable in any large set of measurements, causing ambiguity in locations. In addition, historically the long reaches spanned by the triangulation networks, as shown in Figure 3-18, could be helpful in recalculating certain constants, such as the Earth's curvature (Figure 3-5). Later, satellite-based measurements were used to better estimate other constants, such as the datum origin. The positions of all points in a reference datum are estimated in a network-wide *datum adjustment*. The datum adjustment reconciles errors across the network, first by weeding out blunders or obvious mis-measurements or other mistakes, and also by mathematically minimizing errors by combining repeat measurements and statistically assigning higher influence to consistent or more precise measurements. Note that a given datum adjustment only incorporates

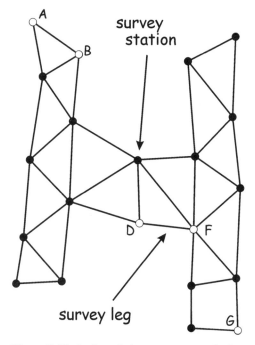

Figure 3-17: A triangulation survey network. Stations may be measured using astronomical (open circles) or surface surveys (filled circles).

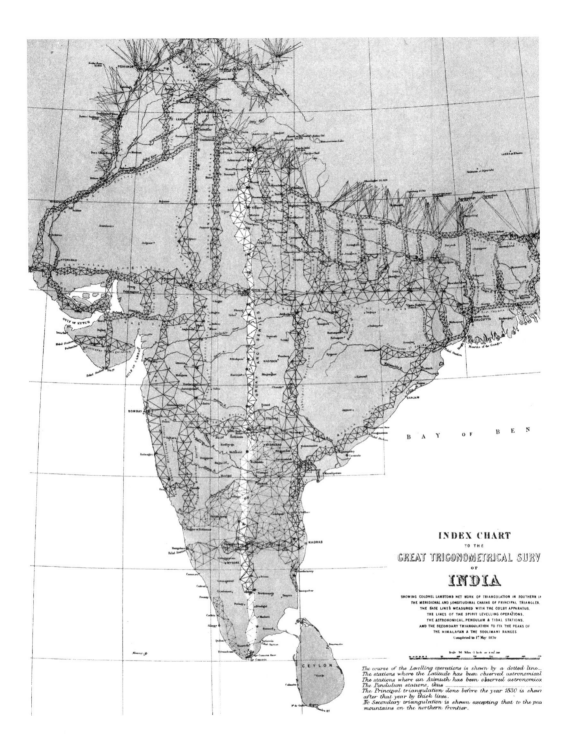

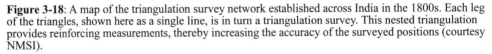

Figure 3-18: A map of the triangulation survey network established across India in the 1800s. Each leg of the triangles, shown here as a single line, is in turn a triangulation survey. This nested triangulation provides reinforcing measurements, thereby increasing the accuracy of the surveyed positions (courtesy NMSI).

measurements up to a given point in time, and may be viewed as our best estimate, at that point, of the measured set of locations.

Periodic datum adjustments result in series of regional or global reference datums. Each datum is succeeded by an improved, more accurate datum. The calculation of a new datum requires that all surveys must be simultaneously adjusted to reflect our current "best" estimate of the true positions of each datum point. Generally a statistical least-squares adjustment is performed, but this is not a trivial exercise, considering the adjustment may include survey data for tens of thousands of old and newly surveyed points from across the continent, or even the globe. Because of their complexity, these continent-wide or global datum calculations have historically been quite infrequent. Computational barriers to datum adjustments have diminished in the past few decades, and so datum adjustments and new versions of datums are now more frequent.

A datum adjustment usually results in a change in the coordinates for all existing datum points, as coordinate locations are estimated for both old and new datum points. The datum points do not move, but our best estimates of the datum point coordinates will change. Differences between the datums reflect differences in the control points, survey methods, and mathematical models and assumptions used in the datum adjustment.

Figure 3-19 illustrates how ellipsoids might change over time, even for the same survey region. Ellipsoid A is estimated with the datum coordinates for pt1 and pt2, with the shown corresponding coordinate axes, origin, and orientation. Ellipsoid B is subsequently fit, after pts 3 through 7 have been collected. This newer ellipsoid has a different origin and orientation for its axis, causing the coordinates for pt1 and pt2 to change. The points have not moved, but the best estimate of their locations, relative to the origin set by the new, more complete set of datum points, will have changed. You can visualize how the latitude angle from the origin to pt1 will change because the origin for ellipsoid A is in a different location than the origin for ellipsoid B. This apparent, but not real, movement is called the datum shift, and is expected with datum adjustments.

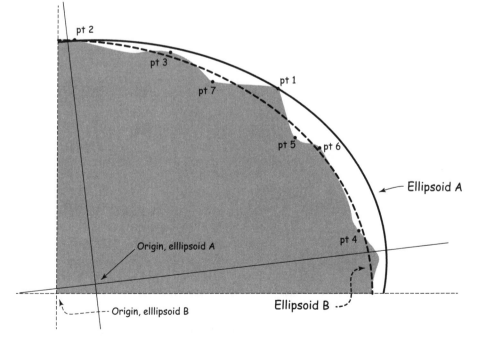

Figure 3-19: An illustration of two datums, one corresponding to Ellipsoid A and based on the fit to pt1 and pt2, and a subsequent datum resulting in Ellipsoid B, and based on a fit of pt1 through pt7.

Commonly Used Datums

Three main series of horizontal datums have been used widely in North America. The first of these is the *North American Datum of 1927* (NAD27). NAD27 is a general least-squares adjustment that included all horizontal geodetic surveys completed at that time. The geodesists used the Clarke Ellipsoid of 1866 and held fixed the latitude and longitude of a survey station in Kansas. NAD27 yielded adjusted latitudes and longitudes for approximately 26,000 survey stations in the United States and Canada.

The *North American Datum of 1983* (NAD83) is the successor datum to NAD27. We place the (1986) after the NAD83 designator to indicate the year, or version, of the datum adjustment. It was undertaken by the NGS to include the large number of geodetic survey points established between the mid-1920s and the early 1980s. Approximately 250,000 stations and 2,000,000 distance measurements were included in the adjustment. The GRS80 ellipsoid was used, and NAD83(1986) is Earth-centered reference, rather than fixing a station as with NAD27. The shifts in estimated coordinate locations between NAD27 and NAD83(1986) were large, on the order of 10's to up to 200 meters in North America. In most instances the surveyed points physically moved very little, e.g., due to tectonic shifts, but our best estimates of point location changed by as much as 200 meters.

Precise GPS data became widely available soon after the initial NAD83(1986) adjustment, and these were often more accurate than NAD83(1986) position estimates. Between 1989 and 2004, the NGS collaborated with other Federal agencies, State and local governments, and private surveyors in creating *High Accuracy Reference Networks* (HARNs), also known as *High Precision Geodetic Networks* (HPGN) in each state and most U.S. territories.

Subsequent NAD83 adjustments have incorporated measurements from the Continuously Operating Reference Station (CORS) network (Figure 3-20). This growing network of satellite observation stations allowed improved datum realizations, including NAD83(CORS93), NAD83(CORS94), NAD83(CORS96), NAD83(2007), and NAD83(2011). The NAD83(2007) datum may be viewed as a successor to the NAD83(HARN). Approximately 70,000 high-accuracy GPS points were adjusted with reference to the NAD83(CORS96) coordinates for the CORS network. NAD83(2011) is a long-observation adjustment based on CORS stations, with coordinates re-estimated for a broad set of bench marks. This datum realization allows surveyors to obtain the coordinates for a widespread set of physical locations, which may then be used as a starting point for subsequent surveys.

Position estimates of locations change by a few centimeters when compared among the NAD83(CORSxx) datums, important improvements for geodesists and extremely precise surveying, but small relative to spatial error budgets for many GIS projects. Differences among current and future NAD83(CORSxx) datums are likely to remain small, on the order of a few centimeters or less in tectonically stable areas, as newer NAD83 datum adjustments are calculated in the future.

The *World Geodetic System of 1984* (WGS84) is a set of datums developed and primarily used by the U.S. Department of Defense (DOD). It was introduced in 1987 based on Doppler satellite measurements of the Earth, and is used in most DOD maps and positional data. The WGS84 ellipsoid is similar to the GRS80 ellipsoid. WGS84 has been updated with more recent satellite measurements and is specified using a version designator. The update based on data collected up to January 1994 is designated as WGS84 (G730). WGS84 datums are not widely used outside of the military because they are not tied to a set of broadly accessible, documented physical points.

There have been several subsequent WGS84 datum realizations. The original datum realization exhibited positional accuracy of key datum parameters to within between one and two meters. Subsequent satellite observations improved accuracies. A re-analysis was conducted on data collected through week 730 of the GPS satellite schedule, resulting in the more accurate WGS84(G730). Successive re-adjustments in weeks 873 and 1150 are known as WGS84(G873) and WGS(G1150), respectively. There will likely be more adjustments in the future.

It has been widely stated that the original WGS84 and NAD83(86) datums were essentially equivalent. Both used the GRS80 ellipsoid, but the defining document for WGS84 notes differences of up to two meters between point locations measured against NAD83(86) versus the original WGS84 datum realizations. You should note that there are positional differences among and between all versions of both NAD83 and of WGS84, and ignoring the datum realiza-

tion may result in positional error. Geodesists at the U.S. National Geodetic Survey have adjusted the NAD83 datum several times since the initial 1986 estimation, and each different from the previous.

Another set of datums used worldwide, known as the *International Terrestrial Reference Frames,* (ITRF), are realizations of the International Terrestrial Reference System (ITRS). A primary purpose for ITRS is to estimate continental drift and crustal deformation by measuring the location and velocity of points, using a worldwide network of measurement locations. Each realization is noted by the year, e.g., ITRF89, ITRF90, ITRF91, and so forth, and includes the X, Y, and Z location of each point and the velocity of each point in three dimensions. The European Terrestrial Reference System (ETRS89 and frequent updates thereafter) is based on ITRF measurements.

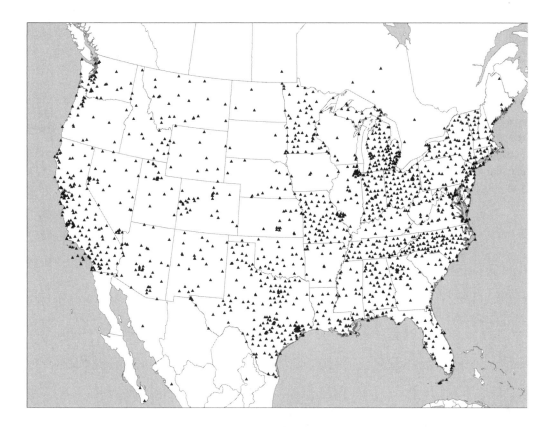

Figure 3-20: Partial distribution of the CORS network, as of 2008. This evolving network is the basis for the NAD83 (1993), NAD83(1994), NAD83(1996), and subsequent U.S. datum adjustments.

As noted earlier, different datums are based on different sets of measurements and ellipsoids, causing the coordinates for bench mark datum points to differ between datums and realizations. Differences are typically largest between legacy pre-satellite datum realizations, and post-satellite measurement datums. For example, the latitude and longitude location of a given bench mark in the NAD27 datum will likely be different from the latitude and longitude of that same bench mark in NAD83 or WGS84 datums by tens of meters, and up to 80 meters. This is described as a *datum shift*.

Figure 3-21 indicates the relative size of datum shifts at an NGS bench mark between NAD27 and NAD83(86) at one point in the eastern U.S., based on estimates provided by the National Geodetic Survey. Notice that the datum shift between NAD27 and

NAD83(86) is quite large, approximately 40 meters (140 feet), typical of the up to 100's of meters of shifts from pre-satellite, regional datums to post-satellite, global datums.

A datum shift does not imply that points have moved. Most monumented points are stationary relative to their immediate surroundings. The locations change over time as the large continental plates move, but these changes are small, on the order of a few millimeters per year, except in tectonically active areas such as coastal California; for most locations it is just our estimates of the coordinates that have changed. As survey measurements improve through time and there are more of them, we obtain better estimates of the true locations of the monumented datum points.

Examples of Datum Shifts
Successive datum transformations for New Jersey control point, Bloom 1

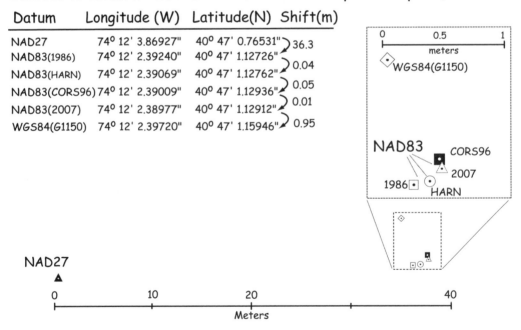

Datum	Longitude (W)	Latitude(N)	Shift(m)
NAD27	74° 12' 3.86927"	40° 47' 0.76531"	36.3
NAD83(1986)	74° 12' 2.39240"	40° 47' 1.12726"	0.04
NAD83(HARN)	74° 12' 2.39069"	40° 47' 1.12762"	0.05
NAD83(CORS96)	74° 12' 2.39009"	40° 47' 1.12936"	0.01
NAD83(2007)	74° 12' 2.38977"	40° 47' 1.12912"	0.95
WGS84(G1150)	74° 12' 2.39720"	40° 47' 1.15946"	

Figure 3-21: Datum shifts in the coordinates of a point for some common datums. Note that the estimate of coordinate position shifts approximately 36 meters from the NAD27 to the NAD83(1986) datum, while the shift from NAD83(1983) to NAD83(HARN) then to NAD83(CORS96) are approximately 0.05 meters. The shift to WGS84(G1150) is also shown, here approximately 0.95 m. Note that the point may not be moving, only our datum estimate of the point's coordinates. Calculations are based on NGS NADCON and HTDP software.

We must emphasize while much data are collected in WGS84 datums using GNSS, most data are converted to a local or national datum before use in a GIS. In the United States, this typically involves GNSS accuracy augmentation, often through a process called differential correction, described in detail in Chapter 5. Corrections are often based on an NAD83 datum, effectively converting the coordinates to the NAD83 reference, but ITRF datums are also commonly used.

For a datum to be practically useful in a a GIS, we typically need the datum coordinates for a widely distributed and uniformly documented set of monumented bench marks. The development of new data through local surveys and image interpretation requires that we tie our new data to this existing network of surveyed points. In the U.S., most spatial data are tied to the widely distributed set of bench marked points reported in the NAD83 (CORSxx) datums, and state, county, and local surveys referenced to these points. The error introduced in ignoring the differences between versions of WGS84 and the NAD83 or other local datums can be quite large, generally up to 2 meters or more (Figure 3-21). Errors in ignoring differences among older datums are larger still, up to 100's of meters. We must use a technique called a datum transformation to combine spatial data measured relative to different datums.

This conversion often happens implicitly when processing the GNSS data. As described in Chapter 6, most precise GNSS data results from correcting field measurements at unknown points against simultaneous GNSS measurements at a known point. This differential correction is usually configured such that the resulting coordinates are expressed in the same datum as the known points. If the correction sources are expressed in NAD83(CORS96) coordinates, corrected positions are initially created in these NAD83(CORS96) coordinates.

There are a few points about datums that must be emphasized. First, different datums mean different coordinate systems. You do not expect coordinates for any physical point to be the same when they are expressed relative to different datums.

Second, the version of the datum is important. NAD83(1986) is a different realization than NAD83(1996). The datum is incompletely specified unless the version is noted. Many GIS software packages refer to a datum without the version, e.g., NAD83. This is indeterminate, and confusing, and shouldn't be practiced. It forces the user to work with ambiguity.

Third, differences between families of datums change through time. The NAD83(86) datum realization is up to two meters different than the NAD83(CORS96), and the original WGS84 differs from the current version by more than a meter over much of the Earth. Differences in datum realizations depend on the versions and location on Earth, and can vary from zero to several hundred meters. This means you should assume all data should be converted to the same datum, via a datum transformation, before combination in a GIS. This rule may be relaxed if the errors due to ignoring the datum differences are small compared to other sources of error, or to the data accuracy required for the intended spatial analysis.

Datum Transformations

Estimating the shift and converting geographic coordinates from one datum to another typically requires a *datum transformation*. A datum transformation provides the latitude and longitude of a point in one datum when we know them in another datum, for example, we can calculate the latitude and longitude of a bench mark in NAD83(HARN) when we know these geographic coordinates in NAD83(CORS96) (Figure 3-22).

Datum transformations are often more complicated when they involve older datums. Many older datums were created piecemeal to optimize fit for a country or continent. The amount of shift between one datum and another often varies across the globe because the errors in measurements may be distributed idiosyncratically. Measurements in one area or period may have been particularly accurate, while in another area or time they may exhibit particularly large errors. Combining them in the datum adjustment affect the local and global differences among datums in their own unique way. Simple formulas often do not exist for transformations involving many older datums, for example from NAD27 to NAD83. Specialized datum transformations may be provided, usually by government agencies, using a number of different methods. As an example, in the United States the National Geodetic Survey has published a number of papers on datum transformations and provided datum transformation software tools, including NADCON to convert between NAD27 and NAD83 datums.

Transformation among newer datums may use more general analytical approaches that apply mathematical transformations

between three-dimensional, Cartesian coordinate systems (Figure 3-22). These Earth or near-Earth centered (geocentric) coordinate systems allow conversion among most GPS and CORS-based NAD83, WGS84 and ITRF systems, and are supported in large part by improved global measurements from artificial satellites, as described in the previous few pages. This three-dimensional approach typically allows for a shift in the origin, a rotation, and a change in scale from one datum to another.

A mathematical geocentric datum transformation is typically a multi-step process. These datum transformations are based on one of a few methods, for example, in past times a *Molodenski transformation* using a system of equations with three or five parameters, or more currently, a *Helmert transformation* using seven parameters (Figure 3-22). First, geographic coordinates on the source datum are converged from longi-

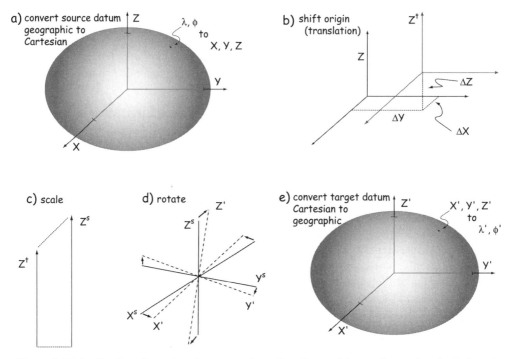

Figure 3-22: Application of a modern datum transformation. Geographic coordinates (longitude, λ, and latitude, ϕ), are transformed to a new datum by a) conversion from geographic to Cartesian coordinates in the old datum (through a set of equations that are not shown), b) applying an origin shift, c) scaling and d) rotating these shifted coordinates, and e) converting these target datum Cartesian coordinates, X', Y', Z', to the longitude and latitude, λ', ϕ', in the target datum.

tude (λ) and latitude (ϕ) to X, Y, and Z Cartesian coordinates. An origin shift (translation), rotation, and scale are applied. This system produces new X', Y', and Z' coordinates in the target datum. These X', Y', and Z' Cartesian coordinates are then converted back to geographic coordinates, longitudes and latitudes (λ' and ϕ'), in the target datum.

More advanced methods allow these 7 transformation parameters to change through time, as tectonic plates shift, for a total of 14 parameters. These methods are incorporated into software that calculate transformations among modern datums, for example, the Horizontal Time Dependent Positioning (HTDP) tool available from the U.S. NGS (www.ngs.noaa.gov/TOOLS/Htdp/ Htdp.shtml) among recent NAD83 datums and most ITRF and WGS84 datums.

Positions also change through time as tectonic plates shift, so that the most precise geodetic measurements refers to the epoch, or fixed time period, at which the point was measured. The HTDP software includes options to calculate the shift in a location due to measuring against different reference datums (e.g., NAD83(CORS96) to WGS84(G1150), the shift due to different realizations of a datum (e.g., NAD83(CORS96) to NAD83(2011)), the shift due to measurements in different epochs (e.g., NAD83(CORS96) epoch 1997.0 to NAD83(CORS96) epoch 2010.0), and the differences due to all three factors. Since most points are moving at velocities less than 0.01 mm per year in the NAD83 reference frame, epoch differences are often ignored for all but precise geodetic surveys.

Datums shifts associated with datum transformations have changed with each successive realization, as summarized in Figure 3-23, and some datums are considered functionally equivalent when combining data from different data layers, or when applying datum transformations. The WGS84(G730) was aligned with the ITRF92 datum, so these may be substituted in datum transformations requiring no better than centimeter level accuracies. Similarly, the

WGS84(G1150) and ITRF00 datums have been aligned, and may be substituted in most transformations.

While differences among the NAD83(CORSXX) and the ITRF/WGS84 datums are commonly over a meter, datum shifts internal to these groupings have become small for recent datums. Differences between NAD83(HARN) and NAD83(CORSxx) datums may be up to 20 cm, but are typically less than 4 cm, so these datum realizations may be considered equivalent if accuracy limits are above 20 cm, and perhaps as low as 4 cm. The differences among NAD83(CORS96) and NAD83(2011) are often on the order of a few centimeters, as are the differences among ITRF realizations, e.g., 91, 94, 00, 05, and 08.

There will be new datum realizations, each requiring additional transformations in the future. The ITRF datums are released every few years, requiring new transformations to existing datums each time. As of this writing, the NGS has released the NAD83(NSRS2007) datum coordinates. This is a re-analysis of state-collected points that were the basis for the NAD83(HARN) network, applying uniform, improved analysis methods. NAD83(2011) is to be released in early 2012, a nationwide adjustment of passive bench mark stations and multi-year observations at GNSS/GPS CORS stations.

Until quite recently, spatial error due to improper datum transformation has been below a detectable threshold in many analyses, so it caused few problems. GNSS receivers can now provide centimeter-level accuracy in the field, so what were once considered small discrepancies often cannot now be overlooked. As data collection accuracies improve, datum transformation errors become more apparent. The datum transformation method within any hardware or software package should be documented and the accuracy of the method known before it is adopted. Unfortunately, both of these recommendations are too often ignored or only partially adopted by software vendors and users.

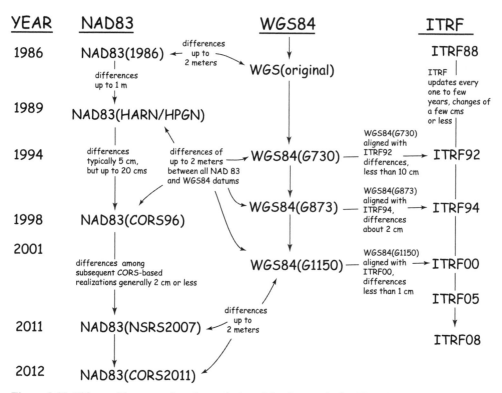

Figure 3-23: This graphic summarizes the evolution of the three main families of datums used in North America. As the datums have been adjusted, horizontal positional differences between bench mark points have varied, within the ranges shown. "Aligned" datums (e.g., WGS84(G1150) and ITRF00) may be considered equivalent for most purposes when applying datum transformations.

The NGS maintains and disseminates a list of control points in the United States (Figure 3-24), including those points used in datum definitions and adjustment. Point descriptions are provided in digital forms, including access via the world wide web (http://www.ngs.noaa.gov). Stations may be found based on a station name, a state and county name, a type of station (horizontal or vertical), by survey order, survey accuracy, date, or coordinate location. These stations may be used as reference points against which to check the accuracy and correctness of any data set, or as a starting point for additional surveys.

These NGS sheets may provide an estimate of the shifts associated with a datum transformation, and so consulting one may give the specific datum shift value in any working area. For example, the values in the datum sheet in Figure 3-24 report bench

mark coordinates in various datum realization that allow datum shift estimates of approximately 57.7 meters from NAD27 to NAD83(86), and 17 cm from NAD(86) to NAD83(CORS96). Similar data from a nearby station allow a calculated datum shift of approximately 1.02 m between NAD83(CORS96) and WGS84(G1150)/ ITRF00. You would expect perfectly accurate data to mis-align by these amounts if the proper datum shifts were not applied.

There are a number of factors that we should keep in mind when applying datum transformations. First, changing a datum changes our best estimate of the coordinate locations of most points. These differences may be small and ignored with little penalty in some specific instances, typically when the changes are smaller than the spatial accuracy required for our analysis. However, many datum shifts are quite large, up to tens

```
National Geodetic Survey,    Retrieval Date = SEPTEMBER 26, 2011
OB0554  DESIGNATION -  CAPE SMALL  OB0554  PID        - OB0554
OB0554  STATE/COUNTY-  ME/SAGADAHOC   USGS QUAD   - PHIPPSBURG (1957)
OB0554
OB0554                      *CURRENT SURVEY CONTROL
OB0554  _____
OB0554* NAD 83(1996)    -  43 46 42.87649(N) 069 50 42.26065(W)    ADJUSTED
OB0554* NAVD 88         -       73.    (meters) 240.    (feet)  SCALED
OB0554  _____
OB0554  LAPLACE CORR-          2.33   (seconds)                   DEFLEC99
OB0554  GEOID HEIGHT-        -25.73   (meters)                    GEOID03
OB0554  HORZ ORDER  -  FIRST
 .
 .
OB0554:                 Primary Azimuth Mark           Grid Az
OB0554:SPC ME W    -  BURNT LEDGE JR 1866             008 26 54.5
OB0554:UTM  19     -  BURNT LEDGE JR 1866             009 15 20.4
OB0554
OB0554|-----------------------------------------------------------------|
OB0554| PID    Reference Object               Distance     Geod. Az  |
OB0554|                                                    dddmmss.s  |
OB0554|
OB0555 BURNT LEDGE JR 1866              APPROX. 2.2 KM 0084015.5 |  OB0554|
OB0531 MT MERRITT 2                     APPROX. 1.3 KM 2083443.8 |
OB0554|-----------------------------------------------------------------|
OB0554
OB0554                      SUPERSEDED SURVEY CONTROL
OB0554
OB0554  NAD 83(1992)-  43 46 42.87431(N)    069 50 42.25948(W) AD( ) 1
OB0554  NAD 83(1986)-  43 46 42.88001(N)    069 50 42.26497(W) AD( ) 1
OB0554  NAD 27      -  43 46 42.57400(N)    069 50 44.10300(W) AD( ) 1
 .
 .
```

Figure 3-24 A portion of a National Geodetic Survey control point data sheet.

of meters. One should know the magnitude of the datum shifts for the area and datum transformations of interest.

Second, datum transformations are estimated relationships which are developed with a specific data set and for a specific area and time. There are spatial errors in the transformations that are specific to the input and datum version. There is no generic transformation between NAD83 and WGS84. Rather, there are transformations between specific versions of each, for example, from NAD83(96) to WGS84(1150).

Finally, GIS projects should not mix datums except under circumstances when the datum shift is small relative to the requirements of the analysis. Unless proven otherwise, all data should be converted to the same coordinate system, based on the same datum. If not, data may mis-align.

Vertical Datums

Just as there are networks of well-measured points to define horizontal position, there are networks of points to define vertical position and *vertical datums*. Vertical datums are used as a reference for specifying heights. Much like horizontal datums, they are established through a set of painstakingly surveyed control points. These point elevations are precisely measured, initially through a set of optical surface measurements, but more recently using GPS, laser, satellite, and other measurement systems. Establishing vertical datums also requires estimating the strength and direction of the gravitational force near the surface of the Earth.

In its simplest definition, a vertical datum is a reference that we use for measuring heights. As noted in the geoid section on page 78, we use a geoid as a reference surface, and specify the orthometric heights as the elevations of points on the Earth's surface above the geoid. We first establish a specific geoid through a set of gravity measurements and then augment this with precise vertical height measurements at points across the globe to establish a set of vertical bench marks, against which we can conveniently measure all other heights. The vertical datum is the set of points, with heights, relative to a specific geoid.

Leveling surveys are among the oldest methods for establishing a vertical point. Distances and elevation differences are precisely measured from an initial point to other points, establishing height differentials. Early leveling surveys were performed with the simplest of instruments, including a plumb bob to establish leveling posts, and a simple liquid level to establish horizontal lines. Early surveys used an approach known as *spirit leveling*. Horizontal rods were placed between succeeding leveling posts across the landscape to physically measure height differences (Figure 3-25).

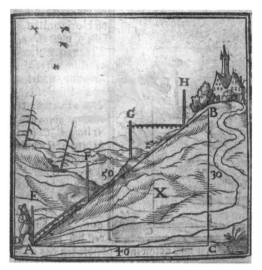

Figure 3-25: Early surveys used level bars placed on vertical posts, simple but effective technology.

The number, accuracy, and extent of leveling surveys increased substantially in the 18th and 19th centuries. Epic surveys that lasted decades were commissioned, such as the Great Arc, from southern India to the Himalayas. These surveys were performed at substantial capital and human expense; in one portion of the Great Arc more than 60% of the field crews died due to illness and mishaps over a six year period. Surface leveling provided most height measurements for vertical datums until the mid to late 20th century, when a variety of satellite-based methods were introduced.

Most leveling surveys from the late 1700s through the mid 20th century employed *trigonometric leveling*. This method uses optical instruments and trigonometry to measure changes in height, as shown in Figure 3-26. Surface distance along the slope was measured to avoid the tedious process of establishing vertical posts and leveling rods. The vertical angle was also measured from a known station to an unknown station. The angle was typically measured with a small telescope fitted with a precisely scribed angle gauge. The gauge could be referenced to zero at a horizontal position, usually with an integrated bubble level, or later, with an electronic level. Surface distance would then be combined with the measured vertical angle to calculate the horizontal and vertical distances. Early surveys measured surface distance along the slope with ropes, metal chains, and steel tapes, but these physical devices have largely been replaced by improved optical methods, or by laser-based methods.

Early national leveling surveys used the concept of mean sea level as a zero, or base height. Sea-level heights vary over time, mostly due to tides, but also due to changes in weather systems, currents, temperature and salinity. Mean sea level at a gauge may be calculated after a sufficient period of time, typically over at least the 19-year cycle of tidal variation. Monumented points were established on rocks, docks, or other oceanside fixed objects near the gauges, and the height of these starting points could then be measured via leveling to the nearby ocean tidal stations. Precise leveling was then extended landward from these oceanside points to measure heights cross-country. All leveled heights could them be tied to a mean sea level through this vertical measurement network.

Note that we said "a" mean sea level, because mean sea level isn't the same everywhere. Mean sea level, even averaged over several decades, varies across the globe due to several factors, for example, persistent differences in water density with temperature and salinity, or regular ocean currents, which may persistently raise or lower the surface in ocean regions. This means the mean sea level is not constant relative to the geoid or ellipsoid, and will be different at Miami than New York. Modern vertical datums do not use mean sea level across many stations as a reference in part because of this variation in mean sea level across the Earth. While most people describe mountain summit elevations or other heights as above mean sea level, geodesists and GIS professionals do not. We use a set of precisely surveyed base bench marks with heights referenced to the Earth's geoid.

As with horizontal datums, the primary vertical datums in use have changed through time as the number, distribution, and accuracy of vertical survey points have increased. Geodetic leveling surveys began in the U.S. in the 1850s, initially focusing on the East Coast and Great Lakes region, and extended across the U.S. between 1877 and 1900. Periodic adjustments harmonized measurements, identified and removed large errors, and distributed small discrepancies

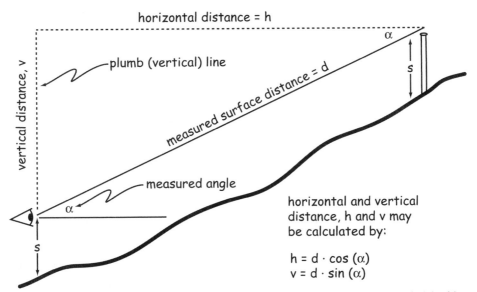

Figure 3-26: Leveling surveys often employ optical measurements of vertical angle (α) with measurements of surface distance (d) and knowledge of trigonometric relationships to calculate horizontal distance (h) and vertical distance (v).

```
National Geodetic Survey,  Retrieval Date = FEBRUARY 12, 2011
FB2737 **************************************************************
FB2737 CBN      - This is a Cooperative Base Network Control Station.
FB2737 DESIGNATION - MITCHELL 2
FB2737 PID      - FB2737
FB2737 STATE/COUNTY- NC/YANCEY
FB2737 USGS QUAD  - MT MITCHELL (1946)
FB2737
FB2737                *CURRENT SURVEY CONTROL
FB2737 _____
FB2737* NAD 83(NSRS2007)- 35 45 53.76569(N) 082 15 54.34377(W)   ADJUSTED
FB2737* NAVD 88     -    2048.27 (meters)   6720. (feet)          LEVELING
FB2737 _____
FB2737 X      -    697,569.032 (meters)           COMP
FB2737 Y      - -5,135,766.495 (meters)           COMP
FB2737 Z      -   3,708,239.126 (meters)          COMP
FB2737 LAPLACE CORR-      -6.76 (seconds)         DEFLEC99
FB2737 ELLIP HEIGHT-    2018.409 (meters)      (02/10/07) GPS OBS
FB2737 GEOID HEIGHT-     -30.00  (meters)         GEOID03
```

Figure 3-27: A portion of a data sheet for a vertical control bench mark.

among stations. These vertical adjustments were conducted in 1899, 1903, 1907, and 1912, relating all measured heights to between five and nine precisely measured tidal gauges.

The first continental vertical datum in North America was the *National Geodetic Vertical Datum* of 1929, also referred to as NGVD29. Vertical leveling was adjusted to 26 tidal gauges, including 5 in Canada, based on local mean sea level at each of the gauges. Geodesists realized that mean sea level varied across the continent, but assumed these differences would be similar or smaller than measurement errors. They wanted to avoid confusion caused by seaside bench marks having heights that differed from mean sea level.

Vertical measurements continued from the 1920s through the 1980s, resulting in the *North American Vertical Datum of 1988* (NAVD88), and many monumented control points have vertical heights reported in NAVD88 (Figure 3-27). The 1988 datum is based on over 600,000 kilometers (360,000 miles) of control leveling performed since 1929, and also reflects geologic crustal movements or subsidence that may have changed bench mark elevation. NAVD88 was fixed relative to only one tidal station, in the town of Rimouski, Quebec, because

improved methods meant measurement errors were much smaller than differences in mean sea level among stations. Surface heights in this datum are not based on mean sea level, because to do so across the set of sea-side benchmarks would require additional warping that would degrade the measurements.

Improved geoid models have been developed concurrently with these newer vertical datums. For example, at this writing, the most current model, GEOID03, integrated nearly 15,000 vertical bench marks to estimate geoidal and orthometric heights. These heights are reported on NGS data sheets for vertical bench marks (Figure 3-27), noting the vertical datum (here NADVD88), the geoid model (GEOID03), the orthometric height (here 2048.27 meters), and the ellipsoidal and geoidal heights.

Dynamic Heights

We must discuss one final kind of height, called a dynamic height, because they are important for certain applications, and are often listed on NGS data sheets and elsewhere. Dynamic heights measure the change in gravitational pull from a given equipotential surface. Dynamic heights are important when interested in water levels and flows across elevations. Points that have the same dynamic heights can be thought of as being at the same water level. Perhaps a bit surprisingly, points with the same dynamic heights often have different orthometric heights (Figure 3-28). To be clear, two distinct points at water's edge on a large lake often do not have the same elevations, that is, they are different orthometric heights above our reference geoid. Since orthometric heights are our bench mark for specifying elevation, water may flow from one point to another, even though those points have the same elevation.

To understand why water may flow between points with the same elevation (orthometric heights), it is important to remember how orthometric heights are defined. An orthometric height is the distance, in the direction of gravitational pull, from the geoid up to a point. But remember, the geoid is a specified gravity value, an "equipotential" surface, where the pull of gravity is at some specified level. As we move up from the geoid toward the surface, we pass through other equipotential sur-

faces, each at a slightly weaker gravitational pull or force, until we arrive at the surface point.

There are two key observations here. First, water spreads out to level across an equipotential surface, absent wind, waves, and other factors. The water level in a still bathtub, pond, or lake is at the same equipotential surface at one end as another. Gravity pulls down on the surface to ensure it conforms to an equipotential surface. Second, the equipotential surfaces are closer together when nearer the mass center of Earth. The equipotential surfaces converge, or become "denser" the closer you are to the center of the Earth.

Because of these two facts, and because the Earth's polar radius is less than the equatorial radius, the orthometric heights of the water surface on large lakes are usually different at the north and south ends. For example, as you move further north in the northern hemisphere, the equipotential surfaces converge due to a decreasing distance from the mass center of the Earth, and the pull of gravity increase (Figure 3-28). An

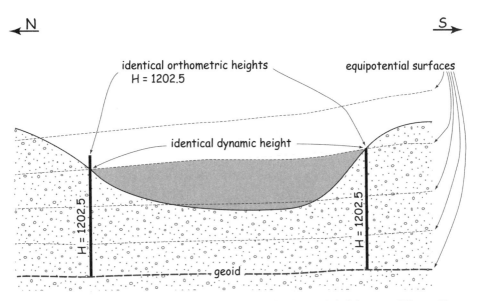

Figure 3-28: An illustration of how dynamic heights and orthometric heights may differ, and how equal orthometric heights may correspond to different heights above the water level on a large lake. Because equipotential surfaces converge, the orthometric height at the water level at the northern extreme of a lake will have different orthometric heights. Dynamic heights and water levels are equal across an equipotential surface.

orthometric height is a fixed height above the geoidal surface, so the northern orthometric height will pass through more equipotential surfaces than the same orthometric height at a more southerly location. Water follows an equipotential surface, so an orthometric height of the water level at the south end of the lake will be higher than at the north end. For example, in the Great Lakes of North America, the orthometric height corresponding to water level at the south end of Lake Michigan is approximately 15 cm higher than the water level at the north end.

Dynamic heights are most often used when we're interested in relative heights for water levels, particularly over large lakes or connected water bodies. Because equal dynamic heights are at the same water level, we can use them when interested in accurately representing hydrologic drop, head, pressure, and other variables related to water levels across distances. But these differences should be confusing when observing benchmark or sea level heights, and underscore that our height reference is not mean sea level, but rather an estimated geoidal surface.

Control Accuracy Specification

In most cases the horizontal datum control points are too sparse to be sufficient for all needs in GIS data development. For example, precise point locations may be required when setting up a GPS receiving station, to georegister a scanned photograph or other imagery, or as the basis for a detailed subdivision or highway survey. It is unlikely there will be more than one or two datum points within any given work area. Because a denser network of known points is required for many projects, datum points are often used as a starting locations for additional surveying. These smaller area surveys increase the density of precisely known points. The quality of the point locations

depends on the quality of the intervening survey.

The Federal Geodetic Control Committee of the United States (FGCC) has published a detailed set of survey accuracy specifications. These specifications set a minimum acceptable accuracy for surveys and establish procedures and protocols to ensure that the advertised accuracy has been obtained. The FGCC specifications establish a hierarchy of accuracy. First order survey measurements are accurate to within 1 part in 100,000. This means the error of the survey is no larger than one unit of measure for each 100,000 units of distance surveyed. The maximum horizontal measurement error of a 5,000 meter baseline (about 3 miles) would be no larger than 5 centimeters (about 2 inches). Accuracies are specified by Class and Orders, down to a Class III, 2nd order point with an error of no more than 1 part in 5,000.

Map Projections and Coordinate Systems

Datums tell us the latitudes and longitudes of a set of points on an ellipsoid. We need to transfer the locations of features measured with reference to these datum points from the curved ellipsoid to a flat map. A *map projection* is a systematic rendering of locations from the curved Earth surface onto a flat map surface. Points are "projected" from the Earth surface and onto the map surface.

Most map projections may be viewed as sending rays of light from a projection source (Figure 3-29). Rays radiate from a source to intersect both the ellipsoid surface and the map surface. The rays specify where each point from the ellipsoid surface is placed on the map surface. In some projections the source is not a single point; however the basic process involves the systematic transfer of points from the curved ellipsoidal surface to a flat map surface.

Distortions are unavoidable when making flat maps, because as we've said, locations are projected from a complexly curved Earth surface to a flat or simply curved map surface. Portions of the rendered Earth sur-

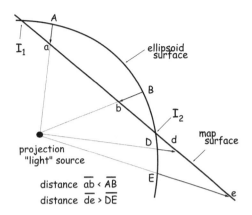

Figure 3-30: Distortion during map projection.

face must be compressed or stretched to fit onto the map. This is illustrated in Figure 3-30, a side view of a projection from an ellipsoid onto a plane. The map surface intersects the Earth at two locations, I_1 and I_2. Points toward the edge of the map surface, such as D and E, are stretched apart. The scaled map distance between d and e is greater than the distance from D to E measured on the surface of the Earth. More simply put, the distance along the map plane is greater than the corresponding distance along the curved Earth surface. Conversely, points such as A and B that lie in between I_1 and I_2 would appear compressed together. The scaled map distance from a to b would be less than the surface-measured distance from A to B. Distortions at I_1 and I_2 are zero.

Figure 3-30 demonstrates a few important facts. First, distortion may take different forms in different portions of the map. In one portion of the map features may be compressed and exhibit reduced areas or distances relative to the Earth's surface measurements, while in another portion of the map areas or distances may be expanded. Second, there are often a few points or lines where distortions are zero and where length, direction, or some other geometric property is preserved. Finally, distortion is usually less near the points or lines of intersection,

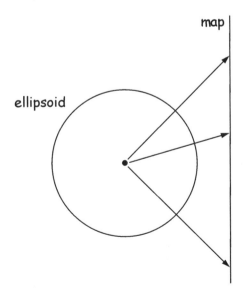

Figure 3-29: A conceptual view of a map projection.

where the map surface intersects the imaginary globe. Distortion usually increases with increasing distance from the intersection points or lines.

Different map projections may distort the globe in different ways. The projection source, represented by the point at the middle of the circle in Figure 3-30, may change locations. The surface onto which we are projecting may change in shape, and we may place the projection surface at different locations at or near the globe. If we change any of these three factors, we will change how or where our map is distorted. The type and amount of projection distortion may guide selection of the appropriate projection or limit the area projected.

Figure 3-31 shows an example of distortion with a projection onto a planar surface. This planar surface intersects the globe at a line of true scale, the solid line labeled as the standard circle shown in Figure 3-31. Distortion increases away from the line of true

scale, with features inside the circle compressed or reduced in size, for a negative scale distortion. Conversely, features outside the standard circle are expanded, for a positive scale distortion. Calculations show a scale error of -1% near the center of the circle, and increasing scale error in concentric bands outside the circle to over 2% near the outer edges of the projected area.

An approximation of the distance distortion may be obtained for any projection by comparing grid coordinate distances to *great circle distances*. A great circle distance is a distance measured on the ellipsoid and in a plane through the Earth's center. This planar surface intersects the two points on the Earth's surface and also splits the spheroid into two equal halves (Figure 3-32). The smallest great circle distance is the shortest path between two points on the surface of the ellipsoid, and by approximation, Earth.

As noted earlier, a straight line between two points on the projected map is likely not

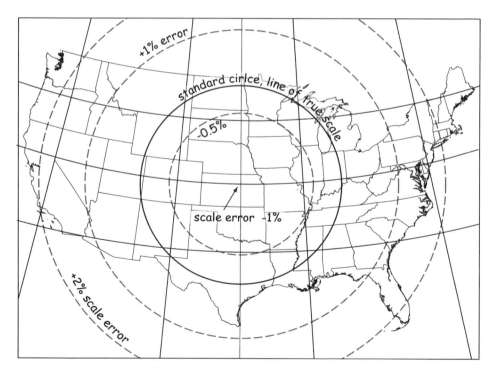

Figure 3-31: Approximate error due to projection distortion for a specific oblique stereographic projection. A plane intersects the globe at a standard circle. This standard circle defines a line of true scale, where there is no distance distortion. Distortion increases away from this line, and varies from -1% to over 2% in this example (adapted from Snyder, 1987).

Great Circle Distance

Consider two points
on the Earth's surface,
A with geographic
coordinates (lat.,lon.)
(ϕ_A, λ_A), and

B, with geographic
coordinates
(ϕ_B, λ_B)

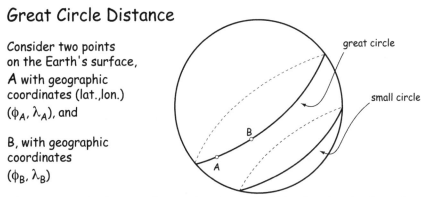

The great circle distance from point A to point B is given by the formula:

$$d = r \cdot \cos^{-1}[(\cos(\phi_A)\cos(\phi_B)\cos(\lambda_A-\lambda_B)\} + \sin(\phi_A)\sin(\phi_B)],$$

where d is the shortest distance on the surface of the Earth from A to B,
and r is the Earth's radius, approximately 6378 km.

This formula may be used to find the distance distortion caused by a
projection between two points, for example, between Ursine and Moab,
Utah, when using UTM Zone 12N coordinates, NAD83?

Great circle distance:
Latitude, longitude of Ursine, Utah = 37.98481°, -114.216944°
Latitude, longitude of Moab, Utah   = 38.57361°, -109.551111°

$$d = 6378 \cdot \cos^{-1}[(\cos(37.98481)\cos(38.57361)\cos(-114.216944 - 109.551111\} +$$
$$\sin(37.98481)\sin(38.57361)]$$
$$= 412.906 \text{ km}$$

Grid distance (UTM Zone 12N coordinates):
Grid coordinates of Ursine, Utah = 217,529.8, 4,208,972.8
Grid coordinates of Moab, Utah = 626,239.2, 4,270,405.9

$$dg = [(X_A - X_B)^2 + (Y_A - Y_B)^2]^{0.5}$$
$$= [(217,529.8 - 626,239.2)^2 + (4,208,972.8 - 4,270,405.9)^2]^{0.5}$$
$$= 413.300 \text{ km}$$

distortion is 412.906 - 413.300 = -0.394 km, or a 394 meter lengthening

Figure 3-32: Example calculation of the distance distortion due to a map projection. The great circle and grid distances are compared for two points on the Earth's surface, the first measuring along the curved surface, the second on the projected surface. The difference in these two measures is the distance distortion due to the map projection. Calculations of the great circle distances are approximate, due to the assumption of a spheroidal rather than ellipsoidal Earth, but are very close.

to be a straight line on the surface of the Earth, and is not the shortest distance between two points when traveling on the surface of the Earth. Conversely, the shortest distance between points when traveling on the surface of the Earth is likely to appear as a curved line on a projected map. The distortion is imperceptible for large scale maps and over short distances, but exists for most lines.

Figure 3-33 illustrates straight line distortion. This figure shows the shortest distance path between Adelaide, Australia, and Tokyo, Japan. Tokyo lies almost due north of Adelaide, and the shortest path approximates a line of longitude, by definition a great circle path. This shortest path is distorted and appears curved by the projection used for this map.

The magnitude of this distortion may be approximated by simple formulas (Figure 3-32). Coordinates may be identified for any

two points in the grid system, and the Pythagorean formula used to calculate distance between the two points. The resulting distance will be expressed in the grid coordinate system, and therefore will include the projection distortion. The distance may also be calculated for a great circle route along the spheroid surface. This calculation will approximate the unprojected distance, measured on the surface of the Earth. This is only an approximation, as we know from the previous section, because the Earth is shaped more like an ellipsoid, and has geoidal undulations. However, the approximation is quite accurate, generally off by less than a few parts per tens of thousands over several hundred kilometers. The great circle and grid coordinate distance may then be compared to estimate the distance distortion (Figure 3-32).

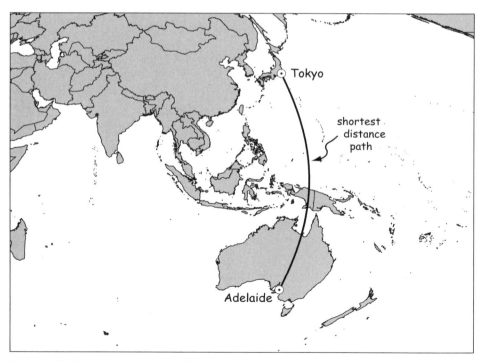

Figure 3-33: Curved representations of straight lines are a manifestation of projection distortion. A great circle path, shown above, is the shortest route when traveling on the surface of the Earth. This path appears curved when plotted on this sinusoidal projection.

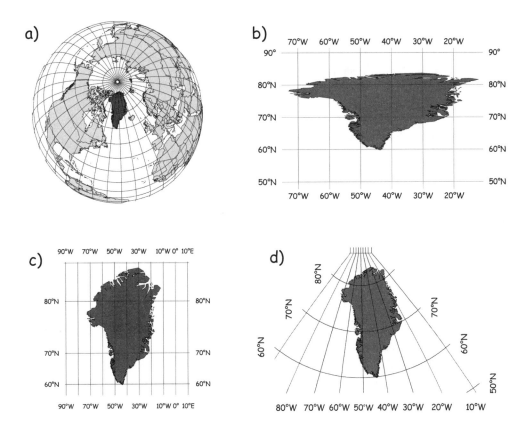

Figure 3-34: Map projections can distort the shape and area of features, as illustrated with these various projections of Greenland, from a) approximately unprojected, b) geographic coordinates on a plane, c) a Mercator projection, and d) a transverse Mercator projection.

Projections may also substantially distort the shape and area of polygons. Figure 3-34 shows various projections for Greenland, from an approximately "unprojected" view from space through geographic coordinates cast on a plane, to Mercator and transverse Mercator projections. Note the changes in size and shape of the polygon depicting Greenland.

Most map projections are based on a *developable surface*, a geometric shape onto which the Earth surface locations are projected. Cones, cylinders, and planes are the most common types of developable surfaces. A plane is already flat, and cones and cylinders may be mathematically "cut" and "unrolled" to develop a flat surface (Figure 3-35). Projections may be characterized according to the developable surface, for

example, as *conic* (cone), *cylindrical* (cylinder), and *azimuthal* (plane). The orientation of the developable surface may also change among projections, for example, the axis of a cylinder may coincide with the poles (equatorial) or the axis may pass through the equator (transverse).

Note that while the most common map projections used for spatial data in a GIS are based on a developable surface, many map projections are not. Projections with names such as pseudocylindrical, Mollweide, sinusoidal, and Goode homolosine are examples. These projections often specify a direct mathematical projection from an ellipsoid onto a flat surface. They use mathematical forms not related to cones, cylinders, planes, or other three-dimensional figures, and may change the projection surface for different

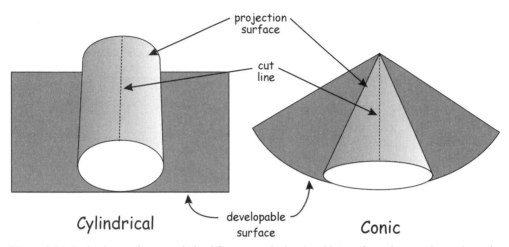

Figure 3-35: Projection surfaces are derived from curved "developable" surfaces that may be mathematically "unrolled" to a flat surface.

parts of the globe. For example, projections such as the Goode homolosine projection are formed by fusing two or more projections along specified line segments. These projections use complex rules and breaks to reduce distortion for many continents.

We typically have to specify several characteristics when we specify a map projection. For example, for an azimuthal projection we must specify the location of the projection center (Figure 3-36) and the location and orientation of the plane onto which the globe is projected. Azimuthal projections are often tangent to (just touch) the ellipsoid at one point, and we must specify the location of this point. A projection center ("light" source location) must also be specified, most often placed at one of three locations. The projection center may be at the center of the ellipsoid (a *gnomonic* projection), at the antipodal surface of the ellipsoid (diametrically opposite the tangent point, a *stereographic* projection), or at infinity (an *orthographic* projection). Scale factors, the location of the origin of the coordinate system, and other projection parameters may be required. Defining characteristics must be specified for all projections, such as the size and orientation of a cone in a conic projection, or the size, intersection properties, and orientation of a cylinder in a cylindrical projection.

Note that the use of a projection defines a projected coordinate system and hence typically adds a third version of North to our description of geography. We have already described magnetic north, towards which a compass points, and geographic north, the pole around which the globe revolves (Figure 3-12). We must add *grid north* to these, defined as the direction of the Y axis in the projection. Grid north is often defined by some meridian in the projection, often known as the central meridian. Grid north is typically not the same as geographic or magnetic north over most of the projected area.

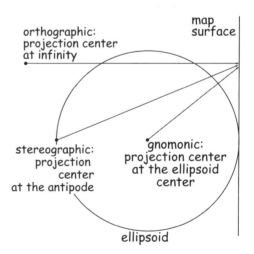

Figure 3-36: The projection center of map projections is most often placed at the center of the ellipsoid, or the antipode, or at infinity.

Common Map Projections in GIS

There are hundreds of map projections used throughout the world, however most spatial data in GIS are specified using projections from a relatively small number of projection types.

The Lambert conformal conic and the transverse Mercator are among the most common projection types used for spatial data in North America, and much of the world (Figure 3-37). Standard sets of projections have been established from these two basic types. The Lambert conformal conic (LCC) projection may be conceptualized as a cone intersecting the surface of the Earth, with points on the Earth's surface projected onto the cone. The cone in the Lambert conformal conic intersects the ellipsoid along two arcs, typically parallels of latitude, as shown in Figure 3-37 (top left). These lines of intersection are known as *standard parallels*.

Distortion in a Lambert conformal conic projection is typically smallest near the standard parallels, where the developable surface intersects the Earth. Distortion increases in a complex fashion as distance from these intersection lines increases. This characteristic is illustrated at the top right and bottom of Figure 3-37. Circles of a constant 5 degree radius are drawn on the projected surface at the top right, and approximate lines of constant distortion and a line of true scale are shown in Figure 3-37, bottom. Distortion decreases towards the standard parallels, and increases away from these lines. Those farther away tend to be more distorted. Distortions can be quite severe, as illustrated by the apparent expansion of southern South America.

Note that sets of circles in an east-west row are distorted in the Lambert conformal conic projection (Figure 3-37, top right). Those circles that fall between the standard parallels exhibit a uniformly lower distortion than those in other portions of the projected map. One property of the Lambert conformal conic projection is a low-distortion band running in an east-west direction between the standard parallels. Thus, the Lambert conformal conic projection is often used for areas that are larger in an east-west than a north-south direction, as there is little added distortion when extending the mapped area in the east-west direction.

Distortion is controlled by the placement and spacing of the standard parallels, the lines where the cone intersects the globe. The example in Figure 3-37 shows parallels placed such that there is a maximum distortion of approximately 1% midway between the standard parallels. We reduce this distortion by moving the parallels closer together, but at the expense of reducing the area mapped at this lower distortion level.

The transverse Mercator is another common map projection. This map projection may be conceptualized as enveloping the Earth in a horizontal cylinder, and projecting the Earth's surface onto the cylinder (Figure 3-38). The cylinder in the transverse Mercator commonly intersects the Earth ellipsoid along a single north-south tangent, or along two *secant* lines, noted as the lines of true scale in Figure 3-38. A line parallel to and midway between the secants is often called the central meridian. The central meridian extends north and south through transverse Mercator projections.

As with the Lambert conformal conic, the transverse Mercator projection has a band of low distortion, but this band runs in a north-south direction. Distortion is least near the line(s) of intersection. The graph at the top right of Figure 3-38 shows a transverse Mercator projection with the central meridian (line of intersection) at 0 degrees longitude, traversing western Africa, eastern Spain, and England. Distortion increases markedly with distance east or west away from the intersection line, for example, the shape of South America is severely distorted in the top right of Figure 3-38. The drawing at the bottom of Figure 3-38 shows lines estimating approximately equal scale distortion for a transverse Mercator projection centered on the USA. Notice that the distortion increases as distance from the two lines of intersection increases. Scale distortion

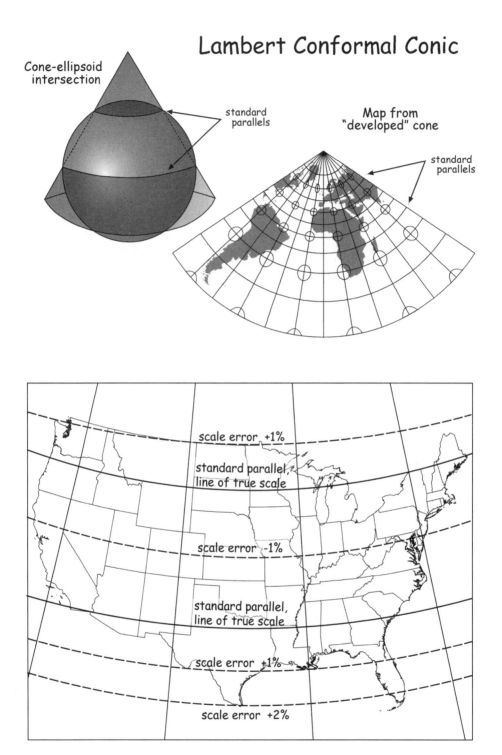

Figure 3-37: Lambert conformal conic (LCC) projection (top) and an illustration of the scale distortion associated with the projection. The LCC is derived from a cone intersecting the ellipsoid along two standard parallels (top left). The "developed" map surface is mathematically unrolled from the cone (top right). Distortion is primarily in the north-south direction, and is illustrated in the developed surfaces by the deformation of the 5-degree diameter geographic circles (top) and by the lines of approximately equal distortion (bottom). Note that there is no scale distortion where the standard parallels intersect the globe, at the lines of true scale (bottom, adapted from Snyder, 1987).

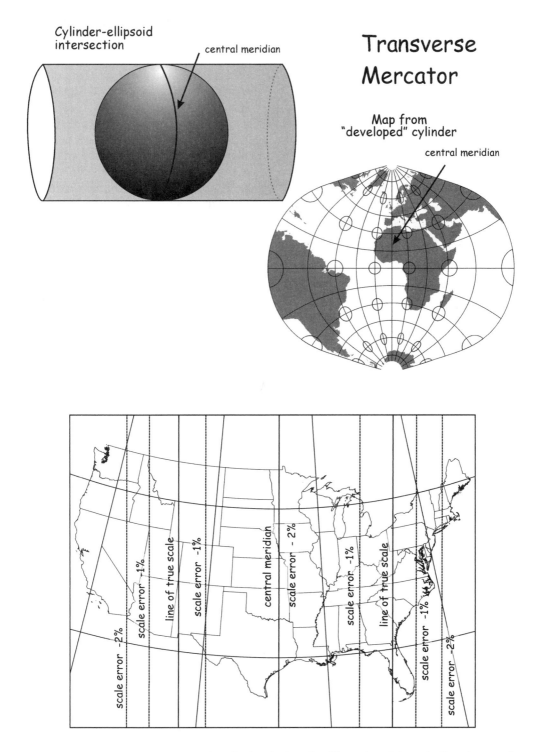

Figure 3-38: Transverse Mercator (TM) projection (top), and an illustration of the scale distortion associated with the projection (bottom). The TM projection distorts distances in an east-west direction, but has relatively little distortion in a north-south direction. This TM intersects the sphere along two lines, and distortion increases with distance from these lines (bottom, adapted from Snyder, 1987).

error may be maintained below any threshold by ensuring the mapped area is close to these two secant lines intersecting the globe. Transverse Mercator projections are often used for areas that extend in a north-south direction, as there is little added distortion extending in that direction.

Different projection parameters may be used to specify an appropriate coordinate system for a region of interest. Specific standard parallels or central meridians are chosen to minimize distortion over a mapping area. An origin location, measurement units, x and y (or northing and easting) offsets, a scale factor, and other parameters may also be required to define a specific projection. Once a projection is defined, the coordinates of every point on the surface of the Earth may be determined, usually by a closed-form or approximate mathematical formula.

The State Plane Coordinate System

The State Plane Coordinate System is a standard set of projections for the United States. The State Plane coordinate system specifies positions in Cartesian coordinate systems for each state. There are one or more zones in each state, with slightly dif-

ferent projections in each State Plane zone (Figure 3-39). Multiple State Plane zones are used to limit distortion errors due to map projections.

State Plane systems greatly facilitate surveying, mapping, and spatial data development in a GIS, particularly when whole county or larger areas are involved. The State Plane system provides a common coordinate reference for horizontal coordinates over county to multi-county areas while limiting distortion error to specified maximum values. Most states have adopted zones such that projection distortions are kept below one part in 10,000. Some states allow larger distortions (e.g., Montana, Nebraska). State Plane coordinate systems are used in many types of work, including property surveys, property subdivisions, large-scale construction projects, and photogrammetric mapping, and the zones and state plane coordinate system are often adopted for GIS.

One State Plane projection zone may suffice for small states. Larger states commonly require several zones, each with a different projection, for each of several geographic zones of the state. For example Delaware has one State Plane coordinate zone, while California has 6, and Alaska has 10 State Plane coordinate zones, each corre-

Figure 3-39: State plane zone boundaries, NAD83.

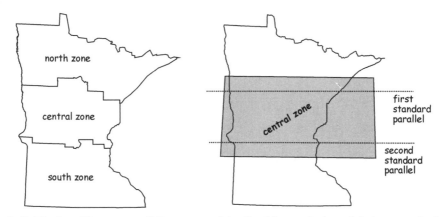

Figure 3-40: The State Plane zones of Minnesota, and details of the standard parallel placement for the Minnesota central State Plane zone.

sponding to a different projection within the state. Zones are added to a state to ensure acceptable projection distortion within all zones (Figure 3-40, left). Zone boundaries are defined by county, parish, or other municipal boundaries. For example, the Minnesota south/central zone boundary runs approximately east-west through the state along defined county boundaries (Figure 3-40, left).

The State Plane coordinate system is based on two types of map projections: the Lambert conformal conic and the transverse Mercator projections. Because distortion in a transverse Mercator increases with distance from the central meridian, this projection type is most often used with states that have a long north-south axis (e.g., Illinois or New Hampshire). Conversely, a Lambert conformal conic projection is most often used when the long axis of a state is in the east-west direction (e.g. North Carolina and Virginia). When computing the State Plane coordinates, points are projected from their geodetic latitudes and longitudes to x and y coordinates in the State Plane systems.

The Lambert conformal conic projection is specified in part by two standard parallels that run in an east-west direction. A different set of standard parallels is defined for each State Plane zone. These parallels are placed at one-sixth of the zone width from the north and south limits of the zone (Figure 3-40,

right). The zone projection is defined by specifying the standard parallels and a central meridian that has a longitude near the center of the zone. This central meridian points in the direction of geographic north, however all other meridians converge to this central meridian, so they do not point to geographic north. The Lambert conformal conic is used to specify projections for State Plane zones for 31 states.

As noted earlier, the transverse Mercator specifies a central meridian. This central meridian defines grid north in the projection. A line along the central meridian points to geographic north, and specifies the Cartesian grid direction for the map projection. All parallels of latitude and all meridians except the central meridian are curved for a transverse Mercator projection, and hence these lines do not parallel the grid x or y directions. The transverse Mercator is used for 22 State Plane systems (the sum of states is greater than 50 because both the transverse Mercator and Lambert conformal conic are used in some states, e.g., Florida).

Finally, note that more than one version of the State Plane coordinate system has been defined. Changes were introduced with the adoption of the North American Datum of 1983. Prior to 1983, the State Plane projections were based on NAD27. Changes were minor in some cases, and major in others, depending on the state and State Plane zone. Some states, such as South Carolina, Nebraska, and California, dropped zones between the NAD27 and NAD83 versions (Figure 3-41). Others maintained the same number of State Plane zones, but changed the projection by the placement of the meridians, or by switching to a metric coordinate system rather than one using feet, or by shifting the projection origin. State Plane zones are sometimes identified by the Federal Information Processing System (FIPS) codes, and most codes are similar across NAD27 and NAD83 versions. Care must be taken when using older data to identify the version of the State Plane coordinate system used because the FIPS and State Plane zone designators may be the same, but the projection parameters may have changed from NAD27 to NAD83.

Conversion among State Plane projections may be additionally confused by the various definitions used to translate from feet to meters. The metric system was first developed during the French Revolution in the late 1700s, and it was adopted as the official unit of distance in the United States, by the initiative of Thomas Jefferson. President Jefferson was a proponent of the metric system because it improved scientific measurements, was based on well-defined, integrated units, reduced commercial fraud, and improved trade within the new nation. The conversion was defined in the United States as one meter equal to exactly 39.97 inches. This yields a conversion for a *U.S. survey foot* of:

$$1 \text{ foot} = 0.3048006096012 \text{ meters}$$

Unfortunately, revolutionary tumult, national competition, and scientific differences led to the eventual adoption of a different conversion factor in Europe and most of the rest of the world. They adopted an *international foot* of:

$$1 \text{ foot} = 0.3048 \text{ meters}$$

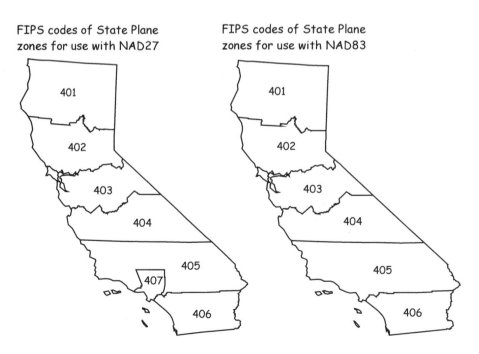

FIPS codes of State Plane zones for use with NAD27

FIPS codes of State Plane zones for use with NAD83

Figure 3-41: State Plane coordinate system zones and FIPS codes for California based on the NAD27 and NAD83 datums. Note that zone 407 from NAD27 is incorporated into zone 405 in NAD83.

The United States definition of a foot is slightly longer than the European definition, by about one part in five million. Both conversions are used in the U.S., and the international conversion elsewhere. The European conversion was adopted as the standard for all measures under an international agreement in the 1950s. However, there was a long history of the use of the U.S. conversion in U.S. geodetic and land surveys. Therefore, the U.S. conversion was called the U.S. survey foot. This slightly longer metric-to-foot conversion factor should be used for all conversions among geodetic coordinate systems within the United States, for example, when converting from a State Plane coordinate system specified in feet to one specified in meters.

Universal Transverse Mercator Coordinate System

The Universal Transverse Mercator (UTM) coordinate system is another standard coordinate, distinct from the State Plane system. The UTM is a global coordinate system, based on the transverse Mercator projection. It is widely used in the United States and other parts of North America, and is also used in many other countries.

The UTM system divides the Earth into zones that are 6 degrees wide in longitude and extend from 80 degrees south latitude to 84 degrees north latitude. UTM zones are numbered from 1 to 60 in an easterly direction, starting at longitude 180 degrees West (Figure 3-42). Zones are further split north and south of the equator. Therefore, the zone containing most of England is identified as UTM Zone 30 North, while the zones containing most of New Zealand are designated UTM Zones 59 South and 60 South. Directional designations are here abbreviated, for example, 30N in place of 30 North.

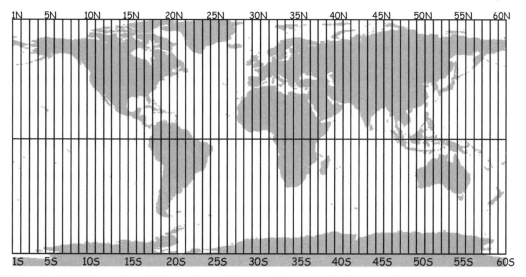

Figure 3-42: UTM zone boundaries and zone designators. Zones are six degrees wide and numbered from 1 to 60 from the International Date Line, 180°W. Zones are also identified by their position north and south of the equator, e.g., Zone 7 North, Zone 16 South.

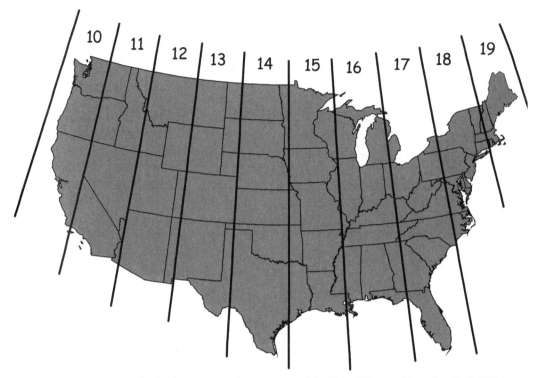

Figure 3-43: UTM zones for the lower 48 contiguous states of the United States of America. Each UTM zone is 6 degrees wide. All zones in the Northern Hemisphere are north zones, e.g., Zone 10 North, 11 North,...19 North.

The UTM coordinate system is common for data and study areas spanning large regions, for example, several State Plane zones. Many data from U.S. federal government sources are in a UTM coordinate system because many agencies manage large areas. Many state government agencies in the United States distribute data in UTM coordinate systems because the entire state fits predominantly or entirely into one UTM zone (Figure 3-43).

As noted before, all data for an analysis area must be in the same coordinate system if they are to be analyzed together. If not, the data will not co-occur as they should. The large width of the UTM zones accommodates many large-area analyses, and many states, national forests, or multi-county agencies have adopted the dominant UTM coordinate system as a standard.

We must note that the UTM coordinate system is not always compatible with regional analyses. Because coordinate values are discontinuous across UTM zone boundaries, analyses are difficult across these boundaries. UTM zone 15 is a different coordinate system than UTM zone 16. The state of Wisconsin approximately straddles these two zones, and the state of Georgia straddles zones 16 and 17. If a uniform, statewide coordinate system is required, the choice of zone is not clear, and either one zones must be chosen, or some compromise projection must be chosen. For example, statewide analyses in Georgia and in Wisconsin are often conducted using UTM-like systems that involve moving the central meridian to near the center of each state.

Distances in the UTM system are specified in meters north and east of a zone origin (Figure 3-44). The y values are known as *northings*, and increase in a northerly direction. The x values are referred to as *eastings* and increase in an easterly direction.

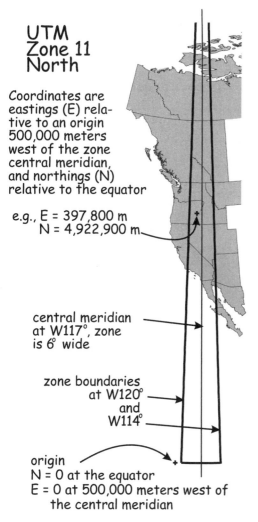

UTM Zone 11 North

Coordinates are eastings (E) relative to an origin 500,000 meters west of the zone central meridian, and northings (N) relative to the equator

e.g., E = 397,800 m
N = 4,922,900 m

central meridian at W117°, zone is 6° wide

zone boundaries at W120° and W114°

origin
N = 0 at the equator
E = 0 at 500,000 meters west of the central meridian

Figure 3-44: UTM zone 11N.

The origins of the UTM coordinate system are defined differently depending on whether the zone is north or south of the equator. In either case, the UTM coordinate system is defined so that all coordinates are positive within the zone. Zone easting coordinates are all greater than zero because the central meridian for each zone is assigned an easting value of 500,000 meters. This effectively places the origin (E = 0) at a point 500,000 meters west of the central meridian.

All zones are less than 1,000,000 meters wide, ensuring that all eastings will be positive.

The equator is used as the northing origin for all north zones. Thus, the equator is assigned a northing value of zero for north zones. This avoids negative coordinates, because all of the UTM north zones are defined to be north of the equator.

University Transverse Mercator zones south of the equator are slightly different than those north of the equator (Figure 3-45). South zones have a *false northing* value added to ensure all coordinates within a zone are positive. UTM coordinate values increase as one moves from south to north in a projection area. If the origin were placed at the equator with a value of zero for south zone coordinate systems, then all the northing values would be negative. An offset is applied by assigning a false northing, a nonzero value, to an origin or other appropriate location. For UTM south zones, the northing values at the equator are set to equal 10,000,000 meters, assuring that all northing coordinate values will be positive within each UTM south zone (Figure 3-45).

Continental and Global Projections

There are map projections that are commonly used when depicting maps of continents, hemispheres, or other large regions. Just as with smaller areas, map projections for continental areas may be selected based on the distortion properties of the resultant map. Sizeable projection distortion in area, distance, and angle are observed in most large-area projections. Angles, distances, and areas are typically not measured or computed from these projections, as the differences between the map-derived and surface-measured values are too great for most uses. Large-area maps are most often used to display or communicate data for continental or global areas.

There are a number of projections that have been widely used for the world. These include variants of the Mercator, Goode, Mollweide, and Miller projections, among others. There is a trade-off that must be made in global projections, between a con-

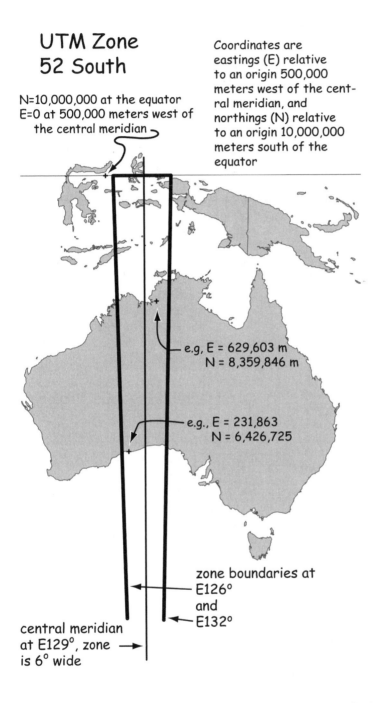

Figure 3-45: UTM south zones, such as Zone 52S shown here, are defined such that all the north-ing and easting values within the zone are positive. A false northing of 10,000,000 is applied to the equator, and a false easting of 500,000 is applied to the central meridian to ensure positive coordi-nate values throughout each zone.

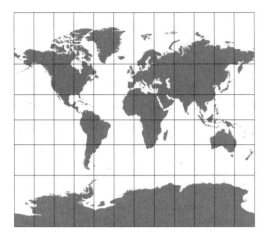

Figure 3-46: A Miller cylindrical projection, commonly used for maps of the world. This is an example of an uninterrupted map surface.

tinuous map surface and distortion. If a single, uncut surface is mapped, then there is severe distortion in some portion of the map. Figure 3-46 shows a Miller cylindrical projection, often used in maps of the world. This projection is similar to a Mercator projection, and is based on a cylinder that intersects the Earth at the equator. Distortion increases towards the poles, although not as much as with the Mercator.

Distortion in world maps may be reduced by using a cut or interrupted surface. Different projection parameters or surfaces may be specified for different parts of the globe. Projections may be mathematically constrained to be continuous across the area mapped.

Figure 3-47 illustrates an interrupted projection in the form of a Goode homolosine. This projection is based on a sinusoidal projection and a Mollweide projection. These two projection types are merged at parallels of identical scale. The parallel of identical scale in this example is set near the mid-northern latitude of $44^\circ\ 40'$ N.

Continental projections may also be established. Generally, the projections are chosen to minimize area or shape distortion for the region to be mapped. Lambert conformal conic or other conic projections are often chosen for areas with a long east-west dimension, for example when mapping the contiguous 48 United States of America, or North America. Standard parallels are placed near the top and bottom of the continental area to reduce distortion across the region mapped. Transverse cylindrical projections are often used for large north-south continents.

None of these worldwide or continental projections are commonly used in a GIS for data storage or analysis. Uninterrupted coordinate systems show too much distortion to be of use in measuring most spatial quantities, and interrupted projections do not spec-

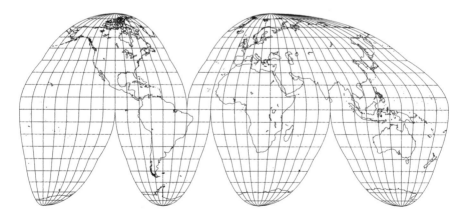

Figure 3-47: A Goode homolosine projection. This is an example of an interrupted projection, often used to reduce some forms of distortion when displaying the entire Earth surface. (from Snyder and Voxland, 1989)

ify a Cartesian coordinate system that defines positions for all points on the Earth surface. Worldwide data are typically stored in geographic coordinates (latitudes and longitudes). These data may then be projected to a specific coordinate system for display or document preparation.

Conversion Among Coordinate Systems

You might ask, how do I convert between geographic and projected coordinate systems? Exact or approximate mathematical formulas have been developed to convert to and from geographic (latitude and longitude) to all commonly used coordinate projections (Figure 3-48). These formulas are incorporated into "coordinate calculator" software packages, and are also integrated into most GIS software. For example, given a coordinate pair in the State Plane system, you may calculate the corresponding geographic coordinates. You may then apply a formula that converts geographic coordi-

nates to UTM coordinates for a specific zone using another set of equations. Since the backward and forward projections from geographic to projected coordinate systems are known, we may convert among most coordinate systems by passing through a geographic system (Figure 3-49, a).

Care must be taken when converting among projections that use different datums. If appropriate, we must insert a datum transformation when converting from one projected coordinate system to another (Figure 3-49, b). A datum transformation is a calculation of the change in geographic coordinates when moving from one datum to another.

Users of GIS software should be careful when applying coordinate projection tools because the datum transformation may be omitted, or an inappropriate datum manually or automatically selected. For some software, the projection tool does not check or maintain information on the datum of the input spatial layer. This will often lead to an inappropriate or no datum transformation, and the output from the projection will be in error. Often these errors are small relative to other errors, for example, spatial imprecision in the collection of the line or point features. As shown in Figure 3-21, errors between NAD83(1986) and NAD83(CORS96) may be less than 10 cm (4 inches) in some regions, often much less than the average spatial error of the data themselves. However, errors due to ignoring the datum transformation may be quite large, for example, 10s to 100s of meters between NAD27 and most versions of NAD83, and errors of up to a meter are common between recent versions of WGS84 and NAD83. Given the sub-meter accuracy of many new GPS and other GNSS receivers used in data collection, datum transformation error of one meter is significant. As data collection accuracy improves, users develop applications based on those accuracies, so datum transformation errors should be avoided in all cases.

Conversion from geographic (lon, lat) to projected coordinates

Given longitude = λ, latitude = ϕ

Mercator projection coordinates are:

$x = R \cdot (\lambda - \lambda_0)$
$y = R \cdot \ln (\tan (90^\circ + \phi/2)$

where R is the radius of the sphere at map scale (e.g., Earth's radius), ln is the natural log function, and λ_0 is the longitudinal origin (Greenwich meridian)

Figure 3-48: Formulas are known for most projections that provide exact projected coordinates, if the latitudes and longitudes are known. This example shows the formulas defining the Mercator projection.

a) From one projection to another - same datum

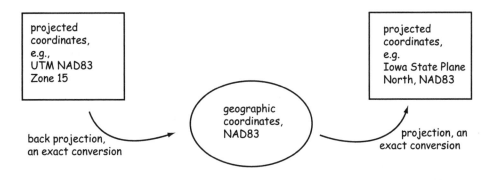

b) From one projection to another - different datums

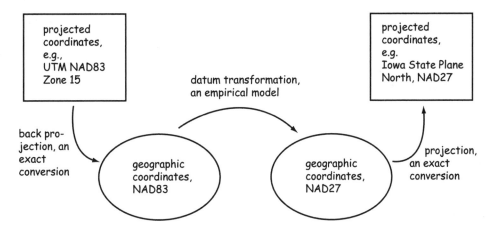

Figure 3-49: We may project between most coordinate systems via the back (or inverse) and forward projection equations. These calculate exact geographic coordinates from projected coordinates (a), and then new projected coordinates from the geographic coordinates. We must insert an extra step when a projection conversion includes a datum change. A datum transformation must be used to convert from one geodetic datum to another (b).

The Public Land Survey System

For the benefit of GIS practitioners in the United States we must cover one final land designation system, known as the *Public Land Survey System*, or PLSS. The PLSS is not a coordinate system, but PLSS points are often used as reference points in the United States, so the PLSS should be well understood for work there. The PLSS is a standardized method for designating and describing the location of land parcels. It was used for the initial surveys over most of the United States after the early 1800s, therefore nearly all land outside the original thirteen colonies uses the PLSS. An approximately uniform grid system was established across the landscape, with periodic adjustments incorporated to account for the anticipated error. Parcels were designated by their location within this grid system.

The PLSS was developed for a number of reasons. First, it was seen as a method to remedy many of the shortcomings of *metes and bounds* surveying, the most common method for surveying prior to the adoption of the PLSS. Metes and bounds describe a parcel relative to features on the landscape, sometimes supplemented with angle or distance measurements. In colonial times a parcel description might state "beginning at the joining of Shope Fork and Coweeta Creek, downstream along the creek approximately 280 feet to a large rock on the right bank, thence approximately northwest 420 feet to a large chestnut oak blazed with an S, thence west 800 feet to Mill Creek, thence down Mill Creek to Shope Fork Road, thence east on Shope Fork Road to the confluence of Shope Fork and Coweeta Creek."

Metes and bounds descriptions require a minimum of surveying measurements, but it was a less than ideal system for describing locations or parcels. These metes and bounds descriptions could be vague, the features in the landscape might be moved or change, and it was difficult to describe parcels when there were few readily distinguished landscape features. Subdivided

6	5	4	3	2	1
7	8	9	10	11	12
18	17	16	15	14	13
19	20	21	22	23	24
30	29	28	27	26	25
31	32	33	34	35	36

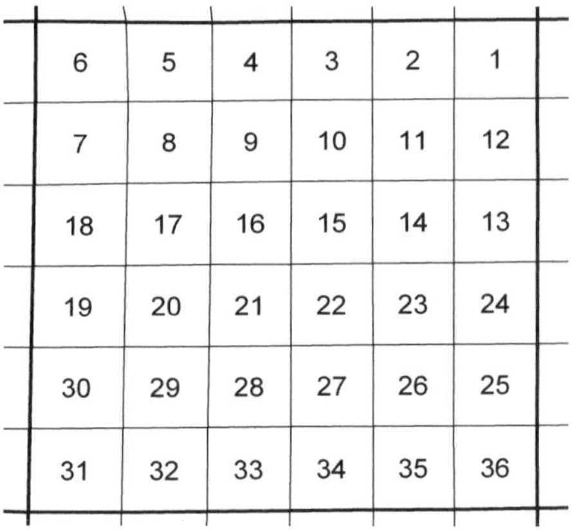

Figure 3-50: Typical layout and section numbering of a PLSS township

parcels were often poorly described, and hence the source of much litigation, ill will, and many questionable real estate transactions.

The U.S. government needed a system that would provide unambiguous descriptions of parcels in unsettled territories west and south of the original colonies. The federal government saw public land sales as a way to generate revenue, to pay revolutionary war veterans, to expand the country, and to protect against encroachment by European powers. Parcels could not be sold until they were surveyed, therefore the PLSS was created. Land surveyed under the PLSS can be found in thirty states, including Alaska and most of the midwestern and western United States. Lands in the original 13 colonies, as well as West Virginia, Tennessee, Texas, and Kentucky were not surveyed under the PLSS system.

The PLSS divided lands by north-south lines running parallel to a principal meridian. New north-south lines were surveyed at six mile intervals. Additional lines were surveyed that were perpendicular to these north-south lines, in approximately east-west directions, and crossing meridian lines, also run at six-mile intervals. These lines form townships that were six miles square. Each township was further subdivided into 36 sections, each section approximately a mile on a side. Each section was subdivided further, to quarter-sections (one-half mile on a side), or sixteenth sections, (one-quarter mile on a side, commonly referred to as quarter-quarter sections). Sections were numbered in a zig-zag pattern from one to 36, beginning in the northeast corner (Figure 3-50).

Surveyors typically marked the section corners and quarter-corners while running survey lines. Points were marked by a number of methods, including stone piles, pits, blaze marks chiseled in trees, and pipes or posts sunk in the ground.

Because the primary purpose of the PLSS survey was to identify parcels, lines and corner locations were considered static on completion of the survey, even if the corners were far from their intended location. Survey errors were inevitable given the large areas and number of different survey parties involved. Rather than invite endless dispute and re-adjustment, the PLSS specifies that boundaries established by the appointed PLSS surveyors are unchangeable, and that township and section corners must be accepted as true. The typical section contains approximately 640 acres, but due in part to errors in surveying, sections larger than 1200 acres and smaller than 20 acres were also established (Figure 3-51).

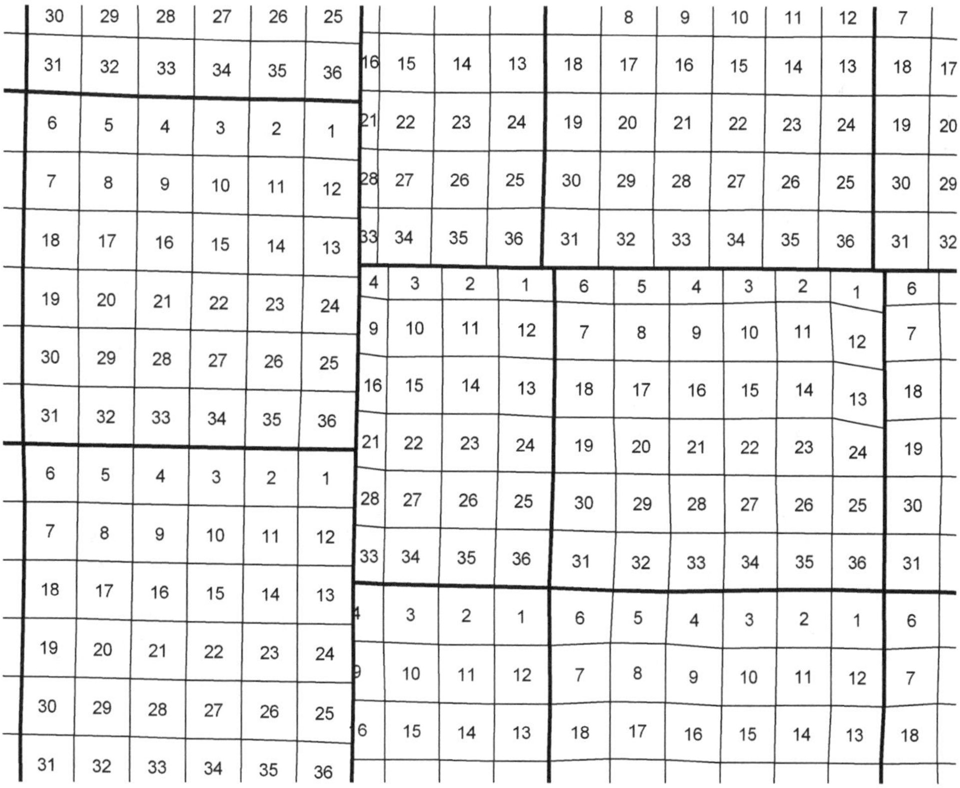

Figure 3-51: Example of variation in the size and shape of PLSS sections. Most sections are approximately one mile square with section lines parallel or perpendicular to the primary meridian, as illustrated by the township in the upper left of this figure. However, adjustments due to different primary meridians, different survey parties, and errors result in irregular section sizes and shapes.

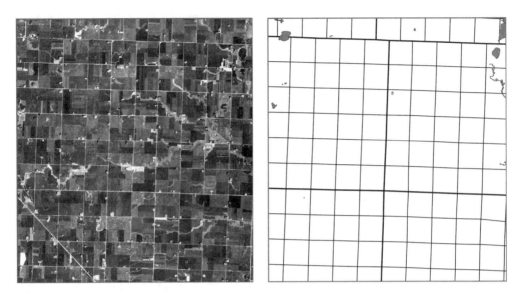

Figure 3-52: PLSS lines are often visible on the landscape. Roads (light lines on the image, above left) often follow the section and township lines (above right).

The PLSS is important today for several reasons. First, since PLSS lines are often property boundaries, they form natural corridors in which to place roads, powerlines, and other public services, so they are often evident on the landscape (Figure 3-52). Many road intersections occur at PLSS corner points, and these can be viewed and referenced on many maps or imagery used for GIS database development efforts. Thus the PLSS often forms a convenient system to co-register GIS data layers. PLSS corners and lines are often plotted on government maps (e.g., 1:24,000 quads) or available as digital data (e.g., National Cartographic Information Center Digital Line Graphs). Further, PLSS corners are sometimes re-surveyed using high precision methods to provide property line control, particularly when a GIS is to be developed (Figure 3-53). Thus these points may be useful to properly locate and orient spatial data layers on the Earth's surface.

Figure 3-53: A PLSS corner that has been surveyed and marked with a monument. This monument shows the physical location of a section corner. These points are often used as control points for further spatial data development.

Summary

In order to enter coordinates in a GIS, we need to uniquely define the location of all points on Earth. We must develop a reference frame for our coordinate system, and locate positions on this system. Since the Earth is a curved surface and we work with flat maps, we must somehow reconcile these two views of the world. We define positions on the globe via geodesy and surveying. We convert these locations to flat surfaces via map projections.

We begin by modeling the Earth's shape with an ellipsoid. An ellipsoid differs from the geoid, a gravitationally-defined Earth surface, and these differences caused some early confusion in the adoption of standard global ellipsoids. There is a long history of ellipsoidal measurement, and we have arrived at our best estimates of global and regional ellipsoids after collecting large, painstakingly-developed sets of precise surface and astronomical measurements. These measurements are combined into datums, and these datums are used to specify the coordinate locations of points on the surface of the Earth.

Map projections are a systematic rendering of points from the curved Earth surface onto a flat map surface. While there are many purely mathematical or purely empirical map projections, the most common map projections used in GIS are based on developable surfaces. Cones, cylinders, and planes are the most common developable surfaces. A map projection is constructed by passing rays from a projection center through both the Earth surface and the developable surface. Points on the Earth are projected along the rays and onto the developable surface. This surface is then mathematically unrolled to form a flat map.

Standard sets of projections are commonly used for spatial data in a GIS. In the United States, the UTM and State Plane coordinate systems define a standard set of map projections that are widely used. Other map projections are commonly used for continental or global maps, and for smaller maps in other regions of the world.

A datum transformation is often required when performing map projections. Datum transformations account for differences in geographic coordinates due to changes in the shape or origin of the spheroid, and in some cases to datum adjustments. Datum transformation should be applied as a step in the map projection process when input and output datums differ.

A system of land division was established in the United States known as the Public Land Survey System (PLSS). This is not a coordinate system but rather a method for unambiguously and systematically defining parcels of land based on regularly spaced survey lines in approximately north-south and east-west directions. Intersection coordinates have been precisely measured for many of these survey lines, and are often used as a reference grid for further surveys or land subdivision.

Suggested Reading

Bossler, J.D. (2002). Datums and geodetic systems, In J. Bossler (Ed.), *Manual of Geospatial Technology.* Taylor and Francis: London.

Brandenburger, A.J. & Gosh, S K. (1985). The world's topographic and cadastral mapping operations. *Photogrammetric Engineering and Remote Sensing*, 51:437-444.

Burkholder, E.F. (1993). Computation of horizontal/level distances. *Journal of Surveying Engineering*, 117:104-119.

Colvocoresses, A.P. (1997). The gridded map. *Photogrammetric Engineering and Remote Sensing*, 63:371-376.

Doyle, F.J. (1997). Map conversion and the UTM Grid. *Photogrammetric Engineering and Remote Sensing*, 63:367-370.

Featherstone, W.E., & Kuhn, M. (2006). Height systems and vertical datums: a review in the Australian context. *Journal of Spatial Science*, 51:21-41.

Habib, A. (2002). Coordinate transformation. In J. Bossler (Ed.), *Manual of Geospatial Technology.* Taylor and Francis: London.

Flacke, W., & Kraus, B. (2005). *Working with Projections and Datum Transformations in ArcGIS: Theory and Practical Examples*. Points Verlag: Norden.

Iliffe, J.C., & Lott, R. (2008). *Datums and Map Projections for Remote Sensing, GIS, and Surveying, 2nd ed.* CRC Press: Boca Raton.

Janssen, V. (2009). Understanding coordinate reference systems, datums, and transformations. *International Journal of Geoinformatics*, 5:41-53.

Keay, J. (2000). *The Great Arc*. Harper Collins: New York.

Leick, A. (1993). Accuracy standards for modern three-dimensional geodetic networks. *Surveying and Land Information Systems,* 53:111-127.

Maling, D.H. (1992). *Coordinate Systems and Map Projections*. George Phillip: London.

Milbert, D. (2008). An analysis of the NAD83(NSRS2007) National Readjustment. Downloaded 9/12/2011 from http://www.ngs.noaa.gov/PUBS_LIB/NSRS2007

National Geospatial-Intelligence Agency (NGA), TR8350.2 World Geodetic System 1984, Its Definition and Relationship with Local Geodetic Systems. http://earth-info.nga.mil/GandG/publications/tr8350.2/tr8350_2.html

NOAA Manual NOS NGS 5 State Plane Coordinate System of 1983 -- http://www.ngs.noaa.gov/PUBS_LIB/ManualNOSNGS5.pdf

Schwartz, C.R. (1989). *North American Datum of 1983, NOAA Professional Paper NOS 2*. National Geodetic Survey: Rockville.

Smith, J. (1997). *Introduction to Geodesy: The History and Concepts of Modern Geodesy*, Wiley: New York.

Sobel, D. (1995). *Longitude*. Penguin Books: New York.

Soler, T. & Snay, R.A.(2004). Transforming positions and velocities between the International Terrestrial Reference Frame of 2000 and the North American Datum of 1983. *Journal of Surveying Engineering*, 130:49-55.

Snay, R.A. & Soler, T. (1999). Modern terrestrial reference systems, part 1. *Professional Surveyor*, 19:32-33.

Snay, R.A. & Soler, T. (2000). Modern terrestrial reference systems, part 2. the evolution of NAD83, *Professional Surveyor*, 20:16-18.

Snay, R.A., & Soler, T. (2000). Modern terrestrial reference systems, part 3. WGS84 and ITRS, *Professional Surveyor*, 20:24-28.

Snay, R.A., & Soler, T. (2000). Modern terrestrial reference systems, part 4, practical considerations for accurate positioning. *Professional Surveyor*, 20:32-34.

Snyder, J. (1993). *Flattening the Earth: Two Thousand Years of Map Projections*. Chicago: University of Chicago Press, Chicago.

Snyder, J. P. (1987). *Map Projections, A Working Manual, USGS Professional Paper No. 1396*. United States Government Printing Office: Washington D.C.

Snyder, J.P., & Voxland, P.M. (1989). *An Album of Map Projections, USGS Professional Paper No. 1453*. United States Government Printing Office: Washington D.C.

Tobler, W.R. (1962). A classification of map projections. *Annals of the Association of American Geographers*, 52:167-175.

U.S. Coast and Geodetic Survey Special Publication 235 The State Coordinate Systems -- http://www.ngs.noaa.gov/PUBS_LIB/publication235.pdf

Van Sickle, J. (2010). *Basic GIS Coordinates, 2nd Edition.* CRC Press: Boca Raton.

Vanicek, P, & Steevens, R.H. (1996). Transformation of coordinates between two horizontal geodetic datums. *Journal of Geodesy,* 70:740-745.

Welch, R., & Homsey, A. (1997). Datum shifts for UTM coordinates. *Photogrammetric Engineering and Remote Sensing,* 63:371-376.

Wolf, P. R., & Ghilani, C.D. (2002). Elementary Surveying (10th ed.). Prentice-Hall: Upper Saddle River.

Yang, Q., Snyder, J.P. & Tobler, W.R. (2000). *Map Projection Transformation: Principles and Applications.* Taylor & Francis: London.

Zilkoski, D., Richards, J. & Young, G. (1992). Results of the general adjustment of the North American Vertical Datum of 1988. *Surveying and Land Information Systems,* 53:133-149.

Study Questions

3.1 - Can you describe how Eratosthenes estimated the circumference of the Earth? What value did he obtain?

3.2 - Assume the Earth is approximately a sphere (not and elllipsoid). Also assume you've repeated Poseidonius' measurements. What is your estimate of the radius of the Earth's sphere given the following distance/angle pairs. Note that the distances are given below meters, and angle in degrees, and calculators or spreadsheets may require you enter trigonometric angles in radians for trigonometric functions (1 radian = 57.2957795 degrees):

> a) angle = $1°$ 18' 45.79558", distance = 146,000 meters
>
> b) angle = $0°$ 43' 32.17917", distance = 80,500 meters
>
> c) angle = $0°$ 3' 15.06032", distance = 6,000 meters

3.3 - Assume the Earth is approximately a sphere (not and elllipsoid). Also assume you've repeated Poseidonius' measurements. What is your estimate of the radius of the Earth's sphere given the following distance/angle pairs. Note that the distances are given below meters, and angle in degrees, and calculators or spreadsheets may require you enter trigonometric angles in radians for trigonometric functions (1 radian = 57.2957795 degrees):

> a) angle = $2°$ 59' 31.33325", distance = 332,000 meters
>
> b) angle = $9°$ 12' 12.77201", distance = 1,020,708 meters
>
> c) angle = $1°$ 2' 12.15566", distance = 115,200 meters

3.4 - What is an ellipsoid? How does an ellipse differ from a sphere? What is the equation for the flattening factor?

3.5 - Why do different ellipsoids have different radii? Can you provide three reasons?

3.6 - Can you define the geoid? How does it differ from the ellipsoid, or the surface of the Earth? How do we measure the position of the geoid?

3.7 - Can you define a parallel or meridian in a geographic coordinate system? Where do the "horizontal" and "vertical" zero lines occur?

3.8 - How does magnetic north differ from the geographic North Pole?

3.9 - Can you define a datum? Can you describe how datums are developed?

3.10 - Why are there multiple datums, even for the same place on Earth? Can you define what we mean when we say there is a datum shift?

3.11 - What is a triangulation survey, and a bench mark?

3.12 - Why do we not measure vertical heights relative to mean sea level anymore?

3.13 - What is the difference between an orthometric height and a dynamic height.

3.14 - Use the NADCON software available from the U.S. NOAA/NGS website (http://www.ngs.noaa.gov/TOOLS/program_descriptions.html) to fill the following table. Note that all of these points are in CONUS, and longitudes are west, but entered as positive numbers.

	NAD27		NAD83(86)		HPGN	
Pnt	latitude	longitude	latitude	longitude	latitude	longitude
1	32°44'15"	117°09'42"	32°44'15.1827"	117°09'45.1202"	32°44'15.1820"	117°09'45.1200"
2	47°27'55"	122°18'06"	47°27'54.3574"	122°18'10.4453"	47°27'54.3642"	122°18'10.4366"
3	43°07'59"	89°20'11"	43°07'58.9806"	89°20'11.4226"		
4	29°58'07"	95°21'31"	29°58'07.7975"	95°21'31.7705"		
5	40°00'00"	105°16'01"			39°59'59.9552"	105°16'02.9712"
6	24°33'30"	81°45'19"			24°33'31.5216"	81°45'18.3362"
7			38°51'10.4052"	77°02'19.9165"	38°51'10.4063"	77°02'19.9041"
8			46°52'0.1524"	68°00'59.0974"	46°52'0.1580"	68°00'59.0995"

3.15 - Use the web version or download and start the HTDP software from the U.S. NOAA/NGS site listed above, and complete the following table. Enter epoch start and stop dates of 1, 1, 1986 and 1,1, 2005, respectively, and specify a zero height or z.

| | NAD83(CORS96) | | WGS(G1150) | | ITRF2005 | |
Pnt	latitude	longitude	latitude	longitude	latitude	longitude
1	32°44'15"	117°09'42"	32°44'15.0321"	117°09'42.0662"	32°44'15.0325"	117°09'45.0663"
2	47°27'55"	122°18'06"	47°27'55.0183"	122°18'06.0583"	47°27'55.0186"	122°18'06.0583"
3	43°07'59"	89°20'11"	43°07'59.0283"	89°20'11.0293"		
4	29°58'07"	95°21'31"	29°58'07.0177"	95°21'31.0293"		
5	40°00'00"	105°16'01"			40°00'00.2143"	105°16'01.422"
6	24°33'30"	81°45'19"			24°33'30.0164"	81°45'19.0145"
7			38°51'1.0288"	77°02'21.0137"	38°51'1.0293"	77°02'21.0136"
8			46°52'0.0363"	68°01'00.0061"	46°52'0.0367"	68°01'01.0061"

3.16 - What is a developable surface? What are the most common shapes for a developable surface?

3.17 - Look up the NGS control sheets for the following points, and record their datums, latitudes and longitudes:

>   DOG, Maine, PID= PD0617.

>   Key West GSL, Florida, PID=AA1645

>   Neah A, Washington, PID=AF882

3.18 - Calculate the great circle distance for the control points, above, from:

>   - DOG to Neah A
>   - Key West to DOG
>   - Neah A to Key West

3.19 - Can you describe the State Plane coordinate system? What type of projections are used in a State Plane coordinate system?

3.20 - Can you define and describe the Universal Transverse Mercator coordinate system? What type of developable surface is used with a UTM projection? What are UTM zones, where is the origin of a zone, and how are negative coordinates avoided?

3.21 - What is a datum transformation? How does it differ from a map projection?

3.22 - Specify which type of map projection you would choose for each country, assuming you could use only one map projection for the entire country, the projection lines of intersection would be optimally-placed, and you wanted to minimize overall spatial distance distortion for the country. Choose from a transverse Mercator, a Lambert conformal conic, or an Azimuthal:

Benin   Bhutan

Slovenia   Israel

3.23 - Specify which type of map projection you would choose for each country, assuming you could use only one map projection for the entire country, the projection lines of intersection would be optimally-placed, and you wanted to minimize overall spatial distance distortion for the country. Choose from a transverse Mercator, a Lambert conformal conic, or an Azimuthal:

Chile   Nepal

Kyrgyzstan   The Gambia

3.24 - Can you describe the Public Land Survey System? Is it a coordinate system? What is its main purpose?

4 Maps, Data Entry, Editing, and Output

Building a GIS Database

Introduction

Spatial data entry and editing are frequent activities for many GIS users. A large number of coordinates is needed to represent features in a GIS, and each coordinate value must be entered into the GIS database. This is often painstakingly slow, even with automated techniques, and spatial data entry and editing take significant time for most organizations.

Most spatial data sources may be categorized as either *hardcopy* or *digital*. Hardcopy forms are any drawn, written, or printed documents, including hand-drawn maps, manually measured survey data, legal records, and coordinate lists with associated tabular data. Most historical spatial data were recorded on maps (Figure 4-1), and although not all maps are suitable for conversion to digital formats, many maps are. Much data were created from hardcopy sources in the early years of GIS via *digitizing*, the process of collecting

Figure 4-1: Maps have served to store geographic knowledge for at least the past 4000 years. This early map of northern Europe shows approximate shapes and relative locations.

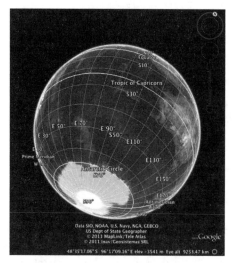

Figure 4-2: An example of commonly produced digital maps.

digital coordinates. Digitizing is a common data entry method today, although primarily from satellite and aerial images.

Digital maps, an electronic, graphic depiction of spatial data, are by far the most common map form today (Figure 4-2). Millions of electronic maps are generated each hour, composed on demand in response to web queries, on automobile navigations sys-

tems, and for commerce and advertising. These maps are flexible, easily customized, inexpensively distributed, and often dynamic.

Most maps, whether digital or hardcopy, contain several components (Figure 4-3). A *data area* or *pane* occupies the largest part of the map, and contains most of the depicted spatial data. A *neatline* is often included to provide a frame around all map elements, and *insets* may contain additional map elements. *Scalebars, legends*, titles, and other graphic elements such as a *north arrow* are often included. All maps have a *map scale*, defined as the ratio of the distance on the map to corresponding distance on the ground. (Figure 4-3).

Maps often depict coordinate lines (Figure 4-4). When the lines represent constant latitude and longitude, a set of coordinate lines is called a *graticule* (Figure 4-4a). These lines may appear curved, depending on the map scale, the map coordinate system, and the location of the area on the Earth's surface. Maps may also depict a *grid* consisting of lines of constant coordinates. Grid lines are typically drawn in both the x

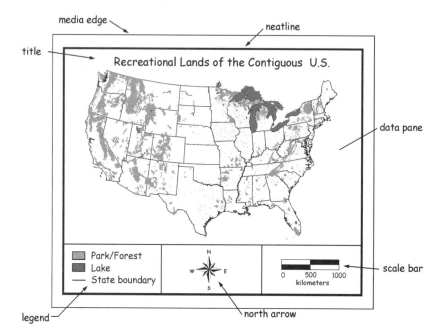

Figure 4-3: An example of a map and its components.

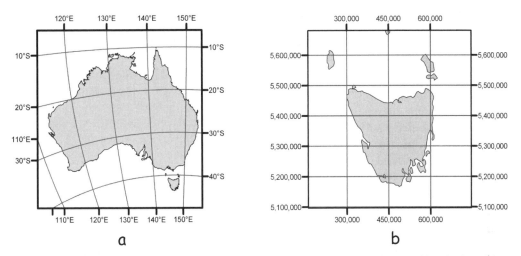

Figure 4-4: Maps often depict lines representing (a) a *graticule* of constant latitude and longitude or (b) a *grid* of constant x and y coordinates.

and y directions, and appear straight on most maps (Figure 4-4b). Graticules and grids are useful because they provide a reference against which location may be quickly estimated. Graticules are particularly useful for depicting the distortion inherent in a map projection, because they show how geographic north or east lines are deformed, and how this distortion varies across the map. Grids may establish a map-projected north, in contrast to geographic north, and may be useful when trying to navigate or locate a position on the map.

Historical and current images are valuable sources of geographic data, and although they are not maps, the line is becoming blurred, as aerial and satellite photographs become common backdrops for digital maps. Photographs do not typically provide an orthographic (flat, undistorted) view, and houses, rivers, or features of interest are not explicitly identified. However, images are a rich source of geographic information, and standard techniques may be used to remove major systematic distortions and extract features, through manual digitizing, described later in this chapter, or through image classification, described in Chapter 6.

Digital spatial data are those provided in a computer-compatible format. These include complete raster and vector data layers, text files, lists of coordinates, and digital images. Files and export formats can be used to transfer them to a local GIS system. Global Navigation Satellite Systems (GNSS), such as the U.S. Global Positioning System (GPS) and coordinate survey devices described in Chapter 5, are direct measurement system that can be used to record coordinates in the field and report them directly into digital formats. Finally, a number of digital image sources are available, such as satellite or airborne images that are collected in a digital raster format, or hardcopy aerial photographs that have been scanned to produce digital images.

Hardcopy data are an important source of geographic information for many reasons. First, most geographic information produced before 1980 was recorded in hardcopy form. Advances in optics, metallurgy, and industry during the 18th and 19th centuries allowed the mass production of precise surveying devices, and by the mid 20th century, much of the world had been plotted on *cartometric* quality maps. Cartometric maps are those that faithfully represent the relative position of objects and thus may be suitable as a source of spatial data.

While much spatial data has been collected from hardcopy sources, data entry from digital sources now dominates. Coordinates are increasingly captured via interpretation of digital image sources (these sources are described in Chapter 6) or collected directly in the field by satellite-based positioning services (Chapter 5).

Our objective in this chapter is to introduce spatial data entry via digitizing and coordinate surveying. We will also cover basic editing methods and data documentation, and rudimentary cartography and output.

Map Types

Many types of maps are produced, and the types are often referred to by the way features are depicted on the map. *Feature maps* are among the simplest, because they map points, lines, or areas and provide nominal information (Figure 4-5, upper left). A

road may be plotted with a symbol defining the type of road or a point may be plotted indicating the location of a city center, but the width of the road or number of city dwellers are not provided in the shading or other symbology on the map. Feature maps are perhaps the most common map form, and examples include most road maps, and standard map series such as the 7.5 minute topographic maps produced by the U.S. Geological Survey.

Choropleth maps depict quantitative information for areas. A mapped variable such as population density may be represented in the map (Figure 4-5, top right). Polygons define area boundaries, such as counties, states, census tracts, or other standard administrative units. Each polygon is given a color, shading, or pattern corresponding to values for a mapped variable, for example, in Figure 4-5, top right, the

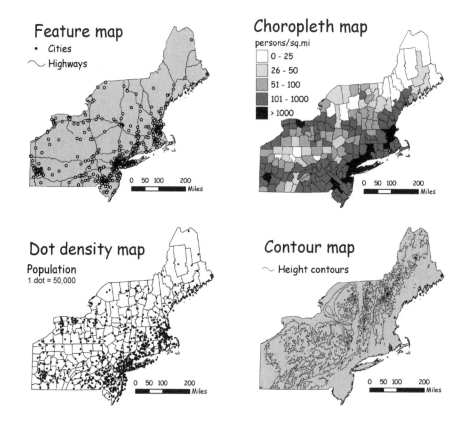

Figure 4-5: Common hardcopy map types depicting New England, in the northeastern United States.

darkest polygons have a population density greater than 1000 persons per square mile.

Dot-density maps are another map form commonly used to show quantitative data (Figure 4-5, bottom left). Dots or other point symbols are plotted to represent values. Dots are placed in the polygon such that the number of dots equals the total value for the polygon. Note that the dots are typically placed randomly within the polygon area. Each dot on the map in the lower left of Figure 4-5 represents 50,000 people, however each point is not a city or other concentration of inhabitants. Note the position of points in the dot-density map relative to the city locations in the feature map directly above it in Figure 4-5.

Isopleth maps, also known as *contour maps*, display lines of equal value (Figure 4-5, bottom right). Isopleth maps are used to represent continuous surfaces. Rainfall, elevation, and temperature are features that are commonly represented using isopleth maps. A line on the isopleth map represents a specified value, for example, a 10°C isopleth defines the position on the landscape that is at that temperature. Lines typically do not cross, in that there cannot be two different temperatures at the same location. However, isopleth maps are commonly used to depict elevation, and cliffs or overhanging terrain do have multiple elevations at the same location. In this case the lower elevations typically pass "under" the higher elevations, and the isopleth is labeled with the tallest height (Figure 4-6). Note that the isopleths are typically estimated surfaces, with the lines drawn based on measurements at a set of point locations; various methods for estimating isopleth lines from points are described in chapter 12.

Map Scale

All maps have a scale, a relationship between a distance on the map and a corresponding distance projected on Earth. Map scale is often reported as a distance conversion, such as one inch to a mile, or as a unitless ratio, such as 1:24,000, indicating a unit

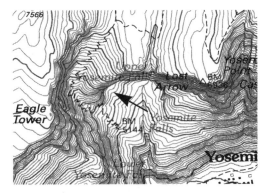

Figure 4-6: Lines on isopleth maps typically do not cross. However, as shown at the arrow in this image, lines may coincide when there is a common value. Here cliffs or overhangs result in converging isopleth lines.

distance on the map is equal to 24,000 units on the Earth's surface (Figure 4-7). Digital maps most often use a third method to report scale, as a bar or line of known distance, labeled on the map.

Note that depicting map scale was unambiguous when only hardcopy maps were produced. A written ratio or conversion, e.g., 1 inch to the mile, was true because the features were fixed on paper or other physical medium. A fixed scale may be erroneous on an electronic document, because the document may be altered by zooming, which changes the magnification on an electronic display. One inch as displayed may not correspond to a mile. Digital documents should most often include a fixed scale bar, depicting an equivalent surface distance, e.g., 1 kilometer, embedded in the map, or some mechanism for re-calculating the scale as the digital map changes size on screen.

The notion of large vs. small scale is often confused because scale implies a ratio, or fraction. A larger ratio signifies a large-scale map, so a 1:24,000-scale map is considered large-scale relative to a 1:100,000-scale map. Many people mistakenly refer to a 1:100,000-scale map as larger scale than a 1:24,000-scale map because it covers a larger area. A 1:100,000-scale map that is 50 cm (20 inches) on a side covers more ground than a 1:24,000-scale map that is 50 cm on a

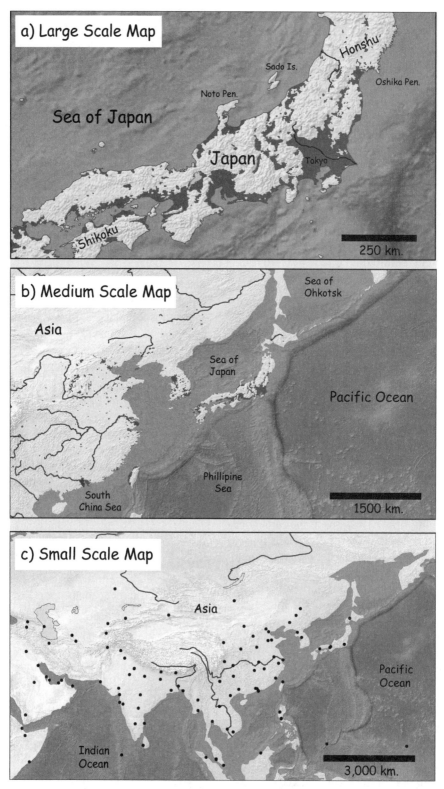

Figure 4-7: Coverage, relative distance, and detail change from larger-scale (top) to smaller-scale (bottom) maps.

side. However, it is the size of the ratio or fraction, and not the area covered that determines the map scale. It is helpful to remember that features appear larger on a large-scale map (Figure 4-7). It is also helpful to remember that large scale maps of a given paper size show more detail, but less area. Notice in Figure 4-7, the larger scale map at the top shows details of Tokyo city. Tokyo shrinks in the successively smaller scale maps, but large additional areas are covered. The larger the ratio (and smaller the denominator), the larger the map scale.

Because maps often report an average scale, and because there are upper limits on the accuracy with which data can be plotted on a map, large-scale maps generally have less geometric error than small-scale maps if the same methods were used to produce them. Small errors in measurement, plotting, printing, and paper deformation are magnified by the scale factor. These errors, which occur during map production, are magnified more on a small-scale map than a large-scale map.

Map Generalization

Maps are abstractions of reality, as are spatial data in a GIS database. This abstraction introduces *map generalization*, the unavoidable approximation of real features when they are represented on a map. Not all the geometric or attribute detail of the physical world are recorded; only the most important characteristics are included. The set of features that are most important is subjectively defined and will differ among users. The mapmaker determines the set of features to place on the map, and selects the methods to collect and represent the shape and location of these features on the map.

The choice of data sources and mapmaking methods will unavoidably set limits on the size and shape of features that may be represented. Consider a project to map lakes, based on image data with a 250 meter cell size (Figure 4-8). The abstraction of the shoreline will not represent bays and peninsulas that are smaller than approximately 250 meters across, by conscious choice of the mapmakers. Small features will be missed, edge detail will be lost, and distances along boundaries will depend on the resolution of the source image.

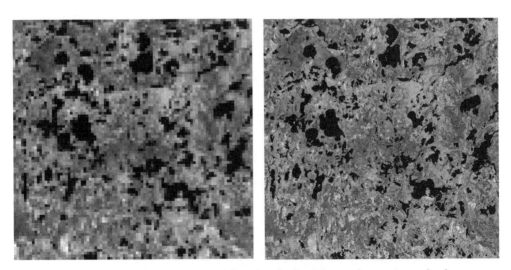

Figure 4-8: A mapmaker chooses the materials and methods used to produce a map, and so imposes a limit on spatial detail. Here, the choice of an input image with a 250 meter resolution (left) renders it impossible to represent all the details of the real lake boundaries (right). In this example, features smaller than approximately 250 meters on a side may not be faithfully represented on the map.

A finer resolution source, such as a 30 meter resolution, may more faithfully depict map detail, but may not be an appropriate choice. The finer resolution may be more expensive, difficult to reproduce, unavailable for the entire mapping area, or inappropriate because it does not show important features, for example, vegetation types or recent developments. Cartographers often must balance several factors in map design, and their choices inevitably lead to some form of map generalization.

Feature generalization is one common form of generalization. Feature generalization is a modification of features when representing them on a map. The geographic aspects of features are generalized because there are limits on the time, methods, or materials available when collecting geographic data. These limits also apply when compiling or printing a map. These feature generalizations, depicted in Figure 4-9, may be classed as:

Fused: multiple features may be grouped to form a larger feature,

Simplified: boundary or shape details are lost or "rounded off",

Displaced: features may be offset to prevent overlap or to provide a standard distance between mapping symbols,

Omitted: Small features in a group may be excluded from the map, or

Exaggerated: standard symbol sizes are often chosen, e.g., standard road symbol widths, which are much larger when scaled than the true road width.

Generalization is present at some level in every map, and should be recognized and evaluated for each map that is used as a source for data in a GIS (Figure 4-10). If generalization results in omission or degradation of data beyond acceptable levels, then the analyst or organization should switch to a larger-scale map if appropriate and available, or return to the field or original source materials to collect data at the required precision.

Map Boundaries and Spatial Data

One final characteristic of maps affects their use as a source of spatial data: hard-copy maps have edges, and discontinuities often occur at these edges. Much digital data have been converted from legacy paper maps, so edge discontinuities have been car-

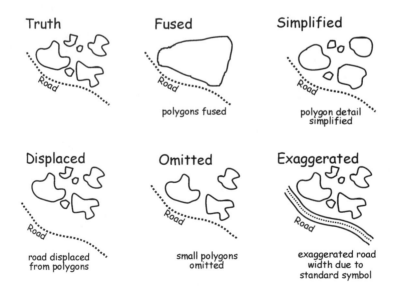

Figure 4-9: Generalizations common in maps.

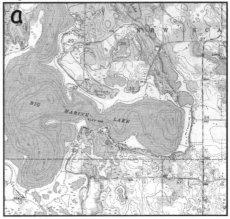

Figure 4-10: Examples of map generalization. Portions are shown for three maps for an area in central Minnesota. Excerpts from a large scale (a, 1:24,000), intermediate scale (b, 1:62,500), and small scale (c, 1:250,000) map are shown. **Note that the maps are not drawn at true scale to facilitate comparison**. The smaller-scale maps (b and c) have been magnified more than a to better show the effects of generalization. Each map has a different level of map generalization. Generalizations increase with smaller-scale maps, and include omissions of smaller lakes, successively greater road width exaggerations, and increasingly generalized shorelines as one moves from maps a through c.

Magnified portion of a 1:62,500-scale map

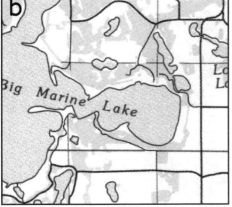

Magnified portion of a 1:250,000-scale map

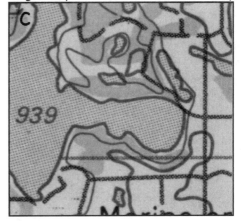

ried to the present. These errors are disappearing as newer data are collected with digital methods, but will be encountered and should be understood.

Large-scale, high-quality maps generally cover small areas. This is because of the trade-off between scale and area coverage, and because of limits on the practical size of a map. Cartometric maps larger than a meter in any dimension have proven to be impractical for most organizations. Maps above this size are expensive and difficult to print, store, or view. Thus, human ergonomics set a practical limit on the physical size of a map.

The fixed maximum map dimension when coupled with a fixed map scale defines the area coverage of the hardcopy map. Larger scale maps generally cover smaller areas. A 1:100,000-scale map that is 18 inches (47 centimeters) on a side spans approximately 28 miles (47 kilometers). A 1:24,000-scale map that is 18 inches on a side represents 9 miles (15 kilometers) on the Earth's surface. Because spatial data in a GIS often span several large-scale maps, these map boundaries may occur in a spatial database. Problems often arise when adjacent maps are entered into a spatial database because features do not align or have mismatched attributes across map boundaries.

Differences in the time of data collection for adjacent map sheets may lead to inconsistencies across map borders. Landscape change through time is a major source of differences across map boundaries. For example, the U.S. Geological Survey has produced 1:24,000-scale map sheets for all of the lower 48 United States of America. The original mapping took place over several decades, and there were inevitable time lags between mapping some adjacent areas. As much as two decades passed between mapping or updating adjacent map sheets. Thus, many features, such as roads, canals, or municipal boundaries, are discontinuous or inconsistent across map sheets.

Different interpreters may also cause differences across map boundaries. Large-area mapping projects typically employ several interpreters, each working on different map sheets for a region. All professional, large-area mapping efforts should have protocols specifying the scale, sources, equipment, methods, classification, keys, and cross-correlation to ensure consistent mapping across map sheet boundaries. In spite of these efforts, however, some differences due to human interpretation occur. Feature placement, category assignment, and generalization vary among interpreters. These

problems are compounded when extensive checking and guidelines are not enforced across map sheet boundaries, especially when adjacent areas are mapped at different times or by two different organizations.

Finally, differences in coordinate registration can lead to spatial mismatch across map sheets. *Registration*, discussed later in this chapter, is the conversion of digitizer or other coordinate data to an earth-surface coordinate system. These registrations contain unavoidable errors that translate into spatial uncertainty. There may be mismatches when data from two separate registrations are joined along the edge of a map.

Spatial data stored in a GIS are not bound by the same constraints that limit the physical dimensions of hardcopy maps. Digital storage enables the production of seamless digital maps of large areas. However, the inconsistencies that exist on hardcopy maps may be transferred to the digital data. Inconsistencies at map sheet edges need to be identified and resolved when maps are converted to digital formats.

Digitizing: Coordinate Capture

Digitizing is the process by which coordinates from a map, image, or other sources are converted into a digital format in a GIS. Points, lines, and areas on maps or images represent real-world entities or phenomena, and these must be recorded in digital forms before they can be used in a GIS. The coordinate values that define the locations and shapes of entities must be captured, that is, recorded as numbers and structured in the spatial database. There is a wealth of spatial data in existing maps and photographs, and new imagery and maps add to this source of information on a nearly continuous basis.

Manual digitization is human-guided coordinate capture from a map or image source. The operator guides an electronic device over a map or image and signals the

capture of important coordinates, often by pressing a button on the digitizing device. Important point, line, or area features are traced on the source materials, and the coordinates are recorded in GIS-compatible formats. Valuable data on historical maps may be converted to digital forms through the use of manual digitizing. On-screen digitizing and hardcopy digitizing are the two most common forms of manual digitization.

On-screen Digitizing

On-screen digitizing, also known as heads-up digitizing, involves manually digitizing on a computer screen, using a digital image as a backdrop. Digitizing software allows the operator to trace the points, lines, or polygons that are identified on the scanned map (Figure 4-11). Digitizing software allows the human operator to specify the type of feature to be recorded, the extent and magnification of the image on screen, the mode of digitizing, and other options to control how data are input. The operator typically guides a cursor over points to be recorded using a mouse, and depresses a button or sequence of buttons to collect the point coordinates. On-screen digitizing can be used for recording information from scanned aerial photographs, digital photographs, satellite images, or other images.

On-screen digitizing offers advantages over hardcopy and scan-digitizing, methods that are described in the following sections. Many data sources are inherently digital, for example, image data collected from aerial photographs and airborne or satellite scanners.These data may be magnified on screen to any desired scale. Converting the image to a paper or other hardcopy form would likely introduce error through the slight deforma-

tion of the paper or printing media, reduce flexibility when digitizing, and add the cost of printing.

On-screen digitizing is often more accurate than manual digitizing because manual map digitization is often limited by the visual acuity and pointing ability of the operator. The pointing imprecision of the operator and digitizing systems translates to a fixed ground distance when manually digitizing a hardcopy map. For example, consider an operator that can reliably digitize a location to the nearest 0.4 millimeters (0.01 inch) on a 1:20,000-scale map. Also assume the best hardcopy digitizing table available is being used, and we know the observed error is larger than the error in the map. The 0.4 millimeter error in precision translates to approximately 8 meters of error on the Earth's surface. The precision cannot be appreciably improved when using a digitizing table, because a majority of the imprecision is due to operator abilities. In contrast, once the map is scanned, the image may be

Figure 4-11: An example of on-screen digitizing. Images or maps are displayed on a computer screen and feature data digitized manually. Buildings, roads, or any other features that may be distinguished on the image may be digitized.

displayed on a computer screen at any map scale. The operator may zoom to a 1:5,000-scale or greater on-screen, and digitizing accuracy and precision improved. While other factors affect the accuracy of the derived spatial data (for example map plotting or production errors, or scanner accuracy), on-screen digitizing may be used to limit operator-induced positional error when digitizing. On-screen digitizing also removes or reduces the need for digitizing tables or map scanners, the specialized equipment used for capturing coordinates from maps.

Hardcopy Map Digitization

Hardcopy digitizing is human-guided coordinate capture from a paper, plastic, or other hardcopy map. An operator securely attaches a map to a digitizing surface and traces lines or points with an electrically sensitized puck (Figure 4-12).The most common digitizers are based on a wire grid

embedded in or under a table. Depressing a button specifies the puck location relative to the digitizer coordinate system. Digitizing tables can be quite accurate, with a resolution of between 0.25 and 0.025 millimeters (0.01 and 0.001 inches).

While once a major method for capturing spatial data, hardcopy map digitizing is diminishing in importance as most paper documents have been converted to digital forms. The tables are large, somewhat expensive, and now little-used. However, because data from hardcopy sources are likely to persist for many decades, and there are still many specialized documents to convert, you should be familiar with the process.

Not all maps are appropriate as a source of information for GIS. The type of map, how it was produced, and the intended purpose must be considered when interpreting the information on maps. Only cartometric maps should be directly digitized, and even though cartometric, a map may not be suit-

Figure 4-12: Manual digitizing on a digitizing table.

able. Consider the dot-density map described in Figure 4-2. Population is depicted by points, but the points are plotted with random offsets or using some method that does not reflect the exact location of the population within each polygon. Before the information in the dot density map is entered into a GIS, the map should be interpreted correctly. The number of dots in a polygon should be counted, this number multiplied by the population per dot, and the population value assigned to the entire polygon.

Maps may be unsuitable for digitizing due to the media. Most hardcopy maps are on paper because it is ubiquitous, inexpensive, and easily printed. Creases, folds, and wrinkles can lead to non-uniform deformation of paper maps.

Characteristics of Manual Digitizing

Manual digitizing, whether from a digital image on screen or from a hardcopy source, is common because it provides sufficiently accurate data for many, if not most, applications. Manual digitizing may be at least the accuracy of most maps or images, so the equipment, if properly used, does not add substantial error. Manual digitizing also requires low equipment investment, often just the software for image display and coordinate capture. The human ability to interpret images or hardcopy maps in poor condition is a unique and important benefit of manual digitizing. Humans are usually better than machines at interpreting the information contained on faded, stained, or poor quality maps and images. Finally, manual digitizing is often best because short training periods are required, data quality may be frequently evaluated, and digitizing equipment is commonly available. For these reasons manual digitization is likely to remain an important data entry method for some time to come.

There are a number of characteristics of manual digitization that may negatively affect the positional quality of spatial data. As noted earlier, map or image scale and res-

olution impacts the spatial accuracy of digitized data. This scale may be the production scale for hardcopy maps, or the display scale for digital images or scanned maps. Table 4-1 illustrates the effects of map scale on data quality. Errors of one millimeter (0.039 inches) on a 1:24,000-scale map correspond to 24 meters (79 feet) on the surface of the Earth. This same one millimeter error on a 1:1,000,000-scale map corresponds to 1000 meters (3281 feet) on the Earth's surface. Thus, small errors in map production or interpretation may cause significant positional errors when scaled to distances on the Earth, and these errors are greater for smaller-scale maps. Errors due to human pointing ability are reduced for on-screen digitizing, because the operator can zoom in to larger scales as needed. However, this does not overcome errors inherent in original images or scanned documents.

Both device precision and map scales should be considered when selecting a digitizing tablet. Map scale and repeatability both set an upper limit on the positional quality of digitized data. The most precise digitizers may be required when attempting to meet a stringent error standard while digitizing small-scale maps.

Table 4-1: The surface error caused by a one millimeter (0.039 inch) map error will change as map scale changes. Note the larger error at smaller map scales.

Map Scale	Error (m)	Error (ft)
1:24,000	24	79
1:50,000	50	164
1:62,500	63	205
1:100,000	100	328
1:250,000	250	820
1:1,000,000	1,000	3,281

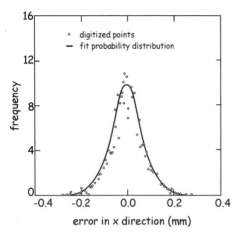

Figure 4-13: Digitizing error, defined by repeat digitizing. Points repeatedly digitized cluster around the true location, and follow a normal probability distribution. (from Bolstad et al., 1990).

The abilities and attitude of the person digitizing (the operator) may also affect the geometric quality of manually digitized data. Operators vary in their visual acuity, steadiness of hand, attention to detail, and ability to concentrate. Some operators will more accurately capture the coordinate information contained in maps. The abilities of any single operator will also vary through time, due to fatigue or difficulty maintaining focus on a repetitive task. Operators should take frequent breaks from digitizing, and comparisons among operators and quality and consistency checks should be integrated into any manual digitization process to ensure accurate and consistent data collection.

The combined errors from both operators and equipment have been well-characterized and may be quite small. One test using a high-precision digitizing table revealed digitizing errors averaging approximately 0.067 millimeters (Figure 4-13). Errors followed a random normal distribution, and varied significantly among operators. These average errors translated to an approximately 1.6 meter error when scaled from the 1:24,000 map to a ground-equivalent distance. This average error is less than the acceptable production error for the map, and is suitable for many spatial analyses.

The Digitizing Process

Manual digitizing involves displaying a digital image on screen or placing a map on a digitizing surface, and tracing the location of feature boundaries. Coordinate data are sampled by manually positioning the puck or cursor over each target point and collecting coordinate locations. This position/collect step is repeated for every point to be captured, and in this manner the locations and shapes of all required map features are defined. Features that are viewed as points are represented by digitizing a single location. Lines are represented by digitizing an ordered set of points, and polygons by digitizing a connected set of lines. Lines have a starting point, often called a *starting node*, a set of *vertices* defining the line shape, and an ending *node* (Figure 4-14). Hence, lines may be viewed as a series of straight line segments connecting vertices and nodes.

Digitizing may be in *point mode*, where the operator must depress a button or otherwise signal to the computer to sample each point, or in *stream mode*, where points are

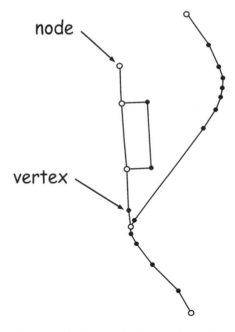

Figure 4-14: Nodes define the starting and ending points of lines. Vertices define line shape.

automatically sampled at a fixed time or distance frequency, perhaps once each meter. Stream mode helps when large numbers of lines are digitized, because vertices may be sampled more quickly and the operator may become less fatigued.

The stream sampling rate must be specified with care to avoid over- or under-sampled lines. Too short a collection interval results in redundant points not needed to accurately represent line or polygon shape. Too long a collection interval may result in the loss of important spatial detail. In addition, when using time-triggered stream digitizing, the operator must remember to continuously move the digitizing puck; if the operator rests the digitizing puck for a period longer than the sampling interval there will be multiple points clustered together. These will redundantly represent a portion of the line and may result in overlapping segments. Pausing for an extended period of time often creates a "rat's nest" of lines that must later be removed.

Minimum distance digitizing is a variant of stream mode digitizing that avoids some of the problems inherent with time-sampled streaming. In minimum distance digitizing a new point is not recorded unless it is more than some minimum threshold distance from the previously digitized point. The operator

may pause without creating a rat's nest of line segments. The threshold must be chosen carefully - neither too large, missing useful detail, nor too small, in effect reverting back to stream digitizing.

Digitizing Errors, Node and Line Snapping

Positional errors are inevitable when data are manually digitized. These errors may be "small" relative to the intended use of the data, for example the positional errors may be less than 2 meters when only 5 meter accuracy is required. However, these relatively small errors may still prevent the generation of correct networks or polygons. For example, a data layer representing a river system may not be correct because major tributaries may not connect. Polygon features may not be correctly defined because their boundaries may not completely close. These small errors must be removed or avoided during digitizing. Figure 4-15 shows some common digitizing errors.

Undershoots and *overshoots* are common errors that occur when digitizing. Undershoots are nodes that do not quite reach the line or another node, and overshoots are lines that cross over existing nodes or lines (Figure 4-15). Undershoots

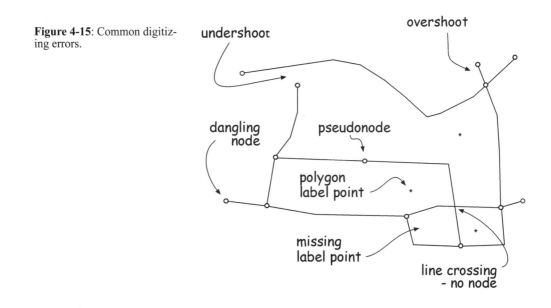

Figure 4-15: Common digitizing errors.

cause unconnected networks and unclosed polygons. Overshoots typically do not cause problems when defining polygons, but they may cause difficulties when defining and analyzing line networks.

Node snapping and *line snapping* are used to reduce undershoots and overshoots while digitizing. Snapping is a process of automatically setting nearby points to have the same coordinates. Snapping relies on a *snap tolerance* or *snap distance*. This distance may be interpreted as a minimum distance between features. Nodes or vertices closer than this distance are moved to occupy the same location (Figure 4-16). Node snapping prevents a new node from being placed within the snap distance of an already existing node; instead, the new node is joined or "snapped" to the existing node. Remember that nodes are used to define the ending points of a line. By snapping two nodes together, we ensure a connection between digitized lines.

Line snapping may also be specified. Line snapping inserts a node at a line crossing and clips the end when a small overshoot is digitized. Line snapping forces a node to connect to a nearby line while digitizing, but only when the undershoot or overshoot is less than the snapping distance. Line snapping requires the calculation of an intersection point on an already existing line. The snap process places a new node at the intersection point, and connects the digitized line to the existing line at the intersection point. This splits the existing line into two new lines. When used properly, line and node snapping reduce the number of undershoots and overshoots. Closed polygons or intersecting lines are easier to digitize accurately and efficiently when node and line snapping are in force.

The snap distance must be carefully selected for snapping to be effective. If the snap distance is too short, then snapping has little impact. Consider a system where the operator may digitize with better than 5 meter accuracy only 10% of the time. This means 90% of the digitized points will be more than 5 meters from the intended location. If the snap tolerance is set to the equivalent of 0.1 meters, then very few nodes will

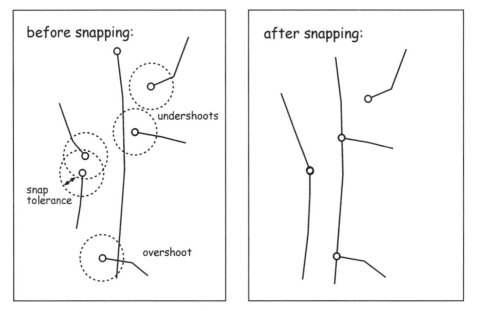

Figure 4-16: Undershoots, overshoots, and snapping. Snapping may join nodes, or may place a node onto a nearby line segment. Snapping does not occur if the nodes and/or lines are separated by more than the snap tolerance.

be within the snap tolerance, and snapping has little effect. Another problem comes from setting the snap tolerance too large. If the snap tolerance in our previous example is set to 10 meters, and we want the data accurate to the nearest 5 meters, then we may lose significant spatial information that is contained in the hardcopy map. Lines less than 10 meters apart cannot be digitized as separate objects. Many features may not be represented in the digital data layer. The snap distance should be smaller than the desired positional accuracy, such that significant detail contained in the digitized map is recorded. It is also important that the snap distance is not below the capabilities of the system used for digitizing. Careful selection of the snap distance may reduce digitizing errors and significantly reduce time required for later editing.

Reshaping: Line Smoothing and Thinning

Digitizing software may provide tools to smooth, densify, or thin points while entering data. One common technique uses *spline* functions to smoothly interpolate curves between digitized points and thereby both smooth and densify the set of vertices used to represent a line. A spline is set of polynomial functions that join smoothly (Figure 4-17). Polynomial functions are fit to successive sets of points along the vertices in a line; for example, a function may be fit to points 1 through 5, and a separate polynomial function fit to points 5 through 11 (Figure 4-17). Constraints force these functions to connect smoothly, usually by requiring the first and second derivatives of the functions to be continuous at the intersection point. This means the lines have the same slope at the intersection point, and the slope is changing at the same rate for both lines at the intersection point. Once the spline functions are calculated they may be used to add vertices. For example, several new vertices may be automatically placed on the line between digitized vertices 8 and 9, leading to the "smooth" curve shown in Figure 4-17.

Data may also be digitized with too many vertices. High densities may occur when data are manually digitized in stream mode, and the operator moves slowly relative to the time interval. High vertex densities may also be found when data are derived from spline or smoothing functions that specify too high a point density. Finally, automated scanning and then raster-to-vector conversion may result in coordinate pairs spaced at absurdly high densities. Many of these coordinate data are redundant and may be removed without sacrificing spatial accuracy. Too many vertices may be a problem in that they slow processing, although this has become less important as computing power has increased. Point thinning algorithms have been developed to reduce the number of points while maintaining the line shape.

Many point thinning methods use a perpendicular "weed" distance, measured from a spanning line, to identify redundant points (Figure 4-18, top). The Lang method exemplifies this approach. A spanning line connects two non-adjacent vertices in a line. A

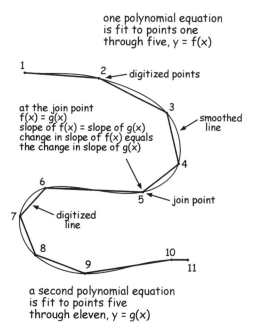

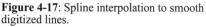

Figure 4-17: Spline interpolation to smooth digitized lines.

pre-determined number of vertices is spanned initially. The initial spanning number has been set to 4 in Figure 4-18, meaning four points will be considered at each starting point. Areas closer than the weed distance are shown in gray in the figure. A straight line is drawn between a starting point and an endpoint that is the 4th point down the line (Figure 4-18a). Any intermediate points that are closer than the weed distance are marked for removal. In Figure 4-18a, no points are within the weed distance, therefore none are marked. The endpoint is then moved to the next closest remaining point (Figure 4-18b), and all intermediate points tested for removal. Again, any points closer than the weed distance are marked for removal. Note that in Figure 4-18b, one point is within the weed distance, and is removed. Once all points in the initial spanning distance are checked, the last remaining endpoint becomes the new starting point, and a new spanning line drawn to connect 4 points (Figure 4-18c, d).

The process may be repeated for successive sets of points in a line segment until all vertices have been evaluated (Figure 4-18e to h). All close vertices are viewed as not recording a significant change in the line shape, and hence are expendable. Increasing the weed distance thins more vertices, and at some upper weed distance too many vertices may be removed. A balance must be struck between the removal of redundant vertices and the loss of shape-defining points, usually through a careful set of test cases with successively larger weed distances.

There are many variants on this basic concept. Some look only at three immediately adjacent points, testing the middle point against the line spanned by its two neighboring points. Others constrain or expand the search based on the complexity of the line. Rather than always looking at four points, as in our example above, more points are scrutinized when the line is not complex (nearly straight), and fewer when the line is complex (many changes in direction).

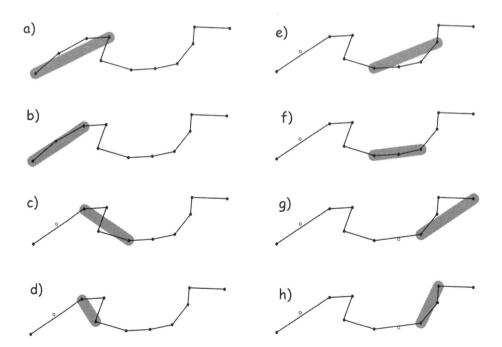

Figure 4-18:The Lang algorithm is a common line-thinning method. In the Lang method vertices are removed, or thinned, when they are within a weed distance to a spanning line (adapted from Weibel, 1997).

Scan Digitizing

Optical scanning is another method for converting hardcopy documents into digital formats (Figure 4-19). Scanners have elements that emit and sense light. Most scanners pass a sensing element over an illuminated map. This device measures both the precise location of the point being sensed and the strength of the light reflected or transmitted from that point. Reflected light intensities are sensed and converted to numbers.

A threshold is often applied to determine if the sensed point is part of a feature. For example, a map may consist of dark lines on a white background. A threshold might be set such that if less than 10% of the light striking the map is returned to the sensor, the sensed point is considered part of a line. If 10% or more of the energy is reflected back to the sensor, the point is considered part of the white space between lines. The scanner then produces a raster representation of the map. Values are recorded where points or lines exist on the map and null or zero values are recorded in the intervening spaces.

Most scanners are either bed or drum designs. Bed scanners provide a flat surface on which the map is placed. A mat or hinged cover is then placed on top of the map, flattening and securing the map to the bed. On some bed scanners an optical train is passed over the map, emitting light and sensing the light reflected back from the map. Sensing arrays are typically used to measure the reflectance so that one to several rows of cells may be scanned simultaneously. A motor then moves the optical train to the adjacent lines and the process is repeated.

Drum scanners differ from flatbed scanners in that they employ a rotating cylinder. A map is fixed onto the surface of this cylinder, and the cylinder set to rotate at a uniform velocity. The angular velocity of a rotating cylinder is easier to control than the straight-line motion of a bed scanner, so many of the early high-precision scanners used drums. Many drum scanners are similar to bed scanners in that they use optical detection of reflected light to sense map elements.

Scanners work best when very clean maps are available. Even the most expensive scanners may report a significant number of spurious lines or points when old, marked, folded, or wrinkled maps are used. These spurious features must be subsequently removed via manual editing, thus negating the speed advantage of scanning over manual digitizing. Scanning also works best when maps are available as map separates, with one thematic feature type on each map. Editing takes less time when maps do not contain writing or other annotation. Strongly contrasting colors are preferred, such as black lines on a white background, rather then dark grey on light grey. Finally, scanning is most advantageous when a large number of cartographic elements is found on the maps.

Scan digitization usually requires some form of *skeletonizing*, or line thinning, particularly if the data are to be converted to a vector data format. Scanned lines are often wider than a single pixel (Figure 4-20). One of several pixels may be selected to specify the position of a given portion of the line. The same holds true for points. A pixel near the "center" of the point or line is typically chosen, with the center of a line defined as the pixel nearest the center of the local perpendicular bisector of the line. Skeletonizing reduces the widths of lines or points to a single pixel.

Figure 4-19: A map scanner (courtesy Calcomp).

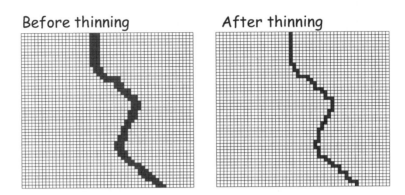

Figure 4-20: Skeletonizing, a form of line thinning that is often applied after scan-digitizing.

Editing Geographic Data

Spatial data may be edited, or changed, for several reasons. Errors and inconsistencies are inevitably introduced during spatial data entry. Undershoots, overshoots, missing or extra lines, missing or extra points or labels are all errors that must be corrected. Spatial data can change over time. Parcels are subdivided, roads extended or moved, forests grow or are cut, and these changes may be entered in the spatial database through editing. New technologies may be developed that provide more accurate positional information, and even though existing data may be consistent and current, the more accurate data may be more useful, leading to data editing.

Identifying errors is the first step in editing. Errors may be identified by printing a map of the digitized data and verifying that each point, line, and area feature is present and correctly located. Plots are often printed both at a similar scale and at a significantly larger scale than the original source materials. The large-scale plots are often paneled with some overlap among panels. Plots at scale are helpful for identifying missing features, and large-scale plots aid in identifying undershoots, overshoots, and small omissions or additions. Operators typically annotate these plots as they are checked systematically for each feature.

Software help operators identify potential errors. Line features typically begin and end with a node, and nodes may be classified as connecting or dangling. A connecting node joins two or more lines, while a dangling node is attached to only one line. Some dangling nodes may be intentional, for example, a cul-de-sac in a street network, while others will be the result of under- or overshoots. Dangling nodes that are plotted with unique symbols can be quickly evaluated, and if appropriate, corrected.

Attribute consistency may also be used to identify errors. Operators note areas in which contradictory theme types occur in different data layers. The two layers are either graphically or cartographically overlain. Contradictory co-occurrences are identified, such as water in one layer and upland areas in a second. These contradictions are then either resolved manually, or automatically via some pre-defined precedence hierarchy.

Many GIS software packages provide a comprehensive set of editing tools (Figure 4-21). Editing typically includes the ability to select, split, update, and add features. Selection may be based on geometric attributes, or with a cursor guided by the operator. Selections may be made individually, by geographic extent (select all features in a box, circle, or within a certain distance of the pointer) or by geometric attributes (e.g., select all nodes that connect to only one

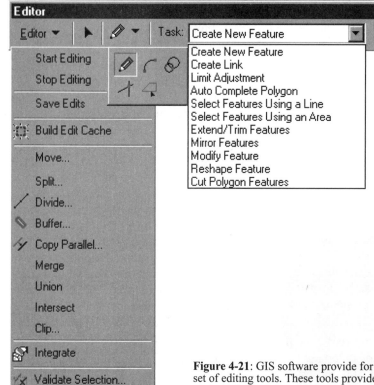

Figure 4-21: GIS software provide for a flexible and complete set of editing tools. These tools provide for the rapid, precise, controlled creation and modification of coordinates and attributes of spatial data (courtesy ESRI).

line). Once a feature is selected, various operations may be available, including erasing all or part of the feature, changing the coordinate values defining the feature, and in the case of lines, splitting or adding to the feature. A line may be split into parts, either to isolate a segment for future deletion, or to modify only a portion of the line. Coordinates are typically altered by interactively selecting and dragging points, nodes or vertices to their best shape and location. Points or line segments are added as needed.

Groups of features in an area may be adjusted through interactive *rubbersheeting*. Rubbersheeting involves fitting a local equation to adjust the coordinates of features. Polynomial equations are often used due to their flexibility and ease of application. Anchor points are selected, again on the graphics screen, and other points are selected by dragging interactively on the screen to match point locations. All lines and points except the anchor points are interactively adjusted. One common application of rubbersheeting involves adjusting linework representing cultural features, such as a road network, when higher geometric-accuracy photo or satellite image data are available. The linework is overlain on an image backdrop and subsequently adjusted.

All edits should be made with due attention to the magnitude of positional change introduced during editing. On-screen editing to eliminate undershoots should only be performed when the "true" locations of features may be identified accurately, and the new features can be confidently placed in the correct location. Automatic removal of "short" undershoots may be performed without introducing additional spatial error in most instances. A short distance for an undershoot is subjectively defined, but typically it is below the error inherent in the source map, or at least a distance that is insignificant when considering the intended use of the spatial data.

Features Common to Several Layers

One common problem in digitizing derives from representation of features that occur on different maps or images. These features rarely have identical locations on each map or image, and often occur in different locations when digitized into their respective data layers (Figure 4-22). For example, water boundaries on soil survey maps rarely correspond exactly to water boundaries found on USGS topographic maps.

Features may appear differently on different maps for many reasons. Perhaps the maps were made for different purposes or at different times. Features may differ because the maps were from different source materials, for example, one map may have been

based on ground surveys while another was based on aerial photographs. Digitizing can also compound the problem due to differences in digitizing methods or operators.

There are several ways to remove this "common feature" inconsistency. One involves re-drafting the data from conflicting sources onto one base map. Inconsistencies are removed at the drafting stage. For example, vegetation and roads data may show vegetation type boundaries at road edges that are inconsistent with the road locations. Both of these data layers may be drafted onto the same base, and the common boundaries fixed by a single line. This line is digitized once, and used to specify the location of both the road and vegetation boundary when digitizing. Re-drafting, although labor intensive and time consuming, forces a resolution of inconsistent boundary locations. Re-drafting also allows several maps to be combined into a single data layer.

A second, often preferable method involves establishing a "master" boundary which is the highest accuracy composite of the available data sets. A digital copy or overlay operation establishes the common features as a base in all the data layers, and this base may be used as each new layer is produced. For example, water boundaries might be extracted from the soil survey and USGS quad maps and these data combined in a third data layer. The third data layer would be edited to produce a composite, high-quality water layer. The composite water layer would then be copied back into both the soils and USGS quad layers. This second approach, while resulting in visually consistent spatial data layers, is in many instances only a cosmetic improvement of the data. If there are large discrepancies ("large" is defined relative to the required spatial data accuracy), then the source of the discrepancies should be identified and the most accurate data used, or new, higher accuracy data collected from the field or original sources.

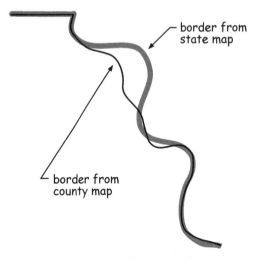

Figure 4-22: Common features may be spatially inconsistent in different spatial data layers.

Coordinate Transformation

Coordinate transformation is a common operation in the development of spatial data for GIS. A coordinate transformation brings spatial data into an Earth-based map coordinate system so that each data layer aligns with every other data layer. This alignment ensures features fall in their proper relative position when digital data from different layers are combined. Within the limits of data accuracy, a good transformation helps avoid inconsistent spatial relationships such as farm fields on freeways, roads under water, or cities in the middle of swamps, except where these truly exist. Coordinate transformation is also referred to as *registration*, because it "registers" the layers to a map coordinate system.

Coordinate transformation is most commonly used to convert newly digitized data from the digitizer/scanner coordinate system to a standard map coordinate system (Figure 4-23). The input coordinate system is usually based on the digitizer or scanner-assigned values. An image may be scanned and coordinates recorded as a cursor is moved across the image surface. These coordinates are usually recorded in pixel, inch, or centimeter units relative to an origin located near the lower left corner of the image. The absolute values of the coordinates depend on where the image happened to be placed on the table prior to scanning, but the relative position of digitized points does not change. Before these newly digitized data may be used with other data, these "inch-space" or "digitizer" coordinates must be transformed into an Earth-based map coordinate system.

Control Points

A set of *control points* is used to transform the digitized data from the digitizer or photo coordinate system to a map-projected coordinate system. Control points are different from other digitized features. When we digitize most points, lines, or areas, we do not know the map projection coordinates for these features. We simply collect the digitizer x and y coordinates that are established with reference to some arbitrary origin on the digitizing tablet or photo. Control points differ from other digitized points in that we

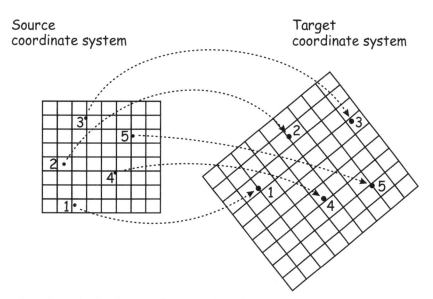

Figure 4-23: Control points in a coordinate transformation. Control points are used to guide the transformation of a source, input set of coordinates to a target, output set of coordinates. There are five control points in this example. Corresponding positions are shown in both coordinate systems.

know both the map projection coordinates and the digitizer coordinates for these points.

These two sets of coordinates for each control point, one for the map projection and one for the digitizer system, are used to estimate the coefficients for transformation equations, usually through a statistical, least-squares process. The transformation equations are then used to convert coordinates from the digitizer system to the map projection system.

The transformation may be estimated in the initial digitizing steps, and applied as the coordinates are digitized from the map or image. This "on-the-fly" transformation allows data to be output and analyzed with reference to map-projected coordinates. A previously registered data layer or image may be displayed on screen just prior to digitizing a new map. Control points may then be entered, the new map attached to the digitizing table, and the map registered. The new data may then be displayed on top of the previously registered data. This allows a quick check on the location of the newly digitized objects against corresponding objects in the study area.

In contrast to on-the-fly transformations, data can also be recorded in digitizer coordinates and the transformation applied later. All data are digitized, including the control point locations. The digitizer coordinates of the control point may then be matched to corresponding map projection coordinates, and transformation equations estimated. These transformation equations are then applied to convert all digitized data to map projection coordinates.

Control points should meet or exceed several criteria. First, control points should be from a source that provides the highest feasible coordinate accuracy. Second, control point accuracy should be at least as good as the desired overall positional accuracy required for the spatial data. Third, control points should be as evenly distributed as possible throughout the data area. A sufficient number of control points should be collected. The minimum number of points

depends on the mathematical form of the transformation, but additional control points above the minimum number are usually collected; this usually improves the quality and accuracy of the statistically-fit transformation functions.

The x, y (horizontal), and sometimes z (vertical or elevation) coordinates of control points are known to a high degree of accuracy and precision. Because high precision and accuracy are subjectively defined, there are many methods to determine control point locations. Sub-centimeter accuracy may be required for control points used in property boundary layers, while accuracies of a few meters may be acceptable for large-area vegetation mapping. Common sources of control point coordinates are traditional transit and distance surveys, global positioning system measurements, existing cartometric quality maps, or existing digital data layers on which suitable features may be identified.

The Affine Transformation

The *affine coordinate transformation* employs linear equations to calculate map coordinates. Map projection coordinates are often referred to as eastings (E) and northings (N), and are related to the x and y digitizer coordinates by the equations:

$$E = T_E + a_1 x + a_2 y \qquad (4.1)$$

$$N = T_N + b_1 x + b_2 y \qquad (4.2)$$

Equations 4.1 and 4.2 allow us to move from the arbitrary digitizer coordinate system to the project map coordinate system. We know the x and y coordinates for every digitized point, line vertex, or polygon vertex. We may calculate the E and N coordinates by applying the above equations to every digitized point.

T_E and T_N are translation changes between the coordinate systems, and can be thought of as shifts in the origins from one coordinate system to the next. The a_i and b_i parameters incorporate the change in scales and rotation angle between one coordinate system and the next. The affine is the most commonly applied coordinate transformation because it provides for these three main effects of translation, rotation, and scaling, and because it often introduces less error than higher-order polynomial transformations.

The affine system of equations has six parameters to be estimated, T_E, T_N, a_1, a_2, b_1, and b_2. Each control point provides E, N, x, and y coordinates, and allows us to write two equations. For example, we may have a control point consisting of a precisely surveyed center of a road intersection. This point has digitizer coordinates of x=103.0 centimeters and y = -100.1 centimeters, and corresponding Earth-based map projection coordinates of E = 500,083.4 and N = 4,903,683.5. We may then write two equations based on this control point:

$$500{,}083.4 = T_E + a_1(103.0) + a_2(-100.1) \quad (4.3)$$

$$4{,}903{,}683.5 = T_N + b_1(103.0) + b_2(-100.1) \quad (4.4)$$

We cannot find a unique solution to these equations, because there are six unknowns (T_E, T_N, a_1, a_2, b_1, b_2) and only two equations. We need as many equations as unknowns to solve a linear system of equations. Each control point gives us two equations, so we need a minimum of three control points to estimate the parameters of an affine transformation. Statistical estimation requires a total of four control points. As with all statistical estimates, more control points are better than fewer, but we will reach a point of diminishing returns after some number of points, typically somewhere between 18 and 30 control points.

The affine coordinate transformation is usually fit using a statistical method that minimizes the *root mean square error,* RMSE. The RMSE is defined as:

$$\text{RMSE} = \sqrt{\frac{e_1^2 + e_2^2 + e_3^2 \ldots + e_n^2}{n}} \quad (4.5)$$

where the e_i are the residual distances between the true E and N coordinates and the E and N coordinates in the output data layer:

$$e = \sqrt{(x_t - x_d)^2 + (y_t - y_d)^2} \quad (4.6)$$

This residual is the difference between the true coordinates x_t, y_t, and the transformed output coordinates x_d, y_d. Figure 4-24 shows examples of this lack of fit. Individual residuals may be observed at each control point location.

A statistical method for estimating transformation equations is preferred because it also identifies transformation error. Control point coordinates contain unavoidable measurement errors. A statistical process provides an RMSE, a summary of the difference between the "true" (measured) and predicted control point coordinates. It provides one index of transformation quality. Transformations are

Figure 4-24: Examples of control points, predicted control locations, and residuals from coordinate transformation.

fit (Figure 4-25). The RMSE will usually be less than the true transformation error at a randomly selected point, because we are actively minimizing the N and E residual errors when we statistically fit the transformation equations. However, the RMSE is an index of accuracy, and a lower RMSE generally indicates a more accurate affine transformation.

Estimating the coordinate transformation parameters is often an iterative process. Control points are rarely exact, and x and y coordinates may not be precisely digitized. Poor eyesight, a shaky hand, fatigue, lack of attention, mis-identification of the control location, or a blunder may result in erroneous x and y values. There may also be errors in the E and N coordinates. Typically, control points are entered, the affine transformation parameters estimated, and the overall RMSE and individual point E and N errors evaluated (Figure 4-24, Figure 4-25). Suspect points are fixed, and the transformation re-estimated and errors evaluated until a final transformation is estimated. The transformation is then applied to all features to convert them from digitizer to map coordinates.

Other Coordinate Transformations

Other coordinate transformations are sometimes used. The conformal coordinate transformation is similar to the affine, and has the form:

$$E = T_E + cx - dy \qquad (4.7)$$

$$N = T_N + dx + cy \qquad (4.8)$$

The coefficients T_E, T_N, c, and d are estimated from control point data. Like the affine transformation, the conformal transformation is also a first-order polynomial. Unlike the affine, the conformal transformation requires equal scale changes in the x

and y directions. Note the symmetry in the equations 4.7 and 4.8, in that the x and y coefficients match across equations, and there is a change in sign for the d coefficient. This results in a system of equations with only four unknown parameters, and so the conformal may be estimated when only two control points are available.

Higher-order polynomial transformations are sometimes used to transform among coordinate systems. An example of a 2nd-order polynomial is:

$$E = b_1 + b_2 x + b_3 y + b_4 x^2 + b_5 y^2 + b_6 xy \qquad (4.9)$$

Note that the combined powers of the x and y variables may be up to 2. This allows for curvature in the transformation in both the x and y directions. A minimum of six control points is required to fit this 2nd-order polynomial transformation, and seven are required when using a statistical fit. The estimated parameters T_E, T_N, a_1, a_2, b_1, and b_2 will be different in equations 4.1 and 4.2 when compared to 4.9, even if the same set of control points is used for both statistical fits. We change the form of the equations by including the higher-order squared and xy cross-product terms, and all estimated parameters will vary.

A Caution When Evaluating Transformations

Selecting the "best" coordinate transformation to apply is a subjective process, guided by multiple goals. We hope to develop an accurate transformation based on a large set of well-distributed control points. Isolated control points that substantially improve our coverage may also contribute substantially to our transformation error.

There are no clear rules on the number of points versus distribution of points trade-off, but it is typically best to strive for the widest distribution of points. We want at least two control points in each quadrant of the working area, with a target of 20% in

Model Fit 1:

E = 1.3325289 * x
+ 0.0058654 * y
- 206851.8

N = - 0.002886 * x
+ 1.3296931 * y
- 1660286

RMSE = 9.36

Examine points 15
& 17, adjust noted
blunders, refit
model

1	518,687.6	5,015,347.0	513,734.1	5,007,087.4	3.07
2	516,907.3	5,013,549.1	511,355.8	5,004,707.2	8.13
3	516,952.2	5,017,965.3	511,438.3	5,010,573.9	4.38
4	518,700.1	5,014,393.4	513,738.9	5,005,831.3	10.99
5	518,099.6	5,013,576.2	512,938.9	5,004,733.6	2.79
6	518,992.6	5,017,306.0	514,144.0	5,009,699.3	8.18
7	519,150.0	5,013,556.6	514,331.9	5,004,709.3	5.66
8	519,259.8	5,013,600.0	514,482.8	5,004,764.0	0.88
9	516,916.8	5,016,528.9	511,378.9	5,008,669.6	4.05
10	516,659.6	5,018,093.8	511,043.8	5,010,744.1	3.37
11	519,474.3	5,018,046.9	514,807.0	5,010,675.2	11.05
12	519,549.2	5,014,375.9	514,873.0	5,005,798.0	2.84
13	518,089.4	5,014,478.2	512,938.6	5,005,931.0	10.36
14	518,087.4	5,014,755.2	512,936.0	5,006,299.0	9.16
15	518,079.9	5,016,484.0	512,912.3	5,008,591.6	19.49
16	516,947.5	5,017,736.1	511,424.5	5,010,277.6	7.16
17	517,016.3	5,014,444.0	511,485.6	5,005,903.6	17.05
18	517,785.1	5,017,492.6	512,542.4	5,009,954.0	9.51
19	519,435.7	5,017,340.7	514,736.0	5,009,735.7	4.46
20	518,710.3	5,016,544.2	513,778.7	5,008,679.7	10.04
21	518,984.0	5,016,548.6	514,127.8	5,008,678.2	9.50
22	516,719.0	5,014,555.9	511,106.7	5,006,028.1	14.96

Model Fit 2:

E = 1.3319386 * x
+ 0.0057193 * y
- 205812.1

N = - 0.002462 * x
+ 1.329962 * y
- 1161855

RMSE = 7.72

Examine points 4
& 22, adjust noted
blunders, refit
model

1	518,687.6	5,015,347.0	513,734.1	5,007,087.4	2.56
2	516,907.3	5,013,549.1	511,355.8	5,004,707.2	7.22
3	516,952.2	5,017,965.3	511,438.3	5,010,573.9	3.21
4	518,700.1	5,014,393.4	513,738.9	5,005,831.3	11.45
5	518,099.6	5,013,576.2	512,938.9	5,004,733.6	1.77
6	518,992.6	5,017,306.0	514,144.0	5,009,699.3	7.79
7	519,150.0	5,013,556.6	514,331.9	5,004,709.3	6.34
8	519,259.8	5,013,600.0	514,482.8	5,004,764.0	1.38
9	516,916.8	5,016,528.9	511,378.9	5,008,669.6	4.62
10	516,659.6	5,018,093.8	511,043.8	5,010,744.1	4.09
11	519,474.3	5,018,046.9	514,807.0	5,010,675.2	11.90
12	519,549.2	5,014,375.9	514,873.0	5,005,798.0	2.79
13	518,089.4	5,014,478.2	512,938.6	5,005,931.0	9.15
14	518,087.4	5,014,755.2	512,936.0	5,006,299.0	8.04
15	518,079.1	5,016,483.3	512,921.1	5,008,596.5	7.71
16	516,947.5	5,017,736.1	511,424.5	5,010,277.6	9.22
17	517,015.8	5,014,443.1	511,495.1	5,005,894.9	6.36
18	517,785.1	5,017,492.6	512,542.4	5,009,954.0	9.62
19	519,435.7	5,017,340.7	514,736.0	5,009,735.7	4.88
20	518,710.3	5,016,544.2	513,778.7	5,008,679.7	8.78
21	518,984.0	5,016,548.6	514,127.8	5,008,678.2	10.08
22	516,719.0	5,014,555.9	511,106.7	5,006,028.1	13.68

Model Fit 3:

E = 1.33118637 * x
+ 0.0056629 * y
- 205490.3

N = - 0.003516 * x
+ 1.3297296 * y
- 1160143

RMSE = 6.78

Examine points,
no more blunders
found.

1	518,687.6	5,015,347.0	513,734.1	5,007,087.4	2.48
2	516,907.3	5,013,549.1	511,355.8	5,004,707.2	5.63
3	516,952.2	5,017,965.3	511,438.3	5,010,573.9	3.84
4	518,699.6	5,014,396.8	513,739.3	5,005,831.0	6.62
5	518,099.6	5,013,576.2	512,938.9	5,004,733.6	2.71
6	518,992.6	5,017,306.0	514,144.0	5,009,699.3	8.40
7	519,150.0	5,013,556.6	514,331.9	5,004,709.3	6.55
8	519,259.8	5,013,600.0	514,482.8	5,004,764.0	1.52
9	516,916.8	5,016,528.9	511,378.9	5,008,669.6	3.30
10	516,659.6	5,018,093.8	511,043.8	5,010,744.1	5.27
11	519,474.3	5,018,046.9	514,807.0	5,010,675.2	11.78
12	519,549.2	5,014,375.9	514,873.0	5,005,798.0	3.54
13	518,089.4	5,014,478.2	512,938.6	5,005,931.0	9.34
14	518,087.4	5,014,755.2	512,936.0	5,006,299.0	8.30
15	518,079.1	5,016,483.3	512,921.1	5,008,596.5	5.67
16	516,947.5	5,017,736.1	511,424.5	5,010,277.6	6.73
17	517,015.8	5,014,443.1	511,495.1	5,005,894.9	3.30
18	517,785.1	5,017,492.6	512,542.4	5,009,954.0	8.86
19	519,435.7	5,017,340.7	514,736.0	5,009,735.7	4.13
20	518,710.3	5,016,544.2	513,778.7	5,008,679.7	9.55
21	518,984.0	5,016,548.6	514,127.8	5,008,678.2	9.53
22	516,717.8	5,014,546.4	511,106.4	5,006,028.8	9.01

Figure 4-25: Iterative fitting of an affine transformation. Control points were examined after each fit, to discover blunders in entry or poor matching of points. Control points with large residuals were examined to determine if the cause for the error may be identified. If so, the control point coordinates may be modified, and transformation re-fit.

each quadrant. This is often not possible. This latter reason is less common with the development of GNSS. The transformation equation should be developed with the following observations in mind.

First, bad control points happen, but we should thoroughly justify the removal of any control point. Every attempt should be made to identify the source of the error, either in the collection or in the processing of field coordinates, the collection of image coordinates, or in some blunder in coordinate transcription. A common error is the misidentification of coordinate location on the image or map, for example, when the control location is placed on the wrong side of a road.

Second, a lower RMSE does not mean a better transformation. The RMSE is a useful tool when comparing among transformations that have the same model form, for example, when comparing one affine to another affine

as in Figure 4-25. The RMSE is not useful when comparing among different model forms, for example, when comparing an affine to a 2nd-order polynomial. The RMSE is typically lower for a 2nd and other higher-order polynomials than an affine transformation, but this does not mean the higher-order polynomial provides a more accurate transformation. The higher-order polynomial will introduce more error than an affine transformation on most orthographic maps, and an affine transformation is preferred. High-order polynomials allow more flexibility in warping the surface to fit the control points. Unfortunately, this warping may significantly deform the non-control-point coordinates, and add large errors when the transformation is applied to all data in a layer (Figure 4-26). Thus, high order polynomials and others should be used with caution.

Finally, independent tests of the transformations make the best comparisons

First order transformation
RMSE = 6.7 m

3rd order transformation
RMSE = 4.2 m

Figure 4-26: An illustration that RMSE should not be used to compare different order transformations, nor should it be used as the sole criterion for selecting the best transformation. Above are portions of a transformed image that was registered to a road network. This area is interstitial to 18 well distributed control points. Because the 3rd-order polynomial is quite flexible in fitting the points and reducing RMSE, it distorts areas between the control points. This is shown by the poor match between image and vector roads, above right. Although it has a higher RMSE, the first order transformation on the left is better overall.

among transformations. A completely independent set of well distributed test points would appear to be ideal, but these rarely exist. The extra points either haven't been collected, or suitable locations do not exist. The best way to test the accuracy of the transformation typically uses a "bootstrap" approach that treats each point as an independent test point. One point is withheld, the transformation estimated, and the error at the withheld point calculated. The point is replaced in the estimation set, and the next point withheld, fitting the same type of transformation. The equations will be slightly different. The error at this second withheld point is then calculated. This process is repeated for each control point, and a mean error calculated.

Control Point Sources: Surveying

Traditional ground surveys based on optical surface measurements are a common, although decreasingly used method for determining control point locations. Modern surveys use complex instruments such as transits and theodolites to precisely measure the relative location of points. If the survey starts from a known point, then the coordinate location of any survey station may be determined via simple trigonometric functions. Federal, state, county, and local governments all maintain a set of accurately surveyed locations (Figure 4-27), and these points may be used as control points or as starting points for additional surveys. Many of these known points have been established using traditional surveying techniques. Indeed, the development of this "control network" infrastructure is one of the first and most important responsibilities of government. These survey points form the basis for distance, location, and area measurements used to define property, political, and municipal boundaries. As a result, this control network underlies most commerce, transportation, and land ownership and management. Coordinates, general location, and descriptions are documented for these control networks, and may be obtained from a

number of government sources. In the United States these sources include county surveyors, state surveyors, departments of transportation, and the National Geodetic Survey (NGS).

The ground survey network is often quite sparse and insufficient for registering many large-scale maps or images. Even when there is a sufficient number of ground-surveyed points in an area, many may not be suitable for use as control points in a coordinate transformation of spatial data. The control points may not be visible on the maps or images to be registered. For example, a surveyed point may fall along the edge of a road. If the control point is at a mapped road intersection, we may use the easting and northing coordinates of the road intersection as a control point during map registration. However, if the surveyed point is along the edge of a road that is not near any mapped feature such as a road intersection, building, or water tower, then it may not be used as a control point. Our control points must have two characteristics to be useful: first, the point must be visible on the map, data layer, or image that we wish to register, and sec-

Figure 4-27: Previous surveys are a common source of control points.

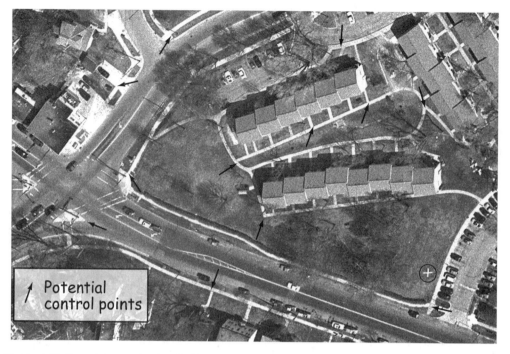

Potential
control points

Figure 4-28: Potential control points, indicated here by arrows, may be extracted from digital reference images. Permanent, well-defined features are identified and coordinates determined from the digital image. Note the white cross, circled in the lower right corner. This is a photogrammetric panel, typically a plastic or painted wooden target placed prior to photo capture, and with precisely surveyed coordinates. These targets are used to create the corrected digital image with a known coordinate system, a process described in Chapter 6.

ond, we must have precise ground coordinates in our target map projection. The first requirement, visibility on the source map or photograph, is often not met for survey-defined control. Therefore, we must often obtain additional control points.

One option for obtaining control points is to perform additional surveys that measure the coordinates of features that are visible on the source materials. Precise surveys are used to establish the coordinate locations of a well-distributed, sufficient set of points throughout the area covered by the source map. While sometimes expensive, new surveys are the chosen method when the highest accuracies are required. Costs were prohibitive with traditional optical surveying methods, however, GNSS positioning technologies allow more frequent, custom collection of control points.

Control Points from Existing Maps and Digital Data

Registered digital image data are common sources of ground control points, particularly when natural resources or municipal databases are to be developed for managing large areas. Digital images often provide a richly detailed depiction of surface features (Figure 4-28). Digital image data may be obtained that are registered to a known coordinate system. Typically, the coordinates of a corner pixel are provided, and the lines and columns for the image run parallel to the easting (E) and northing (N) direction of the coordinate system. Because the pixel dimensions are known, the calculation of a pixel coordinate involves multiplying the row and column number by the pixel size, and applying the corner offset, either by addition or subtraction. In this manner, the image row/column may be converted to an

E, N coordinate pair, and control point coordinates determined.

Existing maps are another common source of control points. Point locations are plotted and coordinates often printed on maps, for example the corner location coordinates are printed on USGS quadrangle maps. Road intersections and other well-defined locations are often represented on maps. If enough recognizable features can be identified, then control points may be obtained from the maps. Control points derived in this manner typically come only from cartometric maps, those maps produced with the intent of giving an accurate, map-projected representation of features on the Earth's surface.

Existing digital data may also provide control points. A short description of these digital data sources are provided here, and expanded descriptions of these and other digital data are provided in Chapter 7. For example, the USGS has produced Digital Raster Graphics (DRG) files that are scanned images of the 1:24,000-scale quadrangle maps. These DRGs come referenced to a standard coordinate system, so it is a simple and straightforward task to extract the coordinates of road intersections or other well-defined features that have been plotted on the USGS quadrangle maps. Vector data of roads are often widely available, and if of sufficient accuracy, may be used as a source of control points at road intersections and other distinct locations.

GNSS Control Points

The global positioning system (GPS), GLONASS, and Galileo are Global Navigation Satellite Systems (GNSS) that allow us to establish control points. GNSS, discussed in detail in Chapter 5, can help us obtain the coordinates of control points that are visible on a map or image. GNSS are particularly useful because we may quickly survey widely-spaced points. GNSS positional accuracy depends on the technology and methods employed; it typically ranges from sub-centimeter (tenths of inches) to a few meters (tens of feet). Most points recently added to the NGS and other government-maintained networks were measured using GNSS technologies.

To sum up: control points are necessary for coordinate transformation, and typically a number of control points are identified for a study area. The x and y coordinates for the control points are obtained from a digitized map or image, and the map projection coordinates, E and N, are determined from survey, GNSS, or other sources (Figure 4-29). These coordinate pairs are then used with a set of transformation equations to convert data layers into a desirable map coordinate system.

Control points

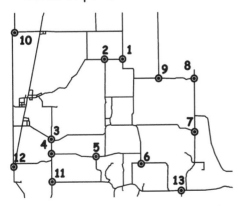

	Digitizer coordinates		Projection coordinates (UTM)	
ID	x	y	E	N
1.0	103.0	-100.1	500,083.4	5,003,683.5
2.0	0.8	-69.1	504,092.3	5,002,499.5
3.0	-20.0	-69.0	504,907.5	5,002,499.5
4.0	-60.0	-47.0	506,493.3	5,001,673.5
5.0	-102.0	-47.2	508,101.3	5,001,651.0
6.0	-101.7	10.8	508,090.1	4,999,384.0
7.0	-86.0	75.8	507,475.9	4,996,849.0
8.0	-40.0	45.7	505,689.2	4,998,022.0
9.0	11.0	36.8	503,679.2	4,998,368.0
10.0	63.0	34.0	501,657.9	4,998,479.5
11.0	63.0	17.7	501,669.1	4,999,116.0
12.0	63.0	64.3	501,680.3	4,997,296.0
13.0	106.0	47.7	500,005.3	4,997,943.5

Figure 4-29: An example of control point locations from a road data layer, and corresponding digitizer and map projection coordinates.

Raster Geometry and Resampling

Data often must be *resampled* when converting between coordinate systems, or changing the cell size of a raster data set (Figure 4-30). Resampling involves reassigning the cell values when changing raster coordinates or geometry. Resampling is required when changing cell sizes because the new cell centers will not align exactly with old cell centers. Changing coordinate systems may change the direction of the x and y axes, and GIS systems often require that the cell edges align with the coordinate system axes. Hence, the new cells often do not correspond to the same locations or extents as the old cells.

Common resampling approaches include the *nearest neighbor* (taking the output layer value from the nearest input layer cell center), *bilinear interpolation* (distance-based averaging of the four nearest cells), and *cubic convolution* (a weighted average of the sixteen nearest cells, Figure 4-30).

An example of a bilinear interpolation is shown in Figure 4-31. This algorithm uses a distance-weighted average of the four nearest cells in the input to calculate the value for the output. The new output location is represented by the black post. Initially, the height, or Z_{out} value, of the output location is unknown. Z_{out} is calculated based on the distances between the output locations and the input locations. The distance in the x direction is denoted in Figure 4-31 by d_1, and the distance in the y direction by d_2. The values in the input are shown as gray posts and are labeled as Z_1 through Z_4. Intermediate heights Z_b and Z_u are shown. These represent the average of the input values when taken in pairs in the x direction. These pairs are, Z_1 and Z_2, to yield Z_u, and Z_3 and Z_4, to yield Z_b. Z_u and Z_b are then averaged to cal-

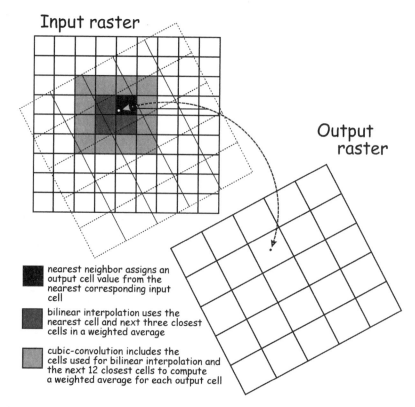

Input raster

Output raster

nearest neighbor assigns an output cell value from the nearest corresponding input cell

bilinear interpolation uses the nearest cell and next three closest cells in a weighted average

cubic-convolution includes the cells used for bilinear interpolation and the next 12 closest cells to compute a weighted average for each output cell

Figure 4-30: Raster resampling. When the orientation or cell size of a raster data set is changed, output cell values are calculated based on the closest (nearest neighbor), four nearest (bilinear interpolation), or sixteen closest (cubic convolution) input cell values.

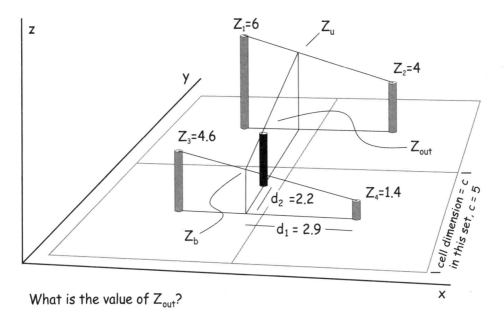

What is the value of Z_{out}?

$$Z_b = Z_4 + \frac{(Z_3 - Z_4)*d_1}{c} \qquad Z_b = 1.4 + \frac{(4.6 - 1.4)*2.9}{5} = 3.26$$

$$Z_u = Z_2 + \frac{(Z_1 - Z_2)*d_1}{c} \qquad Z_u = 4 + \frac{(6 - 4)*2.9}{5} = 5.16$$

$$Z_{out} = Z_b + \frac{(Z_u - Z_b)*d_2}{c} \qquad Z_{out} = 3.26 + \frac{(5.16 - 3.26)*2.2}{5} = 4.1$$

Figure 4-31: The bilinear interpolation method uses a distance weighted average to assign the output value, Z_{out}, based on input values, Z_1 through Z_4.

culate Z_{out}, using the distance d_2 between the input and output locations to weight values at each input location. The cubic convolution resampling calculation is similar, except that more cells are used, and the weighting is not an average based on linear distance.

Map Projection vs. Transformation

Map transformations should not be confused with map projections. A map transformation typically employs a statistically-fit linear equation to convert coordinates from one Cartesian coordinate system to another. A map projection, described in Chapter 3, differs from a transformation in that it is an analytical, formula-based conversion between coordinate systems, usually from a curved, latitude/longitude coordinate system to a Cartesian coordinate system. No statistical fitting process is used with a map projection.

Map transformations should rarely be used in place of map projection equations when converting geographic data between map projections. Consider an example when data are delivered to an organization in Universal Transverse Mercator (UTM) coordinates and are to be converted to State Plane coordinates prior to integration into a GIS database. Two paths may be chosen. The first involves projection from UTM to geographic coordinates (latitude and longitude), and then from these geographic coordinates to the appropriate State Plane coordinates. This is the correct, most accurate approach.

An alternate and often less-accurate approach involves using a transformation to convert between different map projections. In this case a set of control points would be identified and the coordinates determined in both UTM and State Plane coordinate systems. The transformation coefficients would be estimated and these equations applied to all data in the UTM data layer. This new output data layer would be in State Plane coordinates. This transformation process should be avoided, as a transformation may introduce additional positional error.

Transforming between projections is used quite often, inadvertently, when digitizing data from paper maps. For example, USGS 1:24,000-scale maps are cast on a polyconic projection. If these maps are digitized, it would be preferable to register them to the appropriate polyconic projection, and then re-project these data to the desired end projection. This is often not done, because the error in ignoring the projection over the size of the mapped area is typically less than the positional error associated with digitizing. Experience and spe-

cific calculations have shown that the spatial errors in using a transformation instead of a projection are small at these map scales under typical digitizing conditions.

This second approach, using a transformation when a projection is called for, should not be used until it has been tested as appropriate for each new set of conditions. Each map projection distorts the surface geometry. These distortions are complex and nonlinear. Affine or polynomial transformations are unlikely to remove this non-linear distortion. Exceptions to this rule occur when the area being transformed is small, particularly when the projection distortion is small relative to the random uncertainties, transformation errors, or errors in the spatial data. However, there are no guidelines on what constitutes a sufficiently "small" area. In our example above, USGS 1:24,000 maps are often digitized directly into a UTM coordinate system with no obvious ill effects, because the errors in map production and digitizing are often much larger than those in the projection distortion for the map area. However, you should not infer this practice is appropriate under all conditions, particularly when working with smaller-scale maps.

Output: Hardcopy Maps, Digital Data, and Metadata

We create spatial data to use, share, and archive. Maps are often produced during data creation and distribution, as intermediate documents while editing, for analysis, or as finished products to communicate some aspect of our data. To be widely useful, we must also generate information, or "metadata," about the spatial data we've created, and we may have to convert our data to standard forms. This section describes some characteristics of data output. We start with a brief treatment of cartography and map design, by which we produce hardcopy and

digital maps. We then provide a description of metadata, and some observations on data conversion and data transfer standards.

Cartography and Map Design

Cartography is the art and techniques of making maps. It encompasses both mapmaking tools and how these tools may be combined to communicate spatial information. Cartography is a discipline of much depth and breadth, and there are many books, journal articles, conferences, and societies devoted to the science and art of cartography. Our aim in the next few pages is to provide a brief overview of cartography with a particular focus on map design. This is both to acquaint new students with the most basic concepts of cartography, and help them apply these concepts in the consumption and production of spatial information. Readers interested in a more complete treatment should consult the references listed at the end of this chapter.

A primary purpose of cartography is to communicate spatial information. This requires identification of the

- intended audience,

- information to communicate,

- area of interest,

- physical and resource limitations,

in short, the whom, what, where, and how we may present our information.

These considerations drive the major cartographic design decisions we make each time we produce a map. We must consider the:

- scale, size, shape, and other general map properties,

- data to plot,

- symbol shapes, sizes, or patterns,

- labeling, including type font and size,

- legend properties, size, and borders, and

- the placement of all these elements on a map.

Map scale, size, and shape depend primarily on the intended map use. Wall maps for viewing at a distance of a meter or more may have few, large, boldly colored features. In contrast, commonly produced street maps for navigation in metropolitan areas are detailed, to be viewed at short ranges, and have a rich set of additional tables, lists, or other features.

Map scale is often determined in part by the size of the primary objects we wish to display, and in part by the most appropriate media sizes, such as the page or screen size possible for a document. As noted earlier, the map scale is the ratio of lengths on a map to true lengths. If we wish to display an area that spans 25 kilometers (25,000 meters) on a screen that spans 25 centimeters (0.25 meters), the map scale will be near 0.25 to 25,000, or 1:100,000. This decision on size, area, and scale then drives further map design. For example, scale limits the features we may display, and the size, number, and labeling of features. At a 1:100,000 scale we may not be able to show all cities, burgs, and towns, as there may be too many to fit at a readable size.

Maps typically have a primary theme or purpose that is determined by the intended audience. Is the map for a general population, or for a target audience with specific expectations for map features and design? General purpose maps typically have a wide range of features represented, including transportation networks, towns, elevation or other common features (Figure 4-32a). Special purpose maps, such as road maps, focus on a more limited set of features, in this instance road locations and names, town names, and large geographic features (Figure 4-32b).

Once the features to include on a map are defined, we must choose the symbols used to draw them. Symbology depends in part on the type of feature. For example, we have a different set of options when representing continuous features such as elevation or pollution concentration than when representing discrete features. We also must choose among symbols for each of the types of discrete features, for example, the set of symbols for points are generally different from those for line or area features.

Symbol size is an important attribute of map symbology, often specified in a unit called a point. One point is approximately equal to 0.467 mm, or about 1/72 of an inch. A specific point number is most often used to specify the size of symbols, for example, the dimensions of small squares to represent houses on a map, or the characteristics of a specific pattern used to fill areas on a map. A line width may also be specified in points. Setting a line width of two points means we want that particular line plotted with a width of 0.93 mm. It is unfortunate that "point" is both the name of the distance unit and a general property of a geographic feature, as in "a tree is a point feature." This forces us to talk about the "point size" of symbols to represent points, lines, or area fills or patterns, but if we are careful, we may communicate these specifications clearly.

The best size, pattern, shape, and color used to symbolize each feature depends on the viewing distance, the number, density, and type of features, and the purpose of the map. Generally, we use larger, bolder, or thicker symbols for maps to be viewed from longer distances, while we reduce this limit when producing maps for viewing at 50 cm (18 inches). Most people with normal vision under good lighting may resolve lines down to near 0.2 points at close distances, provided the lines show good contrast with the background. Although size limits depend largely on background color and contrast, point features are typically not resolvable at sizes smaller than about one half a point, and distinguishing between shapes is difficult for point features smaller than approximately two points in their largest dimension.

The pattern and color of symbols must also be chosen, generally from a set provided by the software (Figure 4-33). Symbols generally distinguish among feature type by characteristics, and although most symbols are not associated with a feature type, some are, such as, plane outlines for airports, numbered shields for highways, or a hatched line for a railroad.

We also must often choose whether and how to label features. Most GIS software provides a range of tools for creating and placing labels, and in all cases we must choose the label font type and size, location relative to the feature, and orientation. Primary considerations when labeling point

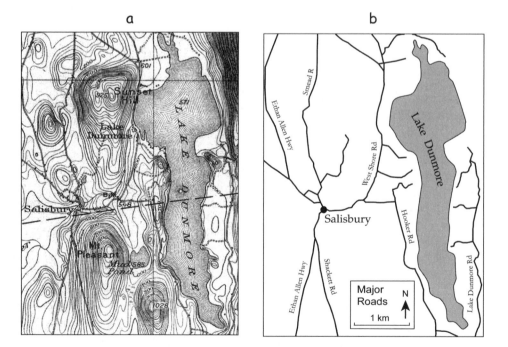

Figure 4-32: Example of a) a detailed, general-purpose map, here a portion of a US Geological Survey map, and b) a specialized map focusing a specific set of selected features, here showing roads. The features chosen for depiction on the map depend on the intended map use.

features are label placement relative to the point location, label size, and label orientation (Figure 4-34). We may also use graduated labels, that is, resize them according to some variable associated with the point feature. For example, it is common to have larger features and label fonts for larger cities (Figure 4-34). Labels may be bent, angled, or wrapped around features to improve clarity and more efficiently use space in a map.

Label placement is very much an art, and there is often much individual editing required when placing and sizing labels for finished maps. Most software provides for automatic label placement, usually specified relative to feature location. For example, one may specify labels above and to the right of all points, or lines labels placed over line features, or polygon labels placed near the

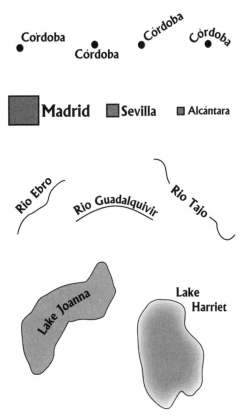

Figure 4-34: Common labeling options, including straight, angled, wrapped text, and graduated labels for points, (top two sets), and angled, wrapped, fronting, and embedded labels for line and polygon features (bottom two sets).

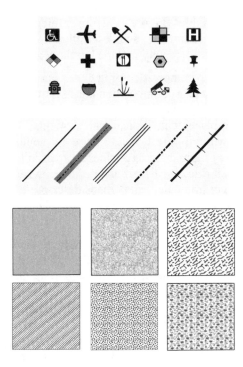

Figure 4-33: Examples of point (top), line (mid), and area (bottom) symbols used to distinguish among features of different types. Most GIS software provide a set of standard symbols for point, line, area, and continuous surface features.

polygon centroid. However, these automatic placements may not be satisfactory because labels may overlap, labels may fall in cluttered areas of the map, or features associated with labels may be ambiguous. Some software provides options for simple to elaborate automatic label placement, including automatic removal or movement of overlapping labels. These often reduce manual editing, but sometimes increase it.

Figure 4-35 shows a portion of a map of southern Finland. This region presents several mapping problems, including the high density of cities near the upper right, an irregular coastline, and dense clustering of islands along the coast. Most labels are placed above and to the right of their corresponding city, however some are moved or angled for clarity. Cities near the coast show both, to avoid labels crossing the water/land boundary where practical. Semi-transparent background shading is added for Parainen and Hanko, cities placed in the island matrix. This example demonstrates the individual editing often required when placing labels.

Most maps should have legends. The legend identifies map features succinctly and describes the symbols used to depict those features. Legends often include or are grouped with additional map information such as scale bars, north arrows, and descriptive text. The cartographer must choose the size and shape of the descriptive symbol, and the font type, size, and orientation for each symbol in the legend. The primary goal is to have a clear, concise, and complete legend.

The kind of symbols appropriate for map legends depends on the types of features depicted. Different choices are available for point, line, and polygon features, or for continuously variable features stored as rasters. Most software provides a range of

legend elements and symbols which may be used. Typically these tools allow a wide range of symbolizations, and a compact way of describing the symbolization in a legend (Figure 4-36).

The specific layout of legend features must be defined, for example the point feature symbol size may be graduated based on some attribute for the points. Successively larger features may be assigned for successively larger cities. This must be noted in the legend, and the symbols nested, shown sequentially, or otherwise depicted (Figure 4-36, top left).

The legend should be exhaustive. Examples of each different symbol type that appears on the map should appear in the legend. This means each point, line, or area symbol is drawn in the legend with some descriptive label. Labels may be next to, wrapped around, or embedded within the features, and sometimes descriptive numbers are added, for example, a range of continuous variables (Figure 4-36, upper left). Scale bars, north arrows, and descriptive text boxes are typically included in the legend.

Map composition or layout is another primary task. Composition consists of determining the map elements, their size, and their placement. Typical map elements shown in Figure 4-3 and Figure 4-4, include one or more main data panes or areas, a legend, a title, a scale bar and north arrow, a grid or graticule, and perhaps descriptive text. These each must be sized and placed on the map.

These map elements should be positioned and sized in accordance with their importance. The map's most important data pane should be largest, and it is often centered or otherwise given visual dominance. Other elements are typically smaller and located around the periphery or embedded within the main data pane. These other elements include map insets, which are smaller data panes that show larger or smaller scale views of a region in the primary data pane. Good map compositions usually group related elements and uses empty space effec-

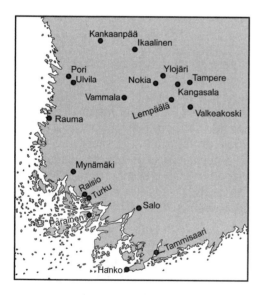

Figure 4-35: Example label placement for cities in southern Finland.

tively. Data panes are often grouped and legend elements placed near each other, and grouping is often indicated with enclosing boxes.

Neophyte cartographers should avoid two tendencies in map composition, both depicted in Figure 4-37. First, it is generally easy to create a map with automatic label and legend generation and placement. The map shown at the top of Figure 4-37 is typical of this automatic composition, and includes poorly placed legend elements and too small, poorly placed labels. Labels crowd each other, are ambiguous, cross water/land or other feature boundaries, and fonts are poorly chosen. You should note that automatic map symbol selection and placement is nearly always sub-optimal, and the novice cartographer should scrutinize these choices and manually improve them.

The second common error is poor use of empty space, those parts of the map without map elements. There are two opposite tendencies: either to leave too much or unbalanced empty space, or to clutter the map in an attempt to fill all empty space. Note that

the map shown at the top of Figure 4-37 leaves large empty spaces on the left (western) edge, with the Atlantic Ocean devoid of features. The cartographer may address this in several ways, either by changing the size, shape, or extent of the area mapped, adding new features, such as data panes as insets, additional text boxes, or other elements, or moving the legend or other map elements to that space. The map shown at the bottom of Figure 4-37, while not perfect, fixes these design flaws, in part by moving the legend and scale bar, and in part by adding labels for the Atlantic Ocean and Mediterranean Sea. The empty space is more balanced in that it appears around the major map elements in approximately equal proportions.

As noted earlier, this is only a brief introduction to cartography, a subject covered by many good books, some listed at the end of this chapter. Perhaps the best compendium of examples is the Map Book Series, by ESRI, published annually since 1984. Examples are available at the time of this writing at www.esri.com/mapmuseum. You should leaf through several volumes in

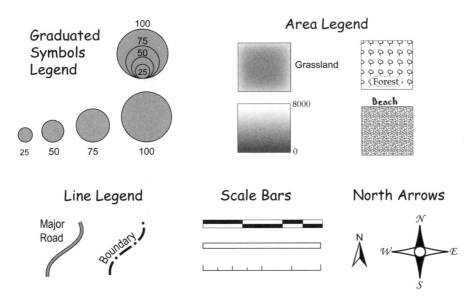

Figure 4-36: Examples of legend elements and representation of symbols. Some symbols may be grouped in a compact way to communicate the values associated with each symbol, e.g., sequential or nested graduated circles to represent city population size, area pattern or color fills to distinguish among different polygon features, line and point symbols, and informative elements such as scale bars and north arrows.

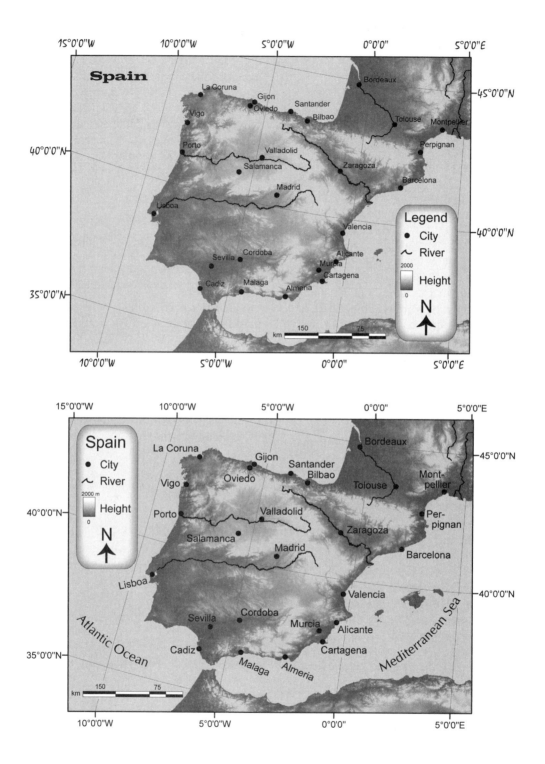

Figure 4-37: An example of poor map design (top). This top panel shows a number of mistakes common for the neophyte cartographer, including small labels (cities) and mismatched fonts (graticule labels, title), poor labeling (city labels overlapping, ambiguously placed, and crossing distinctly shaded areas), unlabeled features (oceans and seas), poorly placed scale bar and legend, and unbalanced open space on the left side of the map. These problems are not present in the improved map design, shown in the lower panel.

this series, with an eye towards critical map design. Each volume contains many beautiful and informative maps, and provides techniques worth emulating.

Digital Data Output

We often must transfer the digital data we create to another organization or user. Given the number of different GIS software, operating systems, and computer types, transferring data is not always a straightforward process. Digital data output typically includes two components, the data themselves in some standard, defined format, and *metadata*, or data about the digital data. We will describe data formats and metadata in turn.

Digital data are the data in some electronic form. As described at the end of the first chapter, there are many file formats, or ways of encoding the spatial and attribute data in digital files. Digital data output often consist of recording or converting data into one of these file formats. These data are typically converted with a utility, tool, or option available in the data development software (Figure 4-38). The most useful of these utilities support a broad range of input and output options, each fully described in the program documentation.

All formats strive for complete data transfer without loss. They must transmit the spatial and attribute data, the metadata, and all other information necessary to effectively use the spatial data. There are many digital data output formats, although many are legacy formats that are used with decreasing frequency.

A common contemporary format is the *Spatial Data Transfer Standard*. This transfer format, also known by the abbreviation SDTS, is a specification first defined by the U.S. Government in 1992. This standard has three basic parts: 1) a logical specification, 2) a description of the types of spatial features supported, and 3) the International Standards Organization (ISO) encoding used. There are four additional parts which define profiles, or descriptions of how vec-

tor, raster, point, and computer aided design data are stored for transfer. Digital data in SDTS formats typically span multiple data files, each holding various data components. At the time of this writing the U.S. Geological Survey was in charge maintaining the SDTS, with the full specification found at mcmcweb.er.usgs.gov/sdts/standard.html.

There are many legacy digital data transfer formats that were widely used before the publication of the SDTS. Among these are several US Geological Survey formats for the transfer of digital elevation models or digital vector data, or software specific formats, such as an ASCII format known as the GEN/UNGEN format that was developed by ESRI. These were useful for a limited set of transfers, but shortcomings in each of these transfer formats led to the development of the SDTS. They will not be discussed further here.

Metadata: Data Documentation

Metadata are information about spatial data. Metadata describe the content, source, lineage, methods, developer, coordinate sys-

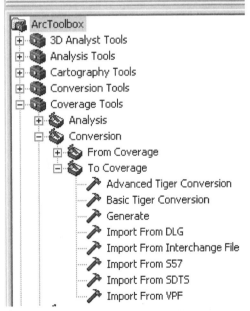

Figure 4-38: An example of a conversion utility, here from the ESRI ArcGIS software. Date may be converted from one of several formats to an ESRI-specific digital data.

tem, extent, structure, spatial accuracy, attributes, and responsible organization for spatial data.

Metadata are required for the effective use of spatial data. Metadata allow the efficient transfer of information about data, and inform new users about the geographic extent, coordinate system, quality, and other data characteristics. Metadata aid organizations in evaluating data to determine if they are suitable for an intended use -- are they accurate enough, do they cover the area of interest, do they provide the necessary information? Metadata may also aid in data updates by guiding the choice of appropriate collection methods and formats for new data.

Most governments have or are in the process of establishing standard methods for reporting metadata. In the United States, the Federal Geographic Data Committee (FGDC) has defined a Content Standard for Digital Geospatial Metadata (CSDGM) to specify the content and format for metadata. The CSDGM ensures that spatial data are clearly described so that they may be used effectively within an organization. The use of the CSDGM also ensures that data may be described to other organizations in a standard manner, and that spatial data may be more easily evaluated by and transferred to other organizations.

The CSDGM consists of a standard set of elements that are presented in a specified order. The standard is exhaustive in the information it provides, and is flexible in that it may be extended to include new elements for new categories of information in the future. There are over 330 different elements in the CSDGM. Some of these elements contain information about the spatial data, and some elements describe or provide linkages to other elements. Elements have standardized long and short names and are provided in a standard order with a hierarchical numbering system. For example, the western-most bounding coordinate of a data set is element 1.5.1.1, defined as follows:

1.5.1.1 West Bounding Coordinate – western-most coordinate of the limit of coverage expressed in longitude.

Type: real

Domain: $-180.0 <= $ West Bounding Coordinate < 180.0

Short Name: westbc

The numbering system is hierarchical. Here, 1 indicates it is basic identification information, 1.5 indicates identification information about the spatial domain, 1.5.1 is for bounding coordinates, and 1.5.1.1 is the western-most bounding coordinate.

There are 10 basic types of information in the CSDGM:

1) identification, describing the data set,

2) data quality,

3) spatial data organization,

4) spatial reference coordinate system,

5) entity and attribute,

6) distribution and options for obtaining the data set,

7) currency of metadata and responsible party,

8) citation,

9) time period information, used with other sections to provide temporal information, and

10) contact organization or person.

The CSDGM is a content standard and does not specify the format of the metadata. As long as the elements are included, properly numbered, and identified with correct values describing the data set, the metadata are considered to conform with the CSDGM. Indentation and spacing are not specified. However, because metadata may be quite complex, there are a number of conventions that are emerging in the presentation of metadata. These conventions seek to ensure that metadata are presented in a clear, logical way to humans, and are also easily ingested by computer software. There is a Standard Generalized Markup Language (SGML) for the exchange of metadata. An example of a portion of the metadata for a 1:100,000 scale digital line graph data set is shown in Figure 4-39.

4. Spatial_Reference_Information:
  4.1 Horizontal_Coordinate_System_Definition:
    4.1.2 Planar:
      4.1.2.2 Grid_Coordinate_System:
        4.1.2.2.1 Grid_Coordinate_System_Name:
                        Universal Transverse Mercator
        4.1.2.2.2 Universal_Transverse_Mercator:
          4.1.2.2.2.1 UTM_Zone_Number: 10-19
      4.1.2.4 Planar_Coordinate_Information:
        4.1.2.4.1 Planar_Coordinate_Encoding_Method:
                        coordinate pair
        4.1.2.4.2 Coordinate_Representation:
          4.1.2.4.2.1 Abscissa_Resolution: 2.54
          4.1.2.4.2.2 Ordinate_Resolution: 2.54
        4.1.2.4.4 Planar_Distance_Units: meters
    4.1.4 Geodetic_Model:
      4.1.4.1 Horizontal_Datum_Name: North American Datum 1927
      4.1.4.2 Ellipsoid_Name: Clark 1866
      4.1.4.3 Semi-major_Axis: 6378206.4
      4.1.4.4 Denominator_of_Flattening_Ratio: 294.98
  4.2 Vertical_Coordinate_System_Definition:
    4.2.1 Altitude_System_Definition:
      4.2.1.1 Altitude_Datum_Name:
              National Geodetic Vertical Datum of 1929
      4.2.1.2 Altitude_Resolution: 1
      4.2.1.3 Altitude_Distance_Units: feet or meters
      4.2.1.4 Altitude_Encoding_Method: attribute values
    4.2.2 Depth_System_Definition:
      4.2.2.1 Depth_Datum_Name: Mean lower low water
      4.2.2.2 Depth_Resolution: 1
      4.2.2.3 Depth_Distance_Units: meters or feet
      4.2.2.4 Depth_Encoding_Method: attribute values

Figure 4-39: Example of a small portion of the FGDC recommended metadata for a 1:100,000 scale derived digital data set.

Metadata are most often created using specialized software tools. Although metadata may be produced using a text editor, the numbering system, names, and other conventions are laborious to type. There are often complex linkages between metadata elements, and some elements are repeated or redundant. Software tools may ease the task of metadata entry by reducing redundant entries, ensuring correct linkages, and checking elements for contradictory information or errors. For example the metadata entry tool may check to make sure the western-most boundary is west of the eastern-most boundary. Metadata are most easily and effectively produced when their development is integrated into the workflow of data production.

Although not all organizations in the United States adhere to the CSDGM metadata standard, most organizations record and organize a description and other important information about their data, and many organizations consider a data set incomplete if it lacks metadata. All U.S. government units are required to adhere to the CSDGM when documenting and distributing spatial data.

Many other national governments are developing metadata standards. One example is the spatial metadata standard developed by the Australia and New Zealand Land Information Council (ANZLIC), known as the ANZLIC Metadata Guidelines. ANZLIC is a group of government, business, and academic representatives working to develop spatial data standards. The ANZLIC metadata guidelines define the core elements of metadata, and describe how to write, store, and disseminate these core elements. Data entry tools, examples, and spatial data directory have been developed to assist in the use of ANZLIC spatial metadata guidelines.

There is a parallel effort to develop and maintain international standards for metadata. The standards are known as the ISO 19115 International Standards for Metadata. According to the International Standards Organization, the ISO 19115 "defines the schema required for describing geographic information and services. It provides information about the identification, the extent, the quality, the spatial and temporal schema, spatial reference, and distribution of digital geographic data".

There is a need to reconcile international and national metadata standards, because they may differ. National standards may require information not contained in international standards, or vice versa. Governments typically create *metadata profiles* that are consistent with the international standard. These profiles establish the correspondence between elements in the different standards, and identify elements of the international profile that are not in the national profile.

Summary

Spatial data entry is a common activity for many GIS users. Although data may be derived from several sources, maps are a common source, and care must be taken to choose appropriate map types and to interpret the maps correctly when converting them to spatial data in a GIS.

Maps are used for spatial data entry due to several unique characteristics. These include our long history of hardcopy map production, so centuries of spatial information are stored there. In addition, maps are inexpensive, widely available, and easy to convert to digital forms, although the process is often time consuming, and may be costly. Maps are usually converted to digital data through a manual digitization process, whereby a human analyst traces and records the location of important features. Maps may also be digitized via a scanning device.

The quality of data derived from a map depends on the type and size of the map, how the map was produced, the map scale, and the methods used for digitizing. Large-scale maps generally provide more accurate positional data than comparable small-scale maps. Large-scale maps often have less map generalization, and small horizontal errors in plotting, printing, and digitizing are magnified less during conversion of large-scale maps.

Snapping, smoothing, vertex thinning, and other tools may be used to improve the quality and utility of digitized data. These methods are used to ensure positional data are captured efficiently and at the proper level of detail.

Map and other data often need to be converted to a target coordinate system via a map transformation. Transformations are different from map projections, which were discussed in Chapter 3, in that a transformation uses an empirical, least-squares process to convert coordinates from one Cartesian systems to another. Transformations are often used when registering digitized data to a known coordinate system. Map transformations should not be used when a map projection is called for.

Cartography is an important aspect of GIS, because we often communicate spatial information through maps. Map design depends on both the target audience and purpose, setting and modes of map viewing, and available resources. Proper map design considers the scale, symbols, labels, legend, and placement to effectively communicate the desired information.

Metadata are the "data about data." They describe the content, origin, form, coordinate system, spatial and attribute data characteristics, and other relevant information about spatial data. Metadata facilitate the proper use, maintenance, and transfer of spatial data. Metadata standards have been developed, both nationally and internationally, with profiles used to cross-reference elements between metadata standards. Metadata are a key component of spatial data, and many organizations do not

consider data complete until metadata have been created.

Suggested Reading

Aronoff, S. (1989). *Geographic Information Systems, A Management Perspective.* WDL Publications: Ottawa.

Bolstad, P., Gessler, P., & Lillesand, T.M. (1990). Positional uncertainty in manually digitized map data. *International Journal of Geographical Information Systems,* 4:399-412.

Burrough, P.A., & Frank, A.U. (1996). *Geographical Objects with Indeterminate Boundaries.*Taylor & Francis: London.

Chrisman, N.R. (1984). The role of quality information in the long-term functioning of a geographic information system. *Cartographica,* 21:79-87.

Chrisman, N.R. (1987). Efficient digitizing through the combination of appropriate hardware and software for error detection and editing. *International Journal of Geographical Information Systems,* 1:265-277.

DeMers, M. (2000). *Fundamentals of Geographic Information Systems* (2nd ed.). Wiley: New York.

Douglas, D.H. & Peuker, T.K. (1973). Algorithms for the reduction of the number of points required to represent a digitized line or its caricature. *Canadian Cartographer,* 10:112-122.

Gesch, D., Oimoen, M., Greenlee, S., Nelson, C,. Steuck, M., & Tyler C., (2002). The National Elevation Dataset. *Photogrammetric Engineering and Remote Sensing,* 68:5-32.

Holroyd, F. & Bell, S.B.M. (1992). Raster GIS: Models of raster encoding. *Computers and Geosciences,* 18:419-426

Joao, E. M. (1998). *Causes and Consequences of Map Generalization.* Taylor & Francis: London.

Laurini, R. & Thompson, D. (1992). *Fundamentals of Spatial Information Systems.* Academic Press: London.

Maquire, D. J., Goodchild, M. F., & Rhind, D. (Eds.). (1991). *Geographical Information Systems: Principles and Applications.*Longman Scientific: Harlow.

McBratney, A.B., Santos, M.L.M., & Minasny, B. (2003). On digital soil mapping. *Geoderma,* 117:3-52.

Muehrcke, P.C. & Muehrcke, J.P. (1992). *Map Use: Reading, Analysis, and Interpretation* (3rd ed.). J.P. Publications: Madison.

Nagy, G. & Wagle, S.G. (1979). Approximation of polygonal maps by cellular maps. *Communications of the Association of Computational Machinery,* 22:518-525.

Peuquet, D.J. (1984). A conceptual framework and comparison of spatial data models, *Cartographica*, 21:66-113.

Peuquet, D.J. (1981). An examination of techniques for reformatting digital cartographic data. Part II: the raster to vector process. *Cartographica*, 18:21-33.

Peuker, T. K. & Chrisman, N. (1975). Cartographic data structures. *The American Cartographer*, 2:55-69.

Shaeffer, C.A., Samet, H., & Nelson R.C. (1990). QUILT: a geographic information system based on quadtrees, *International Journal of Geographical Information Systems*, 4:103-132.

Shea, K.S., & McMaster, R.B. (1989). Cartographic generalization in a digital environment: when and how to generalize. *Proceedings AutoCarto 9*, pp.56-67.

Warner, W. & Carson, W. (1991). Errors associated with a standard digitizing tablet. *ITC Journal*, 2:82-85.

Weibel, R. (1997). Generalization of spatial data: principles and selected algorithms. In van Kreveld, M., Nievergelt, J., Roos, T., & Widmayer, P. (Eds.), *Algorithmic Foundations of Geographic Information Systems*, Springer-Verlag: Berlin.

Wolf, P.R., & C. Ghilani (2002). *Elementary Surveying, an Introduction to Geomatics* (10th ed.). Prentice-Hall: New Jersey.

Zeiler, M. (1999). *Modeling Our World: The ESRI Guide to Geodatabase Design*. ESRI Press: Redlands.

Study Questions

4.1 - Why have so many digital spatial data been derived from hardcopy maps?

4.2 - Which is a larger scale map,
a)1:20,000 or b)1:1,000,000?

c) 1 inch equals 1 mile, or d) 1:100,000

e) 1 mm to 1 kilometer, or f) 1:1,500,000

4.3 - Can you describe three different types of generalization?

4.4 - Identify the kind of generalization at the labeled locations a through d in the map below, left, compared to the "truth" in the image, below right. Categorize the generalizations as fused, simplified, displaced, omitted, or exaggerated.

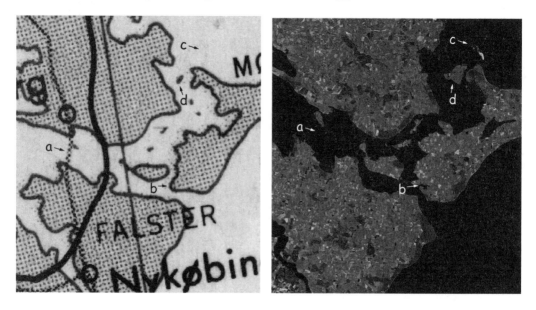

4.5 - Identify the kind of generalization at the labeled locations a through d in the map below, left, compared to the "truth" in the image, below right. Categorize the generalizations as fused, simplified, displaced, omitted, or exaggerated, or if it doesn't fit in one of these categories, then categorize it as "other," and describe the generalization.

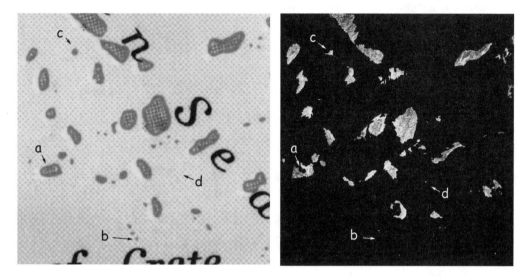

4.6 - What are the most common map media? Why?

4.7 - Is media deformation more problematic with large scale maps or small scale maps? Why?

4.8 - Which map typically shows more detail -- a large-scale map or a small-scale map? Can you give three reasons why?

4.9 - Complete the following table that shows scale measurements and calculations:

Ground distance and units	Correpsonding map distance and units	Map Scale
13,280 feet	6.4 inches	1 : 24,900
126.4 kilometers	25.28 centimeters	
123.6 miles	22.8 inches	
40.7 meters		1 : 502.5
	4.62 inches	1 : 249,685

4.10 - What is snapping in the context of digitizing? What are undershoots and overshoots, and why are they undesirable?

4.11 - Identify a characteristic feature or error in digitizing at each of the labeled letter locations in the drawing below, e.g., node, overshoot, missing label, etc.:

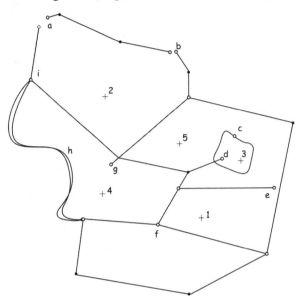

4.12 - Identify a characteristic feature or error in digitizing at each of the labeled letter locations in the drawing below, e.g., node, overshoot, missing label, etc.:

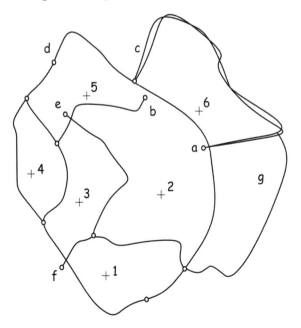

4.13 - Sketch the results of combined node (open circle), vertex (closed circle) and edge (lines) snapping with a snap tolerance of a) a distance of 5 units, and b) a distance of 10 units, as shown snap circles. Note the radius and not the diameter of these circles defines the snapping distance.

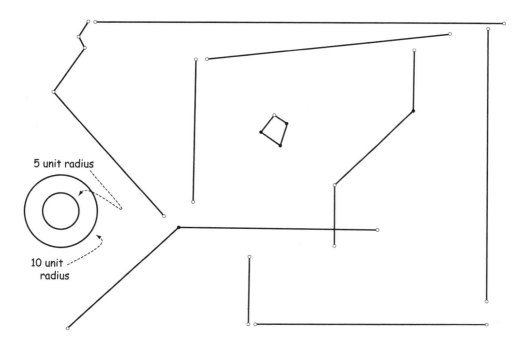

4.14 - What is a spline, and how are they used during digitizing?

4.15 - a) Why is line thinning sometimes necessary?

b) Does increasing the width of the line thinning band tend to increase, decrease, or not affect the number of points removed?

c) Does increasing the number of points initially spanned tend to increase, decrease, or not affect the number of points removed?

4.16 - Can you contrast manual digitizing to the various forms of scan digitizing? What are the advantages and disadvantages of each?

4.17 - What is the "common feature problem" when digitizing, and how might it be overcome?

4.18 - Can you describe the general goal and process of map registration?

4.19 - What are control points, and where do they come from?

4.20 - Can you define an affine transformation, including the form of the equation? Why is it called a linear transformation?

4.21 - What is the root mean square error (RMSE), and how does it relate to a coordinate transformation?

4.22 - Is the average positional error likely be larger, smaller, or about equal to the RMSE? Why?

4.23 - Why are higher order (polynomial) projections to be avoided under most circumstances?

4.24 - Which of the following transformations will likely have the smallest average error at a set of independent test points?

    a) affine, RMSE = 10.23   b) affine, RMSE = 9.8

    c) 2nd order polynomial, RMSE = 4.7   d) 3rd order polynomial, RMSE = 0.45

4.25 - Which of the following transformations will likely have the smallest average error at a set of independent test points?

    a) 1st order polynomial, RMSE = 5.3   b) affine, RMSE = 9.8

    c) 2nd order polynomial, RMSE = 2.9   d) 1st order polynomial, RMSE = 9.9

4.26 - Define and describe metadata. Why are metadata important?

5 Global Navigation Satellite Systems and Coordinate Surveying

---

Introduction

Broadly defined, there are two general ways we measure the locations of geographic features. The first uses field measurements, and are described in this chapter. We travel to a feature and physically occupy a location to measure unknown X, Y, and often Z coordinates. Positioning and measurement systems have become quite sophisticated, incorporating satellite and laser technologies, primarily Global Navigation Satellite Systems (GNSS), as well as traditional ground surveying methods. Field measurements may be accurate to within millimeters (tenths of inches).

The second set of location measurement techniques uses remote data collection, primarily from aerial and satellite images. Coordinate positions may be obtained to within a few centimeters (inches) from properly collected, carefully processed images. Image systems are described in Chapter 6.

GNSS are satellite-based technologies that give precise positional information, day or night, in most weather and terrain conditions (Figure 5-1). GNSS technologies may help navigate and track moving objects large enough to carry a receiver. Receivers shrink in size, weight, and power requirements each year.

Coordinate surveying encompasses traditional surveying through angle and dis-

Figure 5-1: An artist's rendering of a Galileo satellite, part of a planned 30-satellite constellation at the heart of a European Union-led satellite navigation system (courtesy ESA).

tance measurements. A subset of these measurements has been described in Chapter 3, in the section on geodesy. Coordinate surveying is often combined with precise GNSS measurements. Because both measurement methods are important, they will be covered in this chapter, first by describing GNSS and coordinate surveying tools and methods, and then discussing common applications.

GNSS Basics

Because they are inexpensive, accurate, and easy to use, GNSS have significantly changed surveying, navigation, shipping, and other fields, and are also having a pervasive impact in the geographic information sciences. GNSS have become the most common method for field data collection in GIS.

As of 2011 there are two functioning satellite GNSS systems, and two more in advanced stages of development. The NAVSTAR Global Positioning System (GPS) was the first deployed and is the most widely used system. There is an operational Russian system named GLONASS that is increasingly used internationally; at the time of this writing there are 22 operational satellites, giving complete, 24-hour coverage over Russia, and worldwide coverage through more satellites is planned for 2012. A third system, Galileo, is being developed by a consortium of European governments and industries, with a total of 18 satellites in the full constellation, planned for approximately 2015. A fourth system, the Chinese Compass Satellite Navigation System, is also under development. A full constellation of 30 positioning satellites is planned, with global coverage. Although system design and function are little described by the Chinese government, it appears the full constellation should be in operation by mid-decade.

Note that GPS usually refers to the U.S. NAVSTAR system, but is widely used as a

Satellite segment
21+ satellites
6 orbital planes
12 hour return interval for
each satellite

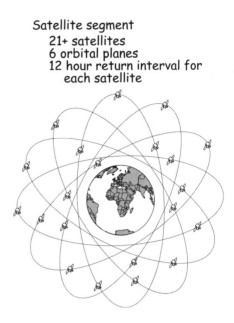

Figure 5-3: Satellite orbit characteristics for the NAVSTAR GPS constellation.

generic term for GNSS because it was the first broadly used satellite navigation system. In the following discussion we use GNSS as a generic term for all four systems, and use GPS to refer specifically to the U.S. NAVSTAR system.

There are three main components, or segments, of any GNSS (Figure 5-2). The first is the *satellite segment*. This segment is a constellation of satellites orbiting the Earth and transmitting positioning signals (Figure 5-3). The second component of any GNSS is a *control segment*. The control segment consists of the tracking, communications, data gathering, integration, analysis, and control facilities. The third part of GNSS is the *user segment*, the set of individuals with one or more receivers. A GNSS receiver is an electronic device that records data transmitted by each satellite, and then processes these data to obtain three-dimensional coordinates (Figure 5-4). There is a wide array of receivers and methods for determining position. Receivers are most often small, handheld devices with screens and keyboards, or elec-

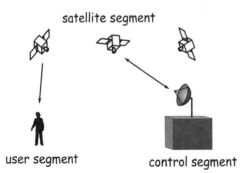

satellite segment

user segment   control segment

Figure 5-2: The three segments that comprise a GNSS.

tronic components mounted on trucks, planes, cargo containers, or cell phones.

The satellite and control segments differ for each GNSS. The NAVSTAR GPS includes a constellation of satellites orbiting the Earth at an altitude of approximately 20,000 kilometers. Initial system design included 21 active GPS satellites and 3 spares, distributed among six offset orbital planes. Every satellite orbits the Earth twice daily, and each satellite is usually above the flat horizon for eight or more hours each day. Experimental and successive operational satellites have outlasted their design life, so there have typically been more than 24 satellites in orbit simultaneously. Between four to eight active satellites are typically visible from any unobstructed viewing location on Earth.

GPS is controlled by a set of ground stations. These are used to observe, maintain, and manage satellites, communications, and related systems. There are five tracking stations in the GPS system, with a Master Control Station in Colorado, USA, and the remaining stations spread across the Earth. Data are gathered from a number of sources by the stations, including satellite health and

status from each GPS satellite, tracking information from each tracking station, timing data from the U.S. Naval Observatory, and earth data from the U.S. Defense Mapping Agency. The Master Control Station synthesizes this information and broadcasts navigation, timing, and other data to each satellite. The Master Control Station also signals each satellite as appropriate for course corrections, changes in operation, or other maintenance.

The GLONASS system is another currently operating GNSS. GLONASS was initiated by the former Soviet Union in the early 1970s. Satellites were first launched in the early 1980s, and the system became functional in the mid 1990s. The GLONASS system was designed for military navigation, targeting, and tracking, and is operated by the Russian Ministry of Defense, with control and tracking stations similar to those for the NAVSTAR GPS system.

GLONASS was designed to include 21 active satellites and 3 spares. New designs have been phased in as older satellites have expired, and system managers have focused on maximizing coverage over Russia. The GLONASS system is stable enough, with a

Figure 5-4: A hand-held GPS receiver (left) and in use (right). (courtesy Topcon)

published renovation and maintenance plan, such that commercial manufacturers have developed dual GPS/GLONASS capable receivers.

Galileo also implements satellite, control, and user segments. There are 30 satellites planned for the complete Galileo constellation, scheduled to be fully operational by 2015. Satellites are arranged in three orbital paths at a 54° orbital inclination, with a satellite altitude near 23,600 km above the Earth. This satellite constellation will provide better coverage of high northern latitudes than the US NAVSTAR GPS system, to better serve northern Europe. Galileo will be managed through two control centers in Europe and 20 Galileo Sensor Stations spread throughout the world to monitor, communicate with, and relay information among satellites and the control centers.

GNSS Broadcast Signals

GNSS positioning is based on radio signals broadcast by each satellite. The NAVSTAR GPS satellites broadcast at a fundamental frequency of 10.23 MHz (MHz = Megahertz, or millions of cycles per second). GPS satellites also broadcast at other frequencies that are integer multiples or divisors of the fundamental frequency (Table 5-1). There are two *carrier signals*, L1 at 1575.42 MHz and L2 at 1227.6 MHz. These carrier signals are modulated to produce two *coded signals*, the C/A code at 1.023 MHz and the P code at 10.23 MHz. The L1 signal carries both the C/A and P codes, while the L2 carries only the P code. Planned upgrades to the GPS system include additional, higher-power L bands that may provide additional information and be easier to receive. These other signals have been added to improve function, for example, the L2C to ease GPS tracking for navigation, L5 for worldwide safety of life applications, and M for enhanced military applications. The coded signals (C/A, P, and M) are sometimes referred to as the *pseudo-random code*, because they appear quite similar to random noise. However, short segments of

Table 5-1 : GPS Signals

Name	Frequency (MHz)
L1, L1C	1575.42
L2, L2CM, L2CL	1227.6
L5	1176.45
P, M	10.23
C/A	1.023

the code are unique in that the signal pattern for each satellite and time is different from other time periods. A receiver may decode each signal to recognize which unique satellite the code came from, when the signal was sent, and where the satellite was at the time the signal was sent. The receiver then uses this information for positioning. The coded signal does repeat, but the repeat interval is long enough to not cause problems in positioning.

Positions based on carrier signal measurements (L1, L2, and L5 frequencies for the NAVSTAR GPS, and sometimes referred to as *carrier phase* measurements) are inherently more accurate than those based on the code signal measurements. The mathematics and physics of carrier measurement are better suited for making positional measurements. However, the added accuracy incurs a cost. Carrier measurements require more sophisticated and expensive receivers. Perhaps a greater constraint on carrier measurements is that carrier receivers must record signals for longer periods of time than C/A code receivers. If the satellite passes behind an obstruction, such as a building, mountain, or tree, the signal may be momentarily lost and carrier phase measurements begun anew. Satellite signals are frequently lost in heavily obstructed environments, substantially reducing the efficiency of carrier phase data collection.

Each GPS satellite also broadcasts data on satellite status and location. Information includes an *almanac*, data used to determine the status of satellites in the GPS constellation. The broadcast also includes *ephemeris data* for the satellite constellation. These ephemerides allow a GPS receiver to accurately calculate the position of the broadcasting satellite and the expected positions of other satellites. Satellite health, clock corrections, and other data are also transmitted.

The other GNSS systems operate in a similar manner, with base, or carrier frequencies, and modulated or coded signals embedded within these carriers (Figure 5-5). GLONASS, the only other fully operational GNSS in 2011, broadcasts L1 and L2 carriers similar to GPS, and an additional L1/L10 and experimental G3 signal at higher and lower frequencies. Galileo plans include the broadcasts of a range of signals on several fundamental carriers, including one that overlaps with the GPS and GLONASS L1 signals at approximately 1575 MHz, with several coded signals planned at lower frequencies.

Substantial effort has been directed at ensuring the NAVSTAR, GLONASS, and Galileo systems do not interfere with each other, and are as compatible as practical. All three systems will broadcast in the 1575 MHz range, with signals coded differently to avoid interference. This simplifies the production of dual-system receivers that are capable of using multiple GNSS signals, improving coverage, accuracy, and efficiency, particularly under difficult data collection conditions. When all GNSS constellations are fully operational, there may be as many as 70 satellites broadcasting in this range, greatly improving coverage.

There is a notable overlap in some frequencies used by the Galileo and Chinese-based Compass system. The Chinese government originally was collaborating with the European consortium developing Galileo, but subsequently elected to produce their own GNSS system. While there is the

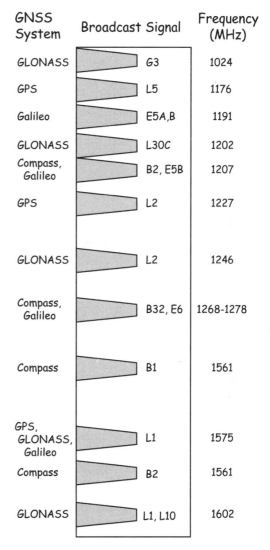

GNSS System	Broadcast Signal	Frequency (MHz)
GLONASS	G3	1024
GPS	L5	1176
Galileo	E5A,B	1191
GLONASS	L30C	1202
Compass, Galileo	B2, E5B	1207
GPS	L2	1227
GLONASS	L2	1246
Compass, Galileo	B32, E6	1268-1278
Compass	B1	1561
GPS, GLONASS, Galileo	L1	1575
Compass	B2	1561
GLONASS	L1, L10	1602

Figure 5-5: Existing and proposed GNSS broadcast signals, frequencies, and positioning services. Signals are spaced to avoid interference, or coded where they overlap. Frequencies are not spaced to scale (courtesy ESA).

possibility for mutual interference in signals, it is likely that the Compass and Galileo systems will be operated to avoid this.

Range Distances

GNSS positioning is based primarily on *range distances* determined from the carrier and coded signals. A range distance, or *range*, is a distance between two objects. For GNSS, the range is the distance between a satellite in space and a receiver on or above the Earth's surface (Figure 5-6). GNSS signals travel at the speed of light. The range distance from the receiver to each satellite is calculated based on signal travel time from the satellite to the receiver:

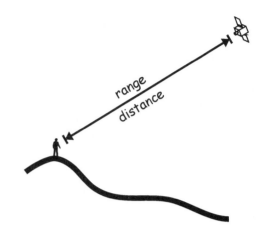

Range = speed of light * travel time   (5.1)

Figure 5-6: A single satellite range measurement.

Coded signals are used to calculate signal travel time by matching sections of the code. Timing information is sent with the coded signal, allowing the GNSS receiver to calculate the precise transmission time for each code fragment. The GNSS receiver also observes the reception time for each code fragment. The difference between transmission and reception times is the travel time, which is then used to calculate a range distance (Figure 5-7). Range measurements can be repeated quite rapidly, typically up to a rate of one per second; therefore several

range measurements may be made for each satellite in a short period of time.

Carrier phase GNSS is also based on a set of range measurements. In contrast to coded signals, the phase of the satellite signal is measured. Each individual wave transmitted at a given frequency is identical, and at any given point in time there is some unknown integer number of waves plus a partial wave that fit in the distance between the satellite and the receiver. Carrier signal observations over extended intervals allow the calculation of wavelength number over

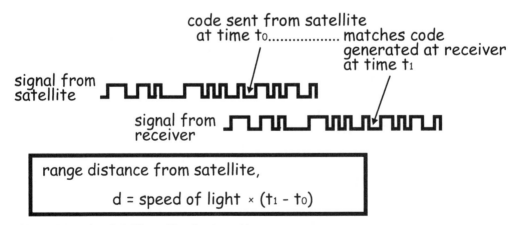

Figure 5-7: A decoded C/A satellite signal provides a range measurement.

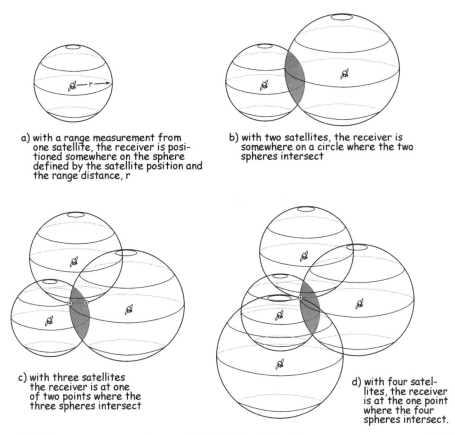

a) with a range measurement from one satellite, the receiver is positioned somewhere on the sphere defined by the satellite position and the range distance, r

b) with two satellites, the receiver is somewhere on a circle where the two spheres intersect

c) with three satellites the receiver is at one of two points where the three spheres intersect

d) with four satellites, the receiver is at the one point where the four spheres intersect.

Figure 5-8: Range measurements from multiple GNSS satellites. Range measurements are combined to narrow down the position of a GNSS receiver. Range measurements from more than four satellites may be used to improve the accuracy of a measured position. (adapted from Hurn, 1989)

the measurement interval, and then the calculation of very precise satellite ranges.

Simultaneous range measurements from multiple satellites are used to estimate a receiver's location. A range measurement is combined with information on satellite location to define a sphere. A range measurement from a single satellite restricts the receiver to a location somewhere on the surface of a sphere centered on the satellite (Figure 5-8a). Range measurements from two satellites identify two spheres, and the receiver is located on the circle defined by the intersection of the two spheres (Figure 5-8b). Range measurements from three satellites define three spheres, and these three spheres will intersect at two points (Figure 5-8c). A sequence of range measurements through time from three satellites will reveal

that one of the points remains nearly stationary, while the other point moves rapidly through space. The second intersection point moves through space because the size and relative geometry of the spheres changes through time as the satellites change position on their orbital paths. If system and receiver clocks were completely accurate it would be possible to determine the position of a stationary receiver by taking measurements from three satellites over a short time interval. However, simultaneous measurements from four satellites (Figure 5-8d) are usually required to reduce receiver clock errors and to allow instantaneous position measurement with a moving receiver, e.g., on a plane, in a car, or while walking. Data may be collected from more than four satellites at a time, and this usually improves the accuracy of position measurements.

Positional Uncertainty

Errors in range measurements and uncertainties in satellite location introduce errors into GNSS-determined positions. Range errors vary substantially even if range measurements are taken just a few seconds apart. Errors in the ephemeris data lead to erroneous estimates of the satellite position, and also result in an offset or positioning error. This causes the intersection of the range spheres to change through time, even when the GNSS receiver is in a fixed location. This results in a band of range uncertainty encompassing the GNSS receiver position (Figure 5-9).

Several methods are used to improve our estimates of receiver position. One common method involves collecting multiple estimates of receiver location while keeping the receiver stationary. Most receivers may estimate a new position, or fix, every second. Receivers collect and average multiple position fixes, yielding a mean position estimate. Multiple fixes enable the receivers to estimate the variability in the position locations, for example a standard deviation, predicted accuracy, or some other figure of merit. While the standard deviation provides no information on the absolute error, it does allow some estimate of the precision of the mean GNSS position fix. The magnitude and

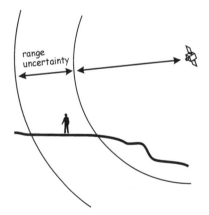

Figure 5-9: Uncertainty in range measurements leads to positional errors in GNSS measurements.

properties of this variation may be useful in evaluating GNSS measurements and the source of errors. However, multiple position fixes are not possible when data collection takes place while moving, for example, when determining the location of an airborne plane. Also, averaging does not remove any bias in the calculated position. Alternative methods for reducing positional error rely on reducing the several sources of range errors.

Sources of Range Error

Ionospheric and *atmospheric delays* are major sources of GNSS positional error. Range calculations incorporate the speed of light. Although we usually assume the speed of light is constant, this is not true. The speed of light is constant only while passing through a uniform electromagnetic field and in a vacuum. The Earth is surrounded by a blanket of charged particles called the ionosphere. These charged particles are formed by incoming solar radiation, which strips electrons from elements in the upper atmosphere. Changes in the charged particle density through space and time result in a changing electromagnetic field surrounding the Earth.

The atmosphere is below the ionosphere, and atmospheric density is significantly different from that of a vacuum. Variation in the atmospheric density is due largely to changes in temperature, atmospheric pressure, and water vapor. Range errors occur because the travel speed of the GNSS signal is altered slightly as it passes through the ionosphere and atmosphere.

We can attempt to reduce ionospheric errors by adjusting the value used for the speed of light when calculating the range distance. Physical models can be developed that incorporate measurements of the ionospheric charge density. Because the charge density varies both around the globe and through time, and because there is no practical way to measure and disseminate the variation in charge density in a timely manner, physical models are based on the average charge. These physical models reduce the

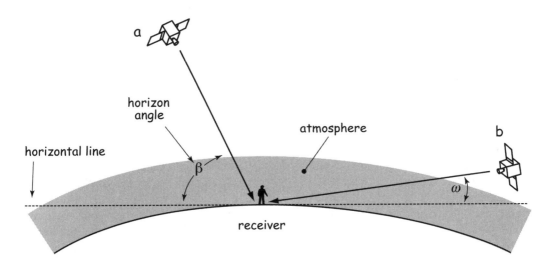

Figure 5-10: GNSS receivers often discard signals from satellites near the horizon. As this image shows, signals from satellites high above the horizon (**a**) with high horizon angles (β) have shorter path lengths in the atmosphere than low-angle satellites (**b**, with angle ω). Low horizon angles increase atmospheric delays and hence range errors for satellites with low horizon angles. Typically a limit or "mask" is set at approximately 15 degrees above the horizon, and satellites are ignored if they are below this limit.

range error somewhat. Alternatively, correction could be based on the observation that the change in the speed of light depends on the frequency of the light. Specialized *dual frequency* receivers collect information on multiple GNSS signals simultaneously, and use sophisticated physical models to remove most of the ionospheric errors. Dual frequency receivers are limited though, because they are typically much more expensive than code-based receivers, and because they must maintain a continuous fix for a longer period of time. Finally, there are no good models for atmospheric effects, thus there is no analytical method to remove range errors due to atmospheric delays.

System operation and delays are other sources of range uncertainty. Small errors in satellite tracking cause errors in satellite positional measurements. Timing and other signals are relayed from globally-distributed monitoring stations to the Master Control Center and up to the satellites, but there are uncertainties and delays in signal transmission, so timing signals may be slightly offset. Atomic clocks on the satellite may be un-synchronized or in error, although this is typically one of the smaller contributions to positional errors.

Receivers also introduce errors into GNSS positions. Receiver clocks may contain biases or may use algorithms that do not precisely calculate position. Signals may reflect off of objects prior to reaching the antenna. These reflected, or *multipath* signals travel a further distance than direct GNSS signals, and so introduce an offset into GNSS positions. Multipath signals often have lower power than direct signals, so some multipath signals may be screened by setting a threshold signal-to-noise ratio. Signals with high noise relative to the mean signal strength are ignored. Multipath signals may also be screened by properly designed antennas. Multipath signals are most commonly a problem in urban settings that have an abundance of corner reflectors, such as the sides of buildings and streets.

Satellite Geometry and Dilution of Precision

The geometry of the GNSS satellite constellation is another factor that affects positional error. Range errors create an area of uncertainty perpendicular to the transmission direction of the GNSS signal. These areas of uncertainty may be visualized as a set of nested spheres, with the true position somewhere within the volume defined by the intersection of these nested spheres (Figure 5-11). These areas of uncertainty from different satellites intersect, and the smaller the intersection area, the more accurate the position fixes are likely to be (Figure 5-12). Signals from widely spaced satellites are complementary because they result in a smaller area of uncertainty. Signals from satellites in close proximity overlap over broad areas, resulting in large areas of positional uncertainty. Widespread satellite constellations provide more accurate GNSS position measurements.

Satellite geometry is summarized in a number called the *Dilution of Precision*, or DOP. There are various kinds of DOPs, including the Horizontal (HDOP), Vertical (VDOP), and Positional (PDOP) Dilution of

Precision. The PDOP is most used and is the ratio of the volume of a tetrahedron created by the four most widespread, observed satellites to the volume defined by the ideal tetrehedron. This ideal tetrahedron is formed by one satellite overhead and three satellites spaced at 120-degree intervals around the horizon. This constellation is assigned a PDOP of one, and closer groupings of satellites have higher PDOPs. Lower PDOPs are better. Most GNSS receivers review the almanac transmitted by the GNSS satellites and attempt to select the constellation with the lowest PDOP. If this best constellation is not available, for example, some satellites are not visible, successively poorer constellations are tested until the best available constellation is found. The receivers typically provide a measurement of PDOP while data are collected, and a maximum PDOP threshold may be specified. Data are not gathered when the PDOP is above the threshold value.

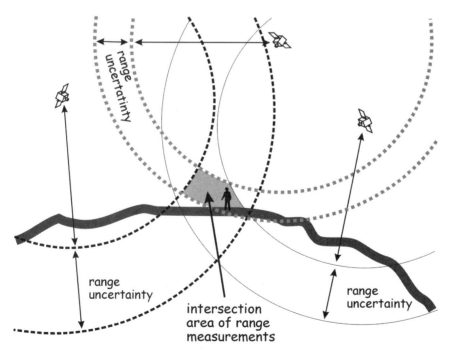

Figure 5-11: Relative GPS satellite position affects positional accuracy. Range uncertainties are associated with each range measurement. These combine to form an area of uncertainty at the intersection of the range measurements.

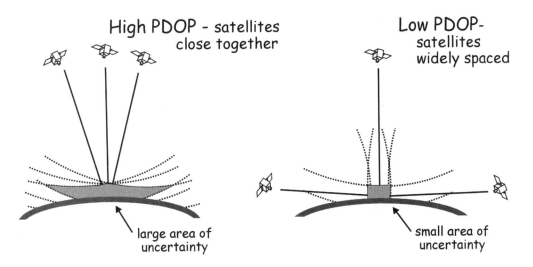

High PDOP - satellites close together

Low PDOP- satellites widely spaced

large area of uncertainty

small area of uncertainty

Figure 5-12: GPS satellite distribution affects positional accuracy. Closely-spaced satellites result in larger positional errors than widely-spaced satellites. Satellite geometry is summarized by PDOP with lower PDOPs indicating better satellite geometries.

Range errors and DOPs combine to affect GNSS position accuracies. There are many sources of range error, and these combine to form an overall range uncertainty for the measurement from each visible GNSS satellite. If more precise coordinate locations are required, then the choices are to use equipment that makes more precise range measurements, and/or to collect data when DOPs are low.

GNSS accuracies depend on the type of receiver, atmospheric and ionospheric conditions, the number of range measurements, the satellite constellation, and the algorithms used for position determination (Figure 5-13). Current C/A code receivers typically provide accuracies between 3 and 30 meters for a single fix. Errors larger than 100 meters for a single fix occur occasionally. Accuracies may be improved substantially, to between 2 and 15 meters, when multiple fixes are averaged. The longer the data collection time the greater the accuracy. Improvements come largely from reducing the impact of rarer, large errors, but average accuracies are rarely below one meter when using a single C/A code receiver.

Accuracies when using carrier phase or similar receivers are much higher, on the order of a few centimeters. These accuracies come at the cost of longer data collection times, and are most often obtained when using differential correction, a process described in the following section.

Differential Correction

The previous sections have focused on GNSS position measurements collected with a single receiver. This operating mode is known as autonomous GNSS positioning.

An alternative method, known as *differential positioning*, employs two or more receivers. Differential positioning measurements are used primarily to remove most of the range

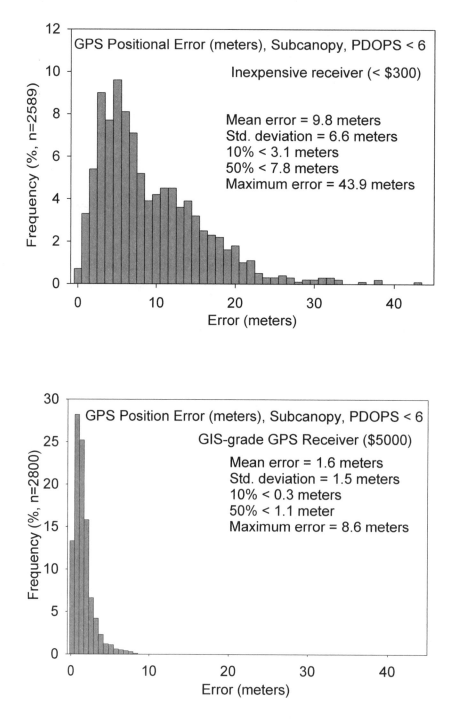

Figure 5-13. Observed GPS error distributions for single fix observations under a forest canopy. Results show a lower accuracy for an inexpensive recreational GPS receiver (top) when compared to an expensive GPS receiver optimized for accurate spatial data collection (bottom). Note the differences in average and maximum errors. Errors will vary with equipment and conditions, and should decrease as technology develops.

errors and thus greatly improve the accuracy of GNSS positional measurements. However, differential positioning is not always employed, because single receiver positioning is accurate enough for some applications, and differential positioning requires more time and greater expense.

Differential GNSS positioning entails establishing a *base station* receiver at a known coordinate point (Figure 5-14). The true coordinate location of the base station is typically determined using high-accuracy surveying methods, for example, repeated astronomical observations, highest-accuracy GNSS, or precise ground surveys, as described in Chapter 3.

We use the base station to estimate a portion of the range measurement errors for each position fix. Remember that GNSS is based on a set of range measurements, and these range measurements contain errors. Some of these errors are due to uncertainty in the measured travel times from the satellite to the receiver. These travel time errors, also known as timing errors, are often among the largest sources of positional uncertainty.

In differential correction we use the known base station position to estimate the timing errors and hence range errors. Each GNSS satellite broadcasts its position along with the ranging signal. The "true" distance from a given satellite to the base station can be calculated because the base station and satellite locations are known. However, note the qualifying quote marks around the "true." We can not exactly define where the satellite is, and the base station coordinates have some (usually small) level of uncertainty associated with them. However, if we are very careful about surveying the location of our base station, then the errors in the base-to-satellite measurement are almost always smaller than the range errors contained in our uncorrected timing measurement.

The difference between the true distance and GNSS-measured distance is used to estimate the timing error for a given satellite at any given second. The timing errors change each second, so they should be measured frequently.

Timing corrections may be applied to the range measurements collected by a roving receiver (Figure 5-15). These roving receivers are used to measure GNSS positions at field locations with unknown coordinates. The timing error, and hence range error, for each satellite observed at a field location is assumed to be the same as the range error observed simultaneously at the base station. We adjust the timing of each satellite measurement made by the rover, then calculate the rover's position in the field. This adjustment usually reduces each range error and substantially improves each position fix taken with the roving field receivers.

The timing errors change across the surface of the Earth, and this places a restriction on the use of differential GNSS correction. Our roving receivers must be "near" our base station for differential correction to work. A substantial portion of the range error is due to atmospheric and ionospheric interference with the GNSS signal. Fortunately, these conditions often vary slowly with distance through the atmosphere, so interaction in one location is likely to be similar to interaction, and thus error, in a nearby location. Therefore, as long as the

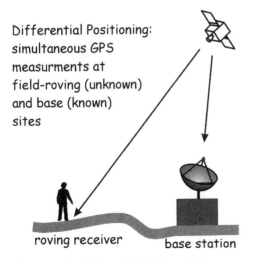

Differential Positioning: simultaneous GPS measurments at field-roving (unknown) and base (known) sites

roving receiver   base station

Figure 5-14: Differential GNSS positioning.

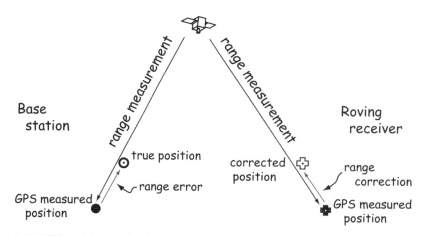

Figure 5-15: Differential correction is based on measuring GPS timing and range errors at a base station, and applying these errors as corrections to simultaneously measured rover positions.

rover is close to the base station we may expect differential correction to improve our position measurements.

Differential correction requires the base station and roving receivers to collect data from a similar set of satellites. We cannot fix a timing error we do not measure. Any four satellites providing acceptable PDOPs will suffice.

The simultaneous viewing requirement limits the distance between the base and roving receivers. The farther apart the receivers, the more likely they will not view the same set of satellites. While differential positioning has been successful at distances of over 1500 kilometers (1000 miles), best performance is achieved when the roving receiver is within 300 kilometers (180 miles) of the base station, and closer is usually better.

Successful differential correction also requires near simultaneity in the base and rover measurements. Errors change rapidly through time. If the base and rover measurements are collected more than a few tens of seconds apart, they do not correspond to the same set of errors, and thus the difference at the base station cannot be used to correct the rover data. Many systems allow data collection to be synchronized to a standard timing signal, thereby ensuring a good match when the error vectors are applied to correct the roving-receiver GNSS data.

Base station data and roving receiver data must be combined for differential correction. A base station correction may be calculated for each fix, but this correction must somehow be joined with the roving receiver data to apply the correction. Many receivers allow large amounts of data to be stored, either internal to the receiver, or in an attached computer. Files may be downloaded from the base station and roving units to a common computer. Software provided by most GNSS system vendors is then used to combine the base and rover data and compute and apply the differential corrections to the position fixes. This is known as post-processed differential correction, as corrections are applied after, or post, data collection (Figure 5-16, top).

Post-processed differential positioning is suitable for many field digitization activities. Road locations may be digitized with a GNSS receiver mounted to the top of a vehicle. The vehicle is driven over the roads to be digitized, and rover data recorded simultaneously with a base station. Base and rover data are downloaded to a process computer and differential corrections computed and applied. These roads data may then be further processed to create a data layer in a GIS.

Post-processing differential positioning has one serious limitation. Because precise positions are not known when the rover is in the field, post-processing technologies are

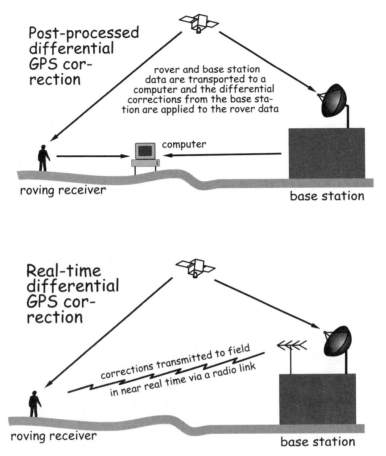

Figure 5-16: Post-processed and real-time differential GPS correction.

useless when precise navigation is required. A surveyor recovering buried or hidden property corners often needs to navigate to within a meter of a position while in the field, so that monuments, stakes, or other markers may be recovered. When using post-processed differential GNSS the field receiver is operating as an autonomous positioning device, and accuracies of a few meters to tens of meters are expected. This is not acceptable for many navigation purposes because too much time will be spent searching for the final location when one is close to the destination. Alternative methods, described in the next section, provide more precise in-field position determination.

Real-Time Differential Positioning

An alternative GNSS correction method, known as *real-time differential correction*, may be appropriate when precise navigation is required. Real-time differential correction requires some extra equipment and there is some cost in slightly lower accuracy when compared to post-processed differential GNSS. However, the accuracy of real-time differential correction is substantially better than autonomous GNSS, and accurate locations are determined while still in the field.

Real-time differential GNSS positioning requires a communications link between the base station and the roving receiver (Figure 5-16, bottom). Typically the base station is connected to a radio transmitter and an antenna. FM radio links are often used due to their longer range and good transmission through vegetation, into canyons or deep valleys, or into other constrained terrain. The base station collects a GNSS signal and calculates range distances. The error is calculated for each range distance. The magnitude and direction of each error is passed to the radio transmitter, along with information on the timing and satellite constellation used. This continuous stream of corrections is broadcast via the base-station radio and antenna.

Roving GNSS receivers are outfitted with a receiving radio, and any receiver within the broadcast range of the base station may receive the correction signal. The roving receiver is also recording GNSS data and calculating position fixes. Each position fix by the roving receiver is matched to the corresponding correction from the base-station radio broadcast. The appropriate correction is then applied to each fix and accurate field locations are computed in real time.

Real-time differential correction requires a broadcasting base station; however, every user is not required to establish a

base station and radio or other communication system. The U.S. Coast Guard has established a set of GPS *radio beacons* in North America that broadcast a standardized correction signal (Figure 5-17). Any roving GPS receiver within the broadcast range of these radio beacons can use the standardized signal for differential correction. These GPS *beacon receivers* typically have an additional antenna and electronics for receiving and decoding the radio beacon signal. The radio beacons were originally intended to aid in ship navigation and control in coastal and major inland waters, so beacons are concentrated near the Atlantic, Pacific, and Gulf Coasts of the United States, the Great Lakes/St. Lawrence Seaway, and along the Mississippi River. This system has become the Maritime Differential GPS systems, and is part of a National Differential GPS system (NDGPS) under development with collaboration of Federal Departments of Transportation, Homeland Security, and others. This will support navigation and positioning in areas distant from the Coast Guard network. Many GPS manufacturers sell beacon

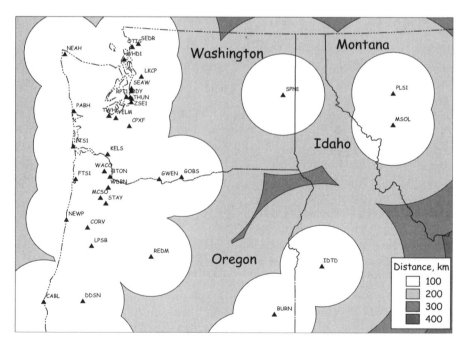

Figure 5-17: The location of GPS radio beacons (triangles) in the Pacific Northwest of the U.S, on June, 2004. Station locations and names and distances from the nearest beacon are shown.

receiver packages that support real-time correction using the beacon signal.

WAAS and Satellite-based Corrections

There are alternatives to ground-based differential correction for improving the accuracy of GNSS observations. One alternative, known as the Wide Area Augmentation System (WAAS), is administered by the U.S. Federal Aviation Administration to provide accurate, dependable aircraft navigation. WAAS is designed to provide real-time accuracies from single fixes to within seven meters or less. Accuracies should be better when multiple fixes are averaged.

WAAS is based on a network of ground reference stations scattered about North America, and it works only with the NAVSTAR GPS. Signals from GPS satellites are received at each station and errors calculated, as with differential GPS. A generalized correction is calculated based on location and this correction is transmitted to a geo-stationary satellite. The correction signal may then be broadcast. The correction is collected by WAAS-compatible roving receivers and applied to each position measurement. Preliminary tests indicate individual errors are less than 7 meters 95% of the time, and average errors when collecting data for 30 minutes are between 1 and 3 meters. This is a substantial improvement over uncorrected C/A code, in which errors above 15 meters are common.

Real Time Kinematic and Virtual Reference Stations

The highest accuracy differential correction is provided with dual-frequency, carrier phase positioning, often called *real-time kinematic* (RTK) GNSS. The amount of ionospheric delay is different for different frequencies, so by comparing signals, such as the GPS carriers L1 and L2, the ionospheric delays may be estimated and removed. The CORS and WAAS differential positioning systems described so far are primarily single frequency, and so less accurate than a rigorous dual-frequency system. While single-frequency positions collected for periods of less than an hour are typically in error by tens of centimeters (a half-foot) or more, dual-frequency GNSS are often accurate to a few centimeters (an inch) or better.

There are disadvantages to RTK GNSS. The receivers are more expensive, often three-times the cost of an a single-frequency GPS system and 30 to 50 times the cost of a common, WAAS-enabled handheld receiver, although prices are dropping. The roving RTK receiver must be closer to the base station for highest accuracies, typically within 10s of kilometers, and usually less than 100s, in comparison to several hundred kilometers for single-frequency differential correction. This requires either a denser network of base stations, or that the RTK users set up and maintain their own base station for each project. Finally, as with all carrier-phase positioning, satellite signals must be continuously tracked for longer periods, although modern receivers have reduced this time to a few to tens of minutes, as opposed to hours required when the technology was first developed.

RTK is such a powerful technology that many state governments are establishing a dense constellation of dual-frequency receivers in a *Virtual Reference Station* network (VRS). Stations are spaced in a network over some region such that a roving receiver is never more than an acceptable distance from a base (Figure 5-18). The systems provide dual-frequency base-station data broadcast in a standard way over radio frequency, along with base station information. A roving dual-frequency receiver may identify the closest or best local receiver, and compare base to collected signals to obtain positions to within a few centimeters, in real time, in the field.

A Caution on Datums

Errors may be easily injected into GNSS data because insufficient attention is paid to datums. This is in part due to the way GNSS

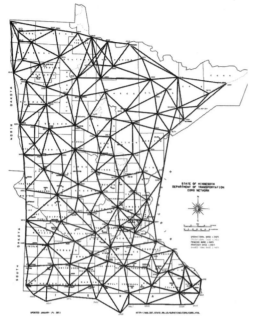

Figure 5-18: The distribution of stations in the Minnesota Department of Transportation VRS network. A station is located at each vertex in the network, ensuring close proximity for any roving receiver within the network (courtesy MNDOT).

satellite systems have been developed, and in part due to design decisions by the GNSS receiver and software manufacturers. One must be cautious in using GNSS data, either directly, or after applying a projection or differential correction because the datum transformation used is often not transparent, and selecting the incorrect transformation may cause positional error. The U.S. NAVSTAR GPS system provides a good example of the confusion that may occur.

GPS satellite locations are reported in the most current WGS84 datum. This datum is rarely used as the basis for GIS data because most older data and thus projects are in national or local datums, for example NAD83(HARN), and because there is not a dense network of points with accurate WGS84 coordinates that are readily found in the field by surveyors and other users. GPS data are typically transformed to a coordinate system that does not use the WGS84 datum. As noted in Chapter 3, ignoring or selecting the wrong datum transformation will introduce error into this process. The

vendors of GPS hardware and software typically provide an option to report data in one of these commonly used coordinate systems, for example, the user may set the GPS receiver to display UTM or State Plane coordinates, and write these to point, line, or area features collected in files. However, the GPS receiver and software vendors often do not specifically identify the datum transformation used. As noted in Chapter 3, early versions of the NAD83 datum that underlie the UTM system were nearly identical to the original WGS84 datum. However, these have diverged over time, and because there is as much as a meter difference between later versions, significant error may be injected into the collected data. Errors may be tens of meters or worse for different transformations.

Confusion may be introduced during differential correction. Here, base station coordinates define time-varying corrections. These coordinates may be based on a datum different from the WGS used for GPS data, hence the data may be reported or corrected across datums. Appropriate transformations between datums must be applied to maintain accuracy. For example, the CORS network of GPS stations is a common source of base data for differential corrections. The coordinates for these base stations are typically reported in the most recent CORS realization of the NAD83 datum, or in an ITRF datum. An appropriate datum transformation must be applied when using these as a base for correction if the highest accuracies are to be maintained. As noted earlier, the differences between the later NAD83(CORS) datums and most recent realizations of WGS84 are typically less than a meter, so introduced errors may be small relative to the accuracy required for intended analyses. However, some projects require submeter accuracy, and for some conditions the errors may be quite large, to tens or hundreds of meters, depending on the datums and projections involved. These errors may be avoided at little cost with the application of appropriate knowledge, typically provided in the vendor's documentation.

Optical and Laser Coordinate Surveying

Historically, coordinate surveys from optical instruments such as transits, theodolites, and electronic distance meters were the primary means of collecting geographic data. While these methods are slowly being replaced by satellite-based positioning and ground-based lasers, they are still quite common, and any competent GIS user should be familiar with optically-based, field surveying methods. Spatial data layers are often produced directly from field surveys, or from field surveys combined with measurements on aerial photographs.

Surveying is particularly common when a highly valued data layer is to be developed or when very precise coordinates are required. Property lines are a good example. Real estate in upscale markets may be valued at hundreds to thousands of dollars per square meter or foot, and buildings, pools, or other improvements add substantial value.

Zoning ordinances often specify the minimum distances between improvements and property boundaries. These factors raise the importance of precisely locating property lines, and justify precise and expensive coordinate surveys (Figure 5-19). Other commonly surveyed features include power lines, fiber-optic cables, sewers, pipelines, and other high valued utilities.

Plane surveying is horizontal surveying based on a planar (flat) surface. The assumption of a flat surface provides a significant computational advantage; the mathematics used to calculate positions in plane surveys are substantially less complicated than those required for geodetic surveys. The flat surface in a plane survey is usually defined by a map projection, with a known point serving as the starting location for the survey.

In plane surveying we typically assume *plumb* lines are perpendicular to the surface

Figure 5-19: Surveying establishes the coordinates for most property lines. Field measurements of distance and direction are used to establish the set of vertices that define property boundary lines. This is the only way to collect these data, as the features are not visible on any other source.

at all points in the survey. A plumb bob or weight is suspended from a string, and is assumed to hang in a vertical direction and intersect the plane surface at a 90° angle. This is a valid assumption when the errors inherent with ignoring the Earth's curvature and density variations are small compared to the accuracy requirements of the survey or to the errors inherent in the survey measurements themselves. The distance error due to assuming a flat rather than curved surface over 10 kilometers (6 miles) is 0.72 centimeters (0.28 inches). Therefore, plane surveys are typically restricted to distances under a few tens of kilometers. This restriction is met in many surveys, and a substantial majority of the lines and points surveyed to date have been measured using plane surveying methods. Plane surveying is sufficient for most subdivisions, public works, construction projects, and property surveys.

Historically, plane surveys have been conducted with optical instruments similar to those described for geodetic surveys. These instruments typically have angle gauges in the horizontal and vertical planes and an optical sight, usually with some degree of telescopic magnification. The instruments go by a number of different names, including, in increasing order of sophistication and capabilities, a level, a transit, a theodolite, and a total station (Figure 5-20).

Distance and angle measurements are the primary field activities in plane surveying. Distances are measured between two *survey stations*, which are points occupied on the ground. The direction is specified by an angle between a standard direction, usually North or South, and the direction of the surveyed line between the two stations (Figure 5-21). The distance is in some standard units, for example, standard international meters.

There are two common ways of specifying angles. The first uses the azimuth. An azimuth angle is measured in a clockwise direction, typically relative to grid or geographic North (Figure 5-21). Angles may also be specified by bearings, which use a

Figure 5-20: A surveying instrument for collecting coordinate geometry data. (courtesy Leica Corporation)

reference direction, an angle direction, and an angular amount (Figure 5-21). The reference direction is either North or South, and the angle direction is East or West. The angle and direction are specified as "N71°E", or "S27°W", as shown in Figure 5-21.

Most surveys have been conducted as *traverses*, a series of connected lines that have a marked beginning and ending point. Traverses typically start at a known control point, or start at a point that has been referenced to a known control point. As described in the preceding sections, the control points are often part of a geodetic control network, or part of a subnetwork established by a municipal surveyor. A distance and angle is measured from the control point to the first survey station. *Coordinate geometry* (COGO) may be used to calculate the station coordinates. Subsequent distance and angle measurements may be taken, and in turn used to calculate the coordinates of subsequent stations. A traverse may be *open*, with a different beginning and ending point, or *closed* with the traverse eventually connecting back to the starting location. Most of the millions of miles of property lines in North America have been established via plane surveys of open and closed traverses.

Coordinate geometry consists of a starting point (a station) and a list of directions (bearings) and distances to subsequent stations. The COGO defines a connected set of points from the starting station to each subsequent station. A sample COGO description follows:

"The starting point is a 1-inch iron rod that is approximately 102.4 feet north and 43.1 feet west of the northeast quarter of the southeast quarter section of section 16 of Township 24 North, Range 16 East, of the 2nd Principal Meridian. Starting from the said point, thence 102.7 feet on a bearing north 72.3 degrees east, to a 1-inch iron pipe; thence 429.6 feet on a bearing south, 64.3 degrees east to a 2-inch iron pipe....."

Basic trigonometric functions are used to calculate the coordinates for each survey station. These stations are located at the vertices that define lines or areas of interest. In the past, these distance and bearing data were manually plotted onto paper maps. Most survey data are now transferred directly to spatial data formats from the surveying instrument or associated software.

Field measurements may be directly entered and coordinate locations derived in the GIS software, or the coordinate calculations may be performed in the surveying instrument first. Many current surveying instruments contain an integrated computer and provide for digital data collection and storage. Coordinates may be tagged with attribute data in the field, at the measurement location. These data are then downloaded directly from a coordinate measuring device to a computer. Specialized surveying programs may be used for error checking and other processing. Many of these surveying packages will then output data in formats designed for import into various GIS software systems.

COGO calculations are illustrated on the left of Figure 5-22. Starting from a known coordinate, x_o, y_o, we measure a distance L and an angle θ. We may then calculate the distances in the x and y directions to another set of coordinates, x_1 and y_1. The coordinates of x_1 and y_1 are obtained by addition of the appropriate trigonometric functions. COGO calculations may then be repeated, using the x_1 and y_1 coordinates as the new starting location for calculating the position of the next traverse station.

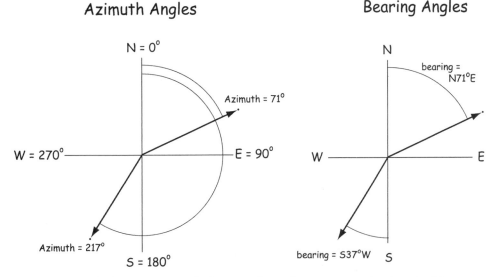

Figure 5-21: Angles in surveys are typically reported one of two ways, as azimuth angles (above left) measured clockwise relative to North, or as bearing angles (above right) measured relative to North or South with the turning direction specified as East or West, e.g., S37°W.

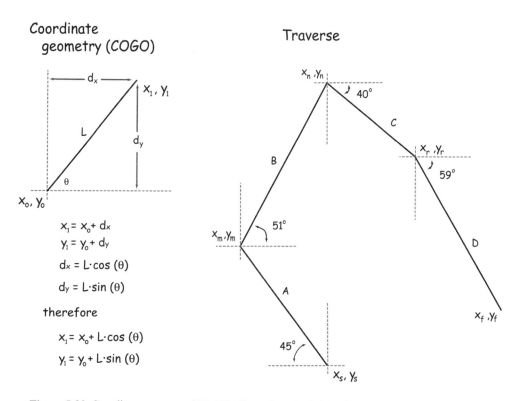

Figure 5-22: Coordinate geometry (COGO) allows the calculation of coordinate locations from open (shown above) or closed traverses. Distance and angle measurements are combined with trigonometric formulas to calculate coordinates.

The right side of Figure 5-22 shows a sequence of measurements for a traverse. Starting at x_s, y_s, the distance **A** and bearing angle, here 45°, are measured to station x_m, y_m. The bearing and distance are then measured to the next station, with coordinates x_n, y_n. Distances and angles are measured for all subsequent stations. Starting with the known coordinates at the starting station, x_s, y_s, coordinates for all other stations are calculated using COGO formulas.

Past survey records of bearings and distances may serve as a source of COGO input. Most of these measurements were recorded for property boundary locations, and are stored in notebooks, on deeds, and in plat books. These paper records must be converted to an electronic format prior to conversion to the coordinate locations. As described above, trigonometric functions may then be used to calculate the position of each station.

Terrestrial, three-dimensional lasers are rapidly becoming common, and although currently used primarily for structure analysis, may be used for GIS data entry. These systems emit narrow, directed laser pulses. By carefully measuring the horizontal and vertical angles relative to the established coordinate system, these laser distance measurements can be converted to three-dimensional coordinates via coordinate geometry. GNSS systems typically provide the location of the laser at the time of data capture, but additional measurements are often necessary to establish and initial or reference pointing direction. These data are then used to generate two- or three-dimensional data layers for spatial data bases.

Terrestrial 3d laser systems collect billions of points, and collections from multiple locations must be combined through three-dimensional reconstructive models to create complete digital representations of real-world objects (Figure 5-23). As software and computer systems improve, three-dimensional terrestrial lasers will become common.

Figure 5-23:Three-dimensional scanning lasers use coordinate geometry to record comprehensive x, y, and z coordinate data for GIS. Here, a scanning laser is shown (foreground) with a portion of a the 3-D laser measurements (left) superimposed bridge on the measured bridge (right). Courtesy Leica Geosystems.

GNSS Applications

Tracking, navigation, field digitizing, and surveying are the main applications of GNSS. Navigation is finding a way or route, and tracking involves noting the location of objects through time. A common example is tracking delivery vehicles in near real time. Large delivery and distribution organizations frequently require information on the location of a fleet of vehicles. Vehicles equipped with a GNSS receiver and a radio broadcast link may report back to a dispatch office every few seconds. In effect, the dispatcher may have a real-time map of the vehicle location. Icons on a digital map are used to represent vehicles, and a quick glance can reveal which vehicle is nearest a delivery or retrieval site, or which driver overly frequents a donut shop.

Navigation is a second common GNSS application. GNSS receivers have been developed specifically for navigation, with digital maps or compasses set into on-screen displays (Figure 5-24). Digital maps may be uploaded to these GNSS receivers from larger databases, and streets, water features, topography, or other spatial data shown as a background. Directions to identified points may be displayed, either as a route on the digital map, or as a set of instructions, for example, directions to turn at oncoming streets. These GNSS receivers and digital maps are extremely specialized GIS systems. These systems are useful when collecting or verifying spatial data, such as to navigate to the approximate vicinity of communications towers for which a set of attribute measurements will be collected.

Field Digitization

Field digitization is a primary application of GNSS in GIS. Data may be recorded directly in the field to update point, line, or area locations. Features are visited or traversed in the field, and an appropriate number of GNSS fixes collected. GNSS receivers have been carried in automobiles, on boats, bicycles, and helmets, or by hand to capture the coordinate locations of points and boundaries (Figure 5-25).

GNSS data are often more accurate than data collected from the highest-quality cartometric maps. For example, differentially corrected C/A code GPS data typically have

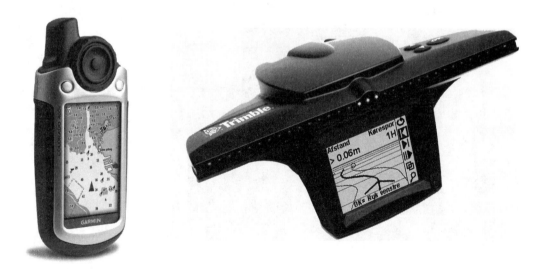

Figure 5-24: A GPS receiver developed for marine (left) and aerial navigation (right), courtesy Garmin Corp. and Trimble Ltd.

Figure 5-25: Line features may be field-digitized via GPS, as in this example of a GIS/GPS system mounted in an automobile. Data display and digitizing software on a portable computer are used to record coordinates collected by a GPS receiver (above). An antenna placed on a pole or rack (right) reduces obstruction (courtesy USDI, left, and G.Johnson, Ducks Unlimited, right).

accuracies better than 5 meters, and often below 2 meters, while accuracies are often near 15 to 20 meters for data collected via manual digitization of 1:24,000-scale images. Precise differential correction of carrier-phase GNSS data often yield centimeter-level accuracies, far better than can be obtained from digitizing all but the largest-scale maps or images.

GNSS is often used to directly digitize new control points. Remember that control points are used to correct and transform map or image data to real-world coordinates. Aerial photographs may be available, and the coordinates may be unknown for features visible on the aerial image. Control points may be difficult or impossible to obtain directly from surveys or from the information plotted on existing maps, particularly when graticule or gridlines are absent. GNSS offers a direct method for measuring the coordinates for potential control points represented on the image or map. Road intersections or other points may be identified and then visited with a GNSS receiver.

GNSS-measured control points are the basis for almost all current projects that perform analytical correction of aerial imagery (see Chapter 6). Most image data are not initially in a map coordinate system, yet images are often particularly useful for developing or updating spatial data. Aerial photographs contain detailed information. However, aerial photographs are subject to geometric distortion. These errors may be analytically corrected through suitable methods (see Chapter 6), but these methods require several control points per image, or at least per project when multiple, overlapping aerial photographs are used. GNSS significantly reduces the cost of control point collection, thereby making single- or multi-photo correction a viable alternative for most organizations that collect spatial data.

Digitization with GNSS often involves the capture of both coordinate and attribute data in the field. Typically the GNSS receiver is activated and detects signals from a set of satellites. A file is opened and position fixes are logged at some fixed rate, such as every two seconds. Attribute data may also be entered, either while the position fixes are being collected, or before or after positional data collection. In some software the position fixes may be tagged or identified. For example, a specific corner may be tagged while digitizing a line. Multiple features may be collected in one file and the identities maintained via attached attributes. Data are processed as needed to improve accuracy, and converted to a format compatible with the GIS system in use. GNSS data

Figure 5-27: Features may be entered and edited in the field using a GPS receiver and appropriate software (courtesy Trimble, Ltd.).

collection and data reduction tools often provide the ability to edit, split, or aggregate collected data, for example, converting multiple fixes into a single point average. These functions may be applied for all position fixes in a file, or for a subset of position fixes embedded in a GNSS file.

GNSS field digitization is most commonly used for collection of point and line features. Lines and points can be unambiguously digitized using GNSS. Multiple position fixes provide higher accuracies, so multiple fixes are often collected for point locations. Multiple position fixes also may be collected for important vertices in line data. However, GNSS data collection for area features suffer from a number of unique difficulties. First, it takes considerable time to traverse an area, so relatively large parcels or many small parcels may be impractical to digitize in the field. Second, the problem of multiple representations of the same boundary occurs when digitizing polygonal features. It is impossible to walk exactly the same line and record the same coordinates when GNSS-digitizing a new polygon that is

adjacent to an already digitized polygon. Attempting to re-trace the common boundary wastes time and provides redundant and conflicting data. The alternative is to digitize only the new lines, and snap to "field-nodes", much as when capturing data using a coordinate digitizer (see Chapter 4). This method is often used, with subsequent editing in a GIS.

Large, field portable displays and advance editing software may be combined with real-time differential correction to improve field digitizing. Tablet computers are available with large color screens (Figure 5-26). Scanned digital images may be displayed with existing digital data. New data may be input via a GNSS receiver, in real time, or via penstrokes on the screen. The operator may digitize new features, edit old ones, or perform some combination of the two while in the field (Figure 5-27). Snapping tolerances, maximum overshoots, and all other digitizing controls may be applied in the field, much like when digitizing on-screen in an office.

GNSS field software are often optimized to streamline the input of attributes that are associated with spatial data. Forms may be developed that provide menus, pick lists, and variable entry boxes in a pre-determined order. These software often improve attribute data accuracy, in part by helping avoid blunders. For example, the entry options for a specific attribute such as fire hydrant color may be restricted to red, green, or yellow from a "pick list," if we know that those are the only possible values. These attribute entry forms also increase completeness, in part by ensuring that every variable is presented to the operator, and these forms also can be configured to show a warning when all variables have not been entered.

Figure 5-26: Ruggedized tablet computers may be carried in the field and used for data entry. Large screens allow efficient display and field editing of spatial data.

Field Digitizing Accuracy and Efficiency

Field GNSS collections are affected by an obstructed sky. Terrain, trees, buildings, or other objects block out portions of the sky, causing temporary interruptions in satellite reception, or forcing the GNSS receiver to estimate position from a constantly changing set of satellites. Maximum collection rates when GNSS digitizing are typically near one fix per second. Obstructions may increase collection intervals to several seconds or minutes.

Obstructions may halt GNSS field digitization entirely if they reduce the number of visible satellites to three or less. Sky obstructions reduce the efficiency of field digitization because more time is spent collecting a given number of fixes, and personnel must wait for the satellite constellation to change when satellites are too few or poorly distributed. Alternately, they may collect fewer positions, thereby reducing positional accuracy.

Reductions in the efficiency of GNSS digitization depend on the nature of the obstruction, the type of equipment, the equipment configuration, and satellite number and position. GNSS signals may pass through foliage when collecting data below a forest canopy, although signals become weaker as they pass through several canopy layers. Satellite signals are blocked by stems and branches, though individual satellites are typically obstructed by stems for relatively short durations. Under dense canopy, the available satellite constellation may change frequently; slightly changing the position of the GNSS antenna, by raising or lowering it, may result in a new constellation of visible satellites. Despite these efforts, efficiency reductions may be substantial, doubling or tripling collection times, but single-fix collection times rarely take longer than a few seconds to minutes when forest canopy is the primary sky obstruction. Collection times will increase correspondingly when multiple fixes are required per feature, as when collecting 100 fixes for each point feature.

Terrain may block satellites, and this becomes a significant problem when the blocked satellites are greater than 15° above the local horizontal plane. Satellites less than 15° above the local horizontal plane are of limited use, even in open conditions, because they exhibit large range errors. Atmospheric interference is magnified at low horizon angles, substantially reducing GNSS accuracy. GNSS receivers designed for GIS data collection typically provide settings that automatically reject satellites below a specified horizon angle.

Terrain obstructions often rise above 15°, such as when mountains, hillslopes, or canyon walls reduce the number of visible GNSS satellites (Figure 5-28). Terrain obstruction often reduces collection efficiencies and accuracies. Because the GNSS signals do not pass through soil, rock, wood, or concrete, any obstructed satellite cannot be used for GNSS positioning. In some instances a short wait may result in a rearrangement of the satellite constellation, such as from point c to point b in Figure 5-28. However, on average, an obstructed sky

results in a reduced constellation of GNSS satellites and lower PDOPs when compared to flat terrain. This problem is particularly vexing in urban settings because the horizon angles change substantially over short distances. This makes it difficult to predict when GNSS satellite coverage will be adequate, and thus plan data collection efforts.

Forest and terrain effects may occur together, further reducing accuracies and decreasing efficiencies. This is a common occurrence in forested, mountainous terrain, and in urban areas with both tall buildings and mature trees. Forest and urban inventories may be substantially hampered by the combination of terrain, building, and canopy obstructions.

The use of a *range pole* is perhaps the easiest, most common, and often most effective method to improve collection efficiency. A range pole is an extendable pole on which a GNSS antenna is mounted. Raising the antenna often provides a better satellite constellation. This may help in non-forest conditions to reduce the horizon angle, for example by raising the antenna above relatively short structures near the collection point.

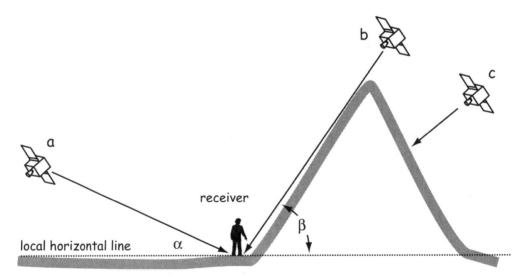

Figure 5-28: GPS satellite signals may be blocked by terrain or built structures. This reduces the constellation of available satellites, increasing error and reducing collection efficiency. Here, satellites a and b are visible with corresponding local horizon angles of α and β. The signal from satellite c is blocked by local terrain.

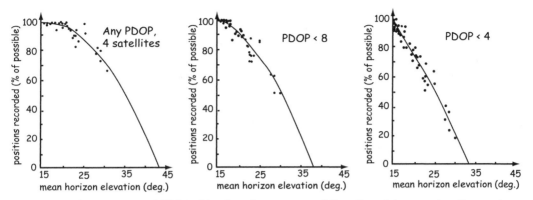

Figure 5-29:The percentage of GPS position fixes that are successfully collected decreases in valleys, or in any other location where the angle to the horizon increases. This may be offset somewhat by allowing poorer (larger) PDOPs, as shown in the leftmost vs. rightmost figures above (adapted from Scrinzi and Floris, 1998).

A range pole is often particularly effective in urban and forested conditions, where canopy gaps and building obstructions vary vertically. A range pole facilitates the search for an acceptable set of satellites. The antenna is raised and lowered during data collection as the satellite constellation changes through time and long pauses are encountered. A range pole is perhaps most useful when digitizing point features or important vertices in line features, when the receiver remains stationary.

Backpack-mounted poles commonly improve efficiency for line and area digitization with GNSS by raising the antenna just a few meters off the ground, avoiding low obstructions such as the body and thick skull of the human carrying the GNSS receiver.

There is often a trade-off between accuracy and efficiency during field digitization, particularly in obstructed locations (Figure 5-29). This series of graphs shows data collected by Scrinzi and Floris (1998), in rough terrain and under forest canopy. As shown in the left-most graph of Figure 5-29, they found that 100% of the possible fixes may be collected when the average horizon angle is near 15°. They also collected data at various points in hilly terrain, where the horizon angle was greater because mountains and ridges block lower portions of the sky. Efficiency dropped to near 70% as average horizon angle increased to near 30° (left-most

graph, Figure 5-29). Collections took about 30% longer or fixes were 30% less frequent when in valley locations compared to flat terrain. However, the left-most graph gives optimistic estimates in that it shows efficiencies when accepting fixes with any PDOP. Since we know accuracy decreases at higher PDOPs, we often set a maximum PDOP threshold. This increases the accuracy and increases collection times when using a GNSS.

The center and right graphs in Figure 5-29 show the increased impact of horizon angle when keeping PDOPs below specified thresholds. The center graph shows that collection efficiency falls off more rapidly when the PDOP threshold is set at eight. Collection efficiency is approximately 50% when horizon angles average 30° -- in other words, it takes approximately twice as long to collect the same amount of data, or approximately one-half the fixes are collected in the same amount of time. These effects are magnified when PDOPs are restricted to less than 4 (Figure 5-29, right graph). Approximately 20% of the possible position fixes were recorded at a 30° average terrain angle and PDOP threshold of 4, suggesting only one in five position fixes will be obtained. While Figure 5-29 was generated with a specific, high quality receiver optimized for field digitization, the general patterns are true for all currently available

GNSS systems -- efficiency decreases in obstructed terrain, and the rate of decrease changes with the allowable PDOP. As GNSS receivers improve, and can measure GPS, GLONASS, and Galileo GNSS simultaneously, efficiencies and accuracies in obstructed environments will substantially increase.

We may improve the efficiency of GNSS digitization by altering PDOP and signal strength thresholds, but this often comes at the expense of decreased positional accuracy. As shown above, accepting higher PDOPs increases the number of fixes recorded, but at lower accuracies. Sophisticated receivers allow multiple settings; a target PDOP, above which the receiver will search for better satellite constellations, and a maximum PDOP, above which data collection will cease. This allows the user to balance the trade-off between accuracy and efficiency.

Some GNSS receivers allow adjustments in the threshold for acceptable signal strength. For example, satellite signals that pass through a forest canopy are weaker. Including these weaker signals improves the number and often the distribution of satellites, thereby increasing collection efficiency and perhaps accuracy. However, weak signals may also result from reflected or multipath transmissions. As described in the section on sources of range error, multi-path signals take a longer than straight-line path to the receiver because they are reflected from nearby solid objects, such as buildings, tree trunks, or the ground. These weaker, reflected signals introduce error through incorrect range measurements. Lowering the threshold for acceptable signal strength is likely to increase positional error, as it increases the likelihood of multi-path measurements. However, some data are often better than none, and lowering the PDOP threshold for collection is sometimes the only way to collect data. In some cases, lowering the PDOP threshold may increase accuracy, if a substantially lower PDOP may be achieved, and the multipath effects are not severe.

GNSS receivers specifically designed for GIS data collection may be fitted with a sensitive antenna that also reduces multipath reception. Manufacturers such as Leica Geosystems, Sokkia, and Trimble have invested substantially in optimizing antenna design and collection systems to control these multiple trade-offs. The availability of specifically optimized antennas is a primary difference between GIS-grade receivers and recreational receivers costing much less. Recreational receivers are substantially less accurate in obstructed terrain, and the user has little control over how data are collected. Recreational receiver thresholds for signal strength or PDOP are often configured for highest efficiency and thus lowest accuracy under obstructed conditions, and these thresholds often may not be adjusted by the GNSS user. Irrespective of the equipment, there is a trade-off between the acceptable

Figure 5-30: A laser rangefinder may substantially improve the efficiency of field data collection with GPS. Here, a system integrates a binocular unit to automatically calculate positions from GPS measurements and distance and angle observations. Also note the rangepole to raise the GNSS antenna above the collector's head (courtesy Leica Geosystems).

signal strength and the introduction of multi-path errors. Setting the PDOP or signal strength thresholds lower will increase efficiency of collection, but often at the cost of increased error.

Rangefinder Integration

There are other limits to GNSS data collection. For example, the need to occupy every vertex and node in the field is a primary drawback of GNSS digitizing. Sometimes it may be dangerous to physically place the GNSS receiver over each point, for example, when a stream to be digitized is in a field full of rutting buffalo. Features may be difficult to reach, costing the user more time in travel than in GNSS data collection. This is particularly common when point features to be digitized are widely dispersed. Features may be numerous, intervisible, but separated by a barrier, for example, a sequence of fence posts or power poles on opposite sides of a limited access highway.

Peripheral measuring devices, such as laser rangefinders, may be attached to GNSS data collectors to substantially improve field data collection (Figure 5-30). These devices typically measure distance with a laser and measure direction with a compass. Measurements are made from each occupied GNSS point to the nearby features of interest. The target coordinate calculations are often automatic because direction is measured with an integrated electronic compass. The rangefinder is pointed at the feature to be digitized. The system calculates the observer's position from the GNSS, and this position is combined with distance and angle measurements in coordinate geometry to calculate the feature coordinates. The person operating the GNSS/laser rangefinder may stand in one location and collect positions for several to tens of features, thereby saving substantial travel time. These systems are most often used to inventory point features such as utility poles, signs, wells, trees, or buildings.

Laser rangefinders are available that can measure features at distances up to 600

meters (2000 feet). Realized accuracies depend on both the quality of the GNSS receiver and the distance measuring subsystem, however, submeter accuracies are possible under open sky conditions.

GNSS Tracking

GNSS tracking of people, vehicles, packages, or animals is an innovative and growing applications of GNSS. GNSS receivers are routinely placed on trucks, ships, buses, boats, or other transport vehicles. These receivers are often part of systems that include information on local conditions, speed of travel, and perhaps the condition of the shipped equipment or cargo.

GNSS tracking for individual or fleets of vehicles typically involves a number of subsystems (Figure 5-31). GNSS receivers and radio transmitters must be placed on each vehicle to record and transmit position. Satellite or ground-based receiver networks

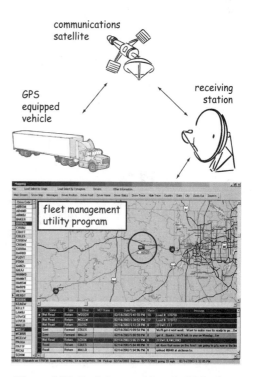

Figure 5-31: Real-time tracking via GPS substantially improves vehicle fleet management, particularly when combined with other data in a GIS.

collect and transmit positional and other data to a computer running a tracking and management program that may be used to display, analyze, and control vehicle movement. Information or instructions may be passed back to vehicles on the road.

GNSS-aided vehicle management may be combined with other spatial data in a GIS framework to add immense value to spatial analyses. Vehicle location can be monitored in real-time, and compared to delivery locations. Delivery planning may be optimized and delivery windows specified with much greater accuracies. This in turn may substantially reduce costs, increase data gathering, and improve profits for participating businesses. Transport may be dispatched more

efficiently, recurring problems analyzed, and solutions more effectively tailored.

GNSS is also increasingly applied to track individual organisms. This is revolutionizing animal movement analysis because of the frequency and density of points that may be collected (Figure 5-32). More position fixes can be collected in a month using GNSS equipment than may be collected in a decade using alternative methods.

Animal movement analysis has long been based on observation of recognizable individuals. Each time a known animal is seen the location is noted. The number of position fixes is often low, however, because some animals are difficult to spot, elusive, or live in areas of dense vegetation or varied terrain. Early alternatives to direct human

Figure 5-32: A wildebeest fit with a GPS tracking collar. The antenna is visible as the white patch on top of the collar, and the power supply and data logging housing is visible at the bottom of the collar. Animal position is tracked day and night, yielding substantially improved information on animal activity and habitat use. (Courtesy Gordon T. Carl, Lotek Wireless Inc.)

observations were based primarily on *radio-telemetry*. Radiotelemetry involves the use of a transmitting and receiving radio unit to determine animal location. A transmitting radio is attached to an animal, and a technician in the field uses a radio receiver to determine the position of the animal. Measurements from several directions are combined and the approximate location of the animal may be plotted.

GNSS animal tracking is a substantial improvement over previous methods. GNSS units are fit to animals, usually by a harness or collars (Figure 5-33). The animals are released, and positional information recorded by the GNSS receiver. Logging intervals are variable, from every few minutes to every few days, and data may be periodically downloaded via a radio link. Systems may be set up with an automatic or radio-activated drop mechanism, so that data may be downloaded and the receiver reused. While only recently developed, GNSS-based animal tracking units are currently in

Figure 5-33: A GPS collar used in tracking animal movement. (courtesy Lotek Wireless, Inc.).

Figure 5-34: GPS tracking data for wildebeest in the Serengeti and Ngorongoro Crater regions of Tanzania. Studies of travel routes and habitat use by migratory animals are substantially improved by GPS data collection (courtesy S. Thirgood, A. Mosser, and M. Borner).

use on all continents in the study of threatened, endangered, or important species (Figure 5-34).

Imaginative uses of GNSS are arising almost daily as this technology revolutionizes positional data collection. GNSS equipment has been interfaced with grain harvesting equipment. Grain production is recorded during harvest, so that yield and grain quality is mapped every few meters in a farm field. This allows the farmer to analyze and improve production on a site-specific basis, for example, by tailoring fertilizing applications for each square meter in the field. The mix of fertilizers may change with position, again controlled by a GNSS receiver and software carried aboard a tractor.

Summary

GNSS is a satellite-based positioning system. It is composed of user, control, and satellite segments, and allows precise position location quickly and with high accuracy.

GNSS is based on range measurements. These range measurements are derived from measurements of a broadcast signal that may be either coded or uncoded. Uncoded, carrier-phase signals are the basis for the most precise position determination, but are of limited use for locating features due to measurement requirements. Code-phase measurements are primarily used for feature collection and entry into GIS. Range measurements from multiple satellites may be combined to estimate position.

GNSS positional estimates contain error due to uncertainties in satellite position, atmospheric and ionospheric interference, multipath reflectance, and poor satellite geometry. These uncertainties vary in time and space.

There are a number of ways to ensure the highest accuracy when collecting GNSS data. Perhaps the greatest improvement comes from differentially correcting GNSS positions. Differential correction is based on simultaneous GNSS measurements at a known base location and at unknown field locations. Errors are calculated for each position fix at the base station, and subtracted to the field collections to improve accuracy. Accuracy may also be improved by collecting with low PDOPs, averaging multiple position fixes for each feature, avoiding multipath or low-horizon signals, and using a GNSS receiver optimized for accurate GIS data collection.

GNSS is most commonly used in GIS to digitize features in the field, either for primary data collection, to update existing data, or for secondary data collection to support orthoimage creation. Terrain, buildings, or tree canopy commonly obstruct the sky, leading to reduced accuracy and efficiencies. Modifying PDOP and signal strength thresholds to account for these obstructions may increase collection efficiencies, but often at the expense of reducing accuracies. Specialized antennas and firmware help, and these are commonly available on GIS-grade receivers, but not on commercial receivers.

GNSS receivers are also used for tracking, navigation, and field surveying. Vehicle tracking applications require GNSS, transmission, and interpretation subsystems, and are becoming widely applied. Animal and human movements are increasingly being tracked via GNSS.

Suggested Reading

Abidin, H. (2002). Fundamentals of GPS signals and data. In Bossler, J. (Ed.). *Manual of Geospatial Science and Technology*. Taylor and Francis: London.

Bergstrom, G. (1990). *GPS in forest management. GPS World*, 10:46-49.

Bobbe, T. (1992). Real-time differential GPS for aerial surveying and remote sensing. *GPS World*, 4:18-22.

Deckert, C.J. & Bolstad, P. V. (1996). Forest canopy, terrain, and distance effects on global positioning system point accuracy. *Forest Science*, 62:317-321.

Dominy, N.J., & Duncan, B. (2002). GPS and GIS in an African rain forest: Applications to tropical ecology and conservation. *Conservation Ecology*, 5:537-549.

Dow, J.M., Neilan, R.E, & Rizos, C. (2009). The International GNSS Service in a changing landscape of Global Navigation Satellite Systems. *Journal of Geodesy* 83:191-198.

Dwolatzky, B., Trengove, E., Struthers, H., McIntyre, J.A., & Martinson, N.A. (2006). Linking the global positioning system (GPS) to a personal digital assistant (PDA) to support tuberculosis control in South Africa: a pilot study. *International Journal of Health Geographics*, 5:34

Gao, J., & Liu, Y.S. (2001). Applications of remote sensing, GIS and GPS in glaciology: a review. *Progress in Physical Geography*, 25:520-540.

Gao, J. (2002). Integration of GPS with remote sensing and GIS: Reality and prospect. *Photogrammetric Engineering & Remote Sensing*, 68:447-453.

Kaplan, E.D., & Hegarty, C. J.(2006). Understanding GPS: Principles and Applications. Artech House, Norwood, MA.

Fix, R. A. & Burt, T. P. (1995). Global Positioning Systems: an effective way to map a small area or catchment. *Earth Surface Processes and Landforms*, 20:817-827.

Jagadeesh, G.R., Srikanthan, T. & Zhang, X.D. (2004). A map matching method for GPS based real-time vehicle location. *Journal of Navigation*, 57:429-440.

Johnson, C.E. & Barton, C.C. (2004). Where in the world are my field plots? Using GPS effectively in environmental field studies. *Frontiers in Ecology & the Environment*, 2:475-482.

Hurn, J. (1989). *GPS, a guide to the next utility*. Trimble Navigation Ltd.: Sunnyvale.

Kennedy, M. (1996). *The Global Positioning System and GIS*. Ann Arbor Press:Ann Arbor.

Mintsis, G., Basbas, S., Papaioannou, P., Taxiltaris, C., & Tziavos, N. (2004). Applications of GPS technology in the land transportation system. *European Journal of Operational Research*, 152:399-409.

Naesset, E., & Jonbeister, T. (2002). Assessing point accuracy of DGPS under forest canopy before data acquisition, in the field, and after processing. *Scandinavian Journal of Forest Research*, 17:351-358.

Scrinzi, G., & Floris, A. (1998). Global Positioning Systems (GPS), una nuova realtà nel rilevamento forestale, Atti del Convegno "Nuovi orizzonti per l'assestamento forestale" 14-56

Small, E.D., Wilson, J.S., & Kimball, A.J. (2007). Methodology for the re-location of permanent plot markers using spatial analysis. *Northern Journal of Applied Forestry*, 24:30-36.

Thirgood, S., Mosser, A., Tham, S. Hopcraft, G. Mwangomo, M. Mlengeya, T. Kilewo,M. Fryxell, J., Sinclair,A.R.E., and Borner, M. (2004). Can parks protect migratory ungulates? The case of the Serengeti wildebeest. *Animal Conservation*, 7:113-120.

Welch, R., Remillard, M., & Alberts, J. (1992). Integration of GPS, remote sensing, and GIS techniques for coastal resource management. *Photogrammetric Engineering and Remote Sensing*, 58:1571-1578.

Wilson, J.P., Spangrud, D.S., Nielsen, G.A.,Jacobsen, J.S., & Tyler, D.A. (1998). GPS sampling intensity and pattern effects on computed terrain attributes. *Soil Science Society of America Journal*, 62:1410-1417.

Zygmont, J. (1986). Keeping tabs on cars and trucks. *High Technology*, 18-23.

Study Questions

5.1 - Can you describe the general components of GNSS, including the three common segments and what they do?

5.2 - What is the basic principle behind GNSS positioning? What is a range measurement, and how does it help you locate yourself?

5.3 - Can you describe the GNSS signals that are broadcast, and the basic difference between carrier and coded signals?

5.4 - How many satellites must you measure to obtain a 3-dimensional position fix?

5.5 - What are the main sources and relative magnitudes of uncertainty in GNSS positioning?

5.6 - How accurate is GNSS positioning? Be sure you specify a range, and describe under what conditions accuracies are at the high and low end of the range.

5.7 - What is a dilution of precision (DOP)? How does it affect GNSS position measurements?

5.8 - Which of the following figures depict the lowest and highest PDOPs, assuming the observer is near the center of the drawn surface?

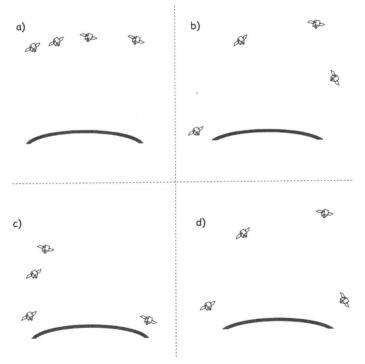

5.9 - Which of the following figures depict the lowest and highest PDOPs, assuming the observer is near the center of the drawn surface?

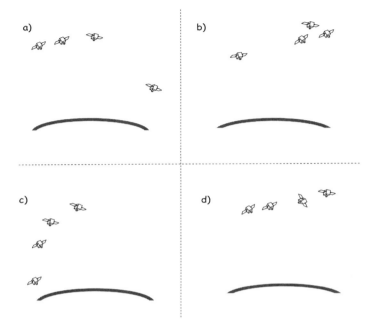

5.10 - Can you describe the basic principle behind differential positioning?

5.11 - What is the difference between post-processed and real-time differential positioning? Can you list three pros and cons of each?

5.12 - How is GNSS accuracy affected by the local terrain horizon? How is it affected by canopy cover or building obstructions? Why does positional accuracy change as these conditions change?

5.13 - How are GNSS data accuracy and efficiency (points collected per given time interval) related when collecting data in obstructed environments? Why? How is this controlled by field personnel?

5.14 - What is WAAS? Is it better or worse than ground-based differential positioning?

5.15 - Why are distance measurements devices and offset used when collecting GNSS data?

5.16 - What is COGO?

5.17 - Complete the table below, calculating missing elements according to the formulas presented in this chapter. Distances are in meters, the angle θ in degrees.

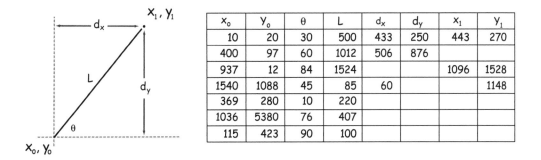

x_0	y_0	θ	L	d_x	d_y	x_1	y_1
10	20	30	500	433	250	443	270
400	97	60	1012	506	876		
937	12	84	1524			1096	1528
1540	1088	45	85	60			1148
369	280	10	220				
1036	5380	76	407				
115	423	90	100				

5.18 - Complete the table below for a traverse with the listed distances and bearings, given as azimuth degrees (drawing a rough sketch may help with the calculations). Note that many spreadsheet, calculator, and trigonometric functions require input in radians (approximately 57.2958 degrees = 1 radian). What is the distance and azimuth from P6 to P0?

Starting point P0, X = 10,128.3, Y = 6,096.4

Point ID	Azimuth	Distance	Delta X	Delta Y	X	Y
P1	32.4	122	65.4	103.0	10,193.7	6,199.4
P2	91.7	207	206.9	-6.1	10,400.6	6,193.3
P3	123.3	305				
P4	212.5	193				
P5	273.9	206				
P6	355.5	145	-60.1	131.9	10,334.9	6,021.6

6 Aerial and Satellite Images

Introduction

Aerial and satellite images are a valuable and common source of data for GIS. These images are data recorded from a distance; thus, photos and satellite images are often referred to as *remotely-sensed* data. Remotely-sensed data come in many forms, however in the context of GIS we usually use the term to describe *aerial images* taken from aircraft using film or digital cameras, or *satellite images* recorded with satellite scanners. Until the 1970s, most mapping images were taken with film and aerial cameras. Digital aerial cameras are now a primary source of images and will soon entirely replace film cameras. In addition, satellite scanners covering a range of resolutions are finding wide use. Whatever their origin, images are a rich source of spatial information and have been used as a basis for mapping for more than seven decades.

Remotely-sensed images are valuable sources of spatial data for many reasons, including:

Large area coverage – images capture data from large areas at a relatively low cost and in a uniform manner (Figure 6-1). For example, it would take months to collect enough ground survey data to accurately produce a topographic map for 10 square kilometers. Images of a region this size may be collected in a few minutes and the topographic data extracted and interpreted in a few weeks.

Extended spectral range – photos and scanners can detect light from wavelengths outside the range of human eyesight. Some kinds of aerial photographs are sensitive to infrared wavelengths, a portion of the light spectrum that the human eye cannot sense. Aerial and satellite scanners sense even broader spectral ranges, up to thermal wavelengths and beyond. This expanded spectral range allows us to detect features or phenomena that appear invisible to the human eye.

Geometric accuracy – remotely-sensed data may be converted to geometrically accurate spatial data. Aerial images are the source of many of our most accurate large-area maps. Under most conditions, aerial images contain geometric distortion due to imperfections in the camera, lens, or film systems, or due to camera tilt or terrain variation in the target area. Satellite scanners may also contain errors due to the imaging equipment or satellite platform. However, distortion removal methods are well established, and provide highly accurate spatial data from images. Cameras and imaging scanners have been developed specifically for the purpose of quantitative mapping. These systems are combined with techniques for identifying and removing most of the spatial error in aerial or satellite images, so spatially accurate data may be collected from images.

Permanent record – an image is fixed in time, so the conditions at the time of the photograph may be analyzed many years hence. Comparison of conditions over multiple dates, or determination of conditions at a specific date in the past are often quite valuable, and remotely-sensed images are often the most accurate source of historical information.

Figure 6-1: Images are a valuable source of spatial data. The upper image, centered on northeastern Egypt, illustrates the broad-area coverage provided by satellite data. The lower image of pyramids in Egypt illustrates the high spatial detail that may be obtained (courtesy NASA, top, and Space Imaging, bottom).

Basic Principles

The most common forms of remote sensing are based on reflected electromagnetic energy. When energy from the sun or another source strikes an object, a portion of the energy is reflected. Different materials reflect different amounts of incoming energy, and this differential reflectance gives objects a distinct appearance. We use these differences to distinguish among objects.

Light is the principal energy form detected in remote sensing for GIS. Light energy is characterized by its *wavelength*, the distance between peaks in the electromagnetic stream. Each "color" of light has a distinctive wavelength, for example, we perceive light with wavelengths between 0.4 and 0.5 micrometers (μm) as blue. Light emitted by the sun is composed of several different wavelengths, and the full range of wavelengths is called the *electromagnetic spectrum*. A graph of this spectrum may be used to represent the amount of incident light energy across a range of wavelengths

(Figure 6-2). The amount of energy in each wavelength is typically plotted against each wavelength value, yielding a curve depicting total electromagnetic energy reaching any object. Notice in Figure 6-2 that the amount of energy emitted by the sun increases rapidly to a maximum between 0.4 to 0.7 μm, and drops off at higher wavelengths. Some regions of the electromagnetic spectrum are named: X-rays have wavelengths of approximately 0.0001 μm, visible light is between 0.4 and 0.7 μm, and near-infrared light is between 0.7 and 1.1 μm.

Our eyes perceive light in the visible portion of the spectrum, between 0.4 and 0.7 μm. We typically identify three base colors: blue, from approximately 0.4 to 0.5 μm, green from 0.5 to 0.6 μm, and red from 0.6 to 0.7 μm. Other colors are often described as a mixture of these three colors at varying levels of brightness. For example, an equal mixture of blue, green, and red light at a high intensity is perceived as "white" light. This same mixture but at lower intensities pro-

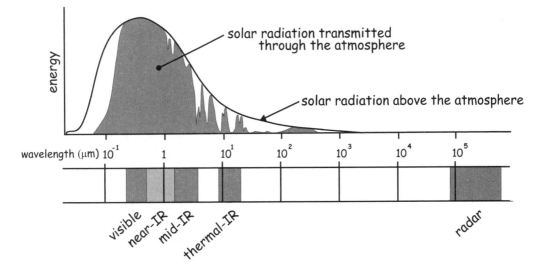

Figure 6-2: Electromagnetic energy is emitted by the sun and transmitted through the atmosphere (upper graph). Solar radiation is partially absorbed as it passes through the atmosphere. This results in variable surface radiation in the visible and infrared (IR) wavelength regions (lower graph).

duces various shades of gray. Other colors are produced with other mixes, for example, equal parts red and green light are perceived as yellow. The specific combination of wavelengths and their relative intensities produce all the colors visible to the human eye.

Electromagnetic energy striking an object is reflected, absorbed, or transmitted. Most solid objects absorb or reflect incident electromagnetic energy and transmit none. Liquid water and atmospheric gasses are the most common natural materials that transmit light energy as well as absorb and reflect it.

Energy transmittance through the atmosphere is most closely tied to the amount of water vapor in the air. Water vapor absorbs energy in several portions of the spectrum, and higher atmospheric water content results in lower transmittance. Carbon dioxide, other gasses, and particulates such as dust also contribute to atmospheric absorption, attenuating radiation in portions of the electromagnetic spectrum (Figure 6-2).

Natural objects appear to be the color they most reflect, for example, green leaves absorb more red and blue light and reflect more green light. Our eyes sense these differences in reflectance properties across a

range of wavelengths to distinguish among objects. While we perceive differences in the visible wavelengths, these differences also extend into other portions of the electromagnetic spectrum that we cannot perceive (Figure 6-3). For example, individual leaves of many plant species appear to be the same shade of green, however some reflect much more energy in the infrared portion of the spectrum, and thus appear to have a different "color" when viewed at infrared wavelengths.

Most remote-sensing systems are *passive*, in that they use energy generated by the sun and reflected off of the target objects. Aerial images and most satellite data are collected using passive systems. The images from these passive systems may be affected by atmospheric conditions in multiple ways. Figure 6-4 illustrates the many paths by which energy reaches a remote sensing device. Note that only one of the energy paths is useful, in that only the surface-reflected energy provides information on the features of interest. The other paths result in no or only diffuse radiation reaching the sensor, and provide little information about the target objects. Most passive systems are not useful during cloudy or extremely hazy periods because nearly all the energy is scattered

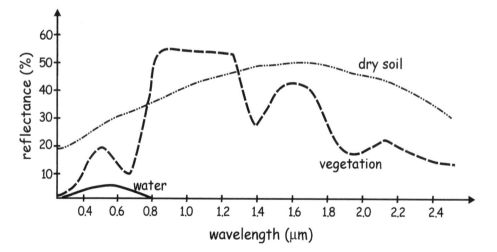

Figure 6-3: Spectral reflectance curves for some common substances. The proportion of incoming radiation that is reflected varies across wavelengths (adapted from Lillesand and Kiefer, 1999).

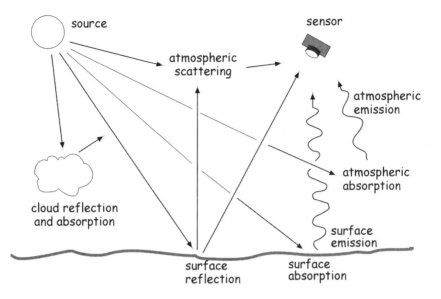

Figure 6-4: Energy pathways from source to sensor. Light and other electromagnetic energy may be absorbed, transmitted, or reflected by the atmosphere. Light reflected from the surface and transmitted to the sensor is used to create an image. The image may be degraded by atmospheric scattering due to water vapor, dust, smoke, and other constituents. Incoming or reflected energy may be scattered.

and no directly reflected energy may reach the sensor. Most passive systems rely on the sun's energy, so they have limited use at night.

Active systems are an alternative for gathering remotely sensed data under cloudy or nighttime conditions. Active systems generate an energy signal and detect the energy returned. Differences in the quantity and direction of the returned energy are used to identify the type and properties of features in an image. Radar (radio detection and ranging) is the most common active remote sensing system, while the use of LiDAR systems (light detection and ranging), is increasing. Radar focuses a beam of energy through an antenna, and then records the reflected energy. These signals are swept across the landscape, and the returns are assembled to produce a radar image. Because a given radar system is typically restricted to one wavelength, radar images are usually monochromatic (in shades of gray). These images may be collected day or night, and most radar systems penetrate clouds because water vapor does not absorb the relatively long radar wavelengths.

Aerial Images

Images taken from airborne cameras are and have historically been a primary source of geographic data. Aerial photography quickly followed the invention of portable cameras in the mid-19th century, and became a practical reality with the development of dependable airplanes in the early 20th century (Figure 6-5). *Photogrammetry*, the science of measuring geometry from images, was well-developed by the early 1930s, and there have been continuous refinements since. Aerial images underpin most large-area maps and surveys in most countries. Digital mapping cameras have been become common in the 21st century, largely supplanting aerial cameras. Aerial images are routinely used in urban planning and management, construction, engineering, agriculture, forestry, wildlife management, and other mapping applications.

Although there are hundreds of applications for aerial images, most applications in support of GIS may be placed into three main categories. First, aerial images are often used as a basis for mapping, to measure and identify the horizontal and vertical locations of objects. Measurements on images offer a rapid and accurate way to obtain geographic coordinates, particularly when image measurements are combined

with ground surveys. In a second major application, image interpretation may be used to categorize or assign attributes to surface features. For example, images are often used as the basis for landcover and infrastructure mapping, and to assess the extent of fire, flood, or other damage. Finally, images are often used as a backdrop for maps of other features, as when photographs are used as a background layer for soil survey maps produced by the U.S. National Resource Conservation Service.

Camera Aircraft, Formats and Systems

Aerial camera systems are most often specifically designed for mapping, so the camera and components are built to minimize geometric distortion and maximize image quality. Mapping cameras have features to reduce image blur due to aircraft motion, enhancing image quality. They maintain or record orientation angles, so distortions can be minimized. These camera systems are precisely made, sophisticated, highly specialized, and expensive, and images suitable for accurate mapping are rarely collected with non-mapping cameras.

Mapping cameras are usually carried aboard specialized aircraft designed for photographic mapping projects (Figure 6-6). These aircraft typically have an instrument bay or hole cut in the floor, through which the camera is mounted. The camera mount and aircraft control systems are designed to maintain the camera optical axis as near vertical as possible. Aircraft navigation and control systems are specialized to support aerial photography, with precise positioning and flight control.

Aerial cameras for spatial data collection are large, expensive, sophisticated devices, but in principle they are similar to simple cameras. A simple camera consists of a lens and a body (Figure 6-7). The lens is typically made of several individual glass

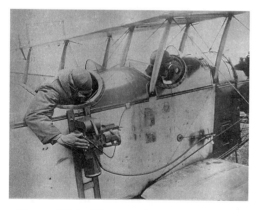

Figure 6-5: Aerial surveys began shortly after the development of reliable airplanes and portable, film-based cameras (courtesy Canadian Government Photographic Archives).

Figure 6-6: Aerial photographs are often taken from specialized aircraft, such as this low altitude airplane, or from helicopters or higher-flying, larger aircraft (courtesy Seabird Ltd.)

elements, with a *diaphragm* or other mechanism to control the amount of light reaching the *sensing media*, the digital sensor or film that records light. These sensors have a characteristic dimensions, **sd**, and for digital senors, a pixel size, that when combined with the flying height (**H**), and focal length (**h**), determine the ground resolution and imaged area. An exposure control, such as a *shutter* within the lens, controls the length of time the film is exposed to light. Cameras also have an *optical axis*, defined by the lens and lens mount. The optical axis is the central direction of the incoming image, and it is precisely oriented to intersect the sensor in a perpendicular direction. Digital sensors are connected to electronic storage, so that successive images may be saved, while film is typically wound on a supply reel (unexposed film) and a take-up reel (exposed film). Images are recorded on a flat stage called the camera's *focal plane*, perpendicular to the optical axis. The time, altitude, and other conditions or information regarding the photographs or mapping project may be recorded by the camera, often as an electronic *header* on digital image files, or on the *data strip* for film cameras, a line of text in the margin of the photograph.

Image scale and *extent* are important attributes of remotely sensed data. Image scale, as in map scale, is defined as the relative distance on the image to the corresponding distance on the ground. For example, one inch on a 1:15,840-scale photograph corresponds to 15,840 inches on the Earth's surface. As shown in Figure 6-7, image scale will be h/H, the ratio of focal length to flying height.

Image extent is the area covered by an image, and depends on the physical size of the sensing area or element (**sd** in Figure 6-7), the camera focal length (**h**), and the flying height (**H**), according to:

$$gd = sd * H / h \qquad (6.1)$$

The extent depends on the physical size of the recording media, sc, (e.g., 5 x 5 cm digital sensor), and the lens system and flying height. For example, a 5 cm sensing element with a 4 cm focal length lens flown at 3000 meters height (about 10,000 ft) results in an extent of approximately 3.75 by 3.75 square kilometers, or 5.1 square miles on the surface of the Earth.

Image resolution is another important concept. The resolution is the smallest object that can reliably be detected on the image.

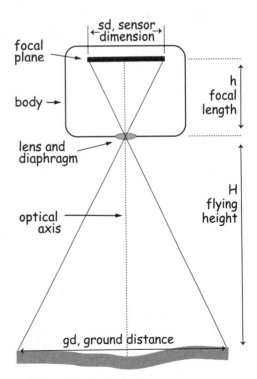

Figure 6-7: A simple camera.

Resolution in digital cameras is often set by the pixel size, the size of individual sensing elements in an array. For example, a 5 x 5 cm array with 7000 cells in each direction will have a cell size of 7.1 x 10^{-6} m, or 7.13 μm.

The realized, or ground resolution on aerial images may be approximately calculated from equation (6.1), substituting cell dimension for sensor dimension, sd. In our example, if the camera has a 5 cm (0.05m) focal length, and is flown at 3000m, the ground resolution is:

$$0.42m = 7.1 \times 10^{-6} * 3000 / 0.05 \quad (6.2)$$

Resolution in aerial photographs is more complicated, and depends on film grain size and exposure properties, and is often tested via photographs of alternating patterns of black and white lines. At some threshold of line width, the difference between black and white lines cannot be distinguished.

Digital Aerial Cameras

Digital aerial cameras are the most common systems used for aerial mapping, and routinely provide high-quality images (Figure 6-8). Film cameras were most common for the 1920s through the mid 1990s, but we are nearing the end of a transition from film to digital cameras. Digital aerial cameras provide many advantages over film cameras, including greater flexibility, easier planning and execution, greater stability, and direct to digital output. While many film cameras are still in use today, camera production has effectively ceased, and film will soon follow.

Digital cameras typically consist of an electronic housing which sits atop a lens assembly (Figure 6-9). The lens focuses light onto charge-coupled devices (CCDs),

Figure 6-8: Digital images may provide image quality equal to or better than film images. This figure shows images collected at 15 cm resolution. Extreme detail is visible, including roof vents, curb locations, and power transmission wires (courtesy Washington County, MN).

or similar electronic scanning elements. A CCD is a rectangular array of *pixels*, or picture elements, that respond to light.

The CCD is comprised of layers of semiconducting material with appropriate reflective and absorptive coatings, insulators, and conducting electrodes (Figure 6-10). Incoming radiation passes through the coatings and into the semiconductors, dislodging electrons and creating a voltage or current. Response may be calibrated and converted to measures of light intensity. Response varies across wavelength, but can be tuned to wavelength regions by manipulating semiconductor composition. Since the pixels are in an array, the array then defines an image.

Digital cameras sometimes us a multilens cluster rather than a single lens, or they may split the beam of incoming light via a prism, diffraction grating, or some other mechanism (Figure 6-11). Since CCDs are typically configured to be sensitive to only a narrow band of light, multiple CCDs may be used, each with a dedicated lens and a specific waveband. Multiple CCDs typically allow more light for each pixel and wave-

Figure 6-9: Digital aerial cameras are superficially similar to film aerial cameras, but typically contain many and more sophisticated electronic components (Courtesy Leica Geosystems).

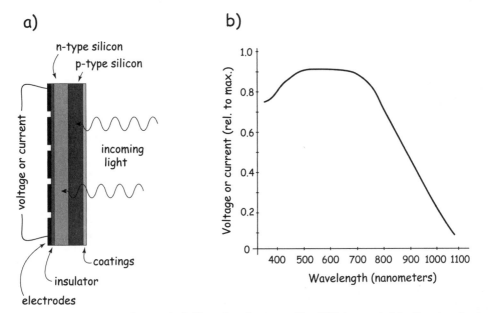

Figure 6-10: CCD response for a typical silicon-based receptor. The CCD is a sandwich of semiconducting layers (a, on left) that generates a current or voltage in proportion to the light received. Response varies over a wavelength region (b, on right).

CCD sensor triplet

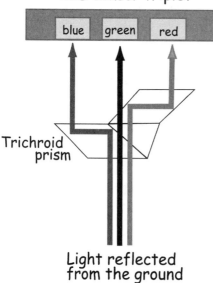

Figure 6-11: Digital cameras often use a prism or other mechanism to separate and direct light to appropriate CCD sensors (adapted from Leica Geosystems).

band, but this increases the complexity of the camera system. If a multi-lens system is used, the individual bands from the multiple lenses and CCDs must be carefully *co-registered*, or aligned, to form a complete multiband image.

Digital cameras most commonly collect images in the blue (0.4-0.5 μm), green (0.5-0.6 μm), or red (0.6-0.7 μm) portion of the electromagnetic spectrum. This provides an image approximately equal to what the human eye perceives. Systems may also record near-infrared reflectance (0.7-1.1 μm), particularly for vegetation mapping. The camera may also have a set of filters that may be placed in front of the lens, e.g., for protection or to reduce haze.

Digital cameras typically have a computer control system, used to specify the location, timing, and exposure, record GPS and aircraft altitude and orientation information, provide data transfer and storage, and allow the operator to monitor progress and image quality during data collection (Figure 6-12).

Digital cameras may have several features to improve data quality. For example, digital cameras may employ electronic image motion compensation, combining information collected across several rows of CCD pixels. This may lead to sharper images (Figure 6-8), while reducing the likelihood of camera malfunction due to fewer moving parts. In addition, digital data may be recorded in long, continuous strips, easing the production of image mosaics.

Film and Film Cameras

While most future aerial images will be collected with digital cameras, there is a vast archive of past aerial images collected with aerial film cameras.These images come in various *formats*, or sizes, usually specified by the edge dimension of the imaged area. Film cameras typically specify their dimensions in physical units, for example, a 240 mm (9-inch) format specifies a square photograph 240 millimeters on a side. Cameras capable of using 240 mm film are considered *large-format* (Figure 6-13), while smaller sizes, e.g., 70mm, were once common. Large-format cameras are most often used to take photographs for spatial data development.

Film consists of a sandwich of light-sensitive coatings spread on a thin plastic sheet.

Figure 6-13: A large-format film camera (courtesy Z/I Imaging Systems).

Film may be black and white, with a single layer of light-sensitive material, or color, with several layers of light-sensitive material. Each film layer is sensitive to a different set of wavelengths (Figure 6-14). These layers, referred to as the *emulsions*, undergo chemical reactions when exposed to light. More light energy falling on the film results in a more complete chemical reaction, and hence a greater film exposure.

Most black and white films are made with a thin emulsion containing silver halide crystals (Figure 6-14). These crystals change when exposed to light. Color films are a more complex combination of several dye layers, each sensitive to a different set of wavelengths. The basic principles of exposure and development apply. Normal color film has blue, green, and red sensitive layers. Color is represented by the differences in densities of color in the dye layers when the film is developed.

Films may be categorized by the wavelengths of light they respond to. Black and white films are sensitive to light in the visible portion of the spectrum, from 0.4 to 0.7 μm and are often referred to as *panchromatic* films. Panchromatic films were widely used for aerial photography because they were inexpensive and could obtain a useful image over a wide range of light conditions. *True color* film is also sensitive to light across the visible spectrum, but in three separate colors.

Infrared films have been developed and were widely used when differences in vegetation type were of interest. These films are sensitive through the visible spectrum and longer infrared wavelengths, up to approximately 0.95 μm.

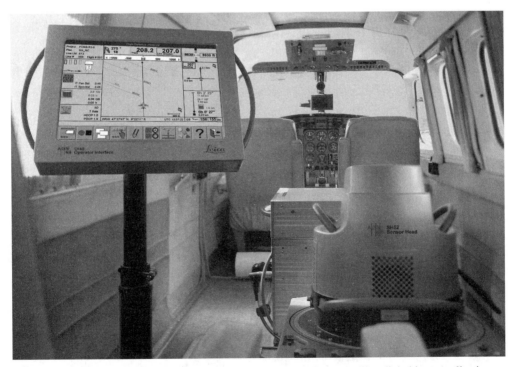

Figure 6-12: An example of the sophisticated system (upper left) for controlling digital image collection, here with a Leica Geosystems ADS40 digital aerial camera (lower right). These systems record and display flight paths and camera stations in real-time, and may be used to plan, execute, and monitor image data collection (courtesy Leica Geosystems).

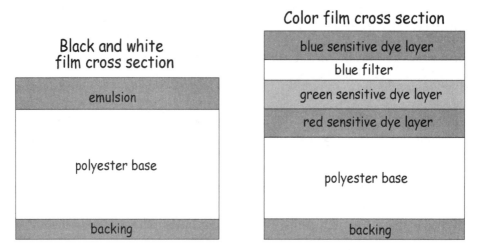

Figure 6-14: Black-and-white and color film are composed of a sandwich of layers. Emulsions on the film change in response to light, resulting in an image. (adapted from Kodak, 1982).

Geometric Quality of Aerial Images

Aerial images are a rich source of spatial information for use in GIS, but most aerial images contain geometric distortion (Figure 6-15). Most geometrically precise maps are orthographic. An orthographic map plots the position of objects after they have been projected onto a common plane, often called a datum plane (Figure 6-16). Objects above or below the plane are vertically projected down or up onto the horizontal plane. Thus, the top and bottom of a building should be projected onto the same location in the datum plane. The tops of all buildings are visible, and all building sides are not (Figure 6-16, left). Except for overhangs, bridges, or similar structures, the ground surface is visible everywhere.

Unfortunately, most aerial images provide a non-orthographic *perspective view* (Figure 6-15 and Figure 6-16, right). Perspective views give a geometrically distorted image of the Earth's surface. Distortion affects the relative positions of objects, and uncorrected data derived from aerial images may not directly overlay data in an accurate

Figure 6-15: Relief distortion is pronounced in vertical aerial images, demonstrated here by buildings that appear to lean outward from the center of the image (courtesy J. Murphy, City of Minneapolis, MN).

orthographic map. The amount of distortion in aerial images may be reduced by selecting the appropriate camera, lens, flying height, and type of aircraft. Distortion may also be controlled by collecting images under proper weather conditions during periods of low wind and by employing skilled pilots and operators. However, some aspects of the distortion may not be controlled, and no camera system is perfect, so there is some geometric distortion in every uncorrected aerial image. The real question becomes "is the distortion and geometric error below acceptable limits, given the intended use of the spatial data?" This question is not unique to aerial images, it applies equally well to satellite images, spatial data derived from GPS and traditional ground surveys, or any other data.

Distortion in aerial images comes primarily from six sources: terrain, camera tilt, film deformation, the camera lens, sensor defects or other camera errors, and atmo-spheric bending. The first two sources of error, terrain variation and camera tilt, are usually the largest sources of geometric distortion when using an aerial mapping camera. The last four are relatively minor when a mapping camera is used, but they may still be unacceptable, particularly when the highest quality data are required. Established methods may be used to reduce the typically dominant tilt and terrain errors and the usually small geometric errors due to lens, camera, and atmospheric distortion.

Camera and lens distortions may be quite large when non-mapping, small-format cameras are used, such as 35 mm or 70 mm format cameras. Small-format cameras can be used for GIS data input, but spatial errors are usually quite large, and great care must be taken in ensuring that geometric distortion is reduced to acceptable levels when using small-format cameras.

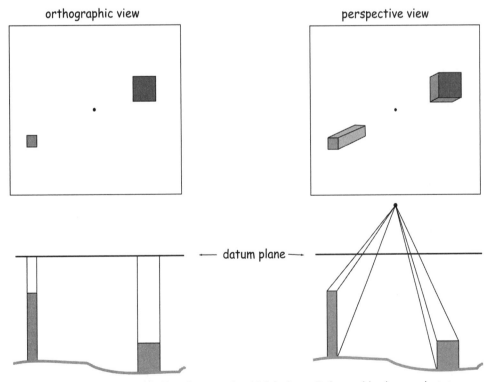

Figure 6-16: Orthographic (left) and perspective (right) views. Orthographic views project at right angles to the datum plane, as if viewing from an infinite height. Perspective views project from the surface onto a datum plane from a fixed viewing location.

Terrain and Tilt Distortion in Aerial Images

Terrain variation, defined as differences in elevation within the image area, is often the largest source of geometric distortion in aerial images. Terrain variation causes *relief displacement*, defined as the radial displacement of objects that are at different elevations.

Figure 6-17 illustrates the basic principles of relief displacement. The figure shows the photographic geometry over an area with substantial differences in terrain. The reference surface (datum plane) in this example is chosen to be at the elevation of the *nadir* point directly below the camera, N on the ground, imaged at n on the photograph. The camera station P is the location of the camera at the time of the photograph. We are assuming a vertical photograph, meaning the optical axis of the lens points vertically below the camera and intersects the reference surface at a right angle at the nadir location.

The locations for points A and B are shown on the ground surface. The corresponding locations for these points occur at A′ and B′ on the reference datum surface. These locations are projected onto the imaging sensor or film, as they would appear in a photograph taken over this varied terrain. In a real camera the sensor is behind the lens, however it is easier to visualize the displacement by showing the sensor in front of the lens, and the geometry is the same. Note that the points a and b are displaced from their reference surface locations, a′ and b′. The point a is displaced radially outward relative to a′, because the elevation at A is higher than the reference surface. The displacement of b is inward relative to b′, because B is lower than the reference datum.

Note that any points that have elevations exactly equal to the elevation of the reference datum will not be displaced, because the reference and ground surfaces coincide at those points.

Figure 6-17: Geometric distortion on an aerial photograph due to relief displacement. P is the camera station, N is the nadir point (adapted from Wolf, 1983, and Lillesand and Kiefer, 1999).

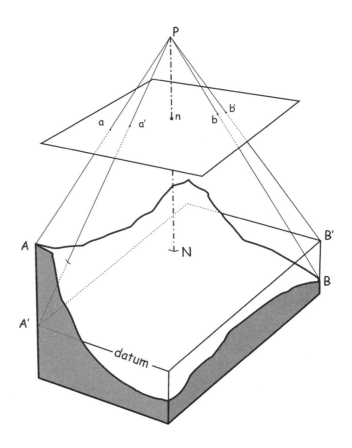

Figure 6-17 illustrates a few key characteristics of terrain distortion in vertical aerial images:

Terrain distortions are radial – higher elevations are displaced outward, and lower elevations displaced inward relative to the center point.

Relief distortions affect angles and distances on an image- relief distortion changes the distances between points, and will change most angles. Straight lines on the ground will not appear to be straight on the image, and areas will expand or shrink.

Scale is not constant on aerial images - scale changes across the photograph and depends on the magnitude of the relief displacement. We may describe an average scale for a vertical aerial photograph over varied terrain, but the true scale between any two points will often differ.

A vertical aerial image taken over varied terrain is not orthographic. We cannot expect geographic data from terrain-distorted images to match orthographic data in a GIS. If the distortions are small relative to digitizing error or other sources of geometric error, then data may appear to match data from orthographic sources. If the relief dis-

Figure 6-18: An example of tilt convergence. Crop bands of equal width appear narrower towards the horizon in this highly tilted image (courtesy USDA).

placement is large, it will add significant errors.

Camera tilt may be another large source of positional error in aerial images. Camera tilt, in which the optical axis points at a non-vertical angle, results in complex *perspective convergence* in aerial images (Figure 6-18). Objects farther away appear to be closer together than equivalently-spaced objects that are nearer the observer (Figure 6-19).

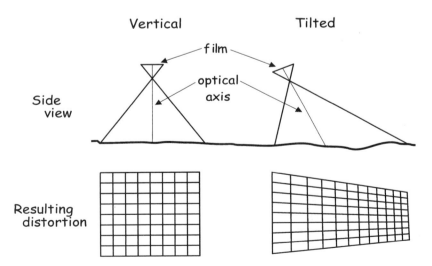

Figure 6-19: Image distortion caused by a tilt in the camera optical axis relative to the ground surface. The perspective distortion, shown at the bottom right, results from changes in the viewing distance across the photograph.

Tilt distortion is zero in vertical photographs, and increases as tilt increases.

Contracts for aerial mapping missions typically specify tilt angles of less than 3 degrees from vertical. Perspective distortion caused by tilt is somewhat difficult to remove, and removal tends to reduce resolution near the edges of the image. Therefore, efforts are made to minimize tilt distortion by maintaining a vertical optical axis when images are collected. Camera mounting systems are devised so the optical axis of the lens points directly below, and pilots attempt to keep the aircraft on a smooth and level flight path as much as possible. Planes have stabilizing mechanisms, and cameras may be equipped with compensating mechanics to maintain an untilted axis. Despite these precautions, tilt happens, due to flights during windy conditions, pilot or instrument error, or system design.

Tilt is often characterized by three angles of rotation, often referred to as omega (ω), phi (ϕ), and kappa (κ). These are angles of rotation about the X, Y, and Z axes that define three dimensional space (Figure 6-20). Rotation about the Z axis alone does not result in tilt distortion, because it occurs around the axis perpendicular with the surface. If ω and ϕ are zero, then there is no tilt distortion. However, tilt is almost always present, even in small values, so all three rotation angles are required to describe and correct it.

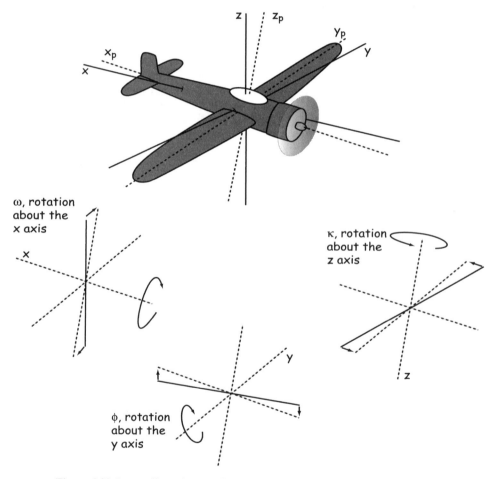

Figure 6-20: Image tilt angles are often specified by rotations about the X-axis (angle ω), the Y-axis (angle ϕ), and the Z-axis (angle κ).

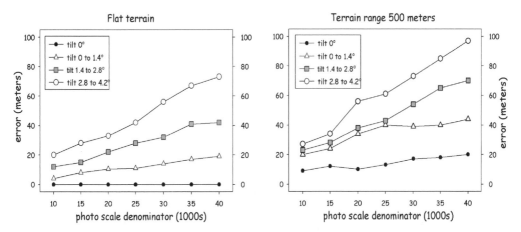

Figure 6-21: Terrain and tilt effects on mean positional error when digitizing from uncorrected aerial images. Distortion increases when tilt and terrain increase, and as photo scale decreases (from Bolstad, 1992).

Tilt and terrain distortion may both occur on aerial images taken over varied terrain. Tilt distortion may occur even on vertical aerial images, because tilts up to 3 degrees are usually allowed. The overall level of distortion depends on the amount of tilt and the variation in terrain, and also on the photographic scale. Not surprisingly, errors increase as tilt or terrain increase, and as photographic scale becomes smaller.

Figure 6-21 illustrates the changes in total distortion with changes in tilt, terrain, and image scale. This figure shows the error that would be expected in data digitized from vertical aerial images when only applying an affine transformation, a standard procedure used to register orthographic maps (see Chapter 4). The process used to produce these error plots mimics the process of directly digitizing from uncorrected aerial images. Note first that there is zero error across all scales when the ground is flat (terrain range is zero) and there is no tilt (bottom line, left panel in Figure 6-21). Errors increase as image scale decreases, shown by increasing errors as you move from left to right in both panels. Error also increases as tilt or terrain increase.

Geometric errors can be quite large, even for vertical images over moderate terrain (Figure 6-21, right side). These graphs clearly indicate that geometric errors will occur when digitizing from vertical aerial images, even if the digitizing system is perfect and introduces no error. Thus the magnitude of tilt and terrain errors should be assessed relative to the geometric accuracy required before digitizing from uncorrected aerial images.

System Errors: Media, Lens, and Camera Distortion

The film, camera, and lens system may be a significant source of geometric error in aerial images. The perfect lens-camera-film system would exactly project the viewing geometry of the target onto the image recording surface, either film or CCD. The relative locations of features on the image in a perfect camera system would be exactly the same as the relative locations on a viewing plane an arbitrary distance in front of the lens. Real camera systems are not perfect and may distort the image. For example the light from a point may be bent slightly when traveling through the lens, or the film may shrink or swell, both causing a distorted image.

Radial lens displacement is one form of distortion commonly caused by the camera system. Whenever a lens is manufactured

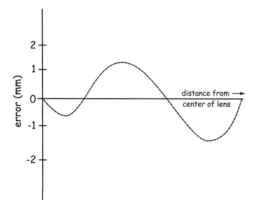

Figure 6-22: An example of a radial lens distortion curve. Negative distortion (inward) and positive distortion (outward) are found at different distances from the center of the lens.

there are always some imperfections in the curved shapes of the lens surfaces. These cause a radial displacement, either inward or outward, from the true image location (Figure 6-22). Radial lens displacement is typically quite small in mapping camera systems, but it may be quite large in other systems. A radial displacement curve is often developed for a mapping camera lens, and this curve may be used to correct radial displacement errors when the highest mapping accuracy is required.

Mapping camera systems are engineered to minimize systematic errors. Lenses are designed and precisely manufactured so that image distortion in minimized. Lens mountings, the platten or CCD, and the camera body are optimized to ensure a faithful rendition of image geometry. Films are designed so that there is limited distortion under tension on the camera spools. This optimization leads to extremely high geometric fidelity in the camera/lens system. Thus, camera and lens distortions in mapping cameras are typically much smaller than other errors, for example, tilt and terrain errors, or errors in converting the image data to forms useful in a GIS.

We also wish to emphasize that camera-caused geometric errors may be quite high when a non-mapping camera is used, such as when photographs are taken with small-for-

mat 35 mm or 70 mm camera system. Lens radial distortion may be extreme, and these systems are likely to have large geometric errors when compared to mapping cameras. That is not to say that mapping cameras must always be used. In some circumstances the distortions inherent in small-format camera systems may be acceptable, or may be reduced relative to other errors, for example, when very large scale photographs are taken, and when qualitative or attribute information are required. However, the geometric quality of any non-mapping camera system should be evaluated prior to use in a mapping project.

Atmospheric distortion is another source of geometric error in aerial images, although it is usually small relative to other errors. Light bends when passing through the atmosphere, and this may result in spatial displacement and hence distortion. Distortion increases with increasing atmospheric depth, and is largest when photographs are taken from extremely high altitudes or at oblique angles through the atmosphere. Under most conditions, atmospheric distortion is quite small relative to other errors, and it is corrected only when the highest spatial accuracies are required.

Stereo Photographic Coverage

As noted above, relief displacement in vertical aerial images adds a radial displacement that depends on terrain heights. The larger the terrain differences, the larger the relief displacement. This relief displacement may be a problem if we wish to produce a map from a single photograph. Photogrammetric methods can be used to remove the distortion. However, if two overlapping photographs are taken, called a *stereopair*, then these photographs may be used together to determine the relative elevation differences. Relief displacement in a stereopair may be used to determine elevation and remove distortion. Many mapping projects collect *stereo photographic coverage* in which sequential photographs in a flight line overlap, called *endlap*, and adjacent flightlines overlap, called *sidelap* (Figure 6-23). Stereo photographs typically have near 65% endlap and 25% sidelap. Some digital cameras collect data in continuous strips and so only collect sidelap.

A *stereomodel* is a three-dimensional perception of terrain or other objects that we see when viewing a stereopair. As each eye looks at a different, adjacent photograph from the overlapping stereopair, we observe a set of parallax differences, and our brain may convert these to a perception of depth. When we have vertical aerial images, the distance from the camera to each point on the ground is determined primarily by the elevations at each point on the ground. We may observe parallax for each point and use this parallax to infer the relative elevation for every point. Stereo viewing creates a

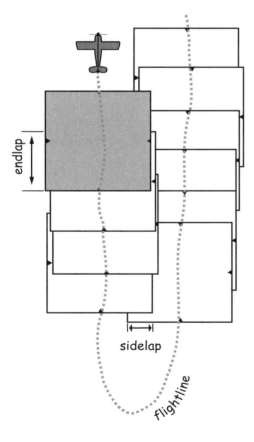

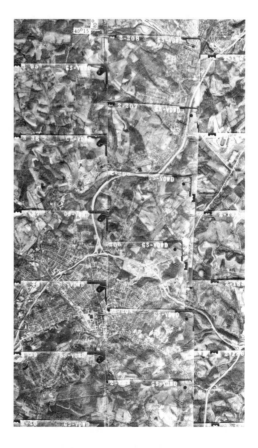

Figure 6-23: Aerial images often overlap to allow three-dimensional measurements and the correction of relief displacement. Sidelap and endlap are demonstrated in the figure (left) and the photomosaic (right).

three-dimensional stereomodel of terrain heights, with our left eye looking at the left photo and our right eye looking at the right photo. The three-dimensional stereomodel can be projected onto a flat surface and the image used to digitized a map. We may also interpret the relative terrain heights on this three-dimensional surface, and thereby estimate elevation wherever we have stereo coverage. We can use stereopairs to draw contour lines or mark spot heights. This has historically been the most common method for determining elevation over areas larger than a few hundred hectares.

Stereomodels are visible in stereopairs due to *parallax*, a shift in relief displacement due to a shift in observer location. Figure 6-24 illustrates parallax. The block (closer to the viewing locations) appears to shift more than the sphere when the viewing location is changed from the left to the right side of the objects. The displacement of any given point is different on the left vs. the right ground views because the relative viewing geometry is different. Points are shifted by different amounts, and the magnitude of the shift depends on the distance from the observer (or camera) to the objects. This shift in position with a shift in viewing location is by definition the parallax, and is the basis of depth perception.

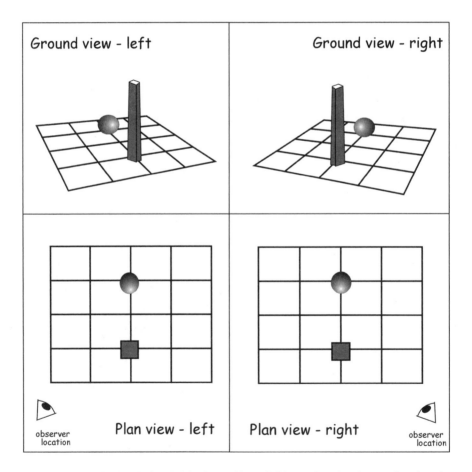

Figure 6-24: Parallax is the relative shift in the position of objects when the observer location changes. The top panels show a ground view and the bottom panels show the plan view (from above) of observer locations. The relative positions of the sphere and the tower change when the observer moves from left to right.

Geometric Correction of Aerial Images

Due to the geometric distortions described above, it should be quite clear that uncorrected aerial images should not be used directly as a basis for spatial data collection under most circumstances. Points, lines, and area boundaries may not occur in their correct relative positions, so length and area measurements may be incorrect. These distortions are a complex mix of terrain and tilt effects, and will change the locations, angles, and shapes of features in the image and any derived data. Worse, when spatial data derived from uncorrected photographs are combined with other sources of geographic information, features may not occur at their correct locations. A river may fall on the wrong side of a road or a city may be located in a lake. Given all the positive characteristics of aerial images, how do we best use this rich source of information? Fortu-

nately, photogrammetry provides the tools needed to remove geometric distortions from photographs (Figure 6-25)

These corrections depend on two primary sets of measurements. First, the location of each image's *perspective center* or *focal center* must be known. This is approximately the location of the camera focal point at the time of imaging. It can be determined from precise GNSS, or deduced from ground measurements. Second, some direct or indirect measurement of terrain heights must be collected. These heights may be collected at a few points, and stereopairs used to estimate all other heights, or they may be determined from another source, e.g., a previous survey, or Radar or LiDAR systems described later in this chapter. Armed with perspective center and height measurements, we may correct our aerial images.

Geometric correction of aerial images involves calculating the distortion at each

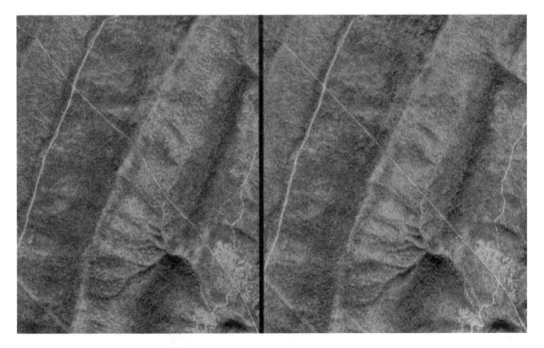

Figure 6-25: An example of distortion removal when creating an orthoimage. A nearly straight pipeline right-of-way spans uncorrected (left) and corrected (right) images, from the lower right to upper left in each image. The path appears bent in the image on the left as it alternately climbs ridges and descends into valleys. Using equations described in this section, these distortions may be removed, resulting in the orthographic image on the right, showing the nearly straight pipeline trajectory (courtesy USGS).

point, and shifting the image location to the correct orthographic position. Consider the tower in Figure 6-26. The bottom of the tower at B is imaged on the photograph at point b, and the top of the tower at point A is imaged on the photograph at point a. Point A will occur on top of point B on an orthographic map. If we consider the flat plane at the base of the tower as the datum, we can use simple geometry to calculate the displacement from a to b on the image. We'll call this displacement d, and go through an explanation of the geometry used to calculate the displacement.

Observe the two similar triangles in Figure 6-26, one defined by the points S-N-C, and one defined by the points a-n-C. These triangles are similar because the angles are equal, that is, the interior angle at n and N are both 90°, the triangles share the angle at C, and the interior angle at S equals the interior angle at a. C is the focal center of the camera lens, and may be considered the location through which all light passes. The film in a camera is placed behind the focal center; however, as in previous figures, the film is shown here in front of the focal center

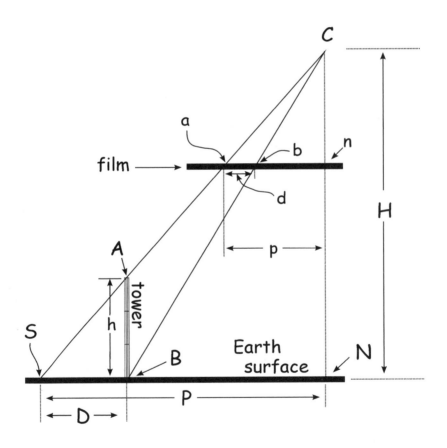

Figure 6-26: Relief displacement may be calculated based on geometric measurement. Similar triangles S-N-C and a-n-C relate heights and distances in the photograph and on the ground. We usually know flying height, H, and can measure d and p on the photograph.

for clarity. Note that the following ratios hold for the similar triangles:

$$D/P = h/H \qquad (6.3)$$

and also

$$d/p = D/P \qquad (6.4)$$

so

$$d/p = h/H \qquad (6.5)$$

rearranging

$$d = p*h/H \qquad (6.6)$$

where:

d = displacement distance

p = distance from the nadir point, n, on the vertical photo to the imaged point a

H = flying height

h = height of the imaged point

We usually know the flying height, and can measure the distance p. If we can get h, the height of the imaged point above the datum, then we can calculate the displacement. We might climb or survey the tower to measure its height, h, and then calculate the photo displacement by Equation (6.6). Relief displacement for any elevated location may be calculated provided we know the height. Heights have long been calculated by measurements from stereopairs, but are increasingly measured using LiDAR, described later in this chapter. These heights and equations are used to adjust the positional distortion due to elevation, "moving" imaged points to an orthographic position.

Equation 6.4 applies to vertical aerial images. When photographs are tilted the distortion geometry is much more complicated, as are the equations used to calculate tilt and elevation displacement. Equations may be derived that describe the three-dimensional projection from the terrain surface to the two-dimensional film plane. These equations and the methods for applying them are part of the science of photogrammetry, and will not be discussed here.

Digital orthophotographs are most often produced using a *softcopy* photogrammetric workstation (Figure 6-27). This method uses digital (softcopy) images, either scanned versions of aerial images or images from a digital aerial camera. *Softcopy photogrammetry* uses mathematical models of photogeometry to remove tilt, terrain, camera, atmospheric, and other distortions from digital images. Control points are identified on sets of photographs, stereomodels developed, and geometric distortions estimated. These distortions are then removed, creating an orthophotograph.

The correction process requires the measurement of the image coordinates and their combination with ground x, y, and z coordinates. Image coordinates may be measured using a physical ruler or calipers, however, they are most often measured using digital

Figure 6-27: A softcopy photogrammetric workstation, used in the production of orthophotographs (courtesy Z/I Imaging Systems).

a)

b)

c)

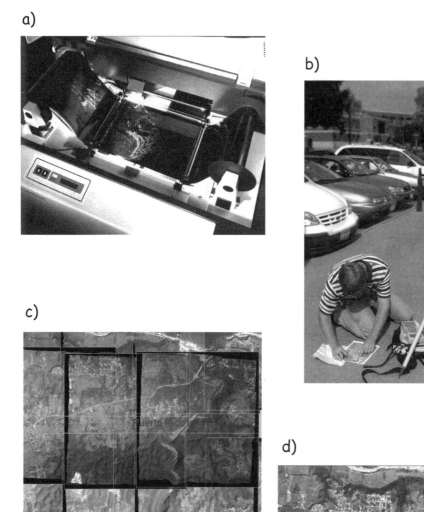

d)

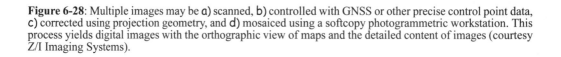

Figure 6-28: Multiple images may be a) scanned, b) controlled with GNSS or other precise control point data, c) corrected using projection geometry, and d) mosaiced using a softcopy photogrammetric workstation. This process yields digital images with the orthographic view of maps and the detailed content of images (courtesy Z/I Imaging Systems).

methods. Typically images are taken with a digital camera, or if taken with a film camera, the images are scanned. Measurements of image x and y are then determined relative to some image-specific coordinate system. These measurements are obtained from one or many images. Ground x, y, and z coordinates come from precise ground surveys.

A set of equations is written that relates image x and y coordinates to ground x, y, and z coordinates. The set of equations is solved, and the displacement calculated for each point on the image. The displacement may then be removed and an orthographic image or map produced. Distances, angles, and areas can be measured from the image. These orthographic images, also known as *orthophotographs* or *digital orthographic images*, have the positive attributes of photographs, with their rich detail and timely coverage, and some of the positive attributes of cartometric maps, such as uniform scale and true geometry.

Multiple images or image strips may be analyzed, corrected, and stitched together into a single mosaic (Figure 6-28).This process of developing photomodels of multiple images at once utilizes interrelated sets of equations to find a globally optimum set of corrections across all images.

Photointerpretation

Aerial images are useful primarily because we may use them to identify the position and properties of interesting features. Once we have determined that the film and camera system meet our spatial accuracy and information requirements, we need to collect the photographs and interpret them. Photo (or image) interpretation is the process of converting images into information. Photo interpretation is a well-developed discipline, with many specialized techniques. We will provide a very brief description of the process. A more complete description may be found in several of the sources listed at the end of this chapter.

Interpreters use the size, shape, color, brightness, texture, and relative and absolute location of features to interpret images (Figure 6-29). Differences in these diagnostic characteristics allow the interpreter to distinguish among features. In the figure, the polygon near the center of the image labeled Pa-C, a pasture, is noticeably smoother than the polygons surrounding it, and the polygon above it labelled As-Y1 shows a finer grained texture, reflecting smaller tree crowns than the polygon labelled NH-M11 above it and to the right. Different vegetation types may show distinct color or texture variations, road types may be distinguished by width or the occurrence of a median strip, and building types may be defined by size or shape.

The proper use of all the diagnostic characteristics requires that the photo-interpreter develop some familiarity with the features of interest. For example, it is difficult to distinguish the differences between many crop types until the interpreter has spent time in the field, photos in hand, comparing what appears on the photographs with what is found on the ground. This "ground-truth" is invaluable in developing the local knowledge required for accurate image interpretation. When possible, ground visits should take place contemporaneously with the photographs. However, this is often not possible, and sites may only be visited months or years after the photographs were collected. The affects of changes through time on the ground-to-image comparison must then be considered.

Photointerpretation most often results in a categorical or thematic map. Identified features are assigned to one of a set of discrete classes. A crop may be corn or soybean, a neighborhood classed as urban or suburban, or a forest as evergreen or deciduous. Mixed classes may be identified, for example, mixed urban-rural, but the boundaries between features of this class and the other finite numbers of categories are discrete.

Photointerpretation requires we establish a target set of categories for interpreted features. If we are mapping roads, we must

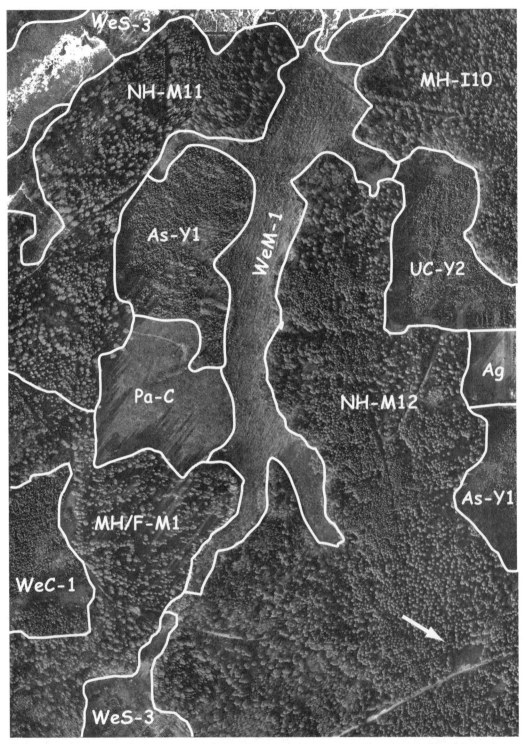

Figure 6-29: Photointerpretation is the process of identifying features on an image. Photointerpretation in support of GIS typically involves digitizing the points, lines, or polygons for categories of interest from a georeferenced digital or hardcopy image. In the example above, the boundaries between different vegetation types have been identified based on the tone and texture recorded in the image. The arrow at the lower right shows an "inclusion area", non-delineated because it is smaller than the minimum mapping unit.

decide what classes to use, for example, all roads will be categorized into one of these classes: unpaved, paved single lane, paved undivided multi-lane, and paved divided multi-lane. These categories must be inclusive, so that in our photos there must be no roads that are multi-lane and unpaved. If there are roads that do not fit in our defined classes, we must fit them into an existing category, or we must create a category for them.

Photointerpretation also requires we establish a *minimum mapping unit*, or MMU. A minimum mapping unit defines the lower limit on what we consider significant, and usually defines the area, length, and/or width of the smallest important feature. The arrow in the lower right corner of Figure 6-29 points to a forest opening smaller than our minimum mapping unit for this example map. We may not be interested in open patches smaller than 0.5 hectares, or road segments shorter than 50 meters long. Although they may be visible on the image, features smaller than the minimum mapping unit are not delineated and transferred into the digital data layer.

Finally, photointerpretation to create spatial data requires a method for entering the interpreted data into a digital form. On-screen digitizing is a common method. Point, line, and area features interpreted on the image may be manually drawn in an editing mode, and captured directly to a data layer. On-screen digitizing requires a digital image, either collected initially, or by scaling a hardcopy photograph.

Another common method consists of interpretation directly from a hardcopy image. The image may be attached to a digitizing board and features directly interpreted from the image during digitizing. This entails either drawing directly on the image or placing a clear drafting sheet and drawing on the sheet. The sheet is removed on completion of interpretation, taped to a digitizing board, and data are then digitized as with a hardcopy map. Care must be taken to carefully record the location of control features on the sheet so that it may be registered.

Satellite Images

In previous sections we described the basic principles of remote sensing and the specifics of image collection and correction using aerial images. In many respects satellite images are similar to aerial images when used in a GIS. The primary motivation is to collect information regarding the location and characteristics of features. However, there are important differences between photographic and satellite-based scanning systems used for image collection, and these differences affect the characteristics and hence uses of satellite images.

Satellite scanners have several advantages relative to aerial imaging systems. Satellite scanners also have a very high perspective, which significantly reduces terrain-caused distortion. Equation 6.4 shows the terrain displacement (d) on an image is inversely related to the flying height (H). Satellites have large values for H, typically 600 kilometers (360 miles) or more above the Earth's surface, so relief displacements are correspondingly small. Because satellites are flying above the atmosphere, their pointing direction (attitude control) is very precise, and so they can be maintained in a near perfect vertical orientation.

There may be a number of disadvantages in choosing satellite images instead of aerial images. Satellite images typically cover larger areas, so if the area of interest is small, costs may be needlessly high. Satellite images may require specialized image processing software. Acquisition of aerial images may be more flexible because a pilot can fly on short notice. Many aerial images have better effective resolution than satellite images. Finally, aerial images are often available at reduced costs from government sources. Many of these disadvantages of using satellite images diminish as more, higher-resolution, pointable scanners are placed in orbit.

Basic Principles of Satellite Image Scanners

Scanners operate by pointing the detectors at the area to be imaged. Each detector has an *instantaneous field of view*, or IFOV, that corresponds to the size of the area viewed by each detector (Figure 6-30). Although the IFOV may not be square and a raster cell typically is square, this IFOV may be thought of as approximately equal to the raster cell size for the acquired image.

The scanner builds a two-dimensional image of the surface by pointing a detector or detectors at each cell and recording the reflected energy. Data are typically collected in the across-track direction, perpendicular

Figure 6-30: A spot scanning system. The scanner sweeps an instantaneous field of view (IFOV) in an across-track direction to record a multispectral response. Subsequent sweeps in an along-track direction are captured as the satellite moves forward along the orbital path.

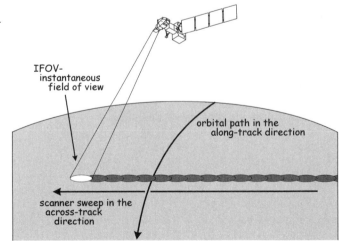

IFOV-instantaneous field of view

orbital path in the along-track direction

scanner sweep in the across-track direction

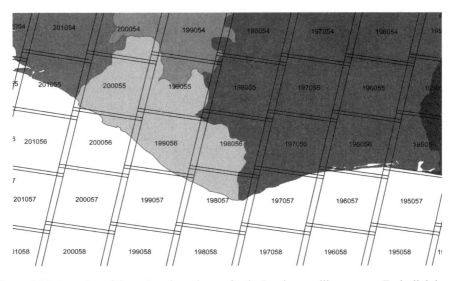

Figure 6-31: A portion of the path and row layout for the Landsat satellite systems. Each slightly overlapping, labelled rectangle corresponds to a satellite image footprint.

to the flight path of the satellite, and in the along-track direction, parallel to the direction of travel (Figure 6-30). Several scanner designs achieve this across- and along-track scanning. Some older designs use a *spot detector* and a system of mirrors and lenses to sweep the spot across track. The forward motion of the satellite positions the scanner for the next swath in the along-track direction. Other designs have a *linear array* of detectors – a line of detectors in the across-track direction. The across-track line is sampled at once, and the forward motion of the satellite positions the array for the next line in the along-track direction. Finally, a *two-dimensional array* may be used, consisting of a rectangular array of detectors. Reflectance is collected in a patch in both the across-track and the along-track directions.

A remote sensing satellite also contains a number of other subsystems to support image data collection. A power supply is required, typically consisting of solar panels and batteries. Precise altitude and orbital control are needed, so satellites carry navigation and positioning subsystems. Sensors evaluate satellite position and pointing direction, and thrusters and other control components orient the satellite. There is a data storage subsystem, and a communications

subsystem for transmitting data back to Earth and for receiving control and other information. All of these activities are coordinated by an onboard computing system.

Several remote sensing satellite systems have been built, and data have been available for land surface applications since the early 1970s. The detail, frequency, and quality of satellite images have been improving steadily, and there are several satellite remote sensing systems currently in operation.

Satellite data are often nominally collected in a path/row system. A set of approximately North-South paths are designated, with approximately East-West rows identified across the paths. Satellite scene location may then be specified by a path/row number (Figure 6-31). Satellite data may also be ordered for customized areas, depending on the flexibility of the acquisition system.

Because most satellites are in near-polar orbits, images overlap most near the poles. Adjacent images typically overlap a small amount near the equator. The inclined orbits are often sun-synchronous, meaning the satellite passes overhead at approximately the same local time.

Sub-meter Satellite Systems

There are a number of operating or planned high-resolution satellite systems. These systems provide resolutions below 1 meter, and record both panchromatic and multispectral data. Commercial systems providing better than 50 centimeter resolution are in the planning and early deployment stages. These resolutions begin to blur the distinction between satellite and photo-based images. While the inherent resolution of many aerial images is much better than 50 centimeters at typical scales, this resolution is often unneeded. Most features in many applications may be accurately identified at 1-meter or coarser resolutions. Spectral range, price, availability, reliability, flexibility, and ease of use may become more important factors in selecting between aerial images and satellite images. Satellite data are replacing aerial images in many projects because satellites may be used to collect data for larger areas more efficiently, satellites can collect data where it is unwise or unsafe

to operate aircraft, and because the data for large areas may be geometrically corrected for less cost and time.

Images from high-resolution satellite systems may provide a suitable source for spatial data in a number of settings. These images provide substantial detail of man-made and natural features, including individual houses, portions of roads, bridges, and other important infrastructure. High resolution systems have been used for construction monitoring, and land use, crop, and fire mapping. All the systems have pointable optics, resulting in short revisit times, on the order of one to a few days.

As of this writing (late 2011) there are five operational systems capable of 1 meter resolution or better: Ikonos, QuickBird and. Worldview-1, Worldview-2, and GeoEye-1. These satellites and related systems are commercial, for-profit enterprises, funded and operated by businesses. The Ikonos satellite was launched in September, 1999, and after a brief test period, began providing commer-

Figure 6-32. A 1-meter resolution image of an urban setting collected by the Ikonos satellite system. Individual streets, small trees, and automobiles may be identified from Ikonos data, rendering it a useful tool for many urban applications (courtesy GeoEye Corp.).

Figure 6-33: A 0.5 meter resolution image from the WorldView-2 satellite (courtesy DigitalGlobe).

cial images (Figure 6-32). The satellite has an orbital altitude of 680 kilometers and an orbital period of 98 minutes. The Ikonos systems provides 1-meter panchromatic and 4-meter visible/infrared images. The swath width is 13 kilometers directly below the satellite, but because the Ikonos system is pointable off nadir, large areas are visible from any satellite track. Revisit times are typically one to three days.

QuickBird was developed by a group of companies, including ITT Industries and Hitachi. QuickBird is similar to Ikonos in that it provides both panchromatic and multispectral data. It differs mainly in the resolution available, down to 0.65 meters panchromatic and 2.44 meters multispectral, and a larger image footprint of 16.5 kilometers, compared to 13 kilometers.

The WorldView-1 high-resolution satellite provides 0.5-meter panchromatic images (Figure 6-33), while the Worldview-2 provides 0.46m panchromatic and multispectral images at 1.8m. Data are collected as often

as a 1.7 day return interval when providing a 1-meter resolution, and 6 days with a 0.5-meter resolution. Images have a swath width of approximately 17 km. Images are collected at approximately 10:30 a.m. local time, a common characteristic of these polar orbiting, sun-synchronous systems.

A number of high-resolution satellite imaging systems have a local focus. The KOMPSAT-2 satellite, launched in 2006, provides 1-meter resolution panchromatic and 4-meter multispectral data primarily for eastern Asia, and the Cartosat-2 satellite, launched in 2007, provides 0.8m resolution data, primarily focuses on south Asia. Resolution will also continue to improve, further blurring the line between aerial images and satellite images. The GEOEYE-2 satellite is planned for launch in 2013 with a planned multispectral resolution of 0.25 meters. The highest resolution data is increasingly dispersed through web-mapping applications.

SPOT

The French Government led the development of the *Systeme Pour l'Observation de la Terre*, or SPOT (Figure 6-34). In February 1986, SPOT-1 was launched, and there have since been four additional SPOT satellites, labeled two through five, placed in orbit. SPOT was the first satellite system designed to serve commercial interests, in a high-volume, production mode.

All SPOT satellites provide panchromatic mode and a high-resolution visible (HRV) mode. The panchromatic mode on SPOT-1 through SPOT-3 contained one image band in the visible wavelengths, and the HRV mode provided one band each in the green, red, and near-infrared portions of the spectrum. The panchromatic mode on these first satellites had a spatial resolution of 10 meters, and the HRV mode has a spectral resolution of 20 meters. SPOT-4 and SPOT-5 increased the panchromatic resolution to between 2.5 and 10 meters, and added a 10 to 20m mid-infrared band to the HRV

mode. This combination provides high resolution over large areas (Figure 6-35), and SPOT data are routinely used in a number of resource management, urban planning, and other applications.

The SPOT scanners have optics pointable to areas up to 27° to either side of the satellite path. This reduces revisit time to between one and five days, and allows the collection of satellite stereopairs suitable for elevation mapping.

SPOT images may also be used to create elevation data, due to the pointable sensors. Just as parallax may be used with aerial images to determine terrain heights, parallax may be measured from a satellite to estimate elevation. Scanners may point forward and back along the flight path, or across separate flight paths to image an area. The relative displacement on the image depends on the relative elevations on the ground. Careful measurements allow us to convert image displacement to elevations with little ground control.

Successor SPOT satellites are under development, with 1.5m resolution panchromatic bands and 6 meter multispectral bands in the blue, green, red, and infrared spectral regions. The system is designed to collect data in a larger swath width than other high-resolution satellites, with two scanners arrayed to provide a 120 km swatch, compared to the 15-20 km swath for sub-meter satellite systems. This improves repeat time at high resolution, and may reduce data costs. The planned launch date is currently late 2012, so system operation details are currently not available. However, new system deployment in a timely manner provides data continuity for users of the current system.

Figure 6-34: The SPOT satellites carry two imaging modules (upper left) as well as solar panels for power (lower right) and communications antennas (center).

Figure 6-35: An example of an image from the SPOT satellite system. The active Mexican volcano Popocatepetl is visible at the image center. This image demonstrates the broad area coverage and fine detail available from the SPOT system (courtesy SPOT Image Corp.).

Landsat

The Landsat system is now effectively in hiatus, but is important because it was the first Earth observing satellite system, there is an image repository spanning five decades, and a majority of these images are available free of charge to anyone with an internet connection, allowing long-term monitoring and analysis. Landsat satellite program began with the launch of the first Landsat satellite on July 23, 1972. Previous satellite systems had focused on observing the weather, oceans, or other features. While earlier systems hinted at the exciting possibilities of land surface remote sensing, these possibilities were not realized until Landsat (Figure 6-36). The success of the first satellite led to the development and launch of six more, with Landsat satellites operating continuously from 1972 to at least 2011.

Landsat satellites have carried three primary imaging scanners. The *Multispectral Scanner* (or MSS) was the first satellite-based land scanner, and it has been carried on board Landsat satellites 1 through 5. The original MSS sensed in four spectral bands, at an 80m resolution: a green, a red, and two infrared bands.

Later Landsat satellites also carried the Thematic Mapper (TM) or Enhanced Thematic Mapper (ETM+), improved over the MSS. TM data contain seven spectral bands (three visible, a near-infrared, two mid-infrared, and a thermal band), and a 28.5m grid-cell resolution for the first six bands. The ETM+ added a 15 meter resolution panchromatic band covering the visible wavelengths. The satellites have had a 16 to 18 day return interval.

The continuous record of Landsat coverage will likely be broken, however, due to funding discontinuity. The current satellite, Landsat 7, has already surpassed its designed lifetime, and provides compro-

mised data over much of the scanning footprint. Data from the ETM+ became degraded in 2003 due to a scanning mirror failure, substantially reducing the utility of approximately the outside 30% of each image. Scenes from the high polar regions are little affected because there is much overlap between paths. However, approximately 30% of area in tropical regions is effectively unsampled by Landsat 7. A Landsat Data Continuity Mission (LDCM) is planned, with a programmed 2012 launch date. The current design has a 15m panchromatic band, 30m visible, near-IR, and shortwave IR, and 100m thermal imaging bands, and an 185km swath width.

Figure 6-36: A portion of a Landsat ETM+ image for an area in South Africa. Vegetation is a mix of savanna and woodland on abandoned grazing land, and agriculture. The image spans approximately 40 km east to west and 30 km north to south.

Resourcesat-1

The Indian Space Research Organization has launched a number of satellites designed for Earth observation, including the Cartosat series, previously described, and the IRS series, beginning with the IRS-1A in 1988. Early satellites were largely experimental and images not widely distributed, but the latest satellite, IRS-6P or Resourcesat-1, provides high-quality, large-area, moderate resolution data over much of the globe (Figure 6-37).

The Resourcesat-1 carries three scanners, LISS-IV with a 5.8m resolution, the LISS-III with a 23.5m resolution, and AWiFS with a 56m resolution. Swath (image) width increases from 70km through 141 km to 740 km for the three instruments, with a 5-day repeat cycle for the AWiFS sensor. The AWiFS is most commonly used outside of India, and provides blue, green, red, and near-infrared sensing bands, with 10-bit data. These images are often used for regional to national analyses because of their large image size and medium resolution, e.g., by the U.S. National Agricultural Statistical Service for annual crop inventories in the U.S.

MODIS and VEGETATION

There are currently two widely-used, coarse resolution sensors, the Moderate Resolution Imaging Sensor, or MODIS, and VEGETATION. MODIS is a NASA research system that collects data at a range of resolutions and wavebands, from visible through thermal infrared bands. Resolutions depend on bands and vary from 250 meters to 1 kilometer, and it has a repeat frequency of every one to two days for the entire Earth's surface when images are sampled at the 1 kilometer resolution. Thirty-six bands are collected when operated in the one kilometer mode, ranging from 0.4 μm to 14.4 μm. Only two bands are collected at the 250 meter resolution, one each in the red and infrared portions of the light spectrum. These are somewhat unique in that the reso-

Figure 6-37: A Resourcesat-1, AWiFS image including Sri Lanka and adjacent India; north is to the left of the image. The Resourcesat-1 provides medium-resolution data over large areas at reasonable costs.

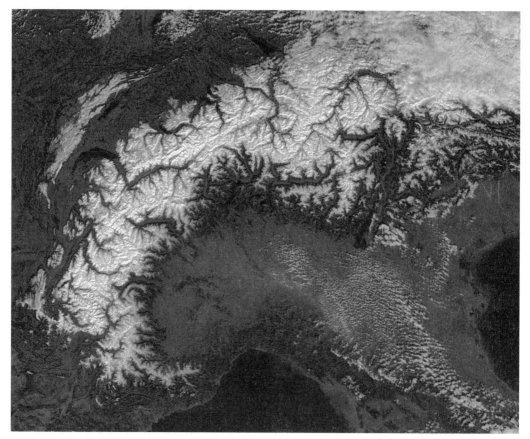

Figure 6-38: A MODIS 250 meter resolution image of northern Italy and Switzerland. The snow-covered Alps cross through the center of this image, north of the Po River Valley in Italy. Small cumulus clouds are visible, as is turbidity in the Mediterranean Sea and variation in land cover (Courtesy NASA).

lution is finer than the 1 kilometer resolution of AVHRR and most MODIS data, but substantially coarser than Landsat, SPOT, and moderate-resolution satellites. Large area coverage is possible at an intermediate level of detail when using MODIS 250 meter data (Figure 6-38).

VEGETATION was developed by the French national space agency, CNES. It collects data in the blue, red, near infrared, and short-wave infrared portions of the spectrum, with a nadir resolution of 1.15 kilometers and a swath width of 2400 kilometers, yielding daily coverage. The scanning system was designed specifically to monitor vegetation and physical environments for continental and larger areas, and it has been

used to monitor crops, fires, storms, and other phenomena (Figure 6-39).

The future of the MODIS system is somewhat unclear. The MODIS satellites were launched in late 1999/mid 2000, with a six-year design life, so they have already outlived their mission, although they continue to function. A limited set of functions may be provided by JPSS, although the primary focus of that system will be weather and climate monitoring and prediction.

MERIS

MERIS is a wide-swath scanner carried aboard the ESA's Envirosat. With 15 selectable bands at a 300m maximum resolution, a 1150 km (750 mile) swath, and 3-day global

Figure 6-39: Dense fog blankets the Central Valley of California in this example image from the VEGETATION sensor (courtesy CNES).

coverage, the system is in many respects an improvement on MODIS. Many of the bands are narrowly specified for a purpose, e.g., several chlorophyll and sediment bands for water quality and vegetation, a yellow band, and bands to aid in atmospheric correction.

The MERIS system is designed in part for oceanographic applications, with specific bands tuned to the low-level returns from ocean targets (Figure 6-40). This leads to improved information extraction from waters, perhaps at the expense in those bands when sensing brighter land-based targets. Response saturation occurs when reflectance reaches the highest values a sensor can record.

Figure 6-40: An example of a MERIS satellite image, showing southern Florida and the western Bahamas. Deeper water penetration is shown in the clear delineation of the sandy shallows surrounding the islands.

LiDAR

A number of laser-based (LiDAR) systems have been developed and are seeing increasing deployment. Lasers are pointed at the Earth's surface, pulses of laser light emitted, and the reflected energy is recorded (Figure 6-41). Like radar, laser systems are active because they provide the energy that is sensed. Unlike radar, lasers have limited ability to penetrate clouds, smoke, or haze.

LiDAR systems have been used to gather data about topography, vegetation, and water quality. A series of studies by NASA has demonstrated that lasers may provide information on vegetation density, height, and structure. Laser pulses are scattered from the canopy and the ground, and the strength and timing of the return may be used to estimate ground height, canopy height, and other canopy characteristics (Figure 6-41). Similar conditions allow multiple measurements from water, including water surface height and returns from various depths, so lasers may be used to measure water clarity.

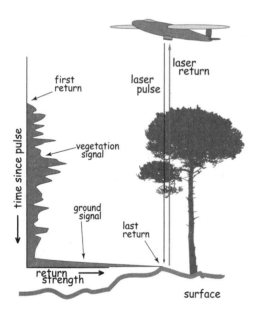

Figure 6-41: Laser mapping systems operate by generating and then sensing light pulses. The return strength used to distinguish between vegetation and the ground, and the travel time may be used to determine heights.

Commercial laser mapping systems are a recent phenomena and have been used primarily for collecting elevation data. A laser pulse is directed downward. Some energy is reflected from vegetation, buildings, or other features above the ground, but under most conditions many signals reach the ground and return to the airborne laser platform. The time interval between laser pulse emission and the ground return may be used to calculate aircraft height above the terrain. If flying height is known, then the terrain elevation may be calculated for each pulse. Pulses may be sent several thousand times a second, so a trace of ground heights may be measured from every few centimeters to a few meters along the ground. Pointable lasers allow the measurement of a swath of elevations.

Discrete-return LiDAR is most common, wherein the system records specific values for each laser pulse downward. Typically the first return from a pulse, last return, and perhaps one to several intermediate returns are recorded. *Waveform LiDAR* collects a continuous record of the pulse returns, the waveform trace shown in Figure 6-41.

Discrete-return LiDAR systems produce point clouds, consisting of X, Y, and Z coordinates and the intensity of the return (Figure 6-42). Modern laser systems often produce densities of several to tens of points per square meter of ground area, and these point clouds must be processed to remove errors, identify ground points, and assign points to feature types such as buildings or vegetation. The point order (first, last, or intermediate) and the location and strength of adjacent returns are used to identify point type.

These airborne laser systems provide detailed elevation and infrastructure mapping over large areas (Figure 6-43). There are a growing number of statewide LiDAR projects, often driven by floodplain mapping or for improved topographic measurements. Ground resolutions of 10 centimeters (4 inches) or better are possible when LiDAR is combined with precise GPS and aircraft orientation measurements. Horizontal and vertical errors less than a few centimeters are

Figure 6-42:An example of a LiDAR point cloud, here a swath through a forested area, displayed over a terrain model. Each point represents a LiDAR return, showing returns from the canopy and taller trees, sub-canopy branches and shrubs, and dense ground returns in canopy gaps.

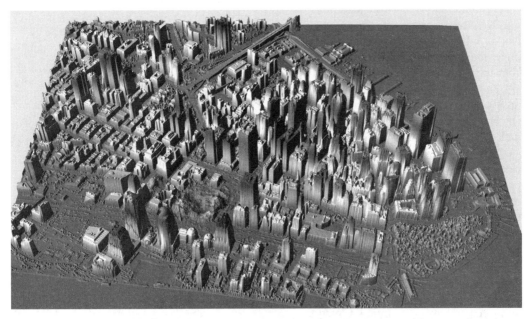

Figure 6-43: Elevation data from lower Manhattan collected with a LiDAR system and processed to create a raster layer. These data are shown in a three-dimensional perspective view. Individual buildings are visible, as are bridges, docks, elevated freeways, and other structures. Data such as these may be collected rapidly over large areas and at high spatial resolution (courtesy NOAA).

possible, allowing the use of airborne lasers to measure building height, floodplain location and extent, forest canopy heights, and urban growth.

Other Systems

There are several other airborne and satellite remote sensing systems that are operational or under development. Although some are quite specialized, each may serve as an important source of data. Some may introduce entirely new technologies, while others replace or provide incremental upgrades to existing systems. Space prevents our offering more than a brief description of these satellite systems here.

Passive optical systems -- there are several remote sensing systems that are based on reflected incident radiation. These include the IRS system deployed by the Indian government, with a 5-meter panchromatic band with a five-day revisit interval, three 24-meter bands that span the green through near-infrared portion of the spectrum, and one 70-meter band in mid-infrared.

A number of radar-based satellite systems have been used as a source of spatial data for GIS. Radar wavelengths are much longer than optical remote-sensing systems, from approximately one to tens of centimeters, and may be used day or night, through most weather conditions. Radar images are panchromatic, because they provide information on the strength of the reflected energy at one wavelength. Radar systems have been successfully used for topographic mapping and some landcover mapping, particularly when large differences in surface texture occur, such as between water and land, or forest and recently clearcut areas. Operational systems include the ERS-1, operated by the European Space Agency, the JERS-1, by the National Space Development Agency of Japan, and the Radarsat system, developed and managed by the Canadian Space Agency.

Satellite Images in GIS

Satellite images have two primary uses in GIS. First, satellite images are often used to create or update landcover data layers. Satellite images are particularly appropriate for landcover classification by virtue of their uniform data collection over large areas. Landcover classes often correspond to specific combinations of spectral reflectance values. For example, forests often exhibit a distinct spectral signature that distinguishes them from other landcover classes (Figure 6-44).

Satellite image classification involves identifying the reflectance patterns associated with each landcover class, and then applying this knowledge to classify all areas of a satellite image. Many techniques have been developed to facilitate landcover mapping using satellite data, as well as techniques for testing the classification accuracy of these landcover data. Regional and statewide classifications are commonly performed, and these data are key inputs in a number of resource planning and management analyses using GIS.

Satellite images are also used to detect and monitor change. The extent and intensity of disasters such as flooding, fires, or hurricane damage may be determined using satellite images. Urbanization, forest cutting, agricultural change, or other changes in land use or condition have all been successfully monitored and analyzed based on satellite data. Change detection often involves the combination of new images with previous landcover, infrastructure, or other information in spatial analyses to determine the extent of damage, to direct appropriate responses, and for long-range planning.

Aerial or Satellite Images in GIS: Which to Use?

The utility of both satellite images and aerial images as data sources for GIS should be clear by now. Several sources are often available or potentially available for a given study area. An obvious question is "Which

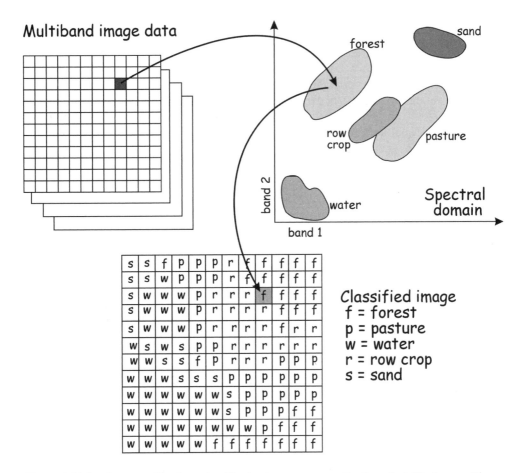

Figure 6-44: Landcover and land use classification is a common application of satellite images. The spectral reflectance patterns of each cover type are used to assign a unique landcover class to each cell. These data may then be imported into a GIS as a raster data layer.

should I use?" A number of factors should be considered when selecting an image source.

First, the image data should provide the necessary spatial resolution. The resolving power of a system depends in part on the difference in color between two adjacent objects, but resolution is generally defined by the smallest high-contrast object that can be detected. Current high-resolution satellite systems have effective spatial resolutions of from one to several meters. Images from mapping cameras, when taken at typical scales and with commonly used film, have maximum resolutions as fine as a few centimeters. Although the gap is narrowing

as higher resolution satellite systems are deployed, aerial images are currently selected in many spatial analyses that require the highest spatial resolution.

Second, the size of the analysis area should be considered. Aerial images are typically less expensive in small-area analyses. Large-scale aerial images are often available from government sources at low cost. Each photograph covers from tens to hundreds of square kilometers, and cost per square kilometer may be quite low. As the size of the study area increases, the costs of using aerial photographs may increase. Multiple photographs must often be placed in a mosaic, and this is typically a time-consuming operation.

Terrain distortion may be significant in mountainous areas, and distortion removal may be more costly with aerial images than with satellite images. Satellites, because of their high view and large area coverage, are less likely to require the creation of mosaics and may have significantly less terrain distortion. Therefore, satellite images may be less expensive for larger study areas.

Satellite scanners may provide a broader spectral range and narrower bands relative to aerial images. As noted earlier, satellite scanners may be designed to detect well beyond the visible and near-infrared portions of the spectrum to which aerial photographs and some aerial scanners are limited. If important features are best detected using these portions of the spectrum, then satellite data are preferred. Specialized sensors may be particularly important under specific conditions, for example radar when clouds prevent the use of aerial images.

Image Sources

National, state, provincial, or local governments are common sources of aerial images. These photographs are often provided at a reduced cost. For example, the National Agriculture Imagery Program (NAIP) provides coverage of much of the lower 48 United States on an annual basis. Images are usually collected in true color, but color infrared images may also be acquired, typically at a resolution of 1 meter or better. Photographs are usually collected mid-growing season. The NAIP program is coordinated through the USDA Farm Services Administration, and so the images are sometimes referred to as FSA or FSA-NAIP photographs. Online and hardcopy indexes are available to aid in identifying appropriate image mosaics.

Aerial images may also be purchased from other government agencies or from private organizations. The USGS and U.S. Forest Service (USFS) routinely take aerial images for specialized purposes. The USFS uses aerial images to map forest type and condition, and often requires images at a

higher spatial resolution and different time of year than those provided by NAIP. The USGS uses aerial images in the development of digital orthophotographs and maps. These organizations are also excellent sources of historical aerial images. Many government agencies contribute to a national archive of aerial images, which may be accessed at the internet addresses listed in Appendix B.

Satellite images may be obtained from various sources. Current Landsat data are available through the NASA and the USGS. SPOT, Ikonos, QuickBird, and other satellite system data may be obtained directly from the managing sources, listed in Appendix B.

Summary

Aerial and satellite images are valuable sources of spatial data. Photos and images provide large-area coverage, geometric accuracy, and a permanent record of spatial and attribute data, and techniques have been well-developed for their use as a data source.

Remote sensing is based on differences among features in the amount of reflected electromagnetic energy. Chemical or electronic sensors record the amount of energy reflected from objects. Reflectance differences are the basis for images, which may in turn be interpreted to provide information on the type and location of important features.

Aerial images are a primary source of coordinate and attribute data. Camera-based mapping systems are well-developed, and are the basis for most large-scale topographic maps currently in use. Camera tilt and terrain variation may cause large errors on aerial images; however, methods have been developed for the removal of these errors. Terrain-caused image displacement is the basis for stereophotographic determination of elevations.

Satellite images are available from a range of sources and for a number of specific purposes. Landsat, the first land remote sensing system, has been in operation for nearly 30 years, and has demonstrated the utility of satellite images. SPOT, AVHRR,

Ikonos, and other satellite systems have been developed that provide a range of spatial, spectral, and temporal resolutions.

Aerial and satellite images often must be interpreted to provide useful spatial information. Aerial images are typically interpreted manually. An analyst identifies features based on their shape, size, texture, location, color, and brightness, and draws boundaries or locations, either on a hardcopy overlay, or on a scanned image. Satellite images are often interpreted using automated or semi-automated methods. Classification is a common interpretation technique that involves specifying spectral and perhaps spatial characteristics common to each feature type.

The choice of photographs or satellite imagery depends on the needs and budgets of the user. Aerial images often provide more detail, are less expensive, and are easily and inexpensively interpreted for small areas. Satellite images cover large areas in a uniform manner, and sense energy across a broader range of wavelengths.

Suggested Reading

American Society of Photogrammetry. (1983). *Manual of Remote Sensing,* (2nd ed.). Falls Church: American Society of Photogrammetry.

Atkinson, P. & Tate, N. (1999). *Advances in Remote Sensing and GIS Analysis.* New York: Wiley.

Avery, T. E. (1973). *Interpretation of Aerial Photographs.* Minneapolis: Burgess.

Befort, W. (1986). Large-scale sampling photography for forest habitat-type identification. *Photogrammetric Engineering and Remote Sensing*, 52:101-108.

Campbell, J. B. (2006). Introduction to Remote Sensing. (4th ed.). New York: Guilford.

Bolstad, P.V. (1992). Geometric errors in natural resource GIS data: the effects of tilt and terrain on aerial photographs. *Forest Science*, 38:367-380.

Dial, G., Bowen, H., Gerlach, F., Grodecki, J., & Oleszczuk, R. (2003). IKONOS satellite, imagery, and products. *Remote Sensing of Environment*, 14:23-36.

Drury, S. A. (1990). *A Guide to Remote Sensing: Interpreting Images of the Earth.* New York: Oxford University Press.

Ehlers, M. (1991). Multisensor image fusion techniques in remote sensing. *Journal of Photogrammetry and Remote Sensing*, 46:19-30.

Elachi, C. (1987). *Introduction to the Physics and Techniques of Remote Sensing.* New York: Wiley.

Goetz, S.J., Wright, R.K., Smith, A.J., Zinecker, E., & Schaub, E. (2003). IKONOS imagery for resource management: Tree cover, impervious surfaces, and riparian buffer analysis in the mid-Atlantic region. Remote Sensing of Environment, 8:195-208.

Hodgson, M.E. & Bresnahan, P. (2004). Accuracy of airborne LiDAR-derived elevation: Empirical assessment and error budget. *Photogrammetric Engineering & Remote Sensing*, 70:331-339.

Kodak.(1982). *Data for Aerial Photography.* Rochester: Eastman Kodak Company: Rochester.

Light, D. (1993). The National Aerial Photography Program as a geographic information systems resource. *Photogrammetric Engineering and Remote Sensing*, 59:61-65.

Lillesand, T.M., Kiefer, R.W., & Chipman, J. (2007). *Remote Sensing and Image Interpretation*, 6th Edition. New York: Wiley.

Meyer, M.P. (1982). Place of small-format aerial photography in resource surveys. *Journal of Forestry*, 80:15-17.

Richards, J.A., & Jia, X. (2005). *Remote Sensing Digital Image Analysis.* New York: Springer.

Ryan, R., Baldbridge, B., Schowengerdt, R.A., Choi, T., Helder, D.L., & Blonski, S. (2003). IKONOS spatial resolution and image interpretability characterization. *Remote Sensing of Environment*, 16:37-52.

Schowengerdt, R.A. (2006). *Remote Sensing: Models and Methods for Image Processing* (3rd ed.). New York: Academic Press.

Nelson, R. & Holben, B. (1986). Identifying deforestation in Brazil using multiresolution satellite data, *International Journal of Remote Sensing*, 1986, 7:429-448.

Teng, W. L. (1990). AVHRR monitoring of U.S. crops during the 1988 drought. *Photogrammetric Engineering and Remote Sensing*, 56:1143-1146.

Warner, W. (1990). Accuracy and small-format surveys: the influence of scale and object definition on photo measurements. *ITC Journal*, 1:24-28.

Welch, R. (1987). Integration of photogrammetric, remote sensing and database technologies for mapping applications. *Photogrammetric Record*, 12:409-428.

Woodcock, C.E., & Strahler, A.H. (1987). The factor of scale in remote sensing. Remote Sensing of Environment, 21:311-332.

Wolf, P.R., & DeWit, B. (2000). *Elements of Photogrammetry with Applications of GIS*. New York: McGraw-Hill.

Yang, C., Everitt, J.H., and Bradford, J.M. (2006). Comparison of QuickBird satellite imagery and airborne imagery for mapping grain sorghum yield. *Precision Agriculture*, 7:33-44.

Study Problems and Questions

6.1 - Can you describe several positive attributes of images as data sources?

6.2 - What is the electromagnetic spectrum, and what are the principle wavelength regions?

6.3 - Define a spectral reflectance curve? Draw typical curves for vegetation and soil through the visible and infrared portions of the spectrum.

6.4 - Can you describe the structure and properties of the main types of photographic film, including their spectral sensitivity curves? Can you describe the structure and properties of digital sensors in digital aerial cameras?

6.5 - What are the basic components of a camera used for taking aerial photographs?

6.6 - Can you describe the most commonly used camera formats for aerial photography, and their relative advantages?

6.7 - What are the major sources of geometric distortion in aerial images, and why? What are other, usually minor, sources of geometric distortion in aerial images?

6.8 - What are typical magnitudes of geometric errors in uncorrected aerial images? How might these be reduced?

6.9 - A tall building is recorded on two vertical aerial photographs, the first photograph at a nominal scale of 1:20,000, the second photograph at a nominal scale of 1:40,000. The building is near the edge of both photographs, and terrain is level throughout the photograph. Which image will show a larger displacement, d, as shown in Figure 6-26?

6.10 - Can you describe stereo photographic coverage, and why it is useful?

6.11 - What is parallax, and why is it useful?

6.12 - Can you describe the basic process of terrain distortion removal?

6.13 - Why do the buildings lean in different directions in the images below?

6.14 - What is photointerpretation, and what are the main photographic characteristics used during interpretation?

6.15 - How are images from satellite scanners different from photographs? How are they similar?

6.16 - Can you describe and contrast the Landsat ETM+, SPOT HRV, Ikonos, and QuickBird satellite imaging systems?

6.17 - What is a LiDAR? What type of information can LiDAR produce?

6.18 - What are some of the criteria used in selecting the type of images for spatial data development?

7 Digital Data

Introduction

Many spatial data currently exist in digital forms. Roads, political boundaries, water bodies, land cover, soils, elevation, and a host of other features have been mapped and converted to digital spatial data for much of the world. Because these data are often distributed at low or no cost, these existing digital data are often the easiest, quickest, and least expensive source for much spatial data (Figure 7-1).

Data are increasingly collected in digital formats. GPS, laser measurements, and satellite scanners all provide primary data in digital forms. They are directly transferable to other digital devices and GIS systems, where they may be further processed. Direct digital collection eliminates data transfer to viewable physical media such as maps or written lists, and then converting these media to a digital form.

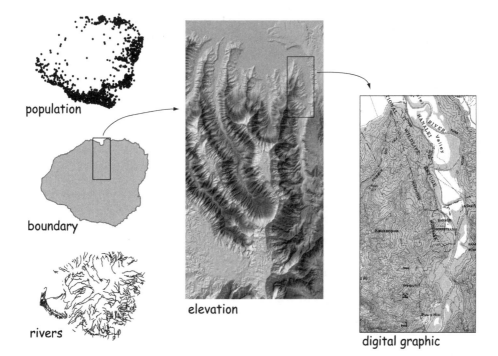

Figure 7-1: Examples of free digital data available at a range of themes, extents, and scales. Vector (left), raster (middle), and georeferenced digital graphic data (right) are shown for Kauai, Hawaii, USA.

Digital data are developed by governments because these data help provide basic public services such as safety, health, transportation, water, and energy. Spatial data are required for disaster planning and management, national defense, infrastructure development and maintenance, and other governmental functions. Many national, regional, and local governments have realized that once these data have been converted to digital formats for use within government, they may also be quite valuable for use outside government. Business, non-profit, education, science, as well as governmental bodies may draw benefit from the digital spatial data, as these organizations benefited in prior times from government-produced paper maps. Some data commonly available throughout the United States and the world are described in this chapter.

Map Services vs. Locally Storable Data

We must distinguish between data that are available for transfer to, storage on, and manipulation in a local computer (locally storable), from those data that are available as a Web Mapping Service (WMS). Digital data were first distributed on physical media, then via the internet, but typically as electronic files that were copied onto a local storage device for use. You maintained a copy on your device, and manipulated those locally-stored data. A WMS eliminates the need for a local copy.

AWMS is a standard way of serving geographic data over the internet. GIS software accesses data via an internet connection, displaying these data on a local machine, although they are "served" from some remote computing system. Image data are most often served, but vector data may also be provided, usually in the form of a georeferenced map backdrop. The data don't reside on the local hard disk, and data are

delivered in response to each pan, zoom, or other change in display.

Web mapping services have both advantages and disadvantages to data stored on a local hard disk. WMS data saves space on the local hard drive, and only the portion of interest from a large dataset need be accessed. A community of users may share the data, e.g., widely used government data, and the most up-to-date data sets provided to a wide set of users. Many different kinds of data may be joined together more easily, as accessing a WMS typically requires a few mouse clicks. However, you may often not manipulate, or change WMS data in any substantial way, and some kinds of analysis may not be supported or allowed. In these cases, local copies of the data are typically required. Map services may also require a fast and reliable internet connection, particularly for high-resolution image data or other large raster datasets.

For the remainder of this chapter, we will discuss data you may download to your local computer storage, as these have the fewest, to no, restrictions on analysis. Through time, many of these data may be offered as WMS, or for both WMS and local storage.

National and Global Digital Data

National governments commonly develop, organize, archive, and distribute national data sets. The standardization of weights and measures is a primary function of most national governments, and spatial data may be viewed as measurements of land, sea, or other national territories. Governments must oversee the planning, construction, and management of public infrastructure such as roads, waterways, and power distribution systems, and these activities, among many others, require spatial data sets that are national in extent.

National digital data often may be obtained from a central source via the internet. For example, Geosciences Australia, a part of the Australian national government, provides a suite of data layers that may be accessed via a website or may be requested in hardcopy form (Figure 7-2). Similar resources are available for Canadian, most European, and many Asian and Latin American countries. A partial list of available data resources is included in appendix B near the end of this book.

Global data sets are also available but are less common than national data sets.

Figure 7-2: National governments often create portals through which digital data may be accessed.

Global data are scarce because few governments collect spatial data in the same way or with the same set of attributes. Different governments specify different datums, standard map projections, data variables, and attributes or have different requirements for survey accuracy or measurement units. Data reduction or documentation methods may be different across national boundaries. There is substantial work in reconciling differences across national boundaries, therefore, global data sets are only occasionally built from a composite of national data sets.

A few global data sets are available that have been collected using a standard set of tools and methods. Global data sets have often been developed using global satellite data at a relatively coarse resolution (e.g., AVHRR or MODIS at one to eight kilometer cell sizes). Using a uniform global data source avoids the problem of reconciling differences among disparately collected data sets, but substantially reduces the number and type of global data sets that may be obtained, because a limited set of data may be derived from satellite images. These data must be visible from coarse-resolution satellites, and there must be an organization interested in collecting and processing global data. NASA provides a large and diverse group of global spatial data sets, due to their leadership in the development and application of satellite images at a range of spatial scales. Global raster data sets available from NASA include elevation, land use, ecosystem type, and a number of measures of vegetation productivity, phenology, structure, and health.

University centers or ad-hoc collaborations are other rich sources of global data. One example is the Center for International Earth Science Information Network, administered by the Earth Institute at Columbia University (www.ciesin.org). It seeks to provide global data to better address environmental problems. Another example is the Global Land Cover Facility at the University of Maryland (http://glcf.umi-acs.umd.edu), a set of earth science data products, primarily derived from NASA satellites. A final example is the collection of Natural Earth datasets (http://www.naturalearthdata.com/), a volunteer collaboration for creating consistent, high-quality data suitable for small-scale mapping (Figure 7-3).

Global spatial data sets are often organized around a theme. For example, the Max Planck Institute in Germany has led an effort to create a gridded data set of historical global precipitation by combining data from 40,000 meteorological stations in 173 countries. These data are compiled, quality checked, and processed to create gridded data sets for normal precipitation. Data sets of annual anomalies, the number of gauges, and systematic error are also provided. This was an expensive and time-consuming undertaking due to the number of different methods used to collect and report precipitation. Considerable time was spent reconciling data collection methods and results. A more complete description of these data is found at http://gpcc.dwd.de.

Given the substantial difficulties in compiling data from disparate global sources, the Global Spatial Dataset Infrastructure (GSDI)

initiative was formed. GSDI is an attempt to coordinate collection and processing methods worldwide to ensure that spatial data are broadly suitable for global-level analysis. The primary goal is to improve the development, use, and sharing of spatial data across the globe. This will be achieved through the adoption of common standards and complimentary policies across governments and regions.

The GSDI initiative began in the late-1990s, and is still a work in progress. Activities during the first few years include identifying participants, developing goals and organizational structure, and identifying and prioritizing early actions. Activities on the GSDI initiatives may be found at www.gsdi.org.

The Global Map is one early GSDI initiative. The Global Map specifies eight thematic layers: boundaries, elevation, land cover, vegetation, transportation, population centers, and drainage. Scale, feature classes, feature types, and feature names are specified, as are attributes, metadata, tiling schemes, and delivery mechanisms. Countries submit data to the Global Map project, which then serves as a distribution node.

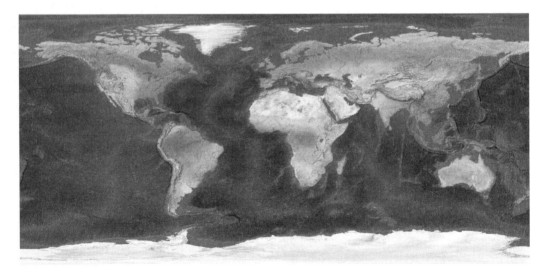

Figure 7-3: A diverse set of global spatial data types is available from various sources, although there is no one central clearinghouse. The United Nations, the European Union, and national government websites are the best sources, although there are specialized compilations by theme, as exemplified by the Natural Earth project, and available at http://www.shadedrelief.com/natural2/index.html.

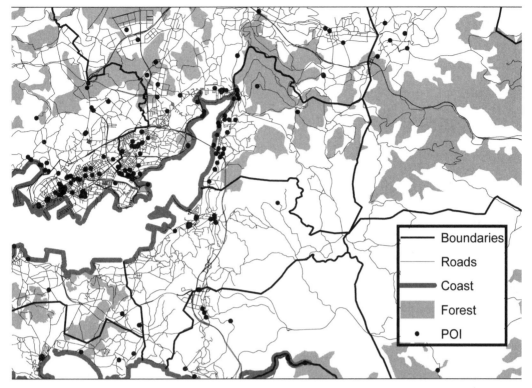

Figure 7-4: An example of OpenStreetMap data for an area in Galicia, in northwestern Spain. Volunteer collaborators created line and attribute data representing important feature layers, at high detail.

OpenStreetMap is one notable effort to develop global data through international volunteer collaboration. Much like Wikipedia, this is an open-access, user generated resource. Individual users register and can check out data sets to modify. Roads and other transportation infrastructure are digitized, typically from image interpretation or via GPS, and submitted for database integration. As with many online collaboratives, there are protocols for review and resolving conflicts, and data may be downloaded in various formats from OpenStreetMap or companion sites.These are often the best data in areas with poorly-developed mapping infrastructure.

While OpenStreetMap provides the best data in many regions, there are potential drawbacks with these data. Because it is a collaborative, quality documentation and uniformity may be lacking. A range of sources, abilities, and methods may be used to develop data, and documentation on these

sources may be unavailable. In addition, data may not be complete, depending on how much volunteer effort has been directed at an area, and the pace of change. Given these drawbacks, the data should be verified for accuracy and completeness, or at least suitability for the intended use, prior to adoption. Although this is true of most data, the burden perhaps falls more heavily on the user with these data.

Another potential drawback may be the method of distribution. Currently the data may be downloaded from the primary website in publicly defined, but rarely-used data formats. Data are available in more standard formats from 3rd-party websites, and the native OpenStreetmap formats are supported by some softwares (e.g., QGIS), and will surely achieve broader support in the future. In spite of these potential drawbacks, these open-source collaboratives have a bright future, and may well become standard for many types of data.

Digital Data for the United States

National Spatial Data Infrastructure

The United States has defined the National Spatial Data Infrastructure (NSDI) as the policies, technologies, and personnel required to ensure the efficient sharing and use of spatial data. The goal of the NSDI is to reduce duplication of effort among agencies, improve quality and reduce the costs of geographic information, to make geographic data more accessible to the public, to increase the benefits of available data, and to establish key partnerships with states, counties, cities, tribal nations, academia and the private sector to increase data availability (www.fgdc.gov).

The NSDI has developed a framework that identifies core data sets commonly used by many organizations. The framework consists of geodetic control, orthoimagery, elevation, transportation, hydrography, cadastral data (property boundaries), and governmental unit boundaries. A primary goal of the NSDI is to foster the efficient development of these core data.

Framework data are to be created and maintained by a diverse set of organizations, both within and outside of governments. Framework data may reside anywhere, but will be submitted to the National Geospatial Data Clearinghouse (NGDC) for certification. These data will be made available for the cost of providing access.

The National Geospatial Data Clearinghouse (NGDC) provides parallel access to many data sets. The NGDC website, found at http://clearinghouse1.fgdc.gov/ is perhaps the best current resource when searching for geospatial data for the United States. The U.S. Geological Survey (USGS), http://www.usgs.gov/, is the best single source of geospatial data from the U.S. federal government, and many of these data will be described in the following sections.

The National Atlas

The U.S. Geological Survey is developing a set of base digital data for the United States, harmonizing the digital spatial data of more than 20 federal agencies and providing them through a single portal. These are generally small-scale, widely-used data of national extent. Map layers may be viewed online, or downloaded in various data formats.

The themes of the National Atlas data vary broadly, and include political boundaries, environmental features, climate, history, biology, and natural hazards. Digital data may be downloaded in both the U.S. Spatial Data Transfer Standard format (STDS) and ESRI shapefile formats. At the time of this writing, description of the National Atlas was found at http://www.nationalatlas.gov.

The U.S. National Map

Digital data are available for most of the U.S. through the "National Map" project, described as a "cornerstone" of U.S. mapping efforts (http://nationalmap.gov). Data are provided on political and civil boundaries, transportation, hydrography, geographic names, structures (e.g., dams, notable buildings, towers, or monuments), elevation, aerial photographs, and landcover (Figure 7-5). Some of these data are available from dedicated projects and web sites, e.g., the National Elevation Datasets (NED) and the National Hydrologic Datasets (NHD). We'll discuss these two datasets in detail in later sections of this chapter, and here focus on a general description of the National Map project, and the additional data available through the national map.

Besides hydrography, the National Map also distributes transportation, structures, and boundaries data in vector formats. Transportation data represent roads, railroads, airports, and other transportation fea-

Figure 7-5: An example of spatial data available through the U.S. National Map, here elevation, road, river, and government building data for an area near Brevard, North Carolina.

tures. National map boundaries data identify national, state, county, and Native American lands, as well as the boundaries for cities and towns. Structures data identify selected man-made facilities, including government centers or service buildings, hospitals, and other important buildings, as well as dams, bridges, and other key physical structures. Limited attribute data are provided with all the National Map vector data.

Data for the National Map come from a variety of sources, including new primary data collections from aerial and satellite images contributed by federal and state agencies, and older data. Much of the National Map data are legacies of USGS hardcopy mapping programs. The USGS began topographic mapping in the U.S. in the 1880s, but the most common detailed map series began in the 1940s, with the production of 1:24,000 scale topographic maps. These paper maps covered 7.5 minutes of arc on a side, and comprised about 55,000 tiles covering the lower 48 states. These data were converted to digital layers, known as

Figure 7-6: An example of a USGS USTopo map.

Digital Line Graphs (DLGs), and much DLG data are still available on legacy websites, although the updated National Map data should be used where available.

The mapping program aimed at paper quadrangle maps ended in the 1990s, and has been replaced by a digital format "US Topo" map (Figure 7-6). Currently, these digital topographic maps are delivered as geographically enhanced postscript document format (PDF) files, with layers for orthoimagery, roads, place names, elevation contours, and rivers, lakes, and other hydrographic features. Layers may be rendered visible or invisible, and the maps displayed with other georeferenced data in appropriate viewers, but these maps are generally not used in a GIS. Complete US Topo maps are not available for all of the 7.5 minute cells for the lower 48 states, but are planned.

As an alternative, older Digital Raster Graphic (DRG) are available for most of the U.S. DRGs are georeferenced raster layers scanned from legacy USGS maps (Figure 7-7). DRGs are available for most of the 1:24,000-scale USGS 7.5 minute quad maps, and also for smaller-scale maps. A UTM coordinate system is used for most 1:24,000-scale USGS maps. The image is resampled to just under 100 dots per centimeter (250 dots per inch), and converted to a compressed GeoTIFF image format.

Figure 7-7: An example from a USGS 1:24,000-scale digital raster graphics (DRG) files.

Digital Elevation Models

Digital elevation models (DEMs) provide elevation data in a raster format and are available for most of the Earth (Figure 7-8). Each raster cell contains a value that corresponds to an elevation at a point on the ground. These DEMs may be used to both depict and quantitatively measure terrain. DEM manipulations and terrain analysis are described in Chapter 11. Here we introduce the sources of DEM data and their basic characteristics.

The United States and other governments have created DEMs at a range of cell sizes and extents, and moderate-resolution data are available for every continent. Worldwide coverage currently exists at various resolutions, older data sets with cell sizes of a kilometer or more, but more recent data have almost complete global coverage with cell sizes near 100 meters. New technologies promise improved elevation data sets, with submeter or better resolutions, at very high accuracies, such as those based on aerial or space-based LiDAR topographic measurements (Chapter 6). At present high-resolution DEMs are common in developed countries.

Ground and aerial surveys are the primary source of original elevation measurements for most DEMs. Traditional distance and angle measurements with surveying equipment were used up until the 1940s to provide precise elevations at specified locations. Because these methods are relatively slow, they provided a sparse network of points, with a dense network suitable for elevation mapping over only small areas. Improved electronic distance meters helped, as did global positioning system technologies, but even with these improvements in survey speed and accuracy, these technologies are too slow to be the sole elevation data collection method over all but the smallest areas.

Aerial images and airborne LiDAR surveys complement field surveying by increasing the number and density of measured elevations. Accurate elevation data may be collected over broad areas with the appropriate selection of aerial mapping technologies. From the 1950s until the late 1990s, most elevation data were compiled using precise mapping aerial cameras, complemented by optical ground surveys. Since the late 1990s, LiDAR mapping has been combined with GNSS to more accurately and rapidly map elevation. These various GNSS and survey methods are discussed in Chapter 5, and aerial images and LiDAR in Chapter 6.

Laser-based elevation mapping and DEM generation are now common, and will be used for the foreseeable future. These

LiDAR systems employ a downward-pointing laser in an aircraft (see Chapter 6). The laser measures height above terrain, and may include pointing optics that can measure the distance from the plane to locations below the aircraft. A precise GNSS system measures flying heights, which are combined with laser distance and angle measurements to estimate terrain heights across the landscape. Heights are then used to create a TIN or grid-based DEM. Statewide LiDAR are becoming common, for example North Carolina completed statewide coverage in

Figure 7-8: Digital elevation models (DEMs) are available at various resolutions and coverage areas for most of the world. This figure near Mono Lake in eastern California was produced using data available from the USGS. A string of lava flows is apparent on the right side of the image, and glacial lakes and valleys to the left.

the mid 2000's (Figure 7-9), and Minnesota is completing statewide coverage in 2013.

Prior to LiDAR, most DEMs were developed using photogrammetry, precision measurements from metric aerial photographs, combined with ground surveys. Photogrammetry has been used since the 1930's to map lines of a constant elevation (contours) and spot heights, and data have been developed for much of the Earth's surface using these techniques.

Completely digital, "softcopy" photogrammetric workstations are common, and although being supplanted by LiDAR, are still used in many projects (see Chapters 4 and 6). Parallax measurements on digital images are combined with camera location and orientation to generate contours. These systems may support full or semi-automated DEM and contour generation, techniques that greatly improve the efficiency of DEM creation in new surveys.

DEMs with 3-, 10-, and 30-meter horizontal sampling frequency are available for much of the United States. The highest resolution DEMs are delivered in a number of formats and for a range of footprints. Currently, the USGS delivers these DEMs as part of the National Elevation Dataset (NED), through the National Map portal (http://seamless.usgs.gov/). NED elevation values are typically provided as decimal meters in an NAD83 datum, in a geographic coordinate system. Data are available at a 30-meter resolution for almost all of the United States, and 10-meter resolution for the lower 48 states and Hawaii, and 3 meter for a large and expanding area.

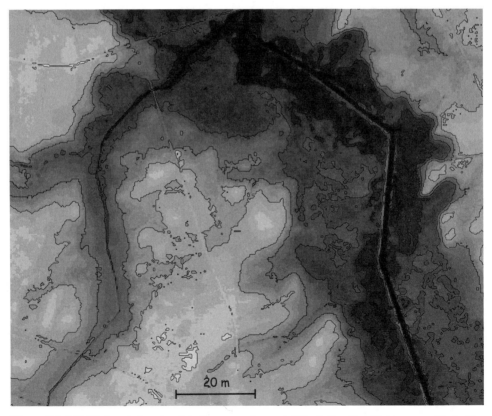

Figure 7-9: LiDAR-based elevation data for an area in coastal North Carolina. Data are provided with a six-meter grid spacing and shown here with a one-meter contour interval. These data are a substantial improvement in resolution and accuracy over previous elevation data sets (courtesy State of North Carolina).

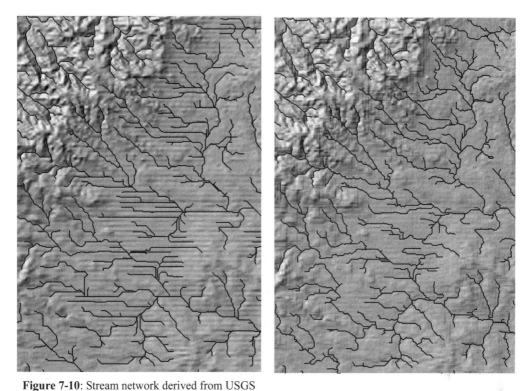

Figure 7-10: Stream network derived from USGS 1:24,000 quad-based DEM (left) and NED data set (right). Notice the parallel streams and right-angle turns derived from the quadrangle data are removed by the error processing in creating the NED data (recreated from USGS data for the Rockypoint, WY quadrangle, based on a USGS example).

GIS users should be cautious in the elevation data they select, because there are many versions of USGS data available on free public and for fee websites. Much of these data are older versions, e.g., 30 meter USGS DEMs that have since been supplanted by more current, higher accuracy, or higher resolution data. The most current USGS data are best accessed through the USGS or related US government websites.

NED is the most current USGS, nation-wide elevation data source, and an improvement data because of higher resolutions, higher accuracies, and fewer artifacts caused by the original data processing (Figure 7-10). DEM creation methods have progressed substantially in the past 20 years. Techniques used for tiled DEMS produced pits, spikes, or linear drainage systems. The NED assembly process is designed to remove most of these artifacts. This improves the

quality of the slope, aspect, shaded-relief, and drainage information that may be derived from the elevation data.

DEM data for the United States are also available with various other sampling frequencies and area coverage. DEMs with a resolution of 3 seconds of arc (100 meters, 330 feet) are available for the entire United States, and were produced by the Defense Mapping Agency (for a subsequent period named the National Imagery and Mapping Agency, NIMA, and now named the National Geospatial Intelligence Agency, NGA). These data are distributed by the USGS in 1 degree by 1 degree blocks.

SRTM DEMs: Radar has also been used to map elevation for much of the Earth's surface, most notably through the Shuttle Radar Topography Mission (SRTM). Elevations were collected from C- and X-band radars over 11 days in February 2000, and pro-

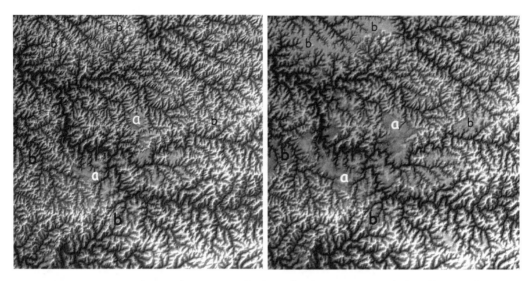

Figure 7-11: The extent of mountain-top removal strip mining in eastern Kentucky is evident in this comparison of older NED data (left) and year 2000 SRTM elevation data (right). Ridgetops are light colored, valley bottoms dark colored. Mines appear as areas of uniform tone on ridgetops. Mine sites labeled a, above, have expanded substantially, while large new mines have also been developed (b).

cessed to yield 1-km and 90-meter resolution elevation data worldwide, and 30-meter resolution data for the United States. The data are generally of high quality, with vertical elevation generally tested to average between 4 and 8 meters absolute error, and lower relative or local error. SRTM and other radar-based data have the advantage of large, seamless data sets that may be downloaded from a single location and in standard formats, balanced by the disadvantage of poor contour representation due to noise in the radar response over short distances. Other radar systems have been used for measuring elevation, including those on board the Radarsat and ERS satellite series.

Because it is a radar product, there are gaps in SRTM coverage in very steep terrain, for example, in canyons of the western United States. However, these gaps occur for a very small portion of the landscape. SRTM elevation data are a useful complement to older NED data collected in the United States during the 1950s and 1960s (Figure 7-11). For much of the rest of the world, SRTM 90 meter (300 feet) data provide 10 times the spatial resolution of the next available source.

Many developed countries have higher-resolution elevation data specific to their areas of control. For example, elevation data on a 20-meter contour or 3-second grid are available for Australia through the Australian National Mapping Division, and data for the United Kingdom are provided at a 50-meter (160 foot) raster cell size by the Ordnance Survey of Britain.

Hydrologic Data

The National Hydrologic Dataset (NHD) contains digital spatial data about surface waters, including rivers, streams, canals, ditches, lakes, ponds, springs, and wells. The NHD combines data from USGS digital line graph data and US Environmental Protection Agency (EPA) river "reach" data. A reach is a segment of a stream, river, or coastline considered homogenous under an EPA classification scheme. NHD data are based on 1:100,000 scale USGS DLG data, but may be improved as new data are developed. Naturally occurring and built features are represented in NHD data (Figure 7-12). These include water bodies, canals, pipelines, dams, and other natural or control structures. Attributes may be provided for these features, for example, a lake type or name, if a dam is earthen or concrete, or ditch type. Features may be points, lines, or polygons.

NHD data also represent network topology, the connection among stream features, and include information on connections and flow directions. Line segments have a designated flow direction, and connections or crossings may be represented as full connections, or noted as a bypass, for example, when a spillway or pipeline crosses a river without the possibility of discharging water into the river. Coding schemes have been developed to identify each reach in the hydrographic network, and to represent network connections among reaches.

NHD data are organized by areas, in a hierarchically nested set of Hydrologic Units, identified by unique codes (HUCs). These units correspond to watersheds, or basins, or logical aggregations or subareas of watersheds. The United States was divided into 21 regions, and these regions further divided into 222 subregions. Subregions were in turn divided, forming a total of 352

hydrologic units, and these are further divided into 2150 hydrologic units. This fourth level division is for the most part along major river basins, outlining distinct watersheds, or intermediate pieces along a the main stem of larger rivers. Each of these divisions is identified by a unique 8-digit code, and so these areas are also known as HUC-8 boundaries. Regions, subregions, and subregion divisions are known as HUC-2, HUC-4, and HUC-6 areas, respectively, providing a nested set of drainage areas for areas larger than the HUC-8 catchments. HUC-8 catchments may in turn be split into smaller HUC-10 and HUC-12 catchments, this last size typically the smallest delineations widely available.

These NHD data are a substantial improvement on previous sources of hydrologic data. These earlier data are digital versions of hydrologic lines and polygons shown on USGS maps. These data are of mixed type, and do not represent accurate topology. Streams are represented by line features and larger rivers and lakes by polygonal features. These typically are not joined

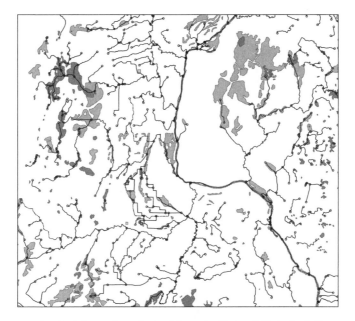

Figure 7-12: An example of sub-basin data obtained from the National Hydrologic Dataset. A number of feature types are represented, including stream segment endpoints (unfilled circles), connected stream networks (solid lines), water bodies (dark polygons), and adjacent wetlands (grey polygons).

to form a network, and flow direction and location is typically ill-defined in a polygonal region. Substantial work is required to connect the lines and generate a connected network through which water flow may be analyzed. These older data are still widely available, but should be used with caution when performing analysis.

The US EPA also provides data on waters and watersheds of various types and formats, organized to correspond to the HUC data at some levels. EPA River Reach Files organize data in a series of versions, from RF1 through the most recent and detailed RF3 data. RF3 data are designed to provide a nationally consistent hydrographic database that records the geography and assigns unique identifiers to all surface water features. It allows the hydrologic ordering of reaches so that larger rivers and segments may be accurately defined, along with river connectedness and flow direction. RF3 data also record the locations and characteristics of additional elements, including gauges, dams, and other hydrologic features (Figure 7-14). RF3 data have been created for more

than 3 million river reaches in the United States.

River reach data are precursors to NHD data, and so contain much of the same base information. RF files are available for most of the contiguous United States. Tabular data on water chemistry and other watershed characteristics are available at http://cfpub.epa.gov/surf/locate/index.cfm.

There are other improved hydrologic data, called NHDPlus. Managers need consistent elevation, stream, and watershed boundary data sets at high resolution to solve many water resource science and planning problems. NHD is a program to generate consistent, high quality hydrographic data with elevation-based catchments, tables of cumulative drainage area data on landcover, mean elevation, slope, and other characteristics, flowline directions, and flow volume and velocity estimates for each stream segment. There was an original version produced in the early 2000s focused on hydrographic data from 1:100,000 scale. Subsequent work has focused on improving the accuracy, consistency, and tools to sup-

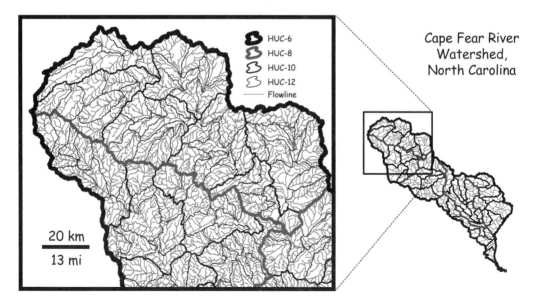

Figure 7-13: An example of nested HUC drainage areas for a portion of the Cape Fear River in North Carolina. Higher HUC designators nest within lower designators, so HUC-12 areas are contained within HUC-8 units.

Figure 7-14: River reach data (RF1 through RF3) contain location and connectivity data for river and stream segments. These data are based on 1:250,000 through 1:100,00 USGS map sources, and have been further enhanced via coding to show river direction, ordering, and other characteristics.

port NHDPlus data, with work on version 2 of NHDPlus beginning in January, 2011.

There is an emerging system for storing, finding, and retrieving hydrologic data associated with the CUAHSI project (www.cuashi.org). CUASHI is a National Science Foundation funded project involving more than 120 universities to support hydrologic science and education. CUASHI-HIS is an internet-based system for sharing hydrologic data via a web service. As noted earlier, a web service is set of protocols that allow communication among computer programs over the internet. In GIS these web services are most often used to streamline data sharing and access. The CUASHI HIS is designed to aid the integration and sharing of disparate hydrologic data, such as stream gauge, precipitation, river location, basin topography or other basin characteristics.

There are many other hydrologic datasets available for the U.S., most often used for analysis or display over larger areas. The U.S.G.S produced digital, nationwide datasets based on paper 1:100,000, 1:250,000, 1:1,000,000, and smaller scale maps (Figure 7-15), and these are available from various state, national, local, and private sources, although somewhat surprisingly, at the time of this writing, not as part of the U.S. National Map. These data show larger rivers and a limited set of attributes for each river, most importantly river names, although names are not provided for all river reaches. These data are also not hydrologically continuous, in that many of the rivers

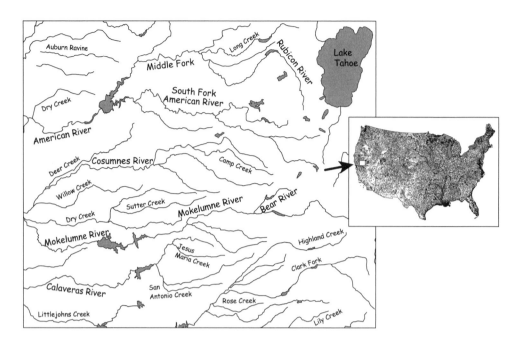

Figure 7-15: An example of U.S.G.S. legacy 1:1,000,000 hydrologic data, with the national data set (inset, right) and a portion in east-central California depicting the level of detail in these data.

do not maintain their connection through water bodies. Despite these limitations, they are often used because they may be more appropriate for statewide or regional analysis involving only the main stems or larger rivers in a region.

Integrated, consistent, continent to worldwide hydrography data are also available from the "Natural Earth" data projects (www.naturalearthdata.com). These data are intended for use in cartography, and not primarily for analysis, as they have been generalized and made consistent primarily for display rather than geographic accuracy. Different hydrographic data are offered, targeted at a range of small scales, for regional through global mapping. A limited set of attributes is available, including reach or river names and cartographic widths.

Digital Images

Digital images are available from a range of sources, including national, state, and county governments, or from private contractors, satellite imaging companies, and resellers. High resolution digital image data are typically collected every five to ten year by the USGS, in partnerships with states or other government agencies. Image archives go back to the 1950s for most of the U.S., with state and countywide coverage as far back as the 1930s not uncommon. Images before the early 2000s were almost always film-based, although many of the images have been scanned and are available for download from the U.S. EROS Data Center. Image sets include the historical black and white aerial photographs, the nation-wide programs of the 1980s and 1990s, high-resolution coastal images, and radar, and other special collections.

The High Resolution Orthophotographs (HROs) are among the highest resolution, widely available image data sets. The HRO images are collected and distributed through the U.S. Geological Survey. These data are often collected at 0.3 m (1 foot) resolution, and at times up to 10 cm (4 inch) resolution. Because they are orthophotographs, object base locations have been corrected for tilt and terrain distortion at ground height. Towers, buildings, bridges, and other tall objects often appear to tilt, as these structure heights above ground are not corrected on the images (Figure 7-16). These images are useful for infrastructure mapping, planning, disaster management, and many other applications. The images are sometimes used for vegetation mapping, but these images are most often collected during leaf-off periods,

Figure 7-16 An example of a High Resolution Orthophotograph (HRO), distributed by the U.S. Geological Survey. Individual persons can be identified near the flagpole in the upper left part of the island, as well as the statue and shadow to the lower right.

Figure 7-17: An example of an historical aerial photograph from the 1940s (left) and 2008 (right), for an area in east-central Minnesota. Early development was restricted to near the lake along the top margin of the 1940 photo, while by 2008 the area had become completely suburbanized. Photographs may be used to map current infrastructure and resources, and their change through time.

and so must often be complemented by images taken during leaf-on periods for most vegetation mapping efforts.

The HRO and other high resolution images are valuable sources of spatial data. These images are typically processed to within a few pixels of the delivered resolution, e.g., typically accurate to within a half-meter for the 0.3m data. Because these and other photos described below record the surface at a fixed point in time, they may be used to create new maps or to monitor change (Figure 7-17).

Legacy Aerial Photographs: SCS, DOQ, NHAP, and NAPP Photos

Earlier photographic programs have provided vertical aerial photographs suitable for mapping. Millions of aerial photographs are available, dating back to at least the 1920s, although only a small portion from before the 1990s have been digitized. The oldest images are available from the U.S. National Archives and Records Administration, Cartographic and Architectural Branch.

Some older photographs and many through the 1990s are available from the USGS, EROS data center aerial photograph archive. Aerial photographs for rural agricultural and forested lands are available from the Farm Services Administration, starting in the 1950s, but as with the national archives, most of these photographs have not been converted to digital forms. Nationwide coverage was provided in the early 1980s via the NHAP initiative, and again in the 1990s via the NAPP program.

Digital orthophoto quadrangles (DOQs), produced in the late 1980s through 1990s, were the first set of widely available, scanned photographic images. Images were orthorectified to remove positional errors, as described in Chapter 6. These corrections yield photographs that are planimetrically correct, and are registered to an Earth coordinate system. These photographs were and may still be a valuable source of spatial information, because they may be used as the basis for recording thematic data layers such as historical roads, vegetation, or buildings.

NAIP Digital Images

The National Aerial Imagery Program (NAIP) acquires photographs during the growing season in the continental United States. NAIP images are distinct from the previous HRO, NHAP, NAP, and DOQ programs because NAIP is primarily for one purpose - to monitor agricultural landscapes. NAIP photographs are typically acquired during the full-leaf period for local crops, so the bulk of the images are collected from June through August, in contrast to HRO, NAPP, NHAP, and DOQ photographs, often taken during leaf-off conditions. In addition, the NAIP photographs typically have a yearly repeat cycle, while other sources are often spaced at five-year or longer intervals. NAIP photographs may be obtained in hardcopy or digital formats.

Images are most often derived from natural color photographs, although sometimes infrared photographs are taken, and digital aerial scanner images will completely replace them at some point. NAIP images are orthorectified and provided at 1- and 2-meter ground resolutions, with corresponding horizontal accuracies at 5 to 10 meters. Images are provided as photomosaics, either as quarter quadrangles (described above for DOQs) or as compressed county mosaics (CCMs). Data are provided in an NAD83 UTM coordinate system corresponding to the image area.

NAIP images are most useful as a base for digitizing, particularly when information on vegetation type or condition is important (Figure 7-18). Leaf-off images are less useful for mapping vegetation, because differences are most often expressed in the color, brightness, and texture of foliage. While the natural color images typically used for NAIP images are inferior to infrared images, substantial information on vegetation can be collected, and sometimes the NAIP images include an infrared band. This, plus the annual image collection cycle, make these images a valuable source of spatial data.

Figure 7-18: An example NAIP image showing fallow farm fields (*A*), grassland (*B*), forest (*C*), and new construction (*D*). These images are particularly useful for agricultural assessments and detecting fine-detail changes at annual time steps.

National Land Cover Data

While land cover is important when managing many spatially distributed resources, data on land cover are quite expensive to obtain over large areas. These data are often scarce, at low categorical or spatial resolution, and rarely available over broad areas. While individual states, counties, metropolitan areas, or private landholders have developed detailed land cover maps, there have been few national efforts to map land cover in a consistent manner. There are two national data sets available, the Land Use Land Cover (LULC) data from the 1970s and early 1980s, and the National Land Cover Data (NLCD) from the 1990s (Figure 7-19), and from 2001 (Figure 7-20). A third national landcover dataset, NLCD 2006, released in 2001, is based on changes in landcover relative to 2001, observed via moderate-resolution satellite images.

LULC data were developed by the USGS by the interpretation of 1970s and early 1980s aerial photographs. These photographs were taken at a range of scales, although scales between 1:40,000 and 1:60,000 were most commonly used. Land use was manually photointerpreted. Photointerpretation involved viewing the photographs and assigning land use and land cover based on differences in the color, tone, shape, and other information contained in the photograph (Chapter 6). Technicians were trained to become familiar with the land cover and land uses common to an area, and to see how these land uses appeared on photographs. The knowledge of image/land use relationships was most often developed through extensive field surveys in areas for which photographs were available. Minimum mapping units of 4 to 16 hectares were applied, depending on the feature type. Maps were then manually digitized. LULC data are available in either vector or raster formats. Data are tiled by 1:100,000 or 1:250,000 USGS map series boundaries.

NLCD are a more recent and detailed source of national land cover information. NLCD are produced in a cooperative effort by a number of U.S. federal government agencies. Their goal is a consistent, current land cover data record for the conterminous United States. NLCD data have been pro-

Figure 7-19: An example of 1992 National Land Cover Data (NLCD) for the state of Virginia. NLCD categorizes land cover into 21 classes, and are provided in a 30-meter raster cell format. In this example forests are shown in darker tones, with urban and agricultural areas in lighter tones of gray.

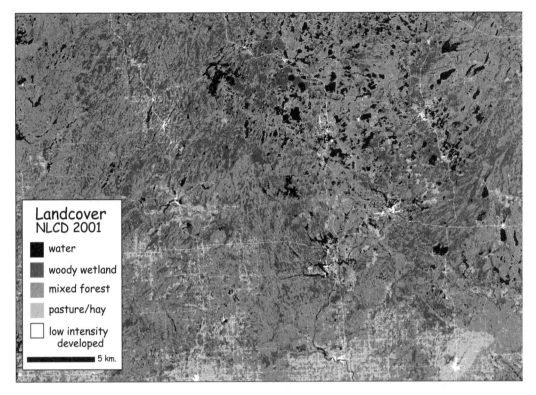

Figure 7-20: An example of NLCD 2001 landcover for a rural region in northern Wisconsin, USA. Note that fine grain features, including road corridors and small lakes, are represented in these data.

duced three times, known respectively as NLCD 1992, NLCD 2001, and NLCD 2006 corresponding to the primary data collection years.

NLCD land cover classifications are based primarily on 30 meter Landsat Thematic Mapper data. NLCD 1992 land cover was assigned to one of 21 classes. Full coverage is obtained from adjacent or overlapping, cloud-free Landsat images. Multiple dates are often acquired in order to improve accuracy and categorical detail through phenologically-driven changes. For example, evergreen forests are more easily distinguished from deciduous forests when both leaf-off and leaf-on images are used. Other spatial data sets are used to improve the accuracy and categorical detail possible through spectral data alone. These data include digital elevation, slope, aspect, Bureau of Census population and housing density data, USGS LULC data, National

Wetlands Inventory data, and STATSGO soils data.

Data are processed in a uniform manner within each state or region, and a national set of categories and protocols is followed. All classifications were subjected to a standardized accuracy assessment, and reported and delivered in a standard format. Accuracy assessments were based on NAPP or other medium- to high-resolution aerial photographs. Areas were stratified based on the photographs, and sampling units were defined. Photointerpretations of land cover were assumed true, and compared to NLCD classification assignments. Errors were noted and reported using standard methods.

NLCD 2001 data differ slightly from NLCD 1992 data. The analysis process for NLCD 2001 was refined to yield more categories, higher accuracy, and a more uniform classification. The classification combined Landsat data from three periods, digital ele-

Table 7-1: NLCD 2001 Land cover classes. NLCD 1992 use a somewhat reduced set.

Water
11 open water
12 perennial ice/snow

Developed
21 open space
22 low intensity
23 medium intensity
24 high intensity

31 bare rock/sand/clay
32 unconsolidate shore

Forested Upland
41 deciduous forest
42 evergreen forests
43 mixed forests

Shrubland
51 dwarf shrub
52 shrub/scrub

Herbaceous Upland Natural
71 grassland/herbaceous
72 sedge/herbaceous
73 lichens
74 moss

Herbaceous Planted/
Cultivated
81 pasture/hay
82 cultivated crop

90 Woody Wetlands
91 palustrine forested wetlands
92 palustrine shrub/scrub wetlands
94 esturine forested wetland

95 Emergent Herbaceous Wetlands
96 palustrine emergent wetlands
97 esturine emergent wetlands
98 palustrine aquatic bed
99 estuarine aquatic bed

vation data, population density, road locations, NLCD 1992, and city lights data. NLCD 2001 are mapped to a set of 29 categories (Table 7-1). The base data were also used to estimate percent impervious surface, and tree canopy density. NLCD are a further improvement on the 2001 data, maintaining categories but improving accuracy and uniformity.

NASS CDL

The National Agricultural Statistical Service (NASS) produces yearly Crop Data Layer (CDL) data, landcover maps that focus on distinguishing major crop types and rotations (Figure 7-21). These data are created from a combination of existing landcover data for non-agricultural lands, multi-date images from mid-resolution satellites such as Landsat and Resourcesat-1, coarser resolution but higher frequency MODIS data for phenological discrimination, and various vector data layers to improve classification accuracy.

CDL landcover classification is based on extensive field surveys conducted by the US Department of Agriculture. Fields are visited, airphotos obtained, and fields, farms, and regions classified by dominant crop types and rotations. Observed crop types are compared to spectral data from satellites, and a classification algorithm developed. Classification methods have changed since 2002, the year nationwide data became available annually. Class assignment accuracies are generally between 85 and 95% for agricultural crops.

CDL data produced annually for most regions, allowing analysis of trends in planting, crop rotations, and harvest. Data may be downloaded for regional, statewide, or sub-state areas, in standard formats and coordinate systems

While NASS-CDL data are the most up to date and accurate landcover classification for agricultural lands, they have limitations. Classification for non-agricultural lands are not as rigorously ground-truthed as agricultural data, and depend on older NLCD classifications. Landcover is classified only for counties with agriculture, although this is a surprisingly large proportion of the country. The 30 to 60 meter cell size is quite good for such a large-area classification, but still too small for field-level assessments, and is better suited to farm-level and larger analyses.

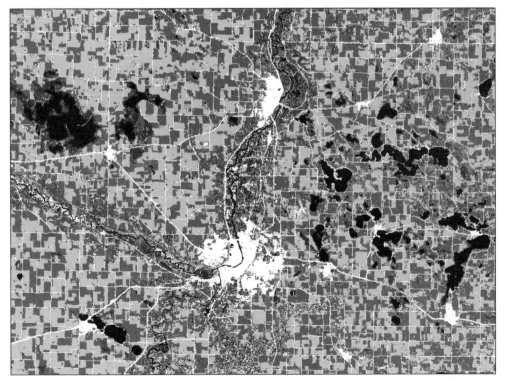

Figure 7-21: An example of NASS landcover data, here for a region in southern Minnesota. Cities and roads are shown in white, lakes in black, wetlands in dark gray, and crops in various shades of gray. The pattern here is dominated by soybeans and corn, large blocks in two mid-shades of gray.

National Wetlands Inventory

Data on the location and condition of wetlands are available for much of the United States through the National Wetlands Inventory (NWI) program. NWI data are produced by the US Fish and Wildlife Service. NWI data portray the extent and characteristics of wetlands, including open water (Figure 7-22), and are available for approximately 90% of the conterminous United States. About 60% of the conterminous United States is available in digital formats. NWI data were produced through the 1970s and 1980s, with an update in the 1990s. Decadal updates are planned.

NWI data were produced through a combination of field visits and airphoto interpretation. Spring photographs at a range of scales and types are used. Color infrared photographs at a scale of 1:40,000 were commonly used, however black and white photographs and scales ranging between 1:20,000 and 1:62,500 have been employed. Spring photographs typically record times of highest water tables and are most likely to record ephemeral wetlands. There is substantial year-to-year variability in surface water levels, and hence there may be substantial wetland omission when photographs are acquired during a dry year.

NWI data provide information on wetland type through a hierarchical classification scheme, with modifiers. Wetlands are categorized as part of a lacustrine (lake), palustrine (pond), or riverine system. Subsystem designators then specify further attributes, to record if the wetland is perennial, intermittent, littoral or deep water. Further class and subclass designators and modifiers provide additional information on wetland characteristics. A shorthand designator is often used to specify the wetland class. A wetland may be designated L1UB2G, as system = lacustrine (L), subsystem = limnetic (1), class = unconsolidated bottom (UB), with subclass = sand (2), and a modifier indicating the wetland is intermittently exposed (G).

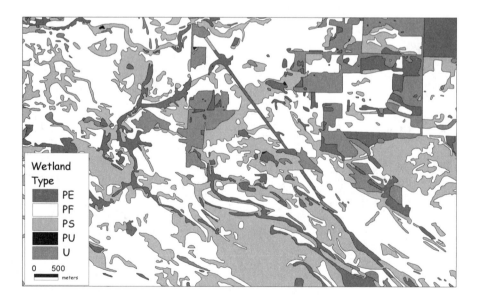

Figure 7-22: An example of national wetlands inventory (NWI) data. Digital NWI data are available for most of the United States, and provide information on the location and characteristics of wetlands.

The *minimum mapping unit* (MMU) is the target size of the smallest feature captured. Features smaller than the MMU are not recorded in these data. NWI data typically specify MMUs of between 0.5 and 2 hectares. MMUs vary by vegetation type, film source, region, and time period. MMUs are typically largest in forested areas and smallest in agricultural or developed areas, because it is more difficult to detect many forested wetlands. MMUs also tend to be larger on smaller-scale photographs. The MMU, scale, and other characteristics of the wetlands data are available in map-specific metadata.

NWI data do not exhaustively define the location of wetlands in an area. Because of the photo scales and methods used, many wetlands are not included. Statutory wetland definition typically includes not only surface water, but also characteristic vegetation or evidence on the surface or in the soils that indicates a period of saturation. Since this saturation may be transient or the evidence may not be visible on aerial photographs, many wetlands may be omitted from the NWI. Nonetheless, NWI data are an effective tool for identifying the location and extent of large wetlands, the type of wetland, and for directing further, more detailed ground surveys.

Digital Soils Data

The Natural Resource Conservation Service (NRCS) of the United States Department of Agriculture has developed three digital soils data sets. These data sets differ in the scale of the source maps or data, and thus in the spatial detail and extent of coverage. The National Soil Geography (NATSGO) data set is a highly generalized soils map for the continental United States, developed from small scale maps. NATSGO data have limited use for most regional or more detailed analyses and will not be further discussed here. State Soil Geographic (STATSGO) data are intermediate in scale and resolution, and Soil Survey Geographic (SSURGO) data provide the most spatial and categorical detail.

SSURGO data are intended for use by land owners, farmers, and planners at the large farm to county level. SSURGO maps indicate the geographic location and extent of the soil map units within the soil survey area (Figure 7-23). Soil map units typically correspond to general grouping, called phases, of detailed soil mapping types. These detailed mapping types are called soil series. There are approximately 18,000 soil series in the United States, and several phases for most series, so there are potentially a large number of map units. Only a small subset of series is likely to occur in a mapped area, typically fewer than a few hundred soil series or series phases. A few to thousands of distinct polygons may occur.

SSURGO data are not intended for use at a site-specific level, such as crop yield predictions for an individual field or septic system location within a specific parcel. SSURGO data are more appropriate for broader-scale application, such as identifying areas most sensitive to erosion, or planning land use and development. SSURGO data and the soil surveys on which they are based are the most detailed soils information available over most of the United States.

SSURGO data are developed from a combination of field and photo-based measurements. Trained soil surveyors conduct a series of field transects in an area to determine relationships between soil mapping units and terrain, vegetation, and land use. Aerial photographs at scales of 1:12,000 to 1:40,000 are used in the field to aid in location and navigation through the landscape. Soil map unit boundaries are then interpreted onto aerial photographs or corresponding orthophotographs or maps. Typical photo scales are 1:15,840, 1:20,000, or 1:24,000. These maps are then digitized in a manner that does not appreciably affect positional accuracy. Soil surveys are often conducted on a county basis, so county mosaics of SSURGO data are common. SSURGO data are reported to have positional accuracy no worse than 13 meters (40 feet) for approximately 90% of the well-defined points when SSURGO data are compiled at 1:24,000-scale.

SSURGO data are linked to a Map Unit Interpretations Record (MUIR) attribute data base (Figure 7-24). Key fields are provided with the SSURGO data, including a unique identifier most often related to a soil map unit, known as the map unit identifier (muid). Tables in the MUIR data base are linked via the muid, and other key fields. Most tables contain the muid field, so a link may be created between the muid value for a polygon and the muid value in another table, such as the Compyld table (Figure 7-24). This creates an expanded table that may be further linked through cropname, clascode, or other key fields. These table structures and linkages are discussed in Chapter 8.

Variables include an extensive set of soil physical and chemical properties. Data are reported for water capacity, soil pH, salinity, depth to bedrock, building suitability, and most appropriate crops or other uses. Most MUIR data report a range of values for each soil property. Ranges are determined from representative field-collected samples for each map unit, or from data collected from similar map units. Samples are analyzed using standardized chemical and physical methods.

STATSGO digital soil maps are smaller scale and cover broader areas than SSURGO soil data. STATSGO data are typically created by generalizing SSURGO data. If SSURGO data are not available, STASGO data may be generated from a combination of topographic, geologic, vegetation, land use, and climate data. Relationships between these factors and general soil groups are used to create STATSGO maps. STASTGO map units are larger, more generalized, and do not necessarily follow the same boundaries as SSURGO map units (Figure 7-25). In addition, STATSGO polygons contain from one to over 20 different SSURGO detailed map units. A SSURGO map unit type is a standard soil type used in mapping. It is often a phase of a map series. Each STATSGO map unit may be made up of thousands of these more detailed SSURGO

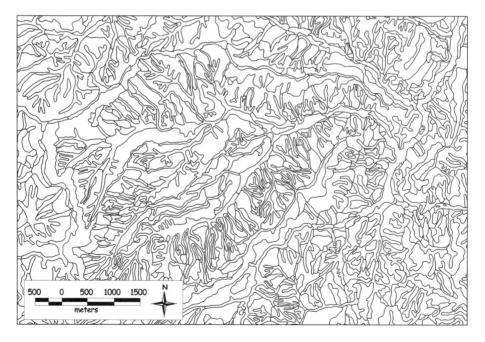

Figure 7-23: An example of SSURGO digital soils data available from the NRCS. Each polygon represents a soil mapping unit of relatively uniform soil properties.

SSURGO Attribute Data Tables

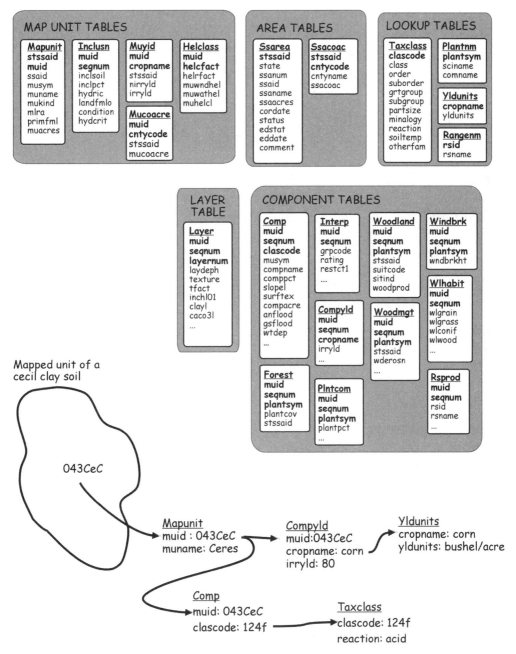

Figure 7-24: The database schema associated with the SSURGO digital soils data. Variables describing soil characteristics are provided in a set of relatable tables. Keys in each table, shown in bold, provide access to items of interest. Codes provided with the digital geographic data, e.g., the muid, provide a link to these data tables. The relation of a mapped soil polygon to attribute data is shown in the example at the bottom. The muid is related from the MAP UNIT and COMP tables, which in turn are used to access other variables through additional keys.

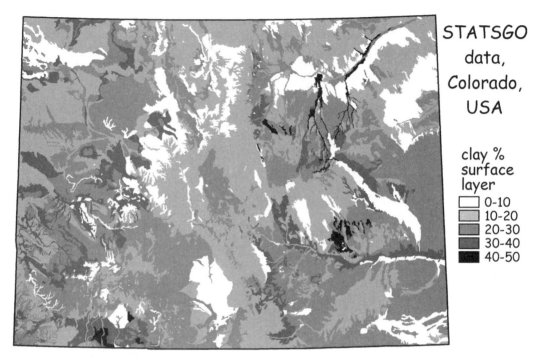

Figure 7-25: An example of STATSGO data for Colorado, USA.

polygons, and many different SSURGO map unit types can be represented within a STATSGO polygon. STATSGO data provide information on some of this variability. Data and properties on multiple components are preserved for each STATSGO map unit.

STATSGO data are developed via a redrafting of more detailed data onto 1:250,000-scale base maps. These data may be existing county soil surveys that form the basis for SSURGO data, or a combination of these and other materials such as previously published statewide soil maps, satellite imagery, or other statewide resource maps. Soil map units are drafted onto mylar sheets overlain on 1:250,000-scale maps. A minimum mapping unit of 625 hectares is specified, and map unit boundaries are edgematched across adjoining 1:250,000-scale maps. These features are then digitized to produce digital soils data for each 1:250,000-scale map. Data may then be joined into a statewide soils layer. Geographic data are then attached to appropriate

attributes. Data are most often digitized in a local zone UTM coordinate system, and converted to a common Alber's equal area projection.

Digital Floodplain Data

Floods cause between $5 billion and $18 billion in damage each year in the United States, and this damage could be reduced with the effective application of GIS. A first step in reducing damage is the mapping of flood prone areas. The Federal Emergency Management Agency (FEMA) develops and disseminates flood hazard maps, commonly known as floodplain maps. Floodplain maps were produced via a variety of methods prior to 1968. These maps located the boundary of areas with a 1% or higher annual chance of flooding, commonly known as 100-year floodplain maps.

FEMA occasionally updates these floodplain maps. The objectives are to develop maps of flood hazard via an improved process, better input data, in a uniform digital format, and to integrate map creation into ongoing local and state government mapping and planning efforts.

Floodplain maps are used for a number of purposes, chief among them setting flood insurance rates. Over 19,000 communities participate in the National Flood Insurance Program (NFIP). The NFIP allows the U.S. federal government to guarantee flood insurance for communities with floodplain management ordinances. These reduce flooding risks for redevelopment and new construction. This helps communities manage risk and avoid losses due to flooding.

Digital floodplains maps are produced, often by cooperating partners or agencies, to define regions within a 100-year floodplain. Boundary accuracy may be challenged, usually by individual landowners, businesses, or

municipalities, and the proposed adjustments evaluated and included in the floodplain map if they improve accuracy.

Maps are most often produced by cooperating technical partners (CTP). Expertise, training, and demonstrated capabilities are required of each CTP. Best technical practices for digital floodplain data development are defined by FEMA, and training is offered to teach best available methods and increase data quality. Guidelines and specifications are defined for the use of LiDAR, datum conversions, and the analysis and mapping of specific landforms. Protocols for verifying and revising maps are defined as part of the data evaluation process.

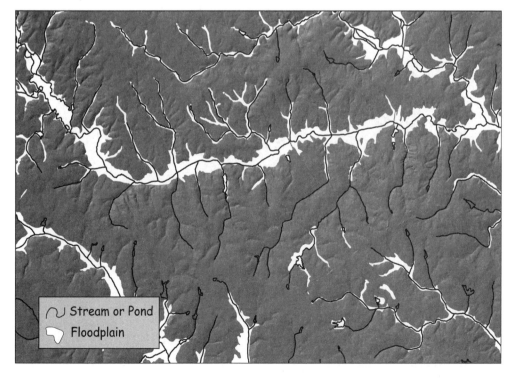

Figure 7-26: An example of FEMA floodplain data for a region near Morgan, Georgia, USA.

Climate, Geology, and Other Environmental Data

Various other spatial environmental data sets are available, including weather, climate, water chemistry, energy resources, and others. Here we provide examples, but there are many types and sources of basic environmental data available.

The National Climatic Data Center (NCDC) maintains historical climate records for the U.S., and provides their data through a web-portal (http://gis.ncdc.noaa.gov/maps). Recording stations may be selected by various criteria, including geography, measured variables, or length of record. Climate data have been converted to spatial fields, and are distributed through the PRISM initiative (prism.oregonstate.edu, see Figure 7-28).

Mineral resources data are available from the US Geological Survey, at http://mrdata.usgs.gov/. These data include maps of basic national geology (Figure 7-27), as well as spatial and tabular data on special-

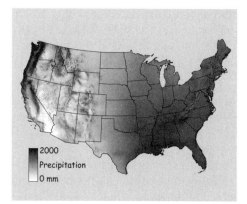

Figure 7-28: U.S. average precipitation, measured from 1971-2000, interpolated from weather stations across the U.S. to create a raster grid (data from the PRISM project).

ized themes such as mineral deposits, mines, claims, smelters and other processing facilities, and energy resources.

Spatial data are available for a range of other environmental parameters, including air pollution, pollutant and contaminant distribution, and some water pollutants through the Environmental Protection Agency, many at www.epa.gov/data/.

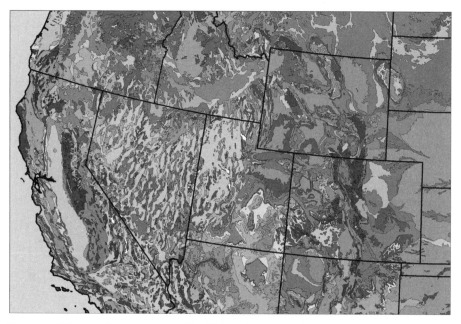

Figure 7-27:A general geologic map of the United States, based on spatial data from the USGS.

Digital Census Data

The U.S. Census Bureau developed and maintains a database system to support the national census. This system is known as the Census TIGER system (Topologically Integrated Geographic Encoding and Referencing). The TIGER system is used to organize areas by state, county, census tract, and other geographic units for data collection and reporting. It also allows the assignment of individual addresses to geographic entities. The census TIGER system links geographic entities to census statistical data on population size, age, income, health, and other factors (Figure 7-29). These entities are typically polygons defined by roads, streams, political boundaries, or other features. The TIGER system is a key government tool in the collection of census data. TIGER also aids in the application of census data during the apportionment of federal government funds, in congressional re-districting, in transportation management and planning, and in other federal government activities.

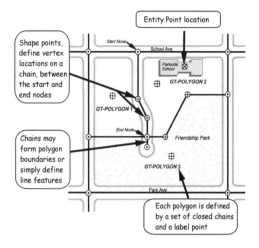

Figure 7-30: TIGER data provide topological encoding of points, lines (chains), and polygons. (from U.S. Dept. of Commerce)

TIGER/Line files are at the heart of the system. They define line, landmark, and polygon features in a topologically integrated fashion. Lines (called chains in the TIGER system) most often represent roads, hydrography, and political boundaries, although railroads, power lines, and pipelines are also represented. Polygon features include census tabulation areas such as census block groups and tracts, and area landmarks such as parks and cemeteries. Point landmarks such as schools and churches may also be represented. Points, lines, and polygons are used to define these features (Figure 7-30).

Nodes and vertices are used to identify line segments. Topological attributes are attached to the nodes and lines, such as the polygons on either side of the line segment, or the line segments that connect to the node. Point landmarks and polygon interior points are other topological elements of TIGER/Line files.

Population change by census tract, 1990-1999

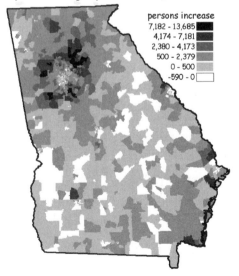

persons increase
7,182 - 13,685
4,174 - 7,181
2,380 - 4,173
500 - 2,379
0 - 500
-590 - 0

Figure 7-29: Digital census data provide spatially referenced demographic and other data for the U.S.

TIGER/Line files contain information to identify street address labels. Starting and ending address numbers are recorded corresponding to starting and ending nodes (Figure 7-31). Addresses may then be assigned within the address range. The system does not allow specific addresses to be assigned to specific buildings. However it does restrict the addresses on a city block to a limited range of numbers, something of great use to field workers responsible for collecting census information.

TIGER/Line files are organized in 17 different record types. A collection of records reports the location and attribute information about a set of census features, including the location, shape, addresses and other census attributes for a county. There is an identifier based on the U.S. Federal Information Processing Standards Code (FIPS) that is used to identify the file and record type. The record types for a county are organized in files that include the unique FIPS code. For example, all records of type 1 for Pulaski County, Georgia, FIPS code 13235, are delivered in a file named tgr13235.rt1. All or part of the 17 files are used to describe

geographic features within the county. Separate records define the end nodes for lines, the vertices for lines, line identifiers, landmark features, names, internal points, ZIP codes, and index and linkage attributes. The index and linkage attributes are used to combine the data across record types. The linkages and indices allow groups of points, lines, or areas to represent features. Identifiers are also included to link areas to Summary Tape (STF) files, Congressional District files, County files, Census Tract/ Block Numbering files, and other data files produced by the U.S. Census Bureau.

Specialized software packages are available to convert TIGER/Line and related census files to data layers in specific GIS formats, and many GIS provide utilities to ingest TIGER/Line and census data. These products are often bundled with census data and enable the selection of areas and attributes based on states, counties, zip codes, or other area units. TIGER/Line files

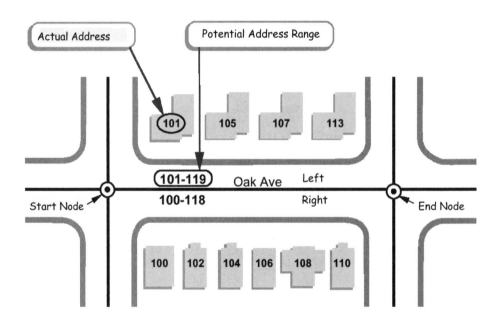

Figure 7-31: TIGER data provide address ranges for line (chain) segments. These ranges may be distributed across the line, giving approximate building locations on a street. (from U.S. Dept. of Commerce)

contain a wealth of attributes for line and area features, including state and county boundaries, school district, city, township, and other minor civil divisions; place names, park locations, road and other infrastructure attributes, as well as summary population data. These data may be extracted in a customized manner.

Many U.S. government data sets are provided with codes compatible to U.S. Census Bureau data. For example, data are delivered in Census-compatible units and codes by the U.S. Department of Education (http://nces.ed.gov/ccd/), the U.S. Department of Transportation (http://www.bts.gov/), and the U.S. Center for Disease Control (http://wonder.cdc.gov). Data from these organizations are delivered with codes needed to link statistics to geography, for example, the average traffic fatality rate from 1997-2006, as shown in Figure 7-32.

Summary

Digital data are available from a number of sources, and provide a means for rapidly and inexpensively populating a GIS database. Most of these data have been produced by government organizations and are available at little or no cost, often via the internet. Data for elevation, transportation, water resources, soils, population, land cover, and imagery are available, and should be evaluated when creating and using a GIS.

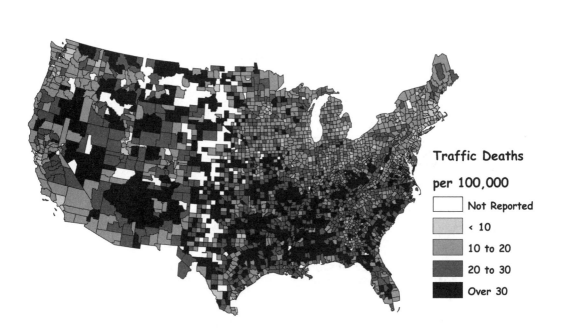

Figure 7-32: Traffic fatality data, showing average number of deaths per 100,000 persons over 1997-2006, derived from data reported by the U.S. Center for Disease Control. These data are published with links to U.S. Census recognized geographies, as are much other data collected by the Federal Government. Here, Federal Information Processing Standard (FIPS) codes were published with the CDC data, and used to link to county boundary files. Note the generally high death rates in some southern and interior western counties, and that New York drivers, counter to reputation, appear to be among the safest on the roads.

Suggested Reading

Broome, F.R. and Meixler, D.B. (1990). The TIGER database structure. *Cartography and Geographic Information Systems*, 17:39-47.

Carter, J.R. (1988). Digital representations of topographic surfaces, *Photogrammetric Engineering and Remote Sensing*, 54:1577-1580.

Chrisman, N.R.(1990). Deficiencies of sheets and tiles: Building sheetless databases, *International Journal of Geographical Information Systems*, 4:157-168.

Decker, D. (2001). *GIS Data Sources*. New York: Wiley.

Elassal, A.A. & Caruso, V.M. (19840. *Digital Elevation Models. Geological Survey Circular 895-B*. Reston: United States Geological Survey.

Gesch, D., Oimoen, M., Greenlee, S., Nelson, C., Steuck, M., & Tyler, D. (2002). The national elevation dataset. *Photogrammetric Engineering and Remote Sensing*, 68:5-11.

Goodchild, M F., Anselin, L., & Deichmann, U. (1993). A framework for the areal interpolation of socioeconomic data. *Environment and Planning*, 25:383-397

Gorokhovich, Y., and Voustianiaouk, A. (2006). Accuracy assessment of the processed SRTM-based elevation data by CGIAR using field data from USA and Thailand and its relation to the terrain characteristics. *Remote Sensing of Environment*, 104:409-415.

Horner, C., Huang, C., Yang, L., Wylie, B, and Coan, M. (2004). Development of a 2001 national land-cover database for the United States. *Photogrammetric Engineering and Remote Sensing*, 70:829-840.

Luzio, M.di, Arnold, J.G., & Srinivasan, R. (2004). Integration of SSURGO maps and soil parameters within a geographic information system and nonpoint source pollution model system. *Journal of Soil and Water Conservation*, 59:123-133.

Lytle, D.J., Bliss, N.B., and Waltman, S.W. (1996). Interpreting the State Soil Geographic Database (STATSGO). In Goodchild, M.F., Steyaert, L.T., Parks, B.O., Johnston, C., Maidment, D., Crane, M., & Glendinning, S. (Eds.). *GIS and Environmental Modeling: Progress and Research Issues*. Fort Collins: GIS World.

Marx R.W. (1986). The TIGER system: automating the geographic structure of the United States Census. *Government Publications Review*, 13:181-201.

Maune, D.F. (2007). *Digital Elevation Model Technologies and Applications: The DEM User's Manual* (2nd ed.). Bethesda: American Society of Photogrammetry and Remote Sensing.

Openshaw, S. & Taylor, P. (1979). A million or so correlation coefficients: three experiments on the modifiable areal unit problem. In N. Wrigle (Ed.), *Statistical Applications in the Spatial Sciences* (pp. 127-144). London: Pion.

Smith, B., & Sandwell, D. (2003). Accuracy and resolution of shuttle radar topography mission data. Geophysical Research Letters, 30:1-20

Taylor, P.J., & Johnston, R.J. (1979). *Geography of Elections*. Hammondsworth: Penguin.

Wilen, B.O., & Bates, K.M. (1995). The US Fish and Wildlife Service's National Wetlands Inventory Project. *Vegetatio*, 118:153-169.

Study Questions

7.1 - What are some advantages and disadvantages of using digital spatial data?

7.2 - What are the most important questions you must ask before using already-developed spatial data?

7.3 - For each of the following data sets, tell us who produces them, what are the source materials, what do the datasets contain, their grain sizes and accuracies, and they how they are delivered: digital raster graphics (DRGs), digital line graphs (DLGs), digital elevation models (DEMs), digital orthophotoquads (DOQs), digital floodplain data, National Wetlands Inventory data (NWI), SSURGO and STATSGO soils data, TIGER census data, and national land cover data (NLCD) sets.

7.4 - What is edge matching and why is it important?

7.5 - Can you identify and describe the characteristics of three different sources of digital elevation data? What are the pros and cons of each source?

7.6 - Visit one of the websites mentioned in this chapter, or in the appendices at the end of this book, and download several data layers of an area of interest. If you have access to a GIS, try to import these data and display them.

8 Attribute Data and Tables

Introduction

We have described how spatial data in a GIS are often split into two components, the coordinate information that describes object geometry, and the attribute information that describes the non-spatial properties of objects. Because these non-spatial data are frequently presented to the user in tables, they are often referred to as tabular data. Tabular data summarize the most important nonspatial characteristics of each carto-graphic object, for example, a table of attributes that describe counties (Figure 8-1). In this example the attributes include the county name, Federal Information Process-ing Standards (FIPS) code, population, area, and population density.

Attribute information in a GIS are typi-cally entered, analyzed, and reported using a *database management system* (DBMS), a specialized computer program for organizing and manipulating data. The DBMS stores the properties of geographic objects and the rela-tionships among the objects. A DBMS incor-porates a specialized set of software tools for managing tabular data, including those for efficient data storage, retrieval, indexing, and reporting. DBMS were initially developed in

Name	FIPS	Pop90	Area	PopDn
Whatcom	53073	128	2170	59
Skagit	53057	80	1765	45
Clallam	53009	56	1779	32
Snohomish	53061	466	2102	222
Island	53029	60	231	261
Jefferson	53031	20	1773	11
Kitsap	53035	190	391	485
King	53033	1507	2164	696
Mason	53045	38	904	42
Gray Harbor	53027	64	1917	33
Pierce	53053	586	1651	355
Thurston	53067	161	698	231
Pacific	53049	19	945	20
Lewis	53041	59	2479	24

Figure 8-1: Data in a GIS include both spatial (left) and attribute (right) components.

the 1960s, and refinements since then have led to robust, sophisticated systems employed by government, businesses, and other organizations. A somewhat standard set of DBMS tools and methods have been developed and are provided by many vendors.

Note that the terms DBMS and database are sometimes used interchangeably. In most cases this is incorrect and in all cases imprecise. A DBMS is a computer program that allows you to work with data. A database is an organized collection of data, often created or manipulated with the help of a DBMS. The database may have a specific form dictated by a DBMS, but it is not the system.

Students often struggle with relational databases at first, and often ask "why bother? Can't we just use a spreadsheet?" Many more people are familiar with spreadsheet forms, programs, and manipulation, and don't see the value added when adopting a DBMS. A short example may help.

Consider the file shown in Figure 8-2, representing business orders. Each row records the purchaser, an order number, and the items ordered. Spreadsheets typically present data like this in a single, "flat file" with two dimensions, rows and columns. Because orders may contain multiple items, we need multiple columns with copies of the item/quantity pair. For example, order number five by Atom Ant includes two items, two B52s and two CR7s, while order number three by Paul Smith has four items, or a total of 8 columns for items. Larger orders would require additional columns.

This storage form has two characteristics. First, we either have to limit the number of items per order (rarely a good thing for a business), or else not know how many columns our database might have, which would complicate programing and management. Second, and more importantly, we can easily have most of the storage in our database contain nothing. We may have thousands of orders with one or two items. However, if we have one order with 50 items, we have to add enough columns to accommodate 50 item/quantity pairs, even in orders with one or two items. As with orders 1, 2, 4, 6, 7, and 8 in the table below, many of the cells will be empty. We can easily have a database that is mostly empty cells. Computer memory has become quite inexpensive, so you may think this a minor disadvantage. However more data means longer processing times, to the extent that the database may not function. This flat file structure is flawed, in both inefficient use of space and slow processing.

There is another obvious disadvantage of this structure. Note that there are two orders from Paul Smith, order #3, and order #8. This means there are redundant sets of information for the same customer. We have

name	surname	address	phone #	order #	item	qty	item	qty	item	qty	item	qty
Leo	Durocher	112 Beal St	5-1307	1	CR7	1						
Rudy	Valentini	1 Hispanola Dr	4-2706	2	F15	1						
Paul	Smith	99 Upstate Ln	0-0000	3	GTO	3	F15	1	B52	1	SR71	1
Adam	Smith	1 Wall St	1-2334	4	626	1						
Atom	Ant	685 Hanbar Rd	4-1222	5	B52	2	CR7	2				
William	Smith	202 Dinkytown	9-9199	6	F111	2						
Alice	Paul	5 Free St.	4-4178	7	SR71	1						
Paul	Smith	99 Upstate Ln	0-0000	8	F15	1						

Figure 8-2: An example of database, in a flat file format.

his first and last name, address, and phone number repeated in both orders. This wastes space and makes editing more cumbersome and error-prone. If Paul Smith changes his phone number, we must search through every line in our database and change every instance of an order that contains Paul Smith's phone number. Difficult or redundant editing is an additional disadvantage of a spreadsheet-like file system.

Functions and programs may be written to address the inefficient use of space, slow processing, and difficulty editing in spreadsheets or other apparently flat file formats. These programs often require specific knowledge of the spreadsheet structure, and so depend on the arrangement and number of columns. While these work-arounds are possible, they are often complicated and require substantial program maintenance. Database management systems were developed to overcome these redundancies and inefficiencies by adding structure to data files in a standard way. While spreadsheets may be used for simple collections of data, DBMS are better for all but the simplest problems.

DBMSs provide other advantages. They provide *data independence*, a valuable characteristic when working with large data sets. Data independence allows us to make changes in the database structure in ways that are transparent to any use or program. This means restructuring the database does not require a user or programmer to modify their procedures. Before data independence became widespread, organizations frequently spent considerable time and money re-writing applications and re-training users with each change of the data structure. Data independence avoids this.

DBMS may also provide for *multiple user views*. Different users may require different information from the database, or the same information delivered in different formats or arrangements. Profiles or forms can be developed that change the way data are provided to each program or user. DBMSs are able to automatically reformat data to meet the viewing preferences of various users. The DBMS eliminates the need to have copies of the data for each user, by changing the presentation to meet each specific need.

A DBMS also allows *centralized control and maintenance* of important data. One "standard" copy of the data may be maintained, and updated on a regular, known basis. These data may be time stamped or provided with a version number to aid in management. These data are then distributed to the various users. A single person or group may be charged with maintaining data currency, quality, and completeness, and with resolving contradictions or differences among various versions of the database.

Adopting a DBMS may come at some cost. Specialized training may be required to develop, use, and maintain a database. Defining the components of a database and relationships among them may be a complex task that may require specialists. Structuring the database for efficient access or creating customized forms will often require significant effort. The software itself may be quite expensive, although free, stable, open-source database management software is available. However, for many users, the value of the DBMS and database development far outweighs these costs.

Database Components and Characteristics

The basic components of a traditional database are *data items* or *attributes*, the indivisible named units of data (Figure 8-3). These items can be identifiers, sizes, colors, or any other suitable characteristic used to describe the features of interest. Attributes may be simple, for example one word or number, or they may be compound, for example an address data item that consists of a house number, a street name, a city, and a zip code.

Items have a *type* and a *domain*, that restrict the values they may take. Types define essential characteristics of an item. Common types include real numbers, integer numbers, both of various lengths, hexadecimal numbers, text fields, hyperlinks, and binary large objects (blobs). Domains define the acceptable values an item may take, for example, integers may be restricted to be larger than 0 but smaller than 10, or there may be a type name "color" that can only take on the values "red", "green", "blue", "yellow", "cyan", or "magenta."

A collection of related data items that are treated as a unit represents an *entity*. In a GIS, the database entities are typically roads, counties, lakes, or other types of geographic features. A specific entity, such as a specific county, is an *instance* of that entity. Entities are defined by a set of attributes and associated geographic data. In our example in Figure 8-3, the attributes that describe a county include the name, a FIPS code, the 1990 population in thousands of persons, area, and the population density. These related data items are often organized as a row or line in a table, called a *record*. A *file* may then contain a collection of records, and a group of files may define the database. Specific database systems often define the terms differently for each of these parts. For example, in the relational database model the record may be called a *row* or an *n-tuple*, and the records are typically organized into a *relational table*.

You should note that the concept of an entity, when referred to in a database, may be slightly different than an entity in a GIS data model. This difference stems from two different groups, geographers and computer scientists, using a word for different but related concepts. An entity in a geographic data model is often used for the real-world item or phenomenon we are trying to represent with a cartographic object. These entities are typically a physical phenomenon, for example, a lake, city, or building, but they may also be a conceptual phenomenon, such as a property boundary. In contrast, computer scientists and database managers often define an entity as the principal data objects about which information will be collected. In the DBMS literature, the entity is the data object that denotes a physical thing, and not the thing itself. Thus properties of entities

Figure 8-3: Components of an attribute data table.

Name	FIPS	Pop90	Area	PopDn
Whatcom	53073	128	2170	59
Skagit	53057	80	1765	45
Clallam	53009	56	1779	32
Snohomish	53061	466	2102	222
Island	53029	60	231	261
Jefferson	53031	20	1773	11
Kitsap	53035	190	391	485

and relationships among entities refer to the structure of the DBMS. These properties and relationships are used to represent the real-world phenomenon. This is a subtle distinction in terminology, but these different definitions can lead to confusion unless the difference in meanings is noted. For the remainder of this chapter we will use the definition of an entity as a data object.

A DBMS typically supports complex structures, primarily to provide data security, stability, and to allow multiple users or programs to access the same data simultaneously. Database users often demand shared access, that is, multiple users or programs may be allowed to open, view, or modify a data set simultaneously. However if each program or user has direct file access, then multiple copies of a database may be open for modification at the same time (Figure 8-4, top). With direct file access, multiple users may try to write to the data file simultaneously, with unforeseen results. The data saved may be the most recent, the first updates, or some mix in between. Because data may be lost or modified in unforeseen ways with direct file access, a DBMS may be designed to manage multi-user access (Figure 8-4, bottom). Some DBMS manage shared files and data, and enforce a pre-determined precedence in simultaneously accessed files. The DBMS may act as an intermediary between the files and the application programs or user. The DBMS may prevent errors due to simultaneous access. Other DBMS programs do not manage simultaneous access, and users of such systems generally must avoid opening multiple copies of the database at one time.

The DBMS is sometimes referred to as a database *server*, and the applications programs as *clients*. The server provides or "serves up" data to the client applications. Clients may be built in by the DBMS vendor, written by the DBMS user, or sold as add-ons by third party software developers. These clients may operate on the same computer as the DBMS server software, or they may provide requests from remote machines

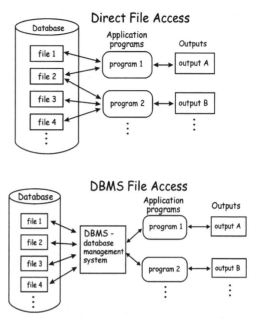

Figure 8-4: Direct and database management system file access (adapted from Aronoff, 1989).

over a network connection. A single server may be configured to respond to many client programs running on separate machines.

The separation of data and functions into multiple levels is often referred to as a *multi-tiered architecture* (Figure 8-5). Data are primarily stored at the lowest tier. These data may be of diverse types, including coordinate data, attributes, text, images, sound, video recordings, or other important, persistent data.

Data sets at the lowest tier may be managed by an individual database system (Figure 8-5). The system or programs that access the first tier, at the bottom of a multi-tier system is often called a *transaction manager*. This transaction manager typically takes requests from higher tiers and searches the relevant portions of the database to identify the requested data, or perform the requested operation.

The next tier in the multi-tier architecture is often referred to as an *applications server* (Figure 8-5). The use of the term server may be a bit confusing, because

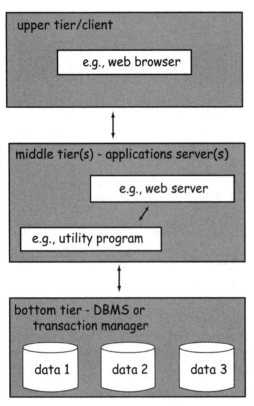

Figure 8-5: An example of a multi-tier architecture. Data are accessed by the bottom tier, and requests made at an upper tier. There may be several middle tiers that act upon the request and translate/process information passed between upper and bottom tiers.

server is also used to describe a computer on which data are stored, and also to describe the database management program in two-tier systems. In the context of a multi-tiered database architecture, an *applications server* is software that passes requests from higher-level tiers to the transaction manager. It converts input from above into a set of instructions the transaction manager below can "understand." A single request from a higher tier may require the applications server to query several databases. For example, a real-estate agent may want to identify all houses in a certain price range, in good school districts, and near rapid transit stations for a prospective buyer. The applications server may generate three different requests -- one to identify houses in a price range, a second to identify good school districts, and a third to find rapid transit stations. The applica-

tions server may then perform the operations to determine where these important criteria are met: to identify the set of houses that are in the given price range, to narrow the search to those that are only in the good school districts, to find the distance for each of these houses to the nearest rapid transit station, and to identify the station by name and station characteristics.

Besides passing requests to lower tiers, the application server may also perform other tasks. For example, it may determine if the real-estate agent has the proper clearance to access the housing data. In addition, the applications server may check to see if the agent has a profile, and so handle requests in a certain way.

The uppermost tier of multi-tier architectures is typically a *user interface* (Figure 8-5). This tier may be a display by a single-purpose program, or a web-based interface with the primary purpose of gathering requests from the user, and presenting information back to the user based on those requests.

Multi-tiered architectures are adopted primarily to insulate the user interface from the processing and data at lower tiers, and to allow access to a more diverse range of data through the lowest tiers. The parts are easier to change when they are isolated, and new, different resources may be more easily integrated. If a company decides to redesign their data entry interface, they may do so easily if the user interface is distinct from the tiers below. They do not have to worry about how the applications server or transaction manager operate. The integration of a new database technology is often easier with multi-tiered architectures. For example, new data types or programs may be incorporated into the system by adding the new data to the database and modifying the transaction manager at the lowest tier, without modifying the applications server or user interface in the tiers above.

Multi-tier architectures are by their nature more complex and variable than two-tier architectures. For example, different implementations of multi-tier architectures may split operations differently between the tiers. In our real-estate example above, one architecture may incorporate all the database query and processing operations into the transaction manager. Another multi-tier architecture may perform the query with the transaction manager, and perform the spatial operations (houses in good school districts, and good houses near transit stations) in the applications server tier. Neither architecture is universally better, rather an organization must adopt an architecture that best suits its needs.

Multi-user access adds substantial overhead and complexity to processing. For example, the server must ensure that when several copies of a database are accessed, changes to the database must be reconciled on resubmission. If two different clients have altered different variables in a database, these two sets of changes must be integrated when the database is stored. If the updates from two clients conflict, such as when one client deletes a record while a second client modifies a value for the same record, the program must resolve the differences; perhaps one user has higher priority, the most recent changes are enforced, or a message is sent to an operator noting the ambiguity.

Physical, Logical, and Conceptual Structures

A database may be viewed as having conceptual, logical, and physical structures. These structures define the entities and their relationships and specify how the data files or tables are related one to another.

The conceptual structure is often represented in a *schema*. This structure may be succinctly described in standard shorthand notations, e.g., using *entity-relationship* diagrams, also known as E-R diagrams. We will not describe E-R diagrams or other conceptual methods here.

A schema is a compact graphical representation of the conceptual model, the entities, and the relationships among them. The relationships may be one-to-one, between one entity and another. They may be one-to-many, or many-to-many, connecting several objects. These relationships are represented by lines connecting the entities, and can indicate if the relationships are between one or many entities.

Databases also have a *logical database design*. Most databases are developed using commercial programs that use a specific database model or type. This constrains the specific way a conceptual model may be implemented, and so the conceptual model must be converted to a specific set of elements that define the structure and interaction of database components. The logical structure influences the specific physical structures that are adopted.

Physical structuring may by achieved in many ways. One common method uses file pointers to connect records in one file with those in other files. Much of this structure is designed to speed access, aid updates, and provide data integrity. This structuring is part of the *physical design* of the database. The design typically strives to physically cluster or link data used together in processes so that these processes may be performed quickly and efficiently.

Relational Databases

Relational databases have grown to become the most common database design since their introduction in 1968. There are a number of reasons relational databases are used widely.

The relational model is more flexible than most other designs. The tables structure does not restrict processing or queries, and the organization is simple to understand, learn, and implement relative to other database designs. It can accommodate a wide range of data types, and it is not necessary to know in advance the kind of queries, sorting, and searching that will be performed on the database.

The tables in a relational database design are also called *relations,* shown for forest and related recreation data in Figure 8-6. Entities are represented by rows in a table. In our forest data example, there may be a forest table with a row for each forest, and other tables representing the trails, trail features, and recreational opportunities. The rows are also called *records* or *tuples.* Columns contain *attributes* that describe entities. Attributes may also be referred to as *items* or *variables.*

Tables are related through *keys*, one or more columns that meet certain requirements and may be used to index the rows. Keys are often columns that uniquely identifies every row in a table. We often assign a unique number or code to be a key, for example, a social security number may be used as a key for a set of people in the U.S. No two people have the same valid social security number, so we can use the number to connect a row of information to a specific person.

Keys are used to join data from one table to associated data in another table (Figure 8-7). Keys are the "key" to the utility and flexibility of relational databases. They allow us to mix and match data from various tables, to display data differently for different projects or audiences, to organize our data in ways that help us more quickly search, select, and update our data, and to isolate our data from calling programs or changes in computer hardware.

Figure 8-7 shows a join (or relation) of our forest and trails data in a relational data structure, with a primary key **Forest-ID** in the **Forests** table. This shows how keys allow us to break our data up into several

Forests

Forest Name	Forest-ID	Location	Size
Nantahala	1	N. Carolina	184,447
Cherokee	2	N. Carolina	92,271

Trails

Trail Name	Forest-ID
Bryson's Knob	1
Slickrock Falls	2
North Fork	1
Cade's Cove	1
Cade's Cove	2
Appalachian	1
Appalachian	2

Characteristics

Trail Name	Feature	Difficulty
Bryson's Knob	Vista	E,M
Bryson's Knob	Ogrth	E,M
Slickrock Falls	Ogrth	M
Slickrock Falls	Wfall	M
North Fork	-	M
Cade's Cove	Ogrth	E
Cade's Cove	Wlife	E
Appalachian	Wfall	M,D
Appalachian	Ogrth	M,D
Appalachian	Vista	M,D
Appalachian	Wlife	M,D
Appalachian	Cmp	M,D

Recreational features

Feature	Description	Activity1	Activity2
Wfall	Waterfall	Photography	Swimming
Ogrth	Old-Growth Forest	Photography	Hiking
Vista	Scenic Overlook	Photography	Viewing
Wlife	Wildlife Viewing	Photography	Birding
Cmp	Camping	Camping	-

Figure 8-6: Forest data in a relational database structure.

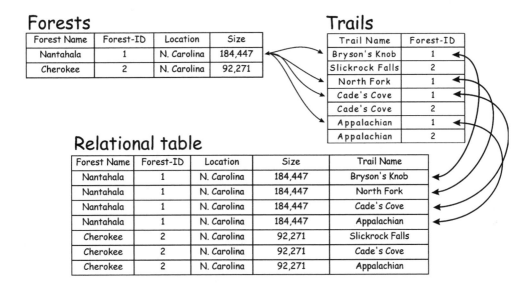

Figure 8-7: Forest and trails data in a relational data structure. Rows hold records associated with an entity, and columns hold items. A key, here **Forest-ID**, is used to join tables.

tables and reap all the benefits described above, while providing a mechanism to link between tables. Note that this linkage of separate tables is usually transparent to the end user, in that all or subsets of the forest and trails data in Figure 8-7 may be displayed on screen or printed as one continuous table. Data from three or more distinct tables in a DBMS are often joined, and columns subsets displayed in what appears to be one table to a user. The joins and columns may be changed, depending on the user.

The general definition of relational databases defines a *relational algebra*. This relational algebra takes relations (tables) as input and returns relations as output. The algebra combines or splits tables, either by rows or columns, to generate subset or expanded tables. The relational algebra may also be used to specify constraints, requirements, and security on a database.

Given their importance, there are some restrictions on keys. For example, null values are typically not allowed to be part of a key. There may be many potential keys (col-

umns that uniquely identify each row), but typically one is chosen for use, called a *primary key*. Most tables in the database will have a primary key, and keys in a table are frequently used to combine tables. Some keys are used to index and add flexibility in selecting data. Too few keys may result in difficulties searching or sorting the database.

Primary Operators

The relational algebra supports eight primary operators: *restrict, project, union, intersection, difference, product* (all combinations of a given set of variables recorded in a database), *join* (combine tables based on matching attribute values), and *divide* (facilitate queries based on condition). These operations are applied in queries to select specific records and items. These operations are depicted graphically in Figure 8-8 and Figure 8-9.

Restrict and project operations select based on rows and columns, respectively, to provide reduced tables. Restrict, also known and described as a *table query*, below, serves up records based on values for given variables. The restrict in Figure 8-8a is specified to restrict the current set to those that have a size that is big or huge -- all other entries in the relation are not selected (remember, tables are called "relations" in a relational database). The restrict then only returns four of the seven records, as shown in Figure 8-8a.

Restrict operations can be compound and complex and involve more than one attribute. Restrict operations most often return a reduced set of rows for a table as output. Examples of more complex restrict operations, or table queries, will be shown later in this chapter.

Project operations return entire columns for a table, in effect subsetting the table vertically, as shown in Figure 8-8b. Database tables may be quite large, and contain hundreds of items. A given analysis may concern only a few of those items, and so the project operation allows only those columns of interest from the table to be subset. This may substantially increase processing speed, reduce the storage space required, and ease viewing and analysis. In the example shown, ID, color, and size are selected from a base relation to create a new relation.

Product operations combine all unique values in one table with all unique values in another table to output a larger table (Figure 8-8c). The values may be single fields, or they may be n-tuples (multiple fields), sets of items taken together in a row. The product defines the complete set of possible combinations that may occur when combining two relations.

The divide operation is often the most confusing of the eight original relational operators, at least to new users. The divide operation is not obviously related to a mathematical divide, at least until the user has accrued some experience using the relational divide. The relational divide is analogous to the mathematical divide in that there is a target (the dividend) that is divided by another table (divisor). However, the confusion comes in that this is done relative, or per, a third table.

Divide operations are generally used in queries that would use the word "all" in a natural language description of the request, for example, "get states with all three major ethnic groups." A dividend is divided by a divisor over a table. For example, the state list is divided by the ethnic group list over a table containing state and ethnic group items. Only those states with all three ethnic groups are returned.

An example of a relational divide operation is shown in Figure 8-8d. The first table with the item named type is divided by the table with the item named size over the table with type and size. This returns a table with the item type and one entry, m. Only those values of type that have records with all values of size are returned. Note that in Figure 8-8d the n is missing size 1,3, and 4, and r is missing size 2 and 4.

The remaining four primary relational operators, as defined by E.F. Codd, typically return records based on membership in two or more tables. These operators are union, intersection, difference, and join, and they are illustrated in Figure 8-9, and described in turn in the following paragraphs.

a) restrict

ID	type	color	size	age
1	a	blue	big	old
2	c	green	big	young
3	a	red	small	mid
4	d	black	big	older
5	x	mauve	tiny	oldest
6	g	dun	huge	young
7	c	ecru	small	mid

restrict →

ID	type	color	size	age
1	a	blue	big	old
4	d	black	big	older
6	g	dun	huge	young
2	c	green	big	young

b) project

ID	type	color	size	age
1	a	blue	big	old
2	c	green	big	young
3	a	red	small	mid
4	d	black	big	older
5	x	mauve	tiny	oldest
6	g	dun	huge	young
7	c	ecru	small	mid

project →

ID	color	size
1	blue	big
2	green	big
3	red	small
4	black	big
5	mauve	tiny
6	dun	huge
7	ecru	small

c) product

No.	Dir.
1	N
2	S

product

App.
Yes
Yes
No

→

No.	Dir.	App.
1	N	Yes
2	S	Yes
1	N	No
2	S	No

d) divide

type
m
n
r

divide by

size
1
2

per

type	size
m	1
m	2
m	3
m	4
n	2
r	1
r	3

→

type
m

Figure 8-8: Relational algebra as originally defined supported eight operators, and the first four are shown here: restrict, project, product, and division, and four are shown in the next figure (modified from C.J. Date, 2004).

a) union

ID	type	color	size	age
1	a	blue	big	old
6	g	dun	huge	young

ID	type	color	size	age
2	c	green	big	young
4	d	black	big	older

union →

ID	type	color	size	age
1	a	blue	big	old
4	d	black	big	older
6	g	dun	huge	young
2	c	green	big	young

b) intersect

ID	color	size
1	blue	big
2	green	big
3	red	small
4	black	big
5	mauve	tiny
6	dun	huge
7	ecru	small

ID	color	size
1	blue	big
5	mauve	tiny
9	ivory	big

intersect →

ID	color	size
1	blue	big
5	mauve	tiny

c) difference

ID	color	size
1	blue	big
2	green	big
3	red	small
4	black	big
5	mauve	tiny
6	dun	huge
7	ecru	small

ID	color	size
1	blue	big
5	mauve	tiny
9	ivory	big

difference →

ID	color	size
2	green	big
3	red	small
4	black	big
6	dun	huge
7	ecru	small

d) join

ID	type
1	a
2	b
3	b
4	a

type	color	size	age
a	blue	big	old
b	dun	tiny	old

join →

ID	type	color	size	age
1	a	blue	big	old
2	b	dun	tiny	old
3	b	dun	tiny	old
4	a	blue	big	old

Figure 8-9: Four of the eight relational algebra operators as originally defined by E.F. Codd: union, intersect, difference, and join.

Before we describe the union, intersect, and difference operations, we must note a limitation in their application. The tables used in these three relational operations must be of the same kind. That means they must have the same set of variables or items before they can be used in these three operations. It makes little sense to find the intersection of two tables when they do not share the same set of items, for example, a table of home addresses and a table of plant species. These tables will always have an empty intersection set, and so this is senseless application of the intersect operation. There are similar problems when the union and difference relational operations are performed on tables that do not have the same set of items. Therefore, these operations are only defined and allowed in the context of tables of the same kind - that is, tables that have exactly the same items, defined in exactly the same way. This doesn't mean the tables are identical -- there will be different values in the various columns. Rather, the columns for both tables must be of the same data type (e.g., integer, text, real), and all columns must be present in both tables.

A union operation combines tables to return records found in either or both tables. As shown in Figure 8-9a, the tables are "stacked" to return a new table with members of both, but it does not show duplicate records for those entries that appear in both tables. As such, the result of a union is at least the size of the largest of the two tables, and no larger than the sum of the two tables.

The intersection operation returns records that occur in both input tables, and omits records found in only one of the two input tables (Figure 8-9b). Note that the records with ID values of 1 and 5 are the only two that are found in both tables, and so they are the only two included in the output table.

The difference operation returns those records that are in the first, but not the second table (Figure 8-9c). This example shows all those records that are not in both tables, in our example 2 through 4, plus 6 and 7.

The order of table input in a union or intersection operation does not change output. However, with the difference operator, order usually matters. The set of records returned from the difference of the first table from the second, shown in Figure 8-9c would be expected to be different from the application of the difference of the second table from the first in Figure 8-9c.

A join operation combines two tables through values found in keys. Values in one or more keys are matched across tables, and the information is combined based on the matching. Figure 8-9d shows an example of a join across two tables, in this case joined through the **type** item. Each **type** entry in the table on the left is matched to the **type** value in the center table, and the data are then joined or related through the values of **type**. The output records to the right of Figure 8-9d are the combined attributes of both tables. Records in the output table with ID values equal to 1 and 4 have **type** values equal to **a** as well as the color, size, and age associated with type **a**. Those records with type **b** have the appropriate IDs (2, 3) and the color, size, and age associated with **type b**.

Hybrid Database Designs in GIS

Data in a GIS are often stored using hybrid designs. Hybrid designs store coordinate data using specialized database structures, and attribute data in a relational database. Thousands to millions of coordinate pairs or cells are typically required to represent the location and shape of objects in a GIS. Even with modern computers, the retrieval of coordinate data stored in a relational database design is often too slow. Therefore, the coordinate data are frequently stored using structures designed for rapid retrieval. This involves grouping coordinates for cartographic objects, for example, storing ordered lists of coordinate pairs to define lines, and indexing or grouping lines to iden-

Attribute data

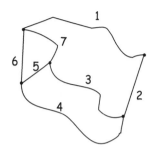

Arc-ID	last-arc	next-arc	type
1	6	2	A
2	1	4	C
3	2	5	F
4	2	5	D
5	6	7	A
6	5	1	F
7	3	6	B

Coordinate Data

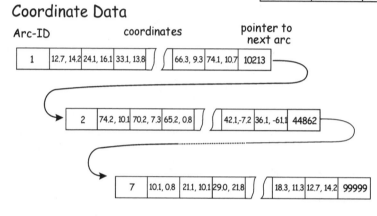

Figure 8-10: A small example of a hybrid database system for spatial data. Attribute data are stored in a relational table, while coordinate data are stored in a network or other structure.

tify polygons. Pointers are used to link related lines or polygons, and unique identifiers link the geographic features (points, lines, or polygons) to corresponding attribute data (Figure 8-10).

Topological relationships may be explicitly encoded to improve analyses or to increase access speed. Addresses to the previous and next data are explicitly stored in an indexing table, and pointers are used to connect coordinate strings. Explicitly recording the topological elements of all geographic objects in a data layer may improve geographic manipulations, includ-

ing determination of adjacencies, line intersection, polygon overlay, and network definition. Coordinates for a given feature or part of a feature may be grouped and these groups indexed to speed manipulation or display.

Hybrid data designs typically store attribute data in a DBMS. These data are linked to the geographic data through unique identifiers or labels that are an attribute in the DBMS. Data may be stored in a manner that facilitates the use of more than one brand of DBMS, and allows easy transport of data from one DBMS to another.

Selection Based on Attributes

The Restrict Operator: Table Queries

Queries are among the most common operations in a DBMS. A query may be viewed as the selection of a subset of records based on the values of specified attributes. Queries may be simple, using one variable, or they may be compound, using conditions on more than one variable, or using multiple conditions on one variable. One might search for all the parcels for which taxes haven't been paid, all census blocks with a size greater than a square mile and less than 200 inhabitants, or all fire hydrants which haven't been pressure tested, are near high-rise buildings, and are farther than 300 meters from the nearest other fire hydrant. In concept, queries are quite simple, but basic query operations may be combined to produce quite complex selections.

Many GIS softwares provide a query builder, a graphical user interface (GUI) that helps in applying selection operations (Figure 8-11). Most GUIs include a list of available fields, operations, and a sample or complete display of values for selected fields. The user constructs queries by alternately clicking on item names, operations, and entering values to build a query expression. This expression may then be applied, and the features matching the query expression are selected. Often you may save complicated or long expressions, to be reused later on different data sets.

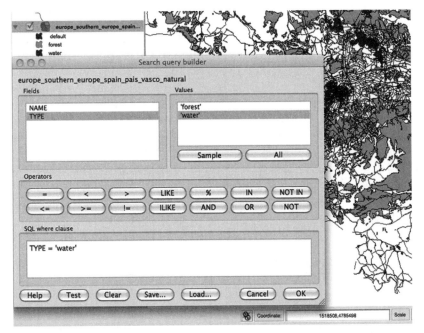

Figure 8-11: A query building GUI of the sort often provided in GIS software, here from QGIS. Selection expressions may be built in the bottom panel by clicking on fields, operators, and values in the upper panels.

The left side of Figure 8-12 demon-strates the selection from a simple query. A single condition is specified, **Area > 20**. The set of selected records is empty at the start of the query. Each record is inspected and added to the selected set if the attribute **Area** meets the specified criteria. Each record in the selected set is shown in gray in Figure 8-12.

The right side of Figure 8-12 demon-strates a compound query based on two attributes. This query uses the **AND** condi-tion to select records that meet two criteria. Records are selected that have a **Landuse** value equal to **Urban**, and a **Municip** value equal to **City**. All records that meet both of these requirements are placed in the selected set, and records that fail to comply with these requirements are in the unselected set. The Boolean operations **AND, OR,** and **NOT** may be applied in combination to select records that meet multiple criteria.

AND combinations typically decrease the number of records in a selected set. They add restrictive criteria, and they provide a more strenuous set of conditions that must be met for selection. In the example on the right side of Figure 8-12, the record with **ID = 7** meets the first criterion, **Landuse = Urban**, but it does not meet the second crite-rion specified in the **AND, Municipality = City**. Thus, the record with **ID = 7** is not selected. **AND**s add restrictions that winnow the selected set.

OR combinations typically increase or add to a selected set in compound queries. **OR** conditions may be considered as inclu-sive criteria. The **OR** adds records that meet a criterion to a set of records defined by pre-vious criteria. In the query on the left side of Figure 8-13, the first criterion, **Area > 20**, results in the selection of records 2, 4, 5, and 6. The OR condition adds any records that satisfy the criterion **Municip = City**, in this case the record with **ID = 1**.

The **NOT** is the negation operation, and may be interpreted as meaning "select those records that do not meet the condition." The right side of Figure 8-13 demonstrates the negation operation. The operation may be viewed as first substituting equals for the **NOT**, and identifying all records. Then the remaining records are placed in the selected set, and the identified records placed in the unselected set.

ANDs, **OR**s, and **NOT**s can have com-plex effects when used in compound condi-tions, and the order or precedence is important in the query. Combinations of

Simple selection:

records with Area > 20.0

ID	Area	Landuse	Municip
1	10.5	Urban	City
2	330.3	Farm	County
3	2.4	Suburban	Township
4	96.0	Suburban	County
5	22.1	Urban	City
6	30.2	Farm	Township
7	4.4	Urban	County

AND selection:

records with (Landuse = Urban) AND (Municip = City)

ID	Area	Landuse	Municip
1	10.5	Urban	City
2	330.3	Farm	County
3	2.4	Suburban	Township
4	96.0	Suburban	County
5	22.1	Urban	City
6	30.2	Farm	Township
7	4.4	Urban	County

Figure 8-12: Simple selection, applying one criterion to select records (left), and compound selection, applying multiple requirements (right).

OR selection:
    records with (Area > 20.0)
    OR (Municip = City)

ID	Area	Landuse	Municip
1	10.5	Urban	City
2	330.3	Farm	County
3	2.4	Suburban	Township
4	96.0	Suburban	County
5	22.1	Urban	City
6	30.2	Farm	Township
7	4.4	Urban	County

NOT selection:
    records with
    Landuse NOT Urban

ID	Area	Landuse	Municip
1	10.5	Urban	City
2	330.3	Farm	County
3	2.4	Suburban	Township
4	96.0	Suburban	County
5	22.1	Urban	City
6	30.2	Farm	Township
7	4.4	Urban	County

Figure 8-13: OR and NOT compound selections.

these three operations may be used to perform very complex selections.

Figure 8-14 shows the results of a complex query, combining AND, OR, and NOT operations. Here, the set of brackets chooses rows with a Landuse values equal to Urban, and Mill Rate values equal to B. Row 5 is the only record satisfying these criteria. The second set of brackets selects those rows that are not in a City, and with a Density greater than 200. This selects Row 3, and the final selected set includes both rows, by the OR operation. Selection operations may get quite complicated, and long, complex selection "sentences" may be saved, that is, the syntax copied in a text file or other repository, and applied when needed.

Complex selection:
    records with [(Landuse = Urban) AND (Mill Rate = B)] OR
            {NOT(Municip = City) AND (Density > 200)}

ID	Area	Landuse	Municip	Density	Mill Rate
1	10.5	Urban	City	1,112.2	A
2	330.3	Farm	County	1.9	C
3	2.4	Suburban	Township	237.5	C
4	96.0	Suburban	County	98.1	A
5	22.1	Urban	City	916.2	B
6	30.2	Farm	Township	3.7	A
7	4.4	Urban	County	153.8	D

Figure 8-14: An example of a complex selection, combining various selection operators.

While database queries are typically applied to tables, we must remember that in a GIS the tables are usually connected in some way to geographic features. Selections of table elements imply the selection of associated geographic elements. It is always a good idea to verify that the selection works as expected. Verification is often easiest by viewing the selection results, either on the table, the geography, or both. Figure 8-15 illustrates the results from three separate selection criteria: a) that county population be greater than 50 persons per square mile, b) that the median age be less than 40 years, and c) that housing vacancy rates be greater than 10%. The rightmost panel shows counties returned from a query specifying that criteria a and b and c all be met. The accuracy of the query may be quickly verified by inspecting maps of the component and final selections, and such an inspection should be conducted whenever possible, but especially when first learning or working with a query system.

Figure 8-16 demonstrates that queries are not generally distributive. For example, if *OP1* and *OP2* are operations, such as AND or NOT, then,

$$\text{OP1 (ConditionA OP2 ConditionB)} \quad (8.1)$$

is not always the same as

$$\text{(OP1 ConditionA)OP2(OP1 ConditionB)} \quad (8.2)$$

For example,

$$\text{NOT } [(\text{Landuse = Urban}) \text{ AND } (\text{Municipality = County})] \quad (8.3)$$

does not yield the same set of records as the expression

$$[\text{NOT (Landuse = Urban)}] \text{ AND } [\text{NOT (Municipality = County)}] \quad (8.4)$$

Parentheses or other delimiters should be used to ensure unambiguous queries.

Relational databases may support a *structured query language* known as SQL (pronounced both sequel and "ess kyou el"). SQL was initially developed by the International Business Machines Corporation but is supported by a number of software vendors. SQL is a non-procedural query language in that the specification of queries does not depend on the structure of the data. The language can be powerful, general, and trans-

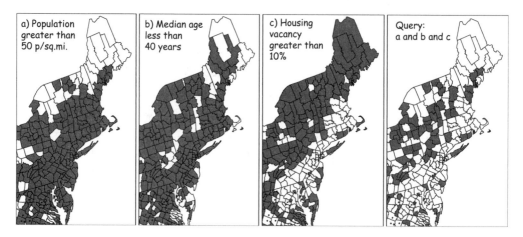

Figure 8-15: Component and composite selection criteria, applied to counties in the northeastern United States. A visual check of the composite against sub-components is often helpful, especially when learning.

NOT [(Landuse = Urban) AND (Municip = County)]

ID	Area	Landuse	Municip
1	10.5	Urban	City
2	330.3	Farm	County
3	2.4	Suburban	Township
4	96.0	Suburban	County
5	22.1	Urban	City
6	30.2	Farm	Township
7	4.4	Urban	County

[NOT (Landuse = Urban)] AND [NOT (Municip = County)]

ID	Area	Landuse	Municip
1	10.5	Urban	City
2	330.3	Farm	County
3	2.4	Suburban	Township
4	96.0	Suburban	County
5	22.1	Urban	City
6	30.2	Farm	Township
7	4.4	Urban	County

Figure 8-16: Selection operations may not be distributed, and the order of application is very important. When the NOT operation is applied after the AND (left) a different set of records is selected than when the NOT operation is applied before the AND (right side). Order of operation is important, and ambiguity should be removed by using parentheses or other delimiters.

ferable across systems, and so has become widely adopted.

SQL provides the capability to both define and manipulate data. Data types may be defined, and tables containing variables of a given type may be specified. Standard operations are used to manipulate data, e.g., to select, delete, insert, and update a database. Long or complicated queries may be saved in text files, or as scripts, that may be debugged, modified, or used later. These scripts may be quite long, and be fairly referred to as programs, given their complexity and capabilities. Utilities can be provided to help write, test, and automate these scripts.

Because SQL as initially defined has limitations for spatial data processing, many spatial operations are not easily represented in SQL. Many more selections may be specified only with complex queries, so various SQL extensions appropriate for spatial data have been developed.

Joining (or Relating) Tables

Relational databases are so powerful in part because we can structure out data in ways that reduce duplication, are easier to maintain, and flexible; much of this flexibility is because we can join tables. A join, also known as a relate, uses columns in one table to match rows based on columns in other tables. Joins were illustrated in Figure 8-7 and Figure 8-9, part d, but additional examples may help you successfully create and apply joins.

Joins are based on joint items, or join fields. In their simplest form, a single column in one table is matched to a column in another table, and a new table "created" by combining rows for matched values. Figure 8-17 shows a simple join between two tables. The simplest joins use a single column in each table as a matching or "join"

item. Here, **Code_A** and **Code_B**, respectively, are used. If we call **Table A** the "target table," and **Table B** the "source table," then our join consists of "copying" the values for a row in the target table to our output table, and then finding the corresponding row in the source table, matched by the join items, and finishing the output row by "copying" the values from the matching row in the source table. The values aren't truly copied, but rather associated and displayed together, so it only appears they've been copied.

If we inspect each **Out Table** row in Figure 8-17, we see this matching by values of the joining items. The **X** value for **Code_B** in the first row of **Table B** matches the **X** for the **Code_A** found in row 1, **Table A**. The first row in **Out Table** is a composite

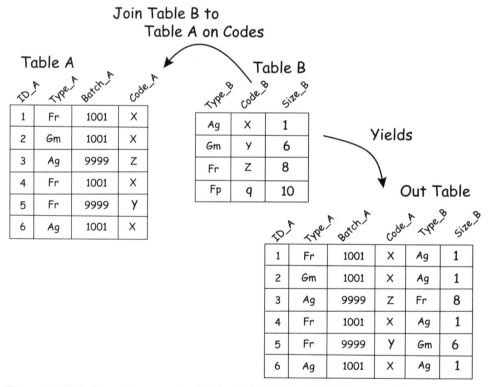

Figure 8-17: This figure illustrates a simple join. Table B is joined to Table A, matching the Code_B values to corresponding Code_A values, to create Out Table.

of row 1 from Table A (ID_A=1, Type_A=Fr, Batch_A=1001) and row 1 from Table B (Type_B=Ag, Size_B=1).

The second row in Out Table is created by "copying" the values in row 2, Table A, and combining it with the "Code" matching row in Table B. Here, the Code_A key for row 2 in Table A has a value X, and this matches the X for Code_B in row 1 of Table B. So, we "join" row 2 in Table A with row 1 in Table B to create row 2 in our Out Table. The third row in Out Table is created by matching the Code_B value, Z, in row 3 of Table B to the Code_A value in row 3 of Table A. This creates the third row in Out Table, and again, note that the variables for both Table A and Table B are "copied" into the output table. Row 4 in Out Table is created by matching the X in Table B, Code_B, row 1, to the X in Table A, Code_A in row 4. Again, the corresponding variables are copied from each table to create the row in the output table. This process continues until we've tried to match all the rows in the Target table.

There are a few things to point out. Non-matches in source table don't always end up in the output table. Inspect the values of Code_B in Table B of Figure 8-17. Note that the value q in row 4 of Table B doesn't match any of the Code_A values in Table A, so data from row 4 of Table B doesn't appear in the output table; in other words, there are no rows with Type_B values of Fp and Size_B values of 10 in the Out Table. Non-matching elements are discarded in this particular join. This is a common way of handling non-matching values, that is, discarding all non-matching data, but it is not the only way. We may specify joins that save some or all of the non-matching elements, but we should be aware of these different join variants.

Figure 8-17 illustrates the most common type of join, often known as an *inner join*, where an output row is created from rows that match on the join items across tables. There are other kinds of joins. For example, an *outer join* saves the information for non-matching rows, placing blank values for the items for rows which don't have a match in target table. We'll show an example of this kind of join later. There are both left- and right-outer joins, depending on whether the source or target non-matches are in the source or target tables. There is also a *natural join*, in which equally-named columns aren't copied, or cross joins, in which all rows in the source table are combined with all rows of the target table, e.g., a cross join of Tables A and B in Figure 8-17 would result in a table with 24 rows (6 rows for A times 4 rows for B).

Mastering the differences between these types of joins are perhaps a bit too much for an introductory GIS course, but I introduce them here because many softwares provide different types of joins as their "standard" or default, and don't identify them by name. These and other different types of joins are covered in depth in most introductory database books and courses, but can be confusing to distinguish and apply without some practice. I introduce them here to:

- warn you of the differences between different types of joins, and to emphasize that different types of joins will usually produce different results, even when applied to the same data, and

- to stress that there are standardized names for different types of joins, although not all GIS softwares use them. You should verify how joins work when first using new software, by comparing source and output tables.

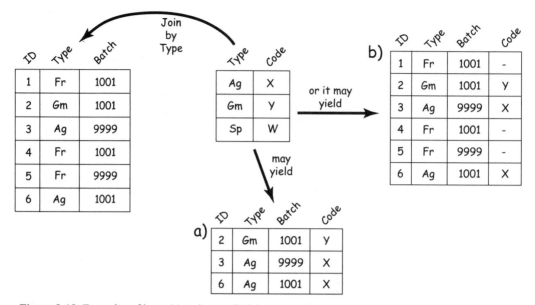

Figure 8-18: Examples of inner (a) and outer (b) joins. Note that an inner join only saves matching rows, while an outer join saves values for both matching and unmatching rows.

Figure 8-18 illustrates a difference between inner and outer joins. The center table is joined to the left-most table on the item **Type**. A resulting inner join is shown in a. Note that only rows 2, 3, and 6 from the "target" table on the left, with values **Ag** and **Gm** for **Type** (our key) recorded in the output table in Figure 8-18a, because those are the only **Type** values found in both tables. Information in rows 1, 4, and 5 is not retained in the output table.

Figure 8-18b show an *outer join*, in which unmatched source table rows are retained. Null or empty values are placed for the non-matching attributes of the target table, as shown by the dashes in the **Code** item for rows 1, 4, and 5 of output table b.

You may have deduced by now that the join items are crucial when joining tables. If the join items are not correctly created, then the joins will likely produce unintended results. We must pay attention to the join items across tables, particularly how many values match for our joining items across the source and target table.

Most joins should involve a one-to-one or a one-to-many relationship between the

source and the target join items. We often run into problems when we have a many-to-one relationship from a source to a target item. If many rows in the source table match a row or rows in the target table, we often can't predict the result. The following paragraphs illustrate these relationships.

A one-to-one relationship means just that, there may be one and only one instance in a join item of a source table that matches one and only one instance of a join item in a target table. The left side of Figure 8-19 illustrates a one to one match for the items **Id1** and **Id2**. Each value of **Id2** matches only one value of **Id1**. Note that not all values of **Id1** have a match in **Id2**.

Tables may also be unambiguously joined if there is a one-to-many relationship between the source join item and the target join item. The join on the right side of Figure 8-19 shows a one-to-many relationship between the source, in item **Id4**, and the target, in item **Id3**. Note there are three instances of **Y** in **Id3**, but they unambiguously match with the one value of **Y** in **Id4**.

We often run into problems when we attempt joins with items that have a many-

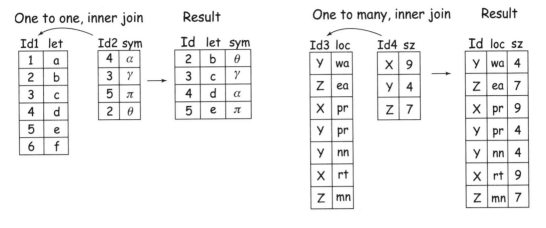

Figure 8-19: An example of a one to one and many to one relationships between tables.

to-one or a many-to-many relationship. These are often considered ill-matching keys, in that results from a join can be indeterminate - you can't predict the results in advance, or they may change due to spurious factors, such as the whims of the programmer's choice of algorithm, or pseudo-random effects of row ordering. Since you're often not sure of the results you'll get, many-to-one or many-to-many relationships between the source and target keys are rarely a good idea. We usually require the source item in a join to be a *key* - a column or set of columns that uniquely identifies the rows of the source table, so that there is a one-to-one or one-to-many relationship in a join.

Figure 8-20 shows an example of an ill-matching join item. The item **Type** in

Source Table is not a key, and this results in a many-to-many join. There are two rows in the **Source Table** with a **Type** value of **Fr**. Both rows may fairly match the **Fr** key values found in the **Target Table**, resulting in an ambiguous assignment for the values of **Code** for those rows. Both **V** and **N** are equally supported, hence our results are uncertain, as shown in the **Output Table**. Such uncertainty is rarely a good thing in table manipulations or analyses. We would have the same ambiguity if there were only one value of **Fr** in the **Target Table**, creating a many to one relationship from **Source** to **Target**. Many-to-one or many-to-many joins should be avoided, except in a constrained set of circumstances.

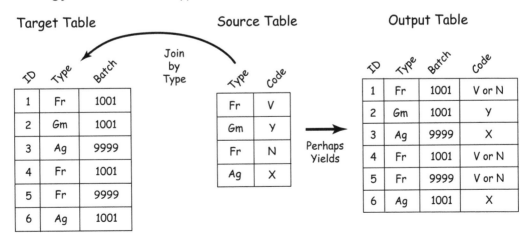

Figure 8-20: An example of a many-to-many table join.

Concatenated Keys

Most examples show joins using a single column as a key. These are most common because you simply need to ensure that all values are unique for the column. This is the idea behind unique identifiers in many databases, for example, an invoice number for a business, unique part ID numbers for a warehouse, or a museum ascension number. These unique IDs allow unique items to be simply identified in a table.

While single-column identifiers are most common, we frequently use multiple columns as keys. These are often used when we have large, multi-table databases that we wish to combine in several different ways. Data from the U.S. Census, or the U.S. SURGO database discussed in the previous chapters use multiple tables, many with two or more columns used in combination as a key.

When multiple columns are used as a key, it is called a *concatenated key*. Concate-

nated keys are typically formed by two columns, and rarely more than three columns.

Figure 8-21 illustrates a concatenated key, here used to uniquely identify U.S. counties. Each U.S. state is assigned a unique Federal Information Processing Standard (FIPS) code, and every county within a state assigned a unique code, but unique only within the state. This allows new codes to be assigned at the state level, e.g., if a new state is added, or new codes to be assigned within a state, e.g., if a county were split to form two new counties. This also allows quick selection of data by state. County FIPS codes (cFIPS) are assigned sequentially, as odd numbers within these states, starting at 1, up to the last county within the state. cFIPS alone can not be used as a key, for example, cFIPS = 1 for both Fairfield County, Connecticut, and Barnstable County, Massachusetts. A concatenated key, using both state and county FIPS numbers, is needed to uniquely distinguish counties across multiple states.

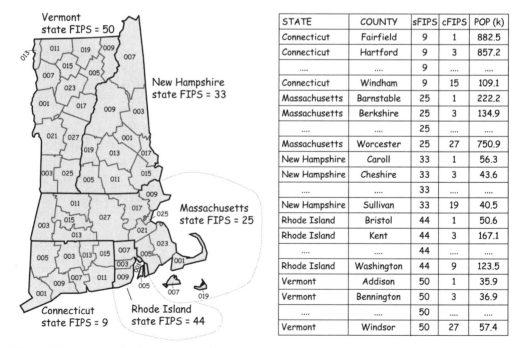

STATE	COUNTY	sFIPS	cFIPS	POP (k)
Connecticut	Fairfield	9	1	882.5
Connecticut	Hartford	9	3	857.2
....	....	9	....	....
Connecticut	Windham	9	15	109.1
Massachusetts	Barnstable	25	1	222.2
Massachusetts	Berkshire	25	3	134.9
....	....	25	....	....
Massachusetts	Worcester	25	27	750.9
New Hampshire	Caroll	33	1	56.3
New Hampshire	Cheshire	33	3	43.6
....	....	33	....	....
New Hampshire	Sullivan	33	19	40.5
Rhode Island	Bristol	44	1	50.6
Rhode Island	Kent	44	3	167.1
....	....	44	....	....
Rhode Island	Washington	44	9	123.5
Vermont	Addison	50	1	35.9
Vermont	Bennington	50	3	36.9
....	....	50	....	....
Vermont	Windsor	50	27	57.4

Figure 8-21: An example of a concatenated key, here the state FIPS (sFIPS) and county FIPS codes (cFIPS)

Multi-table Joins

We may have more than one potential key in a table (remember, the key can consist of one or more columns), but we usually design tables with a main key. We may also join many tables to a single table, often using different target items for each join.

Figure 8-22 shows an example of a multi-table join with distinct keys. School Table may be considered the "foundational" table, and the two tables named County and District are joined to School Table to create an Output Table.

Note that these two joins are based on different items. County Table is joined to School Table based on values in the columns labeled Cty, while the District Table is joined to the School Table based on the values found in the columns labeled DistID. Our output table is shown here without the "copies" of the columns (e.g., Cty and DistID each appear only once in the Output Table), although the "copies" are often displayed.

Note that the joins are one-to-many in both cases; one and only one value in the source columns may match many row values in the target columns. Also note that each of the join items in the source tables are keys - Cty in the County Table uniquely identifies each county, and DistID in the District Table uniquely identifies each district. As noted before, source items in joins are often keys, uniquely identifying the rows in the source table.

County Table

Name	Cty
Washoe	1
Polk	2
Placer	3
Le Seur	4

School Table

Sch#	Name	Cty	Type	DistID
11	Centenial	3	Prm	624
12	Lincoln Union	4	Hgh	624
11	Central Lakes	4	Mdl	113
13	Sierra Gardens	1	Prm	113
10	Richland	2	Prm	95
12	Nevada Union	2	Hgh	95

District Table

DistID	Superintendent
95	M. Skinner
113	D. McGee
624	R. Figgins

Output Table

Sch#	Name	Cty	Type	DistID	Name	Superintendent
11	Centenial	3	Prm	624	Placer	R. Figgins
12	Lincoln Union	4	Hgh	624	Le Seur	R. Figgins
11	Central Lakes	4	Mdl	113	Le Seur	D. McGee
13	Sierra Gardens	1	Prm	113	Washoe	D. McGee
10	Richland	2	Prm	95	Polk	M. Skinner
12	Nevada Union	2	Hgh	95	Polk	M. Skinner

Figure 8-22: An example of a multi-table join, based on different keys. Here, County and District tables are joined to the School Table, to create an Output Table.

Normal Forms in Relational Databases

Keys and Functional Dependencies

The previous sections should point out the need to carefully structure our tables in a relational database, and that keys between tables are especially important. If relational tables are not carefully constructed, they can suffer from serious problems in performance, consistency, redundancy, and maintenance. If all data are stored in one large table there may be large amounts of redundant data or wasted space, and long searches may be needed to select a small set of records. Updates on a large table may be slow, and the deletion of a record may result in the unintended deletion of valuable data from the database. Smaller, carefully constructed tables are often more useful, although proper structuring requires care.

Consider the data in Figure 8-23, in which building records are stored in a single table. Attributes include Parcel-ID, Alderman, Tship-ID, Tship_name, Thall_add, Own-ID, Own_name, and Own_add. Some information is stored redundantly, for example, changing the Alderman for Tship-ID

12 would require changing many rows, and identifying all parcels with Yamane as an owner would require a search of all records for several columns in the table. This storage redundancy is costly both because it takes up disk space and because each extra record adds to the search and access times. A second problem comes with changes in the data. For example, if Devlin, Yamane, and Prestovic sell the parcel they jointly own (first data row), deleting the parcel record for Devlin would purge the database of her address and tax payment history. If these data on Devlin were required later, they would have to be re-entered from an external source.

We may place relational databases in *normal forms* to avoid many of these problems. Data are structured in sequentially higher normal forms to improve correctness, consistency, simplicity, non-redundancy, and stability. There are several levels in the hierarchy of normal forms, but the first three levels, known as the first through third normal forms, are most common. Data are usually structured sequentially, that is, first all tables are converted to first normal forms,

Land Records table, unnormalized form

parcel-ID	Alderman	Tship-ID	Tship_name	Thall-add	Own-ID	Own_name	Own_add
2303	Johnson	12	Birch	15W	122	Devlin	123_pine
618	DeSilva	14	Grant	35E	457	Suarez	453_highland
9473	Johnson	12	Birch	15W	337	Yamane	72_lotus

Own-ID	Own_name	Own_add	Own-ID	Own_name	Own_add
337	Yamane	72_lotus	890	Prestovic	12_clayton
890	Prestovic	12_clayton	231	Sherman	64_richmond
-	-	-	-	-	-

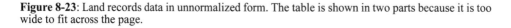

Figure 8-23: Land records data in unnormalized form. The table is shown in two parts because it is too wide to fit across the page.

then converted to 2nd and then 3rd normal forms as needed. Prior to describing normal forms we must introduce some terminology and properties of relational tables.

As noted earlier, relational tables use keys to index data. There are different kinds of keys. A *super key* is one or more attributes that may be used to uniquely identify every record (row) for a table. A subset of attributes of a super key may also be a super key, and is called a *candidate key*. The *primary key* for indexing a table is chosen from the set of candidate keys. There may be many potential primary keys for a given table; however it is usual to use only one primary key per table. The Parcel-ID is a primary key for the table in Figure 8-23, because it uniquely identifies each row in the table.

Functional dependency is another important concept. Attributes are functionally dependent if at a given point in time each value of the dependent attribute is determined by a value of another attribute.

Figure 8-24 illustrates the concept of functional dependency. The table contains a parts list, with ID as the primary key, and a part Name, CNum, Ctype, Thread, and Angle attributes. The ID is unique for each row, and so by definition, all other items are functionally dependent on ID. If we know the value of ID is 1, then we know the part Name is Tec. We denote this as shown,

ID -> Name

We also see that Name is functionally dependent on CNum. If we know a value for CNum, e.g., 2, we know the value of Name will equal Ext. We see that the converse is also true here, CNum is also functionally dependent Name. Note that this is not always true, as shown for Ctype and Thread.

CType -> Thread is true,

but

Thread -> CType is not true

Why? Because for the value of Thread equal to 14, CType may be either E or Er, violating our definition of functional dependence.

In our example in Figure 8-23, we may know that Own_add is functionally dependent on Own_name, and for each Tship_name there is only one Thall-add. In other words, each owner can only have one resident address, so for a given Own_name,

ID	Name	CNum	CType	Thread	Angle
1	Tec	3	M	12	45
2	Cap	1	E	14	20
3	Ext	2	M	12	22
4	Cap	1	M	12	18
5	Tec	3	E	14	20
6	Cap	1	E	14	22
7	Ext	2	Er	14	45

Functional Dependencies: ID -> Name, CNum, Ctype, Thread, Angle
CNum -> Name (or Name -> CNum)
CType -> Thread

Figure 8-24: Example functional dependencies

let's say, **Prestovic**, the **Own_add** is determined. In a similar manner, there is only one Township name, **Tship_name**, for each Town Hall address, **Thall-add**, or

> Own_name -> Own_add
>
> Tship_name -> Thall-add

Remember, these indicate that **Own_add** is functionally dependent on **Own_name**, and **Thall-add** is functionally dependent on **Tship_name**.

Functional dependencies are transitive, so if *A* -> *B*, and *B* -> *C*, then *A* -> *C*. This notation means that if *B* is functionally dependent on *A*, and *C* is functionally dependent on *B*, then *C* is functionally dependent on *A*.

While relational database designs are flexible, the use of keys and functional dependencies places restrictions on relational tables:

- There cannot be repeated records, that is, there can be no two or more rows where all attributes are equal.

- There must be a primary key in a table. This key allows each record to be uniquely identified.

- No member of a column that forms part of the primary key can have a null value. This would allow multiple records which could not be uniquely identified by the primary key.

Land Records table, unnormalized form

parcel-ID	Alderman	Tship-ID	Tship_name	Thall-add	Own-ID	Own_name	Own_add
2303	Johnson	12	Birch	15W	122	Devlin	123_pine
618	DeSilva	14	Grant	35E	457	Suarez	453_highland
9473	Johnson	12	Birch	15W	337	Yamane	72_lotus

Own-ID	Own_name	Own_add	Own-ID	Own_name	Own_add
337	Yamane	72_lotus	890	Prestovic	12_clayton
890	Prestovic	12_clayton	231	Sherman	64_richmond
-	-	-	-	-	-

Land Records table, first normal form (1NF)

Parcel-ID	Alderman	Tship-ID	Tship_name	Thall_add	Own-ID	Own_name	Own_add
2303	Johnson	12	Birch	15W	122	Devlin	123_pine
2303	Johnson	12	Birch	15W	337	Yamane	72_lotus
2303	Johnson	12	Birch	15W	890	Prestovic	12_clayton
618	DeSilva	14	Grant	35E	457	Suarez	453_highland
618	DeSilva	14	Grant	35E	890	Prestovic	12_clayton
618	DeSilva	14	Grant	35E	231	Sherman	64_richmond
9473	Johnson	12	Birch	15W	337	Yamane	72_lotus

Figure 8-25: Relational tables in unnormalized (top) and first normal forms (bottom).

The First and Second Normal Forms

We begin creating tables in normal forms by first gathering all our data, often in a single table. Normal forms typically result in many compact, linked tables, so it is quite common to split tables as the database is *normalized*, or placed in normal forms. After normalization the tables have an indexing system that speeds searches and isolates values for updating.

Tables with repeat groupings, as in the table at the top of Figure 8-25, are *unnormalized*. A repeating group exists in a relational table when an attribute is allowed to have more than one value represented within a row. Owner-ID repeats itself for dwellings with multiple owners.

A table is in first normal form when there are no repeat columns. The Land Records table at the bottom of Figure 8-25 has been *normalized* by placing each owner into a separate row. This is a table in the first normal form (1NF), because each column appears only once in the table definition. A 1NF is the most basic level of table normalization. However, the 1NF table structure still suffers from excessive storage redundancy, inefficient searches, and potential loss of data on updating. First normal forms have an advantage over unnormalized tables because queries are easier to code and implement. Tables in 1NF are usually converted to higher-order normal forms, usually to at least third normal form, 3NF, but it is useful to understand second normal forms before describing 3NF tables.

A table is in second normal form (2NF) if it is in first normal form and every non-key attribute is functionally dependent only on the primary key, or on transitive functional dependencies of the primary key. Remember that functional dependency means that knowing the value for one attribute of a record automatically specifies the value for the functionally dependent attribute. The non-key attributes may be directly dependent on the primary key through some functional dependency, or they may be dependent through a transitive dependency. The Land Records table in 1NF at the bottom of Figure 8-25 has only one possible primary key, the composite of Parcel-ID and Own-ID. No other combination uniquely identifies each row. However, this table is not in second normal form because it has non-key attributes that are not functionally dependent only on the primary key attributes. For example, Tship_name and Thall_add are functionally dependent on Tship-ID.

The Land Records table at the bottom of Figure 8-25 is repeated at the top of Figure 8-26. This table exhibits the primary disadvantages of the first normal form. Parcel-ID, Alderman, and Tship-ID are duplicated when there are multiple owners of a parcel, causing burdensome data redundancy. Each time these records are updated, for example when a new Alderman is elected, data must be changed for each duplicate record. If a parcel changes hands and the seller does not own another parcel represented in the table, then information on the seller is lost.

Some of these disadvantages can be removed by converting the first normal form table to a group of second normal form tables. To create second normal form tables, we make every non-key attribute fully dependent on a primary key in the new tables. Note that the 1NF table will often be split into two or more tables when converting to 2NF, and each new table will have its own key. Any non-key attributes in the new tables will be dependent on the primary keys. The bottom of Figure 8-26 shows our Land Records converted to second normal form. Each of the three tables in second normal form isolates an observed functional dependency, so each table and dependency will be described in turn.

How do we systematically apply this criterion that the non-key attributes be functionally dependent only on the primary key, directly or through a transitive functional

Land records table, first normal form (1NF)

Parcel-ID	Alderman	Tship-ID	Tship_name	Thall_add	Own-ID	Own_name	Own_add
2303	Johnson	12	Birch	15W	122	Devlin	123_pine
2303	Johnson	12	Birch	15W	337	Yamane	72_lotus
2303	Johnson	12	Birch	15W	890	Prestovic	12_clayton
618	DeSilva	14	Grant	35E	457	Suarez	453_highland
618	DeSilva	14	Grant	35E	890	Prestovic	12_clayton
618	DeSilva	14	Grant	35E	231	Sherman	64_richmond
9473	Johnson	12	Birch	15W	337	Yamane	72_lotus

Given functional dependencies:

Parcel-ID ➞ Tship-ID

Tship-ID ➞ Tship_name, Thall_add, Alderman

Own-ID ➞ Own_name, Own_add

Land records tables, second normal form (2NF)

Land Records 1

Parcel-ID	Alderman	Tship-ID	Tship_name	Thall_add
2303	Johnson	12	Birch	15W
618	DeSilva	14	Grant	35E
9473	Johnson	12	Birch	15W

Land Records 2

Own-ID	Own_name	Own_add
122	Devlin	123_pine
337	Yamane	72_lotus
890	Prestovic	12_clayton
457	Suarez	453_highland
231	Sherman	64_richmond

Land Records 3

Parcel-ID	Own-ID
2303	122
2303	337
2303	890
618	457
618	890
618	231
9473	337

Figure 8-26: Ownership data, converted to second normal form.

dependency? We must 1) specify the primary key, 2) identify the main functional dependencies, and 3) project the 1NF table across the key and dependency columns.

First, we must identify the primary key. In our example here, the simplest primary key is the (concatenated) key that is the combination of Parcel-ID and Owner-ID. If our primary key is a single item then the table is already in 1NF by definition, because all non-key attributes will depend on the primary key. However, if our primary key is more than one column, we may have further work to convert to 2NF, focusing on dependence on the components of the primary key.

Our second step is to identify the functional dependencies. We know that parcels occur in only one township, and that each township has a unique Tship-ID, a unique Tship_name, a unique Thall_add, and one Alderman. This means that if we have identified a parcel by its Parcel-ID, the Alderman, Tship-ID, Tship_name, and Thall_add are known. We assign a unique identifier to each parcel of land, and the Alderman, Tship_name, and Thall_add are all dependent on this identifier. This means if we know the parcel identifier, we know these remaining values. This is the definition of functional dependency. We represent these functional dependencies by:

Parcel-ID -> Alderman

Parcel-ID -> Tship-ID

Parcel-ID -> Tship_name

Parcel-ID -> Thall_add

These functional dependencies are incorporated in the table named Land Records 1 in Figure 8-26.

Second, note that once Own-ID is specified, the Own_name and Own_add are determined. Each owner has a unique identifier and only one name (aliases not allowed). Also, each owner has only one permanent home address. Own_name and Own_add are functionally dependent on Own-ID. The functional dependencies are:

Own-ID -> Own_name

Own-ID -> Own_add

The Parcel-ID and Own-ID are called partial functional dependencies, because while both are dependent on the primary key, they aren't dependent on each other. If I have a unique Parcel-ID, I know additional information about some of the columns for any row in the table, but not all of the columns. If I know the Own-ID, I also know the values of a set of columns, but again, not all. When we have a concatenated key, we must identify these in our data, and they guide us in how to further split our table.

How do I get to 2NF? By projecting the 1NF table across the primary key and functional dependencies. Remember, project is just a way of saying we subset the columns, here guided by the functional dependencies. These partial functional dependencies are represented in the tables Land Records 1 and Land Records 2 in Figure 8-26.

Finally, note that we need to tie the owners to the parcels. These relationships are presented in the table Land Records 3 in Figure 8-26. Note that some parcels are jointly owned, and so there are multiple owner IDs for each parcel.

The three tables Land Records 1 through 3 satisfy the conditions of a second normal form. Second normal form eliminates some of the redundancies associated with the 1NF. Note that the redundancy in storing the information on Alderman, Tship-ID, Tship_name, and Thall_add have been significantly reduced, and the minor redundancy in Own_name has also been removed. Editing the tables becomes easier; for example, changes in Alderman entail modifying fewer records. Finally, deletion of a parcel does not have the side effect of deleting the information on the owner, Own-ID, Own_name, and Own_add.

The Third Normal Form

The 2NF still contains problems, although they are small compared to a table in 1NF. They can still suffer from transitive functional dependencies. If a transitive functional dependency exists in a table, then there is a chain of dependencies. A transitive dependency occurs in our example table named **Land Records 1** (Figure 8-26). Note that Parcel-ID specifies Tship-ID, and Tship-ID specifies Tship_name and Thall_add. In our notation of functional dependencies:

Parcel-ID - > Tship-ID

and

Tship-ID -> Tship_name, Thall_add, Alderman

This causes a problem when we delete a parcel from the database. To delete a parcel we remove the parcel from tables **Land Records 1** and **Land Records 3**. In so doing we might also lose the relationship between Tship-ID, Tship_name, Thall_add, and Alderman. To avoid these problems we need to convert the tables to the third normal form.

A table is in the third normal form (3NF) if and only if for every functional dependency A -> B, *A is a super key, or* B *is a member of a candidate key.* This requirement means we must identify transitive functional dependencies and remove them, typically by splitting the table that contains them. The tables **Land Records 2** and **Land Records 3** in Figure 8-26 are already in the 3NF, because the keys for these tables are super keys. **Owner-ID** uniquely identifies the rest of the row in **Land Records 2**, and the concatenated key of Parcel-ID and Tship-ID are the rows in **Land Records 3**.

However the table **Land Records 1** in Figure 8-26 is not in 3NF because the functional dependencies for table **Land Records 1** are:

Parcel-ID - > Tship-ID

Tship-ID -> Tship_name, Thall_add, Alderman

Tship-ID is not a super key for the table, nor are Tship_name and Thall_add members of a primary candidate key for that table. Removing the transitive functional dependency by splitting the table will create two new tables, each of which satisfies the criteria for the 3NF. Figure 8-27 contains the tables **Land Records 1a** and **Land Records 1b**, both of which now satisfy the 3NF criteria, and preserve the information contained in the 1NF table in Figure 8-25. Note that Parcel-ID is now a super key for **Table 1a** and Tship-ID is a super key for **Table 1b**, so the 3NF criteria are satisfied.

A general goal in defining a relational database structure is to have the fewest tables possible that contain the important relationships, and have all tables in at least 3NF. Normal forms higher than three have been described and provide further advantages, however these higher forms are often more limited in their application and depend on the intended use of the database.

While relational tables in normal forms have certain useful characteristics, they may suffer from relatively long access times for specific queries. Databases may be organized around usage, or *denormalized* for the most common processes. These denormalizations typically add extra columns or permanent joins to the database structure. This may add redundancy or move a table to a lower normal form, but these disadvantages often allow significant gains in processing speed. The need to denormalize tables has diminished with improvements in computing power. However, denormalization may be required for extremely large databases, or where access speed is of primary importance.

Land records, third normal form

Land Records 1a

FD: Parcel-ID → Tship-ID

Parcel-ID	Tship-ID
2303	12
618	14
9473	12

Land Records 1b

FD: Tship-ID → Tship_name, Thall_add, Alderman

Tship-ID	Tship_name	Thall_add	Alderman
12	Birch	35W	Johnson
14	Grant	35E	DeSilva

Land Records 2

FD: Own-ID → Own_name, Own_add

Own-ID	Own_name	Own_add
122	Devlin	123_pine
337	Yamane	72_lotus
890	Prestovic	12_clayton
457	Suarez	453_highland
231	Sherman	64_richmond

Land Records 3

No Functional Dependencies

Parcel-ID	Own-ID
2303	122
2303	337
2303	890
618	457
618	890
618	231
9473	337

Figure 8-27: Ownership data in third normal form, with the functional dependencies (FD) noted at the top of the table.

Trends in Spatial DBMS

Computer database technologies continue to evolve, and one of the most striking evolutions in recent years is the widespread adoption of a multi-tiered database architecture. In the past, data were often housed on one large computer and accessed via one program. This might be considered a single-tiered system. Shared databases evolved, as described earlier in this chapter, with a server program providing data as requested by client programs, perhaps running on different computers, in a two-tiered system. Multi-tiered systems are becoming common. A third tier may be added, e.g., a general analysis program, that may spawn a request to a database access program, which in turn queries a database server. Internet applications often interface with databases in this way, passing requests as required to a server.

Distributed database systems are also increasingly common phenomena. A distributed database system maintains multiple data files across different sites on a computer network. Different users can access the database without interfering with or substantially reducing the response of each other. New nodes, often computers, may be added to the network, to add storage and improve performance. The database must be periodically synchronized across the network to reflect any changes by the various users. Distributed databases are becoming more common as we realize the power of combining data from disparate database systems, for example, a DBMS containing information on potential customers and demand may be combined with one for businesses to manage supply of goods or services. These may be tied together in a distributed database system, often with a multi-tier approach.

Summary

Attribute data are an important component of spatial data in a GIS. These data may be organized in several ways, but data structures that use relational tables have become the most common method for organizing and manipulating attribute data in GIS.

Selections, or queries, are among the most common analyses conducted on attribute data. Queries mark a subset of records in a table, often as a precursor to subsequent analyses. Queries may use AND, OR, and NOT operations, among others, alone or in combination.

Relational tables are often placed in normal forms to improve correctness and consistency, to remove redundancy, and to ease updates. Normal forms seek to break large tables into small tables that contain simple functional dependencies. This significantly improves the maintenance and integrity of the database. Normal forms may cause some cost in speed of access, although this is a diminishing problem as computer hardware improves.

Object-relational database systems have been developed that incorporate the strong typing and domains of object-oriented models with the flexibility, logic, and ubiquity of relational data models. These evolutionary improvements to the relational approach will continue as database technologies are extended across networks of computers and the world wide web.

Suggested Reading

Adam, N., & Gangopadhyay, A. (1997). *Database Issues in Geographic Information Systems*. Dordrecht: Kluwer Academic Publishers.

Adler, D.W. (2000). *IBM DB2 Spatial Extender - Spatial data within a RDBMS*. Proceedings 27th International Conference on Very Large Databases. Rome Italy.

Arctur, D., & Zeiler, M. (2004). *Designing Geodatabases: Case Studies in GIS Data Modeling*. ESRI Press: Redlands.

Bhalla, N. (1991). Object-oriented data models: a perspective and comparative review. *Journal of Information Science*, 17:145-160.

Date, C. J. (2004). An Introduction to Database Systems (8th ed.). Pearson/Addison-Wesley: Boston.

Frank, A.U. (1988). Requirements for a database management system for a GIS. *Photogrammetric Engineering and Remote Sensing*, 54:1557-1564.

Lorie, R.A. & Meier, A. (1984). Using a relational DBMS for geographical databases. *Geoprocessing*, 2:243-257.

Milne, P., Milton, S., & Smith, J.L. (1993). Geographical object-oriented databases: a case study. *International Journal of Geographical Information Systems*, 7:39-55.

Rigaux, P., Scholl, M., and Voisard, A. (2002). *Spatial Databases With Applications To GIS*. Morgan Kaufman: San Francisco.

Teorey, T. J. (1999). *Database Modeling and Design* (3rd ed.). Morgan Kaufmann: San Francisco.

Ullman, J.D., & Widom, J. (2008). *A First Course in Database Systems*. Prentice Hall: New York.

Zeiler, M. (1999). *Modeling Our World: An ESRI Guide to Geodatabase Design*. ESRI Press: Redlands.

Study Questions

8.1 - What are the main components of a database management system?

8.2 - What are the primary functions of a database management system?

8.3 - Can you describe the difference between single and multiple user views?

8.4 - What is a one-to-one relationship between tables? A many-to-one relationship?

8.5 - Which single columns in the following table may serve as keys?

PID	Osel	Clr	NumT	SpLm
1	B	or	1	55
3	D	gr	2	55
5	A	rd	11	55
7	C	ye	23	55
9	G	az	1	65
null	X	bl	9	65

8.6 - Which single columns in the following table may serve as keys?

CID	TStmp	Osel	Clr	NumT	Xerr
1	10:12	B	rd	1	110
3	11:44	D	gr	-5	220
5	11:44	A	rd	11	220
7	16:58	C	gr	23	110
9	22:11	F	bl	0	110
Null	23:59	H	bl	-2	220

8.7 - Why have relational database structures proven so popular?

8.8 - What are the eight basic operations formally defined by E.F. Codd for the relational model?

8.9 - What is the primary reason that hybrid database models are used for spatial data?

8.10 - Does an or condition result in more, fewer, or the same number of records than the component parts, e.g., is the set from

> condition A or condition B

> the same, bigger, or smaller than the set from condition A alone, or condition B alone?

> Does an and condition result in more, fewer, or the same number of records as the component parts, e.g., is the set from

> condition A and condition B

> the same, bigger, or smaller than the set from condition A alone, or condition B alone?

8.11 - Identify the states meeting each of the following selection criteria, based on the table below:

a) Smokers < 20%

b) Smokers > 20% and illiteracy < 10

c) Not (Non-federal taxes > 10)

d) Illiteracy < 7 or income > 22,000

e) Get more federal aid than paid in taxes, or non-federal taxes < 8.

f) [Firearms deaths < 10 and income > 21,000] and not {smokers > 20}

FIPS	Name	Smokers (%)	Income ($/person)	Illiteracy (%)	Firearm deaths / 100,000	Non-Federal Tax Rate (%)	Fed. Taxes / Fed. Aid
01	Alabama	22,1	18,189	15	16.2	8.6	1.71
02	Alaska	21.5	22,660	9	20	6.4	1.87
12	Florida	17.5	21,557	20	11.1	7.4	1.02
13	Georgia	19.5	21,154	17	13.4	9.9	0.96
19	Iowa	18.8	19,674	7	6.7	9.7	1.11
27	Minnesota	17.6	23,198	6	6	10.2	0.69
40	Oklahoma	24.7	17,646	12	13.1	9.8	1.48
55	Wisconsin	19.9	21,271	7	8.1	10.2	0.82

8.12 - Identify the countries from the following table that meet the following criteria.

a) Per capita energy use > 4000 and population < 20,000,000

b) Infant mortality < 7 and life expectancy > 79.0

c) Per capita energy use < 4,000 or ((population > 40 million) and (car theft < 1))

d) [Per capita energy use < 4,000 or (population > 40 million)] and (car theft < 1)

e) not (population > 40,000,000)

f) Population < 20,000,000 and not (car theft > 1.5)

Country	Population (millions)	Energy Use (bl.oil/per)	Infant Mortality (per 1000)	Life expect. (years)	Car Theft (%)
Australia	19.9	5,668	4	79.2	2.2
Britain	59.3	5,945	5	77.5	2.6
Finland	5.2	6,456	4	78.0	0.5
France	59.7	4,350	4	79.2	1.8
Japan	127.2	4,071	3	81.6	0.1
Netherlands	16.2	5,993	5	78.3	0.5
Norway	4.6	6,019	4	78.9	1.5
South Africa	45.3	3,703	52	46.5	2.4
Spain	41.1	2,945	5	78.3	0.5
U.S.A.	291.0	8,066	7	77.3	0.5

8.13 - What are normal forms in relational databases? Why are they used, and what are the advantages of putting data in higher normal forms?

8.14 - Sketch the output table resulting from an inner join shown below:

Id1	pos
Y	wa
Z	ea
A	rt
Y	pr
Y	nn
R	rt
Q	mn

Id2	tm
X	5
Y	1
A	6
Q	4
N	3
L	2

8.15 - Sketch an outer join for the table shown in the previous problem.

8.16 - Define the basic differences between first, second, and third normal forms.

8.17 - Give an example of a functional dependency.

8.18 - List the functional dependencies in the following table, using the arrow notation described in this chapter.

ID	Size	Shape	Color	Age	Source
1	large	round	blue	10	A
2	medium	round	green	5	B
3	small	round	red	10	C
4	medium	knobbed	green	5	D
5	medium	knobbed	green	5	E
6	large	round	blue	10	F
7	large	round	blue	10	A

8.19 - What is the object-relational data model? How does it differ from a relational model?

9 Basic Spatial Analyses

'Introduction

Spatial data analysis is the application of operations to coordinate and related attribute data, often to solve a problem, for example, to identify high crime areas, or generate a list of road segments that need repaving. There are hundreds of *spatial operations* or *spatial functions* used in spatial analysis, and all involve calculations with coordinates or attributes.

The terms spatial operations and spatial functions are often used interchangeably. Some practitioners insist an operation does not necessarily produce any output, while a function does. However, in keeping with many authors, GIS practitioners, and software vendors, we will use the terms spatial function and operation interchangeably.

Spatial operations can be applied sequentially to solve a problem. A chain of spatial operations is often applied, with the output of each spatial operation serving as the input of the next (Figure 9-1). Part of the challenge of geographic analysis is selecting appropriate spatial operations, and applying them in the appropriate order.

The table manipulations we described in Chapter 8 are included in our definition of a spatial operation. Indeed, the selection and modification of attribute data in spatial data layers are included at some time in nearly all complex spatial analyses. Many operations incorporate both the attribute and coordinate data, and the attributes must be further selected and modified in the course of a spatial analysis. Some might take issue with our inclu-

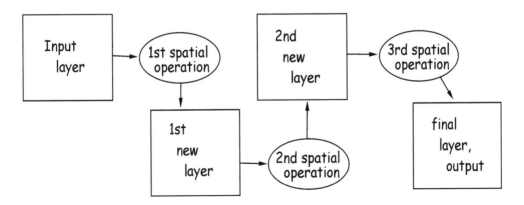

Figure 9-1: A sequence of spatial operations is often applied to obtain a desired final data layer.

sive definition of a spatial operation, in that an operation might be applied only to the "nonspatial" attribute data stored in the database tables. However, attribute data are part of the definition of spatial objects, and it seems artificial to separate operations on attribute data from operations that act on only the coordinate portion of spatial data.

The discussion in the present chapter will expand on rather than repeat the selection operations treated in Chapter 8. This chapter describes spatial data analyses that involve sort, selection, classification, and spatial operations that are applied to both coordinate and associated attribute data.

Input, Operations, and Output

Spatial data analysis typically involves using data from one or more layers to create output (Figure 9-2). The analysis may consist of a single operation applied to a data layer, or many operations that integrate input data from many layers to create the desired output. Many operations require a single data layer as input and generate a single output data layer. Vector to raster conversion is an example of an operation with a one-to-one correspondence between input and output layers (Figure 9-3).

There are also operations that generate several output data layers from a single input. Terrain analysis functions may take a raster grid of elevations as an input data layer and produce both slope (how steep each cell is) and aspect (the slope direction). In this case, two outputs are generated for each input elevation data set.

Operations may also take several input layers to generate a single output layer. A layer average is an example of the use of multiple input layers to produce a single output layer. For example, mean annual grain production might be stored for 10 separate years in raster data layers. To calculate the average grain production over the 10-year period, the annual layers would be averaged on a cell-by-cell basis. This

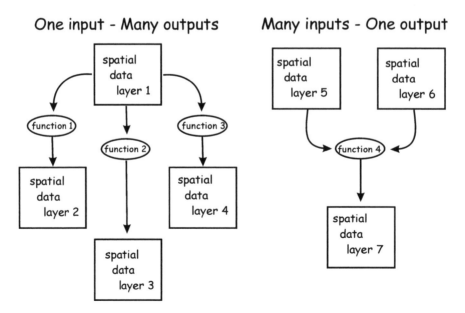

Figure 9-2: Much basic spatial data analysis consists largely of spatial operations. These operations are applied to one or more input data layers to produce one or more output data layers.

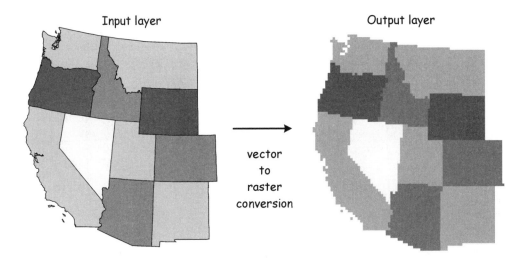

Input layer   Output layer

vector
to
raster
conversion

Figure 9-3: Vector to raster conversion, an example of a spatial data operation.

operation results in a single output data layer. Finally, there are some spatial operations that require many input layers and generate many output layers.

The output from a spatial operation may be spatial, in that a new data layer is produced, or the output may be non-spatial, in that the spatial operation may produce a scalar value, a list, or a table, with no explicit geometric data attached. A layer mean function may simply calculate the mean cell value found in a raster data layer. The input is a spatial data layer, but the output is a scalar.

Other operations create aspatial output. For example, a list of all landcover types may be extracted from a data layer. The list indicates all the different types of landcover that can be found in the layer, but is non-spatial, because it does not attach each landcover to specific locations on the surface of the Earth. One might argue that the list is referenced in a general way to the area covered by the spatial data layer, but each landcover class in the list is not explicitly related to any specific polygons

or points of interest in the data layer. Thus, a spatial input is passed through this "occurrence" operator and provides a list output.

Scope

Spatial data operations may be characterized by their *spatial scope*, the extent or area of the input data that are used in determining the values at output locations (Figure 9-4). There is often a direct correspondence between input data at a location and output data at that same location. The geographic or attribute data corresponding to an input location are acted on by a spatial operation, and the result is placed in a corresponding location in the output data layer. Spatial operations may be characterized as local, neighborhood, or global, to reflect the extent of the source area used to determine the value at a given output location (Figure 9-4).

Local operations use only the data at one input location to determine the value at a corresponding output location (Figure 9-4, top). Attributes or values at adjacent locations are not used in the operation.

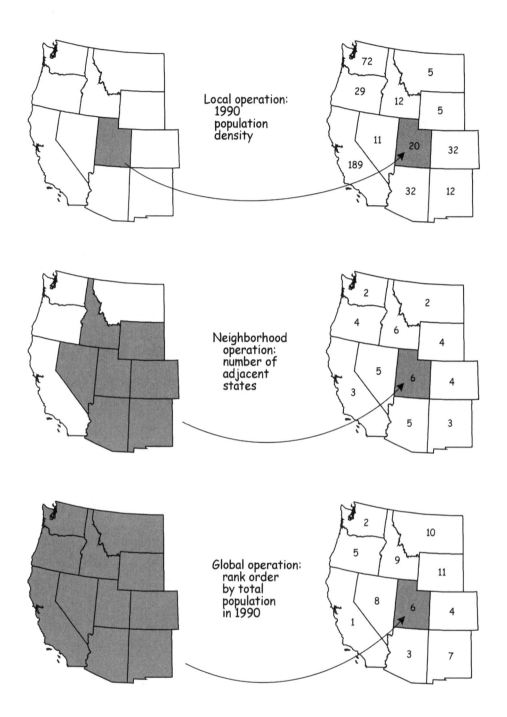

Figure 9-4: Local, neighborhood, and global operations. Specific input and output regions are shown for Utah, the shaded area on the right side of the figure. Shaded areas on the left contribute to the values shown in the shaded area on the right. Local operation output (top right) depends only on data at the corresponding input location (top left). Neighborhood operation output (middle right) depends on input from the local and surrounding areas (middle left). Global operation output (bottom right) depends on all features in the input data layer (bottom left).

Neighborhood operations use data from both an input location plus nearby locations to determine the output value (Figure 9-4, center). The extent and relative importance of values in the nearby region may vary, but the value at an output location is influenced by more than just the value of data found at the corresponding input location.

Global operations use data values from the entire input layer to determine each output value. The value at each location depends in part on the values at all input locations (Figure 9-4, bottom).

The set of available spatial operations depends on the data model and type of spatial data used as input. Some operations may be easily applied to raster or vector data. While the details of the specific implementation may change, the concept of the operation does not. Other operations may be possible in only one data model.

Characteristics of a data model will determine how the concept of any given operation is applied. The specific implementation of many operations, for example, multi-layer addition, depends on the specific data model. A raster operation may produce a different outcome than a vector operation, even if the themes are meant to represent the same features. In a like manner, the specific set and sequence of operations in a spatial analysis will depend on the data model used and the specific operations available in the GIS software.

Operation scope provides a good example of this influence of data models. Cells in a raster data set have uniform size and shape. Therefore, a local operation applied to a raster data layer collects information from a well defined area. In con-

trast, polygons usually vary in size and shape. A local operation for a vector polygon data set is likely drawn from this variable size, shape, and location. In Figure 9-4 the local operation follows a state boundary. Therefore, the operation applies to a different size and shape for each state.

Neighborhood analyses are affected by the shape of adjacent states in a similar manner. Summary values such as populations of adjacent states may be greatly influenced by changes in neighborhood size, so great care must be taken when interpreting the results of a spatial operation. Knowledge of the algorithm behind the operation is the best aid to interpreting the results.

While most operations might be conceptually compatible with most spatial data models, some operations are easier to apply in some models. Most neighborhood operations are quite easy to program when using raster data models, and quite difficult when using vector data models. The reverse is true for network operations, which are generally easier to apply in vector models. In many instances it is more efficient to convert the data between data models and apply the desired operations, and if necessary, convert the results back to the original data model.

Selection and Classification

Selection operations identifies features that meet one to several conditions or criteria. In these operations, attributes or geometry of features are checked against criteria, and those that satisfy the criteria are selected. These selected features may then be written to a new output data layer, or the geometry or attribute data may be manipulated in some manner.

Figure 9-5 shows an example of a selection operation that involves the attributes of a spatial data set. Two conditions are applied and the features that satisfy both conditions are included in the selected set.This example shows the selection of those states in the "lower 48" United States that are a) entirely north of Arkansas, and b) have an area greater than 84,000 square kilometers. The complete set of features that will be considered is shown at the top of the figure. This set is comprised of the lower 48 states, with the state of Arkansas indicated by shading. The next two maps of Figure 9-5 shows those states that match the individual criteria. The second map from the top shows those states that are entirely north of Arkansas, while the third map shows all those states that are greater than 84,000 square kilometers. The bottom part of Figure 9-5 shows those states that satisfy both conditions. This figure illustrates two basic characteristics of selection operations. First, there is a set of features that are candidates for selection, and second, these features are selected based singly or on some combination of the geographic and attribute data.

The simplest form of selection is an *on-screen query*. A data layer is displayed, and features are selected by a human operator. The operator uses a pointing device to locate a cursor over a feature of interest and sends a command to select, often via a mouse click or keyboard entry. On-screen (or interactive) query is used to gather information about specific features, and is often used for interactive updates of

attribute or spatial data. For example, it is common to set up a process such that when a feature is selected the attribute information for the feature is displayed. These attribute data may then be edited and the changes saved.

Queries may also be specified by applying conditions solely to the aspatial components of spatial data. These selections are most often based on the attribute data tables for a layer or layers. These selection operations are applied to a set of features in a data layer or layers. The attributes for each feature are compared to a set of conditions. If the attributes match the conditions, they are selected; if the attributes fail to match the conditions, they are placed in an unselected set. In this manner, selection splits the data into either the selected set or the unselected set. The selected data are then typically acted on in some way, often saved to a separate file, deleted, or changed in some manner.

Selection operations on tables were described in general in Chapter 8. The description here expands on that information and draws attention to specific characteristics of selections applied to spatially-related data. Table selections have spatial relevance because each record in a table is associated with a geographic feature. Selecting a record in a table simultaneously selects the associated spatial features: cells, points, lines, or areas. Spatial selections may be combined with table selections to identify a set of selected geographic features.

Set Algebra

Selection conditions are often formalized using *set algebra*. Set algebra uses the operations less than (<), greater than (>), equal to (=), and not equal to (< >). These selection conditions may be applied either alone or in combination to select features from a set.

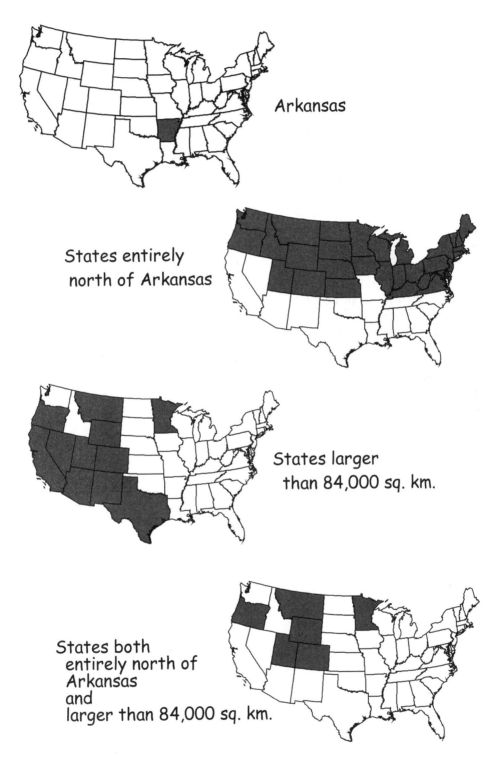

Figure 9-5: An example of a selection operation based on multiple conditions.

Figure 9-6 shows four set algebraic expressions and the selection results for a set of counties in the northeastern U.S. The upper two selections show equal to (=) and not equal to (< >) selections. The upper-left shows all counties with a value for the attribute **state** that equals **Vermont**, while the upper right shows all counties with a value for **state** that are not equal to **New York**. The lower selections in Figure 9-6 show examples of ordinal comparisons. The left figure shows all counties with a size greater than or equal to (> =) 1000 square miles, while the right side shows all counties with a population density less than (<) 250 persons per square mile.

The set algebra operations greater than (>) or less than (<) may not be applied to nominal data, because there is no implied order in nominal data. Green is not greater than yellow, and red is not less than blue. Only the set algebra operations equal to (=) and not equal to (< >) apply to these or other nominal variables. All set algebra operations may be applied to ordinal data, and all are often applied to interval/ratio data.

Boolean Algebra

Boolean algebra uses the conditions OR, AND, and NOT to select features. Boolean expressions are most often used to combine set algebra conditions and create compound spatial selections. The Boolean expression consists of a set of Boolean operators, variables, and perhaps constants or scalar values.

Boolean expressions are evaluated by assigning an outcome, true or false, to each condition. Figure 9-7 shows three example Boolean expressions. The first is an expression using a Boolean AND, with two arguments for the expression. The first argument specifies a condition on a variable named **area**, and the second argument a condition on a variable named **farm_income**. Features are selected if they satisfy both arguments, that is, if their **area** is larger than 100,000 AND **farm_income** is less than 10 billion.

Expression 2 in Figure 9-7 illustrates a Boolean NOT expression. This condition specifies that all features with a variable **state** which is not equal to **Texas** will return a true value, and hence be selected. NOT is also often known as the negation operator. This is because we might interpret the application of a NOT operation as

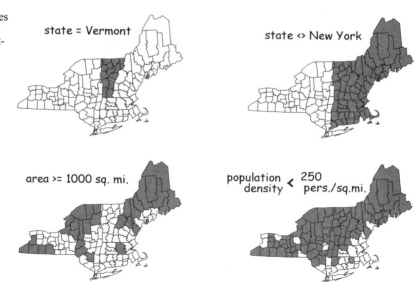

Figure 9-6: Examples of expressions in set algebra and their outcome. Selected features are shaded.

state = Vermont

state <> New York

area >= 1000 sq. mi.

population density < 250 pers./sq.mi.

Boolean expressions

1. (area > 100,000)
   AND
   (farm_income < 10 billion)

2. NOT (state = Texas)

3. [(rainfall > 1000)
     AND
     (taxes = low)]
   OR
   [(house_cost < 65,000)
     AND
     NOT (crime = high)]

Figure 9-7: Example Boolean expressions.

exchanging the selected set for the unselected set. The argument of expression 2 in Figure 9-7 is itself a set algebra expression. When applied to a set of features, this expression will select all features for which the variable **state** is equal to the value **Texas**. The **NOT** operation reverses this, and selects all features for which the variable **state** is not equal to **Texas**.

The third expression in Figure 9-7 shows a compound Boolean expression, combining four set algebra expressions with **AND**, **OR**, and **NOT**. This example shows what might be a naive attempt to select areas for retirement. Our grandparent is interested in selecting areas that have high rainfall and low taxes (a gardener on a fixed income), or low housing cost and low crime.

The spatial outcomes of specific Boolean expressions are shown in Figure 9-8. The figure shows three overlapping circular regions, labeled **A**, **B**, and **C**. Areas may fall in more than one region, for example, the center, where all three regions overlap, is in **A**, **B**, and **C**. As shown in the figure, Boolean **AND**, **OR**, or **NOT** may be used to select any combination or portions of these regions.

OR conditions return a value of true if either argument is true. Areas in either region **A** or region **B** are selected at the top center of Figure 9-8. **AND** requires the conditions on both sides of the operation be met; an **AND** operation results in a reduced selection set (top right, Figure 9-8). **NOT** is the negation operator, and flips the effect of the previous operations, that is, it turns true to false and false to true. The **NOT** shown in the lower left portion of Figure 9-

Figure 9-8: Examples of expressions in Boolean algebra, and their outcomes. Sub-areas of three regions are selected by combining **AND**, **OR**, and **NOT** conditions in Boolean expressions. Any sub-area or group of sub-areas may be selected by the correct Boolean combination.

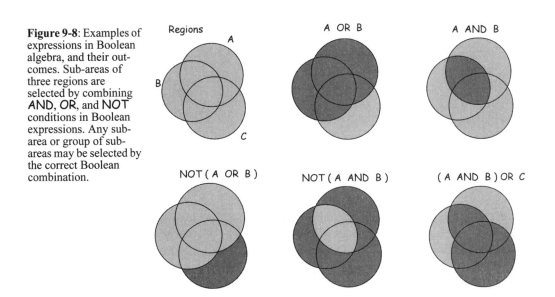

(County = Rice)

AND

(Wshed = Canon)

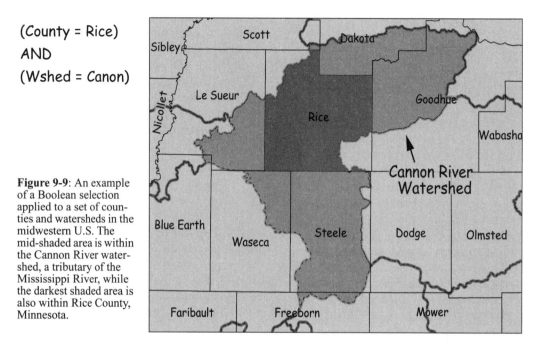

Figure 9-9: An example of a Boolean selection applied to a set of counties and watersheds in the midwestern U.S. The mid-shaded area is within the Cannon River watershed, a tributary of the Mississippi River, while the darkest shaded area is also within Rice County, Minnesota.

8 returns the area that is only in region *C*. Note that this is the converse, or opposite set that is returned when using the comparative *OR*, shown in the top center of Figure 9-8. The *NOT* operation is often applied in combination with the *AND* Boolean operator, as shown at the bottom center of Figure 9-8. Again, this selects the converse (or complement) of the corresponding *AND*. Compare the bottom center selection to the top right selection in Figure 9-8. *NOT*s, *AND*s, and *OR*s may be further combined to select specific combinations of areas, as shown in the lower right of Figure 9-8.

Note that as with table selection discussed in Chapter 8, the order of application of these Boolean operations is important. In most cases you will not select the same set when applying the operations in a different order. Therefore, parentheses, brackets, or other delimiters should be used to specify the order of application. The expression *A AND B OR C* will give different results when interpreted as (*A AND B*) *OR C*, as shown in Figure 9-8, than when interpreted as *A AND* (*B OR C*). Verify this as an exercise. Which areas does the second Boolean expression select?

Figure 9-9 shows a real-world example of a Boolean selection. County agents often must identify areas for treatment, in this case a portion of the Cannon River, a tributary of the Mississippi River, targeted for pollution reduction. Counties are labeled, with boundaries shown as thick solid lines. The Cannon River watershed is shown in darker shades of gray. A Boolean *AND* operation was applied to a data layer containing both watershed and county boundaries, selecting the areas that are both within the Cannon River watershed and within Rice County.

Spatial Selection Operations

Many spatial operations select sets of features. These operations are applied to a spatial data layer and return a set of features that meet a specified condition. Adjacency and containment are commonly used spatial selection operations.

Adjacency operations are used to identify those features that "touch" other features. Features are typically considered to touch when they share a boundary, as when two polygons share an edge. A target or key set of polygon features is identified, and all features that share a boundary with the target features are placed in the selected set.

Figure 9-10a shows an example of a selection based on polygon adjacency. The state of Missouri is shaded on the left side of Figure 9-10a, and states adjacent to Missouri are shaded on the right portion of Figure 9-10a. States are selected because they include a common border with Missouri.

There are many ways the shared border may be detected. With a raster data layer an exhaustive cell-by-cell comparison may be conducted to identify adjacent pairs with different state values. Vector adjacency may be identified by observing the topological relationships (see Chapter 2 for a discussion of topology). Line and polygon topology typically record the polygon identifiers on each side of a line. All lines with Missouri on one side and a different state on the other side may be flagged, and the list of states adjacent to Missouri extracted.

a)

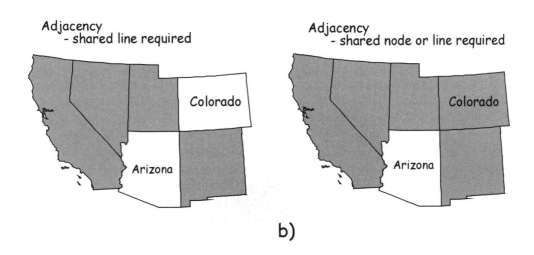

b)

Figure 9-10: Examples of selections based on adjacency. a) Missouri, USA is shown on the left and all states adjacent to Missouri shown on the right. b) Different definitions of adjacency result in different selections. Colorado is not adjacent to Arizona when line adjacency is required (left), but is when node adjacency is accepted (right).

Figure 9-11: An example of a selection based on containment. All states containing a portion of the Mississippi River or its tributaries are selected.

Adjacency is defined in Figure 9-10a as sharing a boundary for some distance greater than zero. Figure 9-10b shows how a different definition of adjacency may affect selection. The left of Figure 9-10b shows the state of Arizona and a set of adjacent (shaded) western states. By the definition of adjacency used in Figure 9-10a, Arizona and Colorado are not adjacent, because they do not share a boundary along a line segment. Arizona and Colorado share a border at a point, called Four Corners, where they join with Utah and New Mexico. When a different definition of adjacency is used, with a shared node qualifying as adjacent, then Colorado is added to the selected adjacent set (right, Figure 9-10b). This is another illustration of an observation made earlier; there are often several variations of any single spatial operation. Care must be taken to test the operation under controlled conditions until the specific implementation of a spatial operation is well understood.

Containment is another spatial selection operation. Containment selection identifies all features that contain or surround a set of target features. For example, the California Department of Transportation may wish to identify all counties, cities, or other governmental bodies that contain some portion of Highway 99, because they wish to improve road safety. A spatial selection may be used to identify these governmental bodies.

Figure 9-11 illustrates a containment selection based on the Mississippi River in North America. We wish to identify states that contain some portion of the river and its tributaries. A query is placed, identifying the features that are contained, here the Mississippi River network, and the target features that may potentially be selected. The target set in this example consists of the lower 48 states of the United States. All states that contain a portion of the Mississippi River or its tributaries are shaded as part of the selected set.

Classification

Classification is a spatial data operation that is often used in conjunction with selection. A classification, also known as a *reclassification* or *recoding*, will categorize geographic objects based on a set of conditions. For example, all the polygons larger than one square mile may be assigned a size value equal to Large, all polygons from 0.1 to 1 square mile may be assigned a size equal to Mid, and all polygons smaller than 0.1 square miles may be assigned a size equal to Small (Figure 9-12). Classifications may add to or modify the attribute data for each geographic object. These modified attributes may in turn be used in further analyses, such as for more complex combinations in additional classification.

Classification may be used for many other purposes. One common end is to group objects for display or map production. These objects have a common property, and the goal is to display them with a uniform color or symbol so the similar objects are identified as a group. The display color and/or pattern is typically assigned based upon the values of an attribute or attributes. A range of display shades may be chosen, and corresponding values for a specific attribute assigned. The map is then displayed based on this classification.

A classification may be viewed as an assignment of features from an existing set of classes to a new set of classes. We identify features that have a given set of values, for example, parcels that are above a certain size, and assign them all a classification value, in this case the class "large". Parcels in another range of sizes may be assigned different class values, for example, "mid" and "small". The attribute that stores the parcel area is used as a guide to assigning the new class value for size.

The assignment from input attribute values (area) to new class values (here, size) may be defined manually, or the assignment may be defined automatically. For manual classifications, the class transitions are specified entirely by the human analyst.

Classifications are often specified by a table or array. The table identifies the input class or values, as well as the output class for each of this set of input values. Figure 9-13 illustrates the use of a classification table to specify class assignment. Input values of A or B lead to an output class value of 1, an input value of E leads to an output value of 2, and an input value of I leads to an output value of 3. The table provides a complete specification for each classification assignment.

Figure 9-13 illustrates a classification based on a manually defined table. A human analyst specifies the In items for the source data layer via a classification table, as well as the corresponding output value for each In variable. Out values must be specified for each input value or there will be undefined features in the output layer. Manually defining the classification table provides the greatest control over class assignment. Alternatively, classifica-

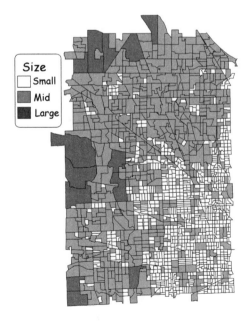

Figure 9-12: Land parcels re-classified by area.

tion tables may be automatically assigned, in that a number of classes may be specified and some rule embodied in a computer algorithm used to assign output classes for each of the input classes.

A *binary classification* is perhaps the simplest form of classification. A binary classification places objects into two classes: 0 and 1, true and false, **A** and **B**, or some other two-level classification. A set of features is selected and assigned a value, and the complement of the set, all remaining features in the data layer, is assigned the different binary value.

A binary classification is often used to store the results of a complex selection operation. A large number of Boolean and set algebra expressions may be used to select a set of features. A specific target attribute is identified for the selected set of features. The value of this target attribute is assigned a unique value. The target

attribute for all features in the unselected set to a different value. In this way create a permanent record of the selected set in the target attribute.

For example, we may wish to select states at least partially west of the Mississippi River as an intermediate step in an analysis (Figure 9-14). We may be using this classification in many subsequent spatial operations. Thus, we wish to store this characteristic, whether the state is west or east of the Mississippi River. States are selected based on location and reclassified. We record this classification by creating a new attribute and assigning a binary value to this attribute, 1 for those parcels that satisfy the criteria, and 0 for those that do not (Figure 9-14). The variable is_west records the state location relative to the Mississippi River. Additional selection operations may be applied, and the created binary variable preserves the information generated in the initial selection.

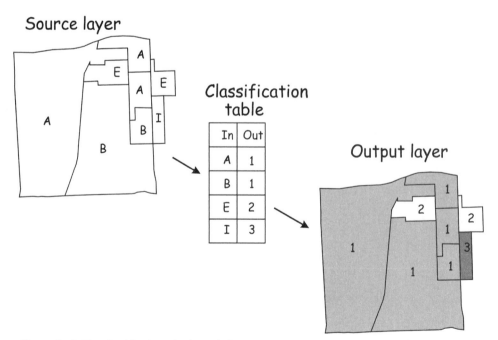

Figure 9-13: The classification of a thematic layer. Values are given to specific attributes in a classification table, which is used, in turn, to assign classes in an output layer.

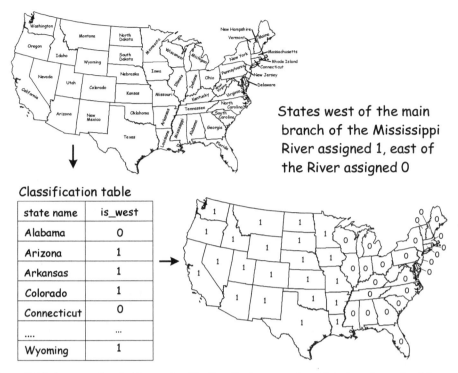

States west of the main branch of the Mississippi River assigned 1, east of the River assigned 0

Classification table

state name	is_west
Alabama	0
Arizona	1
Arkansas	1
Colorado	1
Connecticut	0
....	...
Wyoming	1

Figure 9-14: An example of a binary classification. Features are placed into two classes in a binary classification, west (1) and east (0) of the Mississippi River. The classification table codifies the assignment.

Manually defining the classification table may not always be necessary, and may be tedious or complex. Suppose we wish to assign a set of display colors to a set of elevation values. There may be thousands of distinct elevation values in the data layer, and it would be inconvenient at best to assign each color manually. Automatic classification methods are often used in these instances.

An automatic classification uses some rule to specify the input class to output class assignments. The input and output class boundaries are often based on a set of parameters used to guide class definition.

A potential drawback from an automated class assignment stems from our inability to precisely specify class boundaries. A mathematical formula or algorithm defines the class boundaries, and so specific classes of interest may be split. Thus, the analyst sacrifices precise control over

class specification when an automated classification is used. Considerable time may be saved by automatic class assignment, but the analyst may have to manually change some class boundaries as the only way to achieve a desired classification.

Figure 9-15 describes a data layer we will use to illustrate automatic class assignment. The figure shows a set of "neighborhoods" with populations that range from 0 to 5133. We wish to display the neighborhoods and populations in three distinct classes, high, medium, and low population. High will be shown in black, medium in gray, and low in white. We must decide how to assign the categories - what population levels define high, medium, and low? In many applications the classification levels are previously defined. There may be an agreed-upon standard for high population, and we would simply use this level. However, in many instances the classes are not defined, and we must choose them.

Figure 9-15 includes a bar graph depicting the population frequency distribution; this type of bar graph is commonly called a histogram. The frequency histogram shows the number of neighborhoods that are found in each bar (or "bin") of a set of very narrow population categories. For example, we may count the number of neighborhoods that have a population between 3000 to 3100. If a neighborhood has 3037, 3004, 3088, or any other number between 3000 and 3100, we add one to our frequency sum for the category covering 3000 and 3100. We review all neighborhoods in our area, and we plot the percentage of neighborhoods that have a population between 3000 and 3100 as a vertical bar on the histogram. Approximately 8.4 percent of the neighborhoods have a population in this range, so a vertical bar corresponding to 8.4 units high is plotted. We count and plot the histogram

values for each of our narrow categories (e.g., the number from 0 to 100, from 100 to 200, from 200 to 300), until the highest population value is plotted.

Our primary decision in class assignment is where to place the class boundaries. Should we place the boundary between the low and medium population classes at 1000, or at 1200? Where should the boundary between medium and high population classes be placed? The location of the class boundaries will change the appearance of the map, and also the resulting classification.

One common method for automatic classification specifies the number of output classes and requests equal interval classes over the range of input values. This *equal-interval* classification simply subtracts the lowest value of the classification variable from the highest value, and defines equal-width boundaries to fit the desired number of classes into the range.

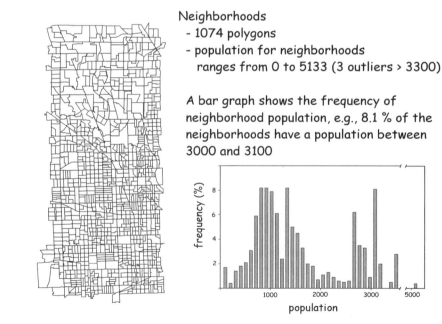

Neighborhoods
- 1074 polygons
- population for neighborhoods
  ranges from 0 to 5133 (3 outliers > 3300)

A bar graph shows the frequency of neighborhood population, e.g., 8.1 % of the neighborhoods have a population between 3000 and 3100

Figure 9-15: Neighborhood polygons and population levels used in subsequent examples of classification assignment. The populations for these 1074 neighborhoods ranges from 0 to 5133. The histogram at the lower right shows the frequency distribution. Note that there is a break in the chart between 3500 and 5000 to show the three "outlier" neighborhoods with populations above 3300.

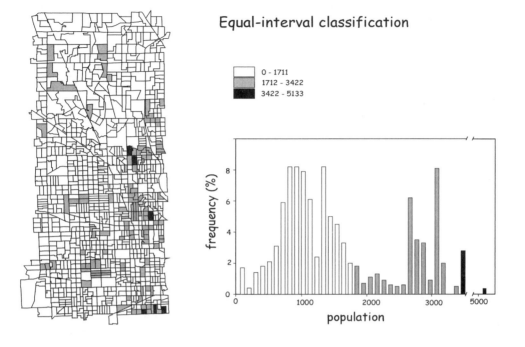

Figure 9-16: An equal interval classification. The range 0 to 5133 is split into three equal parts. Colors are assigned as shown in the map of the layer (left), and in the frequency plot (right). Note the relatively few polygons assigned to the high classes in black. A few neighborhoods with populations near 5000 shift the class boundaries upward.

Figure 9-16 illustrates an equal-interval classification for the population variable. Three classes assigned over the range of 0 to 5133 are specified. Each interval is approximately one-third of this range. This range is evenly divided by 1711. The small class extends from 0 to 1711, the medium class from 1712 to 3422, and the large class from 3423 to 5133. Population categories are shown colored accordingly on the map and the bar graph, with the small (white), medium (gray), and large (black) classes shown.

Note that the low population class shown in white dominates the map; most of the neighborhoods fall in this population class. This often happens when there are features that have values much higher than the norm. There are a few neighborhoods with populations above 5000 (to the right of the break in the population axis of the bar graph), while most neighborhoods have

populations below 3000. The outliers shift the class boundaries to higher values, 1711 and 3422, resulting in most neighborhoods falling in the small population category.

Another common method for class assignment results in an *equal-area* classification (Figure 9-17). Class boundaries are defined to place an equal proportion of the study area into each of a specified number of classes. This usually leads to a visually-balanced map because all classes have approximately equal extents. Equal-area classes are often desirable, for example, when resources need to be distributed over equal areas, or when equally-sized overlapping sales territories may be specified.

Note that the class width may change considerably with an equal-area classification. An equal-area classification sets class boundaries so that each class covers approximately the same area. A class may consist of a few or even one large polygon. This results in a small range for the large-polygon classes. Classes also tend to have a narrow range of values near the peaks in the histogram. Many polygons are represented at the histogram peaks, and so these may correspond to large areas. Both of these effects are illustrated in Figure 9-17. The middle class of the equal-area classification occurs at population values between 903 and 1223. This range of populations is near the peak in the frequency histogram, and these population levels are associated with larger polygons. This middle class spans a range of approximately 300 population units, while the small and large classes span near 900 and 4000 population units, respectively.

Note that equal-area assignments may be highly skewed when there are a few polygons with large areas, and these polygons have similar values. Although not occurring in our example, there may be a relationship between the population and area for a few neighborhoods. Suppose in a data set similar to ours there is one very large neighborhood dominated by large parks. This neighborhood has both the lowest populations and largest area. An equal-area classification may place this neighborhood in its own class. If a large parcel also occurs with high population levels, we may get three classes: one parcel in the small class, one parcel in the high population class, and all the remaining parcels in the medium population class. While most equal-area classifications are not this extreme, unique parcels may strongly affect class ranges in an equal-area classification.

We will cover a final method for automated classification, a method based on *natural breaks*, or gaps, in the data. Natural breaks classification looks for "obvious"

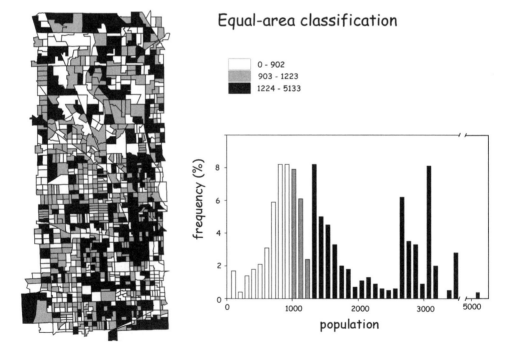

Figure 9-17: Equal-area classification. Class boundaries are set such that each class has approximately the same total area. This often leads to a smaller range when groups of frequent classes are found. In this instance the medium class spans a small range, from 903 to 1223, while the high population class spans a range that is almost 10 times broader, from 1224 to 5133.

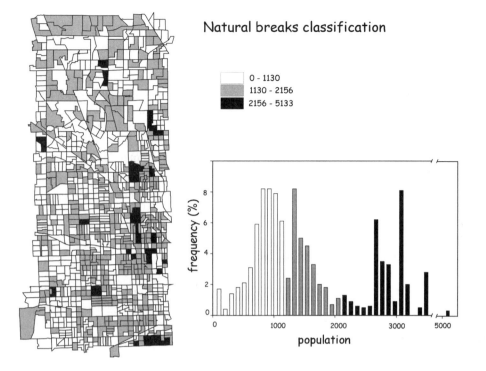

Figure 9-18: Natural breaks classification. Boundaries between classes are placed where natural gaps or low points occur in the frequency distribution.

breaks. It attempts to identify naturally occurring clusters of data, not clusters based on the spatial relationships, but rather clusters based on an ordering variable.

There are various methods used to identify natural breaks. Large gaps in an ordered list of values are one common method. The values are listed from lowest to highest, and the largest gaps in values selected. Barring gaps, low points in the frequency histogram may be identified. There is usually an effort to balance the need for relatively wide and evenly distributed classes and the search for natural gaps. Many narrow classes and one large class may not be acceptable in many instances, and there may be cases where the specified number of gaps does not occur in the data histogram. More classes may be requested than obvious gaps, so some natural break

methods include an alternative method, for example, equally-spaced intervals, for portions of the histogram where no natural gaps occur.

Figure 9-18 illustrates a natural break classification. Two breaks are evident in the histogram, one near 1300 and one near 2200 persons per neighborhood. Small, medium, and large populations are assigned at these junctures.

Figure 9-16 through Figure 9-18 strongly illustrate an important point: you must be careful when producing or interpreting class maps, because the apparent relative importance of categories may be manipulated by altering the starting and ending values that define each class. Figure 9-16 suggests most neighborhoods are low population, Figure 9-17 suggests that high population neighborhoods cover the largest areas, and that these are well mixed with areas of low and medium population, while

Figure 9-18 indicates the area is dominated by low and medium population neighborhoods. Precisely because there are no objectively defined population boundaries we have great flexibility in manipulating the impression we create. The legend in class maps should be scrutinized, and the range between class boundaries noted. A histogram and the maximum/minimum values help when interpreting the legend.

The Modifiable Areal Unit Problem

When there are no objectively recognizable boundaries, polygons may be reclassified and grouped in many ways.The aggregate values for polygon variables such as population, age, and income will depend on the size, shape, and location of the aggregated polygons (Figure 9-19). This general phenomenon, known as the modifiable areal unit problem or MAUP, has been exploited by politicians in the U.S. and elsewhere to redraw political boundaries to one party or another's--but generally not the country's--advantage. The process of aggregating neighborhoods to create majority blocks for political advantage was named gerrymandering after Massachusetts governor Elbridge Gerry, when he crafted a political district shaped like a salamander.

There are two primary characteristics of the MAUP that may be manipulated to affect aggregate polygon values. The first is the *zoning effect*, that aggregate statistics may change by the shape of the units, and the second is the *size effect*, that aggregate statistics may change with the area of the units. For example, the mean income of a unit will change when the boundaries of a unit change, either because of a change in zone or a change in size.

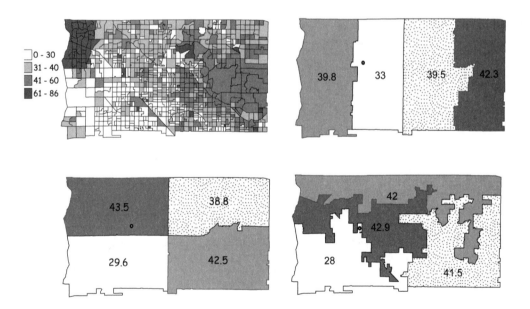

Figure 9-19: An example of the modified areal unit problem (MAUP). Census blocks (upper left) have been aggregated in various ways to produce units that show different mean population age. All units are approximately equal in size. Note that the frequency of units with a mean age over 40 can be 1 (upper right), two (lower left), or three (lower right), and that an individual block (small circle, upper left quadrant) may be in a polygon with a mean age in the 20's, 30's, or 40's, depending on unit shape.

MAUP effects may substantially influence the values for each unit, and they affect many analyses based on the collected data. Openshaw and Taylor published results in 1979 that illustrate MAUP dependencies particularly well. They analyzed the percentage of elderly voters and the number of Republican voters for the 99 counties in Iowa. They showed that the correlation between the elderly voters and republican voter numbers ranged from 0.98 to -0.81 by varying the scale and aggregation units that grouped counties. Additional work has shown that multivariate statistical models based on aggregate data are similarly dependent on the aggregation units, leading to contradictory results in developing multivariate predictions.

Numerous studies of the MAUP have resulted in a number of recommendations to identify and/or reduce the primary negative impacts of the zoning and size effects. The primary recommendation is to work with the basic units of measure. In our census example, this would be to collect and maintain information on the individual person or household. This is often not possible, for example, aggregation is specifically required to maintain the anonymity of the census respondents. However, many efforts allow recording and maintaining data on the primary units within a GIS framework, and this should be implemented when possible.

A second way to address the MAUP is based on optimal zoning. Zones are designed to maximize variation between zones while minimizing variation within zones. Optimal zones are difficult to define for more than one variable, because variables often do not change in concert. For example, an optimal set of zones for determining traffic densities may not be an optimal set of zones for average age. Old people are no more nor less likely to live near busy areas than near roads with little traffic. Optimum zoning approaches are best applied when interest in one variable predominates.

Another approach to solving the MAUP involves conducting a set of sensitivity analyses. Units are aggregated and rezoned across a range of sizes and shapes and the analyses performed for each set. Changes in the results are observed, and the sensitivity to zone boundaries and sizes noted. These tests may identify the relative sensitivities of different variables to size and zoning effects. Robust results may be identified, for example, average age may not change over a range of sizes and unit combinations, yet may change substantially over a narrow range of sizes in some areas. A high computational burden is the primary drawback of this approach, because it depends on replicated runs for each set of variables, zone levels, and shapes. This often overwhelms the available computing resources for many problems and agencies.

Dissolve

A *dissolve* function is primarily used to combine similar features within a data layer. Adjacent polygons may have identical values for an attribute. For example, a wetlands data layer may specify polygons with several sub-classes, such as wooded wetlands, herbaceous wetlands, or open water. If an analysis requires we identify only the wetland areas vs. the upland areas, then we may wish to dissolve all boundaries between adjacent wetlands. We are only interested in preserving the wetland/ upland boundaries.

Dissolve operations are usually applied based on a specific "dissolve" attribute associated with each feature. A value or set of values is identified that belongs in the same grouping. Each line that serves as a boundary between two polygon features is assessed. The values for the dissolve attribute are compared across the boundary line. If the values are the same, the boundary line is removed, or dissolved away. If the values for the dissolve attribute differ across the boundary, the boundary line is left intact.

Figure 9-20 illustrates the dissolve operation that produces a binary classification. This classification places each state of the contiguous United States into one of two categories, those entirely west of the Mississippi River (1) and those east of the Mississippi River (0). The attribute named **is_west** contains values indicating location. A dissolve operation applied on the variable **is_west** removes all state boundaries between similar states. This reduces the set from 48 polygons to two polygons.

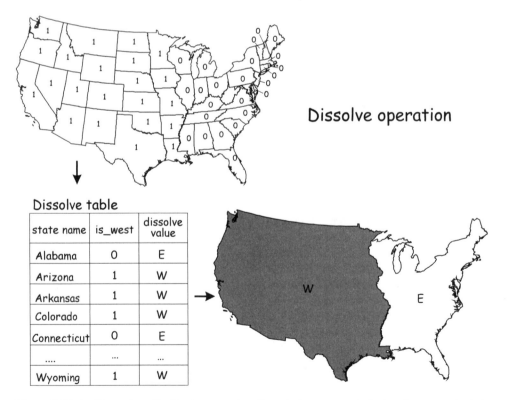

Figure 9-20: An illustration of a dissolve operation. Boundaries are removed when they separate states with the same value for the dissolve attribute **is_west**.

Before dissolve

After dissolve

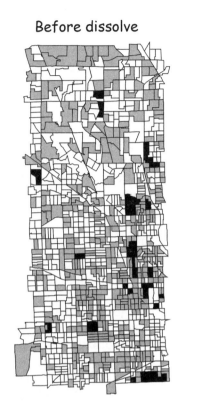 

Figure 9-21: An example of a dissolve operation. Note the removal of lines separating polygons of the same size class. This greatly reduces the number of polygons.

Dissolve operations are often needed prior to applying an area-based selection in spatial analysis. For example, we may wish to select areas from the natural breaks classification shown on the left of Figure 9-21. We seek polygons that are greater than three square miles in area and have a medium population. The polygons may be composed of multiple neighborhoods. We typically must dissolve the boundaries between adjacent, medium-sized neighborhoods prior to applying the size test. Otherwise two adjacent, medium population neighborhoods may be discarded because both cover approximately two square miles. Their total area is four square miles, above the specified threshold, yet they will not be selected unless a dissolve is applied first.

Dissolves are also helpful in removing unneeded information. After the classification into small, medium, and large size classes, many boundaries may become redundant. Unneeded boundaries may inflate storage and slow processing. A dissolve has the advantage of removing unneeded geographic and tabular data, thereby simplifying data, improving processing speed, and reducing data volumes.

Figure 9-21 illustrates the space saving and complexity reduction common when applying a dissolve function. The number of polygons is reduced approximately ninefold by the dissolve, from 1074 on the left to 119 polygons on the right of Figure 9-21.

Proximity Functions and Buffering

Proximity functions or *operations* are among the most powerful and common spatial analysis tools. Many important questions hinge on proximity, the distance between features of interest. How close are schools to an oil refinery, what neighborhoods are far from convenience stores, and which homes will be affected by an increase in freeway noise? Many questions regarding proximity are answered through spatial analyses in a GIS.

Proximity functions modify existing features or create new features that depend in some way on distance. For example, one simple proximity function creates a raster of the minimum distance from a set of features (Figure 9-22). The figure shows a distance function applied to water holes in a wildlife reserve. Water is a crucial resource for nearly all animals, and the reserve managers may wish to ensure that most of the area is within a short distance of water. In this instance point features are entered that represent the location of permanent water. Water holes are represented by individual

points, and rivers by a group of points set along the river course. A proximity function calculates the distance to all water points for each raster cell. The minimum distance is selected and placed in an output raster data layer (Figure 9-22). The distance function creates a mosaic of what appear to be overlapping circles. Although the shading scheme shows apparently abrupt transitions, the raster cells contain a smooth gradient in distance away from each water feature.

Distance values are calculated based on the Pythagorean formula (Figure 9-23). These values are typically calculated from cell center to cell center when applied to a raster data set. Although any distance is possible, the distances between adjacent cells change in discrete intervals related to the cell size. Note that distances are not restricted to even multiples of the cell size, because distances measured on diagonal angles are not even multiples of the cell dimension. There may be no cells that are

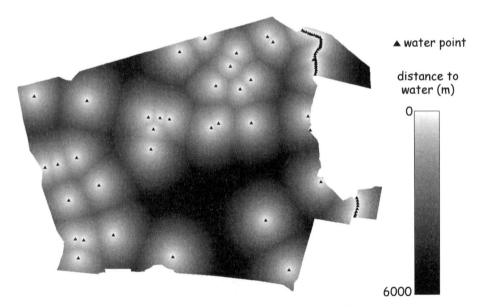

Figure 9-22: An example of a distance function. This distance function is applied to a point data layer and creates a raster data layer. The raster layer contains the distance to the nearest water feature.

exactly some fixed distance away from the target features; however there may be many cells less than or greater than that fixed distance.

Buffers

Buffering is one of the most commonly used proximity functions. A *buffer* is a region that is less than or equal to a specified distance from one or more features (Figure 9-24). Buffers may be determined for point, line, or area features, and for raster or vector data. Buffering is the process of creating buffers. Buffers typically identify areas that are "outside" some given threshold distance compared to those "inside" some threshold distance.

Buffers are used often because many spatial analyses are concerned with distance constraints. For example, emergency planners might wish to know which schools are within 1.5 kilometers of an earthquake fault, a park planner may wish to identify all lands more than 10 kilometers from the nearest highway, or a business owner may wish to identify all potential customers within a given radius of her store. All these questions may be answered with the appropriate use of buffering.

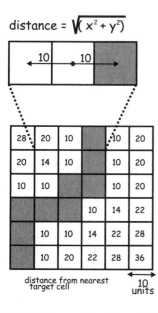

$$distance = \sqrt{(x^2 + y^2)}$$

distance from nearest target cell

10 units

Figure 9-23: A distance function applied to a raster data set.

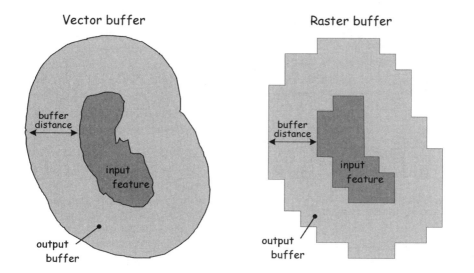

Figure 9-24: Examples of vector and raster buffers derived from polygonal features. A buffer is defined by those areas that are within some buffer distance from the input features.

Raster Buffers

Buffer operations on raster data entail calculating the distance from each source cell center to all other cell centers. Output cells are assigned an *in* value whenever the cell-to-cell distance is less than the specified buffer distance. Those cells that are further than the buffer distance are assigned an *out* value (Figure 9-25).

Raster buffers combine a minimum distance function and a binary classification function. A minimum distance function calculates the shortest distance from a set of target features and stores this distance in a raster data layer. The binary classification function splits the raster cells into two classes: those with a distance greater than the threshold value, and those with a distance less than or equal to a threshold value.

Buffering with raster data may produce a "stair-step" boundary, because the distance from features is measured between cell centers. When the buffer distance runs parallel and near a set of cell boundaries, the buffer boundary may "jump" from one row of cells to the next (Figure 9-25). This phenomenon is most often a problem when the raster cell size is large relative to the buffer distance. A buffer distance of 100 meters may be approximated when applied to a raster with a cell size of 30 meters. A smaller cell size relative to the buffer distance results in less obvious "stair-stepping". The cell size should be small relative to the spatial accuracy of the data, and small relative to the buffer distance. If this rule is followed, then stairs-stepping should not be a problem, because buffer sizes should be many times greater than the uncertainty inherent in the data.

Vector Buffers

Vector buffering may be applied to point, line, or area features, but regardless of input, buffering always produces an output set of area features (Figure 9-26). There are many variations in vector buffering. *Simple buffering*, also known as *fixed distance buffering*, is the most common form of vector buffering (Figure 9-26). Simple buffering identifies areas that are a fixed distance or greater from a set of input features. Simple buffering does not distinguish between regions that are close to one feature from those that are close to more than one feature. A location is either within a given distance from any one of a set of features, or farther away.

Simple buffering uses a uniform buffer distance for all features. A buffer distance of 100 meters specified for a roads layer may be applied to every road in the layer, irrespective of road size, shape, or location. In a similar manner, buffer distances for all points in a point layer will be uniform, and buffer distances for all area features will be fixed.

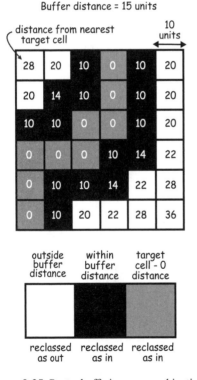

Buffer distance = 15 units

Figure 9-25: Raster buffering as a combination of distance and classification. Here, cells less than 15 units from the target cells are identified.

Buffering on vector point data is based on the creation of circles around each point in a data set. The equation for a circle with an origin at x=0, y=0 is:

$$r = \sqrt{x^2 + y^2} \qquad (9.1)$$

where r is the buffer distance. The more general equation for a circle with a center at x_1, y_1, is:

$$r = \sqrt{(x - x_1)^2 + (y - y_1)^2} \qquad (9.2)$$

Equation (9.2) reduces to equation (9.1) at the origin, where $x_1 = 0$, and $y_1 = 0$. The general equation creates a circle centered on the coordinates x_1, y_1, with a buffer distance equal to the radius, r. Point buffers are created by applying this circle equation successively to each point feature in a data

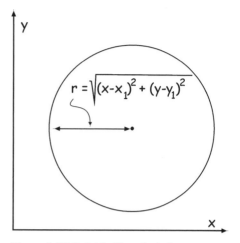

Figure 9-27: Point buffer calculations.

layer. The x and y coordinate locations of each point feature are used for x_1 and y_1, placing the point feature at the center of a circle (Figure 9-27).

Buffered circles may overlap, and in simple buffering, the circle boundaries that occur in overlap areas are removed. For example, areas within 10 kilometers of hazardous waste sites may be identified by creating a buffer layer. We may have a data

Vector buffers

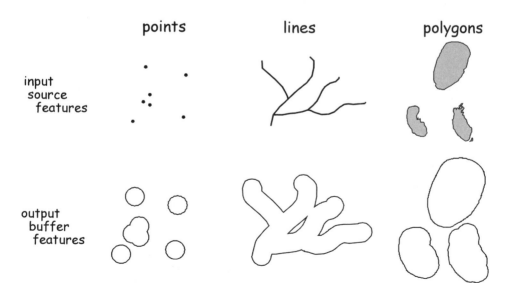

Figure 9-26: Vector buffers produced from point, line, or polygon input features. In all cases the output is a set of polygon features.

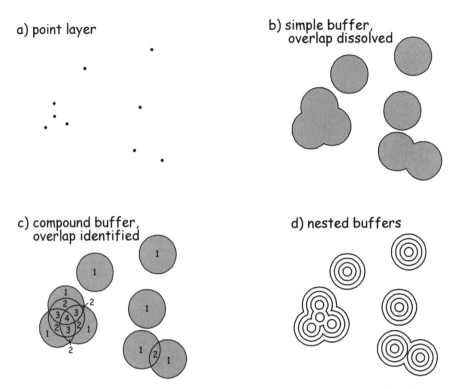

Figure 9-28: Various types of point buffers. Simple buffers dissolve areas near multiple features, more complex buffers do not. Multi-ring buffers provide distance-defined zones around each feature.

layer in which hazardous waste sites are represented as points (Figure 9-28a). A circle with a 10 kilometer radius is drawn around each point. When two or more circles overlap, internal boundaries are dissolved, resulting in non-circular polygons (Figure 9-28b).

More complex buffering methods may be applied. These methods may identify buffer areas by the number of features within the given buffer distance, or apply variable buffer distances depending on the characteristics of the input features. We may be interested in areas that are near multiple hazardous waste sites. These areas near multiple hazardous sites may entail added risk and therefore require special monitoring or treatment. We may be mandated to identify all areas within a buffer distance of a hazardous waste site, and the number of sites. In most applications, most of the dangerous areas will be close to one

hazardous waste site, but some will be close to two, three, or more sites. The simple buffer, described above, will not provide the required information.

A buffering variant, referred to here as *compound buffering*, provides the needed information. Compound buffers maintain all overlapping boundaries (Figure 9-28c). All circles defined by the fixed-radius buffer distance are generated. These circles are then intersected to form a planar graph. For each area, an attribute is created that records the number of features within the specified buffer distance.

Nested (or multi-ring) buffering is another common buffering variant (Figure 9-28d). We may require buffers at multiple distances. In our hazardous waste site example, suppose threshold levels have been established with various actions required for each threshold. Areas very close to hazardous waste sites require evac-

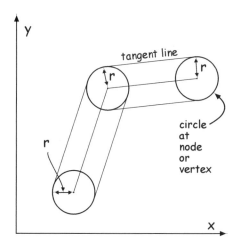

Figure 9-29: The creation of a line buffer at a fixed distance *r*.

Note that three different types of areas may be found when creating simple line buffers (Figure 9-29).The first type of area is within the buffer distance of the line features. An example of this area is labeled **inside buffer** in Figure 9-30. The second type of area is completely outside the buffer. This area is labeled **outside buffer** in Figure 9-30. The third type of area is labeled **enclosed area** in Figure 9-30. This type of area is farther than the buffer distance from the input line data, but completely enclosed within a surrounding buffer polygon. These enclosed areas occur occasionally when buffering points and polygons, but enclosed areas are most frequent when buffering line features.

Variable-distance buffers (Figure 9-31) are another common variant of vector buffering. As indicated by the name, the buffer distance is variable, and may change among features. The buffer distance may increase in steps, for example, we may have one buffer distance for a given set of features, and a different buffer distance for the remaining features. In contrast, the

uation, intermediate distance require remediation, and areas farther away require monitoring. These zones may be defined by nested buffers.

Buffering on vector line and polygon data is also quite common. The formation of line buffers may be envisioned as a sequence of steps. First, circles are created that are centered at each node or vertex (Figure 9-29). Tangent lines are then generated. These lines are parallel to the input feature lines and tangent to the circles that are centered at each node or vertex. The tangent lines and circles are joined and interior circle segments dissolved.

The location of the tangent lines and their intersections with the circles are based on a complicated set of rules and algebra. These operations identify the buffer segments that define the boundaries between **in** and **out** areas. These segments are saved to create a buffer for an individual line. Buffers for separate lines may overlap. With simple line buffering, the internal boundaries in the overlap areas are dissolved. Polygon buffering follows similar steps, and depending on the buffering function, may identify the area internal to the closed polygon and dissolving internal boundaries.

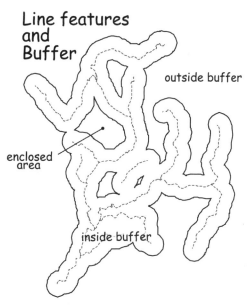

Figure 9-30: Buffers may split geographies into at least three types of features.

buffer distance may vary smoothly; for example, the buffer distance around a city may be a function of the population density in the city.

There are many instances for which we may require a variable-distance buffer. Public safety requires different zones of protection that are dependant on the magnitude of the hazard. We may wish to specify a larger buffer zone around large fuel storage facilities when compared to smaller fuel storage facilities. We often require more stringent protections further away from large rivers than for small rivers, and give large landfills a wider berth than small landfills.

Figure 9-31 illustrates the creation of buffers around a river network. These buffers may be used to analyze or restrict land use near rivers. We may wish to increase the buffer distance for larger rivers. The increase in distance may be motivated by an increased likelihood of flooding downstream, or an increased sensitivity to pollu-

tion, or a higher chance of bank erosion as river size increases. We may specify a buffer distance of 50 kilometers for small rivers, 75 kilometers for intermediate size rivers, and 100 kilometers for large rivers. There are many other instances when variable distance buffers are required, for example, larger distances from noisier roads, smaller areas where travel is difficult, or bigger buffers around larger landfills.

The variable buffer distance is often specified by an attribute in the input data layer. This is illustrated in Figure 9-31. A portion of the attribute table for the river data layer is shown. The attribute table contains the river name in river_identifier and the buffer distance is stored in buffdist. The attribute buffdist is accessed during buffer creation, and the size of the buffer adjusted automatically for each line segment. Note how the buffer size depends on the value in buffdist.

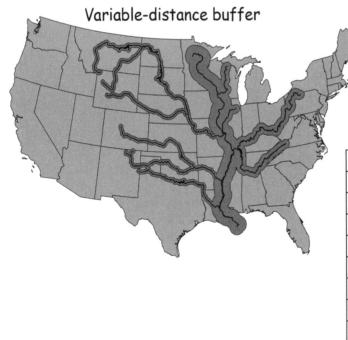

river_identifier	buffdist
mississippi	100
missouri	50
arkansas	50
ohio	75
tennessee	75
st. croix	75
illinois	75
wisconsin	75

Figure 9-31: An illustration of a variable-distance buffer. A line buffer is shown with a variable buffer distance based on a river_identifier. A variable buffer distance, buffdist, is specified in a table and applied for each river segment.

Overlay

Overlay operations are powerful spatial analysis tools, and were an important driving force behind the development of GIS technologies. Overlays involve combining spatial and attribute data from two or more spatial data layers, and they are among the most common and powerful spatial data operations (Figure 9-32). Many problems require the overlay of thematically different data. For example, we may wish to know where there are inexpensive houses in good school districts, where whale feeding grounds overlap with proposed oil drilling areas, or the location of farm fields that are on highly erodible soils. In the latter example a soils data layer may be used to identify highly erodible soils, and a current land use layer may be used to identify the locations of farm fields. The boundaries of erodible soils will not coincide with the boundaries of the farm fields

in most instances, so these soils and land use data must somehow be combined. Overlay is the primary means of providing this combination.

An overlay operation requires that data layers use a common coordinate system. Overlay uses the coordinates that define each spatial feature to combine the data from the input data layers. The coordinates for any point on the Earth depend on the coordinate system used (Chapter 3). If the coordinate systems used in the various layers are not exactly the same, the features in the data layers will not align correctly.

Overlay may be viewed as the vertical stacking and merger of spatial data (Figure 9-32). Features in each data layer are set one "on top" another, and the points, lines, or area feature boundaries are merged into a single data layer. The attribute data are also combined, so that the new data layer includes the information contained in each input data layer.

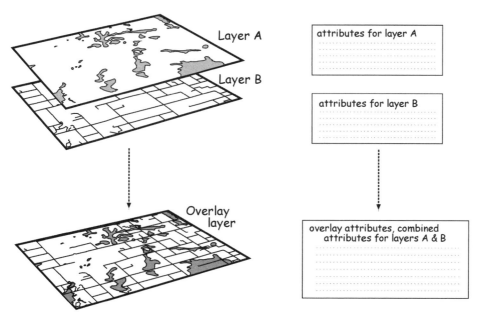

Figure 9-32: Spatial data overlay. Overlay combines both the coordinate information and the attribute information across different data layers.

Raster Overlay

Raster overlay involves the cell-by-cell combination of two or more data layers. Data from one layer in one cell location correspond to a cell in another data layer. The cell values are combined in some manner and an output value assigned to a corresponding cell in an output layer.

Raster overlay is typically applied to nominal or ordinal data. A number or character stored in each raster cell represents a nominal or ordinal category. Each cell value corresponds to a category for a raster variable. This is illustrated in the input data sets shown at the left and center of Figure 9-33. Input **Layer A** represents soils data. Each raster cell value corresponds to a specific soil value. In a similar manner, input **Layer B** records land use, with values **1**, **2**, and **3** corresponding to particular land uses. These data may be combined to create areas combining the two input layers-- cells with values for both soil type and land use.

There are as many potential output categories as there are possible combinations of input layer values. In Figure 9-33 there are two soil types in **Layer A**, and three land use types in **Layer B**. There may be **3 × 2**, or six different combinations in the output layer. Not all combinations will nec-essarily occur in the overlay, as shown in Figure 9-33. In this example only four of the six overlay combinations occur. Unique identifiers must be generated for each observed combination, and placed in the appropriate cell of the output raster layer.

The number of possible combinations is important to note because it may change the number of binary digits or bytes required to represent the output raster data layer. A raster cell typically contains a number or character, and may be a one-byte integer, a two-byte integer, or some other size. Raster data sets typically use the smallest required data size. As discussed in Chapter 2, one unsigned byte may store up to 256 different values. Raster overlay may result in an output data layer that requires a higher number of bytes per cell. Consider the overlay between two raster data layers, one layer that contains 20 different nominal classes, and a second layer with 27 different nominal classes. There is a total of 20 times 27, or 540 possible output combinations. If more than 256 combinations occur, the output data will require more than one byte for each cell. Typically two bytes will be used. This causes a doubling in the output file size. Two bytes will hold more than 65,500 unique combinations; if more categories are required, then four bytes per cell are often used.

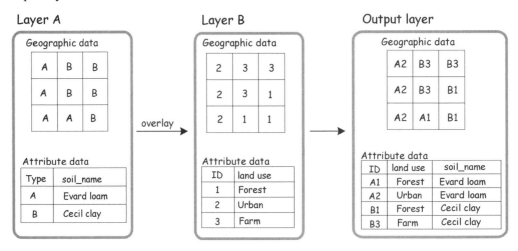

Figure 9-33: Cell-by-cell combination in raster overlay. Two input layers are combined in raster over-lay. Nominal variables for corresponding cells are joined, creating a new output layer. In this example a soils layer (**Layer A**) is combined with a land use layer (**Layer B**) to create a composite **Output layer**.

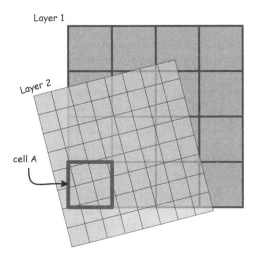

Layer 1

Layer 2

cell A

Figure 9-34: Overlain raster layers should be compatible to ensure unambiguous overlay. In the overlay depicted here it is not clear which cells from Layer 2 should be combined with cell A in Layer 1.

Raster overlay requires that the input raster systems be compatible. This typically means they should have the same cell dimension and coordinate system, including the same origin for x and y coordinates. If the cell sizes differ, there will likely be cells in one layer that match parts of several cells in the second input layer (Figure 9-34). This may result in ambiguity when defining the input attribute value. Overlay may work if the cells are integer multiples with the same origin, for example, the boundaries of a 1 by 1 meter raster layer may be set to coincide with a 3 by 3 raster layer, however this rarely happens. Data are normally converted to compatible raster layers before overlay. This is most often done using a resampling, as described in Chapter 4. In our example, we might choose to resample Layer 2 to match Layer 1 in cell size and orientation. Values for cells in Layer 2 would be combined through a nearest neighbor, bilinear interpolation, cubic convolution, or some other resampling formula to create a new layer based on Layer 2 but compatible with Layer 1.

Vector Overlay

Overlay when using a vector data model involves combining the point, line, and polygon geometry and associated attribute data. This overlay creates new geometry. Overlay involves the merger of both the coordinate and attribute data from two vector layers into a new data layer. The coordinate merger may require the intersection and splitting of lines or areas and the creation of new features.

Figure 9-35 illustrates the overlay of two vector polygon data layers. This overlay requires the intersection of polygon boundaries to create new polygons. The overlay combines attribute data during polygon overlay. The data layer on the left is comprised of two polygons. There are only two attributes for Layer 1, one an identifier (ID), and the other specifying values for a variable named class. The second input data layer, Layer 2, also contains two polygons, and two attributes, ID and cost. Note that the two tables have an attribute with the same name, ID. These two ID attributes serve the same function in their respective data layers, but they are not related. A value of 1 for the ID attribute in Layer 1 has nothing to do with the ID value of 1 in Layer 2. It simply identifies a unique combination of attributes in the output layer.

Vector overlay of these two polygon data layers results in four new polygons. Each new polygon contains the attribute information from the corresponding area in the input data layers. For example, note that the polygon in the output data layer with the ID of 1 has a class attribute with a value of 0 and a cost attribute with a value of 10. These values come from the values found in the corresponding input layers. The boundary for the polygon with an ID value of 1 in the output data layer is a composite of the boundaries found in the two input data layers. The same holds true for the other three polygons in the output data layer. These polygons are a composite of geographic and attribute data in the input data layers.

The topology of vector overlay output will likely be different from that of the input data layers. Vector overlay functions typically identify line intersections during overlay. Intersecting lines are split and a node placed at the intersection point. Thus topology must be re-created if it is needed in further processing.

Any type of vector feature may be overlain with any other type of vector feature, although some overlay operations rarely provide useful information and are performed infrequently. In theory, points may be overlain on point, line, or polygon feature layers, lines on all three types, and polygons on all three types. Point-on-point or point-on-line overlay rarely results in intersecting features, and so are rarely applied. Line-on-line overlay is sometimes required, for example, when we wish to identify the intersections of two networks such as road and railroads. Overlays involving polygons are the most common by far.

Overlay output typically takes the lowest dimension of the inputs. This means point-in-polygon overlay results in point output, and line-in-polygon overlay results in line output. This avoids problems when multiple lower-dimension features intersect with higher-dimension features.

Figure 9-36 illustrates an instance where multiple points in one layer fall within a single polygon in an overlay layer. Output attribute data for a feature are a combination of the input data attributes. If polygons are output (Figure 9-36, right, top) there is ambiguity regarding which point attribute data to record. Each point feature has a value for an attribute named *class*. It is not clear which value should be recorded in the output polygon, the *class* value from point *A*, point *B*, or point *C*. When a point layer is output (Figure 9-36, right, bottom), there is no ambiguity. Each output point feature contains the original point attribute information, plus the input polygon feature attributes.

One method for creating polygon output from point-in-polygon overlay involves recording the attributes for one point selected arbitrarily from the points that fall

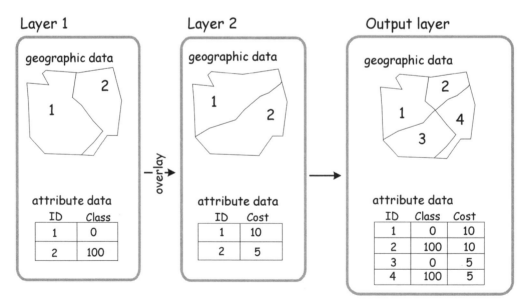

Figure 9-35: An example of vector polygon overlay. In this example output data contain a combination of the geographic (coordinate) data and the attribute data of the input data layers. New features may be created with topological relationships distinct from those found in the input data layers.

within a polygon. This is usually not satisfactory because important information may be lost. An alternative involves adding columns to the output polygon to preserve multiple points per polygon. However this would still result in some ambiguity, such as, what should be the order of duplicate attributes? It may also add a substantial number of sparsely used items, thus increasing file size inefficiently. Forcing the lower order output during overlay avoids these problems, as shown in the lower right of Figure 9-36.

Note that the number of attributes in the output layer increases after each overlay. This is illustrated in Figure 9-36, with the combination of a point and polygon

layer in an overlay. The output point attribute table shown in the lower right portion of the figure contains four items. This output attribute table is a composite of the input attribute tables.

Large attribute tables may result if overlay operations are used to combine many data layers. When the output from an overlay process is in turn used as an input for a subsequent overlay, the number of attributes in the next output layer will usually increase. As the number of attributes grows, tables may become unwieldy, and there may be a need to delete redundant attributes.

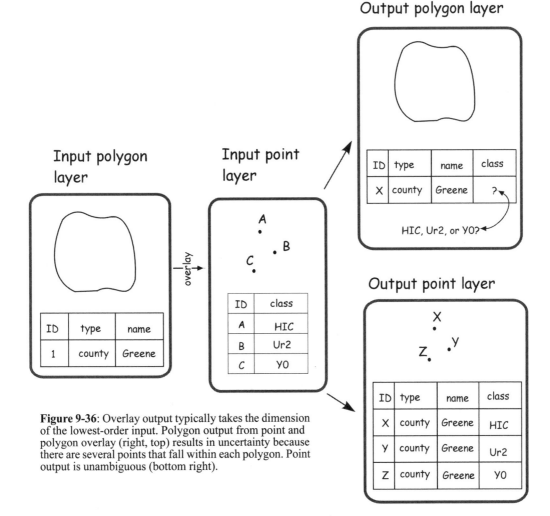

Figure 9-36: Overlay output typically takes the dimension of the lowest-order input. Polygon output from point and polygon overlay (right, top) results in uncertainty because there are several points that fall within each polygon. Point output is unambiguous (bottom right).

Figure 9-37 illustrates the most common types of vector overlay. Each row in the figure represents an overlay operation with point, line, or polygon input. The input data layers are arranged in the left and middle columns and the output data layer is in the right-hand column. Rather than show the complete attribute tables, labels are used to represent the combination of attributes. The bottom row illustrates this for polygon overlay. The two polygons in the first input data layer are labeled 100 and 200. The two polygons in the second input layer (middle panel) are labeled R and S. The resultant composite polygons in the output layer are shown on the left with labels 100R, 100S, 200R, and 200S to represent the combination of attributes from both input data layers.

Common characteristics of various types of overlay are apparent in Figure 9-37. Point-on-line overlay is shown in the top row. Point-on-line overlay is rare, because only in special circumstances do point features intersect with line features (e.g., points representing accident locations). Points have no dimension, and lines have zero width. If the features do not intersect, then the attributes for one of the data layers are not recorded in the output. In our example, the overlay results in joining line data for only one point, 2, which in the overlay layer is labeled 2B. The other points do not intersect a line, as indicated by the minus in the labels denoting attributes, e.g., 3-. This indicates the attributes for point 3 are recorded in the output, but there are no attributes for line features associated with this point

Point features result from point-in-polygon overlay, as shown in the second row of Figure 9-37. Points take the attributes of the coincident polygon. Point location is not changed nor are any geographic data from the polygon features typically incorporated into output point features.

Vector line-on-line overlay results in a line data layer (middle row, Figure 9-37). A planar graph is produced, meaning nodes (open circles) are placed at each line intersection. Each respective line segment maintains the attributes from the source data layer. Node attribute tables, if they exist, may contain information that originated from lines in each of the input data layers.

Line-on-polygon overlay is shown in the fourth row of Figure 9-37. This type of overlay typically produces a vector line output layer. Each line in the output data layer contains attributes from both the original input line data layer and the coincident polygon attribute layer. Line segments are split where they cross a polygon boundary, e.g., the line segment labeled 10 in Input layer 1 is split into two segments in the Output layer. Each segment of this line exhibits a different set of attributes: 10R and 10S. Note that not all line segments may contain a complete set of attributes derived from polygons. Line segments falling outside all polygons may lack polygon attributes, such as the segment at the lower right of line-on-line output panel. This segment is labeled 12-. The minus (-) denotes that polygon attributes are not recorded. These "outside" line segments typically contain null or flag values in the attribute table for the polygon items.

A polygon-on-polygon overlay is shown in the bottom row of Figure 9-37. The combination of polygon features has been discussed on page 379 and will not be described here.

Overlays that include a polygon layer are most common. We are often interested in combination of polygon features with other polygons, or in finding the coincidence of point or line features with polygons. What counties include hazardous waste sites? Which neighborhoods does one pass through on E Street? Where are there shallow aquifers below cornfields? All these examples involve the overlay of area features, either with other area features, or with point or line features.

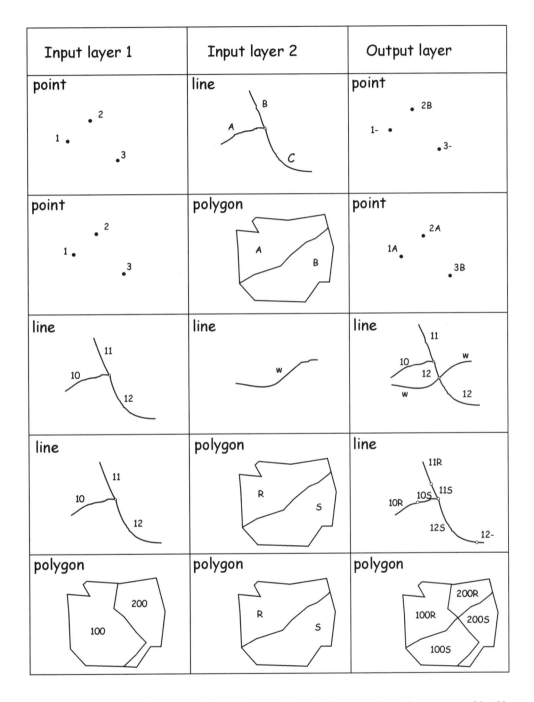

Figure 9-37: Examples of vector overlay. In this example, point, line, or polygon layers are combined in an inclusive overlay. The combination results in an output layer that contains the attribute and geographic data from input layers. The output data are typically the minimum order of the input, e.g., if point data are used as input, the output will be point. If line and line or line and polygon data are used, the output will be a line data layer.

Clip, Intersect, and Union: Special Cases of Overlay

There are three common ways overlay operations are applied: as a *clip*, an *intersection*, or a *union* (Figure 9-38). The basic layer-on-layer combination is the same for all three. They differ in the geographic extent for which vector data are recorded, and in how data from the attribute layers are combined. Intersection and union are derived from general set theory operations. The intersection operation may be considered in some ways to be a spatial AND,

while the union operation is related to a spatial OR. The clip operation may be considered a combination of an intersection and an elimination. All three are common and supported in some manner as standalone functions by most GIS software packages.

A *clip* may be considered a "cookie-cutter" overlay. A bounding polygon layer is used to define the areas for which features will be output. This bounding polygon layer defines the clipping region. Point, line, or polygon data in a second

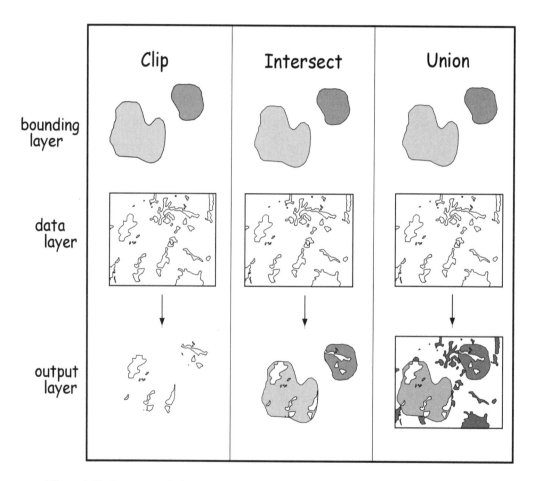

Figure 9-38: Common variations of overlay operations. Clip, intersect, and union operations are used to combine data from separate layers. The clip preserves information only from the data layer and only for the area of the bounding layer. An intersect is restricted to the area of the bounding layer, but combines data from both input layers. The union combines data from both input layers over the combined extent of both layers.

layer are "clipped" with the bounding layer. In most versions of the clip function the attributes for the clipping layer are not included in the output data layer.

An example of a clip is shown on the left side of Figure 9-38. The bounding data layer consists of two polygons, and the data layer contains many small wetland boundaries. The presence of polygon attributes in the bounding layer is indicated by the different shades for the two polygons in that layer. The output from the clip consists of those portions of wetlands in the area contained by the bounding layer polygons. Note that the polygon boundaries defining the bounding layer are not included in the output data layer.

An *intersection* may be defined as an overlay that combines data from both layers but only for the region where both layers contain data. This is illustrated in the central panel of Figure 9-38. Features from the bounding layer and the data layer are combined at the bottom of the central panel. Note that all or parts of polygons in the data layer that are outside the bounding layer are clipped and discarded. A spatial intersection operation differs from a clip in that data from the bounding layer are also included in the output layer. Each polygon in the output layer includes attributes defined in the bounding layer and the data layer.

A *union* is an overlay that includes all data from both the bounding and data layers. A union for our example is shown on the right of Figure 9-38. No geographic data are discarded in the union operation, and corresponding attribute data are saved for all regions. New polygons are formed by the combination of coordinate data from each data layer.

Many software packages support additional variants of overlay operations. Some support a complement to the clip function, in which the areas covered by the input layer are "cut out" or erased from the bounding layer. Other software packages support other variants on unions or inter-

sections. Most of these specialized overlay operations may be created from the application of union or overlay operations in combination with selection operations.

Vector overlay is often a time-consuming computational process, due to the large number of lines that must be compared. A vector overlay typically requires repeated tests of line intersection, a relatively simple set of calculations (Figure 9-39), but there is often a large number of line segments in a data set. Each line segment must be checked against every other line segment, requiring perhaps billions of tests for line intersection.

A Problem in Vector Overlay

Polygon overlays often suffer when there are common features that are represented in both input data layers. We define a common feature as a different representation of the same phenomenon. Figure 9-40 illustrates this problem. A county boundary may coincide with a state boundary. However, different versions of the state and county boundaries may be created independently from two adjacent states, using different source materials, at different times, and using different systems. Thus, these two representations may differ even though they identify the same boundary on the Earth surface.

In most data layers the differences will be quite small, and will not be visible except at very large display scales, for example, when the on-screen zoom is quite high. The differences are shown in the larger-scale inset in Figure 9-40. When the county and state data layers are overlain, many small polygons are formed along the boundary. These polygons are quite small, but they are often quite numerous.

These "sliver" polygons cause problems because there is an entry in the attribute table for each polygon. One-half or more of the polygons in the output data layer may be these slivers. Slivers are a burden because they take up space in the attribute table but are not of any interest or

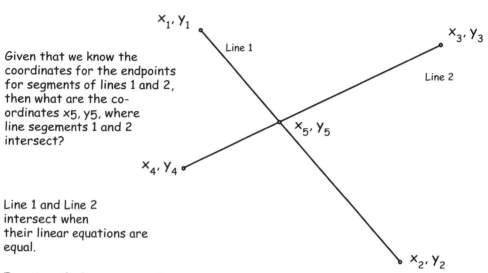

Given that we know the coordinates for the endpoints for segments of lines 1 and 2, then what are the co-ordinates x5, y5, where line segments 1 and 2 intersect?

Line 1 and Line 2 intersect when their linear equations are equal.

Equation of a line: $y = mx + b$

m is the slope, the rise over the run, which is equal to the change in y over a change in x. We may calculate the slope of any line segment from the end points:

$m = \dfrac{rise}{run}$; m_1, the slope of line 1 is: The intercept, b_1, of line 1 is:

$$m_1 = \frac{y_2 - y_1}{x_2 - x_1}$$ $$b_1 = y_1 - m_1 x_1$$

We know x_1, x_2, y_1, and y_2, so we can calculate m_1 and b_1 by the equations, above. We may also calculate the slope and intercept for line 2:

$$m_2 = \frac{y_3 - y_4}{x_3 - x_4}$$ $$b_2 = y_3 - m_2 x_3$$

$y_5 = m_1 x_5 + b_1$ and $y_5 = m_2 x_5 + b_2$

so $m_1 x_5 + b_1 = m_2 x_5 + b_2$

rearranging yields

$$x_5 = \frac{b_2 - b_1}{m_1 - m_2}$$

we may then calculate y_5 by

$$y_5 = m_1 x_5 + b_1$$

Note that before we go to the trouble of searching for an intersection point, we can apply a test to see if the line segments possibly intersect. A common test involves sorting the x and y coordinates by value. In our example, the orders would be:

x_4, x_1, x_2, x_3, and y_2, y_4, y_3, y_1.

Lines cannot intersect if the coordinate values are in order by line. If the x's for each line are together, then the line segments do not intersect. In our example, if x_1 and x_2 were in sequence, and x_3 and x_4 in sequence, and the same held for the ys, then the lines could not intersect. This test is commonly applied in GIS software, for example, when performing a vector overlay to avoid un-neccesary computations.

Figure 9-39: Line intersection is a common operation in vector data overlay. While computationally simple, this operation is repeated many times, hence vector overlay may require substantial time or computing resources.

use. Analyses of large data sets are hindered because all selections, sorts, or other operations must treat all polygons, including the slivers. Processing times often increase exponentially with the number of polygons.

There are several methods to reduce the occurrence of these slivers. One method involves identifying all common boundaries across different layers. The boundary with the highest coordinate accuracy is substituted into all other data layers, replacing the less accurate representations. Replacement involves considerable editing, and so is most often used as a strategy when developing new data layers.

Another method involves manually identifying and removing slivers. Small polygons may be selected, or polygons with two bounding arcs, as commonly

occurs with sliver polygons. Bounding lines may then be adjusted or removed. However, manual removal is not practical for many large data sets due to the high number of sliver polygons.

A third method for sliver reduction involves defining a snap distance during overlay. Much as with a snap distance used during data development (described in Chapter 4), this forces nodes or lines to be coincident if they are within a specified proximity during overlay. As with data entry, this snap distance should be small relative to the spatial accuracy of the input layers and the required accuracy of the output data layers. If the two representations of a line are within the snap distance then there will be no sliver polygons. In prac-

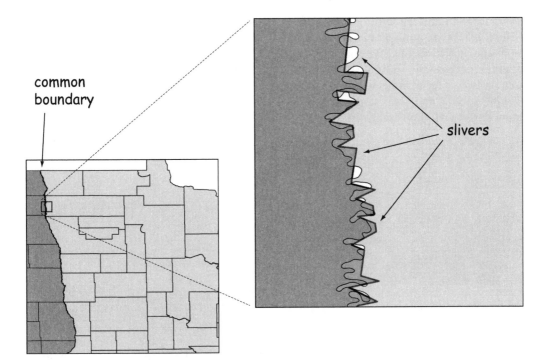

Figure 9-40: Sliver polygons may occur when two representations of a feature are combined. A common boundary between two features has been derived from different sources. The representations differ slightly. This results in small gaps and "sliver" polygons along the margin between these two layers. While slivers may be removed during initial data development, they often occur as a result of overlay operations. Some form of automated or manual sliver removal is often required after multi-layer overlay.

tice, not all sliver polygons are removed, but their numbers are substantially reduced, thereby reducing the time spent on manual editing.

Automatic sliver detection and removal should be applied carefully, as they may delete valuable data. Only small slivers should be removed, with small defined as smaller than an area, length, or width that is worth tracking. This distance may be set by the accuracy of the data collection, or by the requirements of the analysis. If polygon edge locations are only digitized to within one meter of their true position, it makes little sense to maintain polygons that are less than a meter in any dimension. However, if slivers are removed that are substantially wider and longer than a meter, some valuable information may be lost.

An Example Spatial Analysis

Figure 9-41 and the two following briefly illustrate an application of basic spatial analysis functions. We seek to identify suitable areas to investigate wind farms based on two criteria: areas with high average wind speeds, and areas with a low population density. High average winds are preferred because the energy produced at a site increases as a cubic function of wind speed. Low population densities are preferred because land is less expensive and there are fewer neighbors to bother, or to complain.This admittedly simple example does not include obvious additional factors, such as the distance to power lines, protected lands, or the difficulties of building offshore vs. onshore, but it does illustrate how data may be combined in a set of simple spatial functions to answer a question.

Wind data were obtained from the U.S. Department of Energy, and population data from the U.S. Census Bureau. Wind data were reclassed to those values (4 or greater) that provided suitable potential energy (Figure 9-41), and population to suitably sparse areas (Figure 9-42). Reclassed layers were then combined and selected to identify areas that are onshore with a low population density and high windspeed, or offshore with a high wind speed (Figure 9-43).

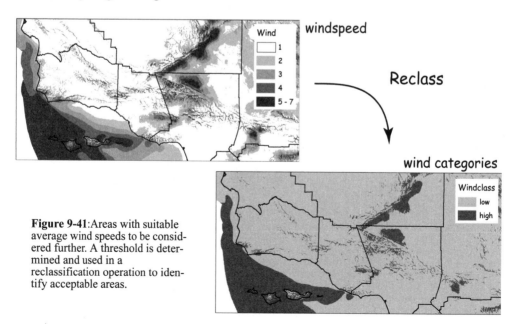

Figure 9-41:Areas with suitable average wind speeds to be considered further. A threshold is determined and used in a reclassification operation to identify acceptable areas.

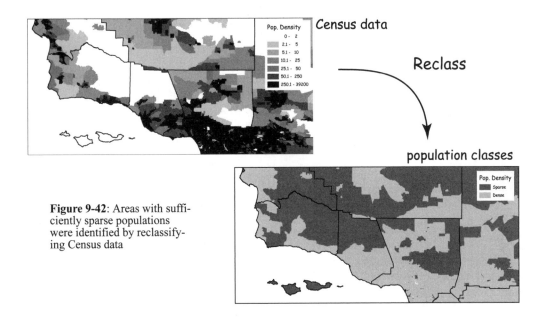

Figure 9-42: Areas with sufficiently sparse populations were identified by reclassifying Census data

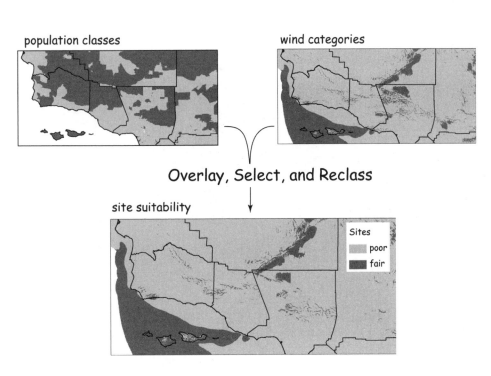

Figure 9-43: These intermediate layers are combined in an overlay operation, and then areas selected based on criteria for each resultant geographic unit.

Network Analysis

Networks are common in our lives. Roads, powerlines, telephone and television cables, and water distribution systems are all examples of networks we utilize many times each day (Figure 9-44). As networks are crucial to civilization, they need to be effectively managed. These networks also represent substantial investments, and their expansion and management merits considerable attention. Spatial analysis tools have been developed to help us use and maintain networks.

A network may be defined as a set of connected features, often termed *centers*. These features may be centers of demand, centers of supply, or both (Figure 9-45). Centers are connected to at least one and possibly many *network links*. Links interconnect and provide paths between centers. Traveling from one center to another often requires traversing many separate links.

Network analyses, also known as network models, are used to represent and analyze the cost, time, delivery, and accumulation of resources along links and between the connected centers. Resources flow to and from the centers through the networks. In addition, resources may be generated or absorbed by the links themselves.

The links that form the networks may have attributes that affect the flow. For example, there may be links that slow or speed up the flow of resources, or a link may allow resources to flow in only one direction. Link attributes are used to model flow characteristics of the real network, for example, travel on some roads is slower than others, or cars may legally move in only one direction on a one-way street.

The concept of a *transit cost* is key to many network analysis problems. A transit cost reflects the price one pays to move a resource through a segment of the network. Transit costs are typically measured in time, distance, or monetary units, for example, it costs 10 seconds to travel through a link.

Figure 9-44: Networks in a GIS are used to represent roads, pipelines, power transmission grids, rivers, and other connected systems through which important resources flow.

Costs may be constant such that it always takes 10 seconds to traverse the link regardless of direction or time of day. Alternatively, costs may vary by time of day or direction, so it may take 15 seconds to traverse an arc during morning and evening rush hours, but 10 seconds otherwise, or it may take twice as long to travel north to south than to travel south to north.

We will discuss three types of problems that are commonly analyzed using networks: route selection, resource and territory allocation, and traffic modeling. There are many types of networking problems; however these three are among the most common and provide an indication of the methods and breadth of network analyses.

Route selection involves identifying a "best" route based on a specified set of criteria. Route selection is often applied to find the least costly route that visits a number of centers. Two or more centers are identified

within a network, including starting and ending centers. These centers must all be visited by traversing the network. There are usually a very large number of alternative routes, or pathways, which may be used to visit all centers. The best route is selected based on some criteria, usually the shortest, quickest, or least costly route. Further restrictions may be placed on the route, for example, the order in which centers are visited may be specified.

Route selection may be used to improve the movement of public transportation through a network. School buses are often routed using network analyses. Each bus must start and finish at a school (a center) and pick up children at a number of stops (also centers). The shortest path or time route may be specified. Alternate routes are analyzed, and the "best" route selected.

Selection of the best route involves an algorithm that recursively follows a least-

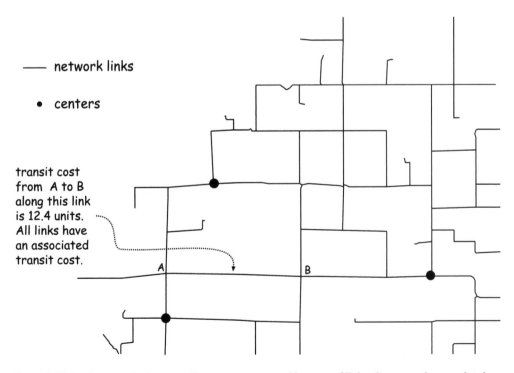

— network links

• centers

transit cost from A to B along this link is 12.4 units. All links have an associated transit cost.

Figure 9-45: Basic network elements. Centers are connected by a set of links. Costs may be associated with traversing the links. Network analysis typically involves moving resources or demands among centers.

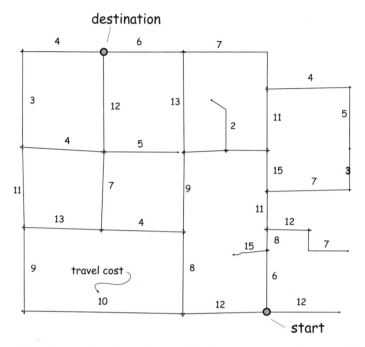

Figure 9-46: An example network. Start and destination centers and costs for link traversal are shown.

cost set of arcs, beginning at the current node. A set of interconnected network links is identified, as well as start and destination centers (Figure 9-46). The route from start to destination locations is typically built iteratively. One route-finding algorithm adds the least-cost link at each step. Multiple paths are tested until a path connects the start and destination centers.

This simple method begins at the start center. Paths are extended by adding the link that gives the lowest total cost for all paths currently pursued. The initial set of candidate links consists of all those connecting to the starting point. The lowest cost link is added, as shown in Figure 9-47a. The link with a value of six is chosen. Now the set of candidate links consists of any link connected to this selected link (the two links with costs of 15 and 8, respectively), plus any connected to the starting point. All paths are examined, and the link added which gives the lowest total path length. In Figure 9-47b, two links are added. Note that the

links added are not connected to the initially selected link. This would have given a total cost of 14 (6 plus 8) or 21 (6 plus 15), while the selected links give a lower path cost of 12. Now, the candidate links are those connected to any of the selected links or to the start point. Since all links from the start point have been selected, only those connected to candidate links are examined. Of these, the lowest cost path is added. The link with a cost of 8 that is attached to the initially selected link is chosen (Figure 9-47c). The candidate set expands accordingly, and is evaluated again. Verify that the links shown in Figure 9-47d and Figure 9-47e should be the next, cumulative low-cost paths selected. This method is used until the destination is reached, and the least-cost path identified (Figure 9-48).

Many different pathfinding algorithms have been developed, most of which are much more sophisticated than the one described above. Note that the described pathfinding algorithm has a rapidly expanding number of links to evaluate at each step. Computational burdens increase accordingly. A subset of all possible candidate paths may be examined because it becomes too computationally time-consuming to examine all possible paths. Most pathfinding algorithms periodically review the total accumulated cost thus far for each candidate path and stop following the highest cost or least promising paths.

There are many variations on this route-finding problem. There may be multiple centers that must be visited in a specific order, and carriers defined to transport specific amounts to or from centers. Centers may add to or subtract from a carrier, for example, some centers might represent houses with children, other centers may represent schools, and carriers represent buses that transport children. Houses must be visited to pick up children, but a bus has a fixed capacity. These children must be transported to the school, and there may be time constraints, for example, children cannot be picked up before 7 a.m. and must be at school by 7:55

a.m. Network-based route selection has been successfully used to solve these and related problems.

Resource allocation problems involve the apportionment of a network to centers. One or more allocation centers are defined in a network. Territories are defined for each of these centers. Territories encompass links or non-allocation centers in the network. These links or non-allocation centers are assigned to only one allocation center. The features are usually assigned to the nearest center, where distance is measured in time, length, or monetary units.

Resource allocation algorithms may be similar to route finding algorithms in that the distance out from each center is calculated along each path. Each center or arc is assigned to the nearest or least-cost center. The route finding method is exhaustive in resource allocation, in that all routes are pursued, not just the least-cost route. The routes are measured outward from each allocation center. (Figure 9-49).

Creating the least cost path

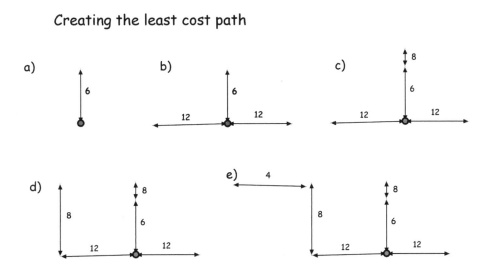

Figure 9-47: Steps in the identification of the least-cost path.

Variations on resource allocation include setting a center capacity. The center capacity sets an upper limit on resources that may be encompassed by a territory. Links are assigned to the nearest center, but once the capacity is reached, no more are added. Maximum distance also serves to limit the range of the territory from the center. Both of these restrictions may result in some unassigned areas, that is, portions of the network that are not allocated to a center.

Resource allocation analyses are used in many disciplines. School districts may use resource allocation to assign neighborhoods to schools. The type and number of dwellings in a district may be included as nodes on a network. The number of children along each link is added until the school capacity is reached. Resource allocation may also be

used to define sales territories, or to determine if a new business should be located between existing businesses. If enough customers fall between the territories of existing business centers, a new business between existing business centers may be justified.

Traffic modeling is another oft-applied network analysis. Streets are represented by a network of interconnected arcs and nodes. Attributes associated with arcs define travel speed and direction. Attributes associated with nodes identify turns and the time or cost required for each turn. Illegal or impossible turns may be modeled by specifying an infinite cost. Traffic is placed in the network, and movement modeled. Bottlenecks, transit times, and under-used routes may be identified, and this information used to improve

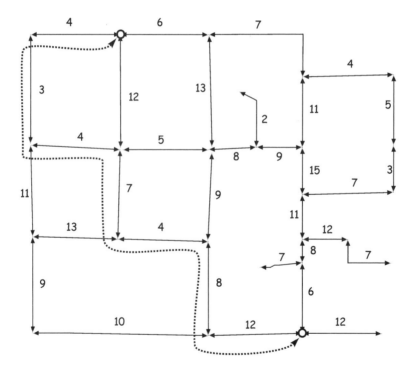

Figure 9-48: Least cost path for the example route-finding algorithm described in the text.

allocation center

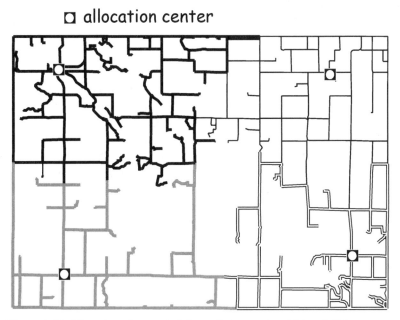

Figure 9-49: Allocation of network links to distinct centers. Network links or resources are assigned to the "nearest" center, where distance may be determined by physical distance, or by cost, travel time, or some other factor.

traffic management or build additional roads.

Traffic modeling through networks is a sub-discipline in its own right. Due to the cost and importance of transportation and traffic management, a great deal of emphasis has been placed on efficient traffic managment. Transportation engineers, computer scientists, and mathematicians have been modeling traffic via networks for many years. An in-depth discussion of network analyses for traffic management may be found in literature listed at the end of this chapter.

Geocoding

Geocoding, also known as *linear referencing*, is another common application of spatial data networks. Geocoding is the process of spatially referencing point features based on the address of the feature and knowledge of an address range for the linear network. Geocoding is commonly applied for business sales, marketing, in vehicle dis-

patch and delivery operations, and for organizing censuses and other government information gathering and dissemination activities.

Geocoding requires a set of addresses associated with a set of linear features. Typically at least the starting and ending addresses for links in a network are known. These starting and ending addresses define an address range, and the range is assumed to linearly span the connecting line. Points on the line may be "geographically coded" (hence the name geocoding), in that given an address, we may calculate approximately where the address should occur on the network link (Figure 9-50).

Geocoded addresses are typically assumed to vary linearly along the link. The starting and ending address are assumed to be at the ends of the link. The estimated location of the geocoded address is based on a linear interpolation, beginning at the starting address and adding a length proportional to the address divided by the address range

(Figure 9-50). The estimated location may be placed within the block or line segment.

Because geocoding only estimates address locations, these locations may contain substantial error. These errors may be larger than the error associated with the linear features along which the geocoded addresses are placed. Figure 9-51 illustrates some sources of error. Geocoding typically involves a regular, linear interpolation of an address across an address range. Address ranges are usually assigned ordinally, while the geocode is an interval estimate. In Figure 9-51a address 250 is not halfway between 200 and 300, and address 240 takes up an entire block. This ordinal/interval mismatch may be particularly bad in rural areas, where development over a long time period may result in substantial nonlinear address arrangements. Figure 9-51b illustrates this, with address 1007 almost opposite address 1026, and numerous inconsistent intervals, for example, the 22 address units between 1071 and 1093 are separated by a shorter distance than the 12 address units between

995 and 1007. These nonlinear addresses can cause substantial confusion, so any application of geocoded data must allow for these inconsistencies, or the data must be evaluated and corrected.

Geocoding is often combined with network analyses to determine shortest path or time travels to a set of locations. Delivery locations may be generated from a list of orders to a business. The locations of these addresses are generated via geocoding. The locations may then be entered into a network search algorithm and the optimal route planned. Businesses save millions of dollars each year applying these basic spatial analyses.

Geocoding: the address 321 M.L. King Drive is placed at the location that is
(321-301)/(359-301) = 0.34
of the distance from the 301 location toward the 359 location, between Third and Fourth streets. Coordinate values are estimated to be approximately

$$X_{321} = X_{301} + 0.34 \cdot (X_{359} - X_{301})$$
$$Y_{321} = Y_{301} + 0.34 \cdot (Y_{359} - Y_{301})$$

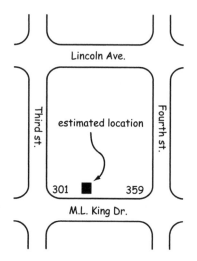

Figure 9-50: Geocoding is the process of estimating the location of an address based on knowledge of an address range along a linear feature. Here an address location is linearly interpolated along a city block, giving the approximate location of a building.

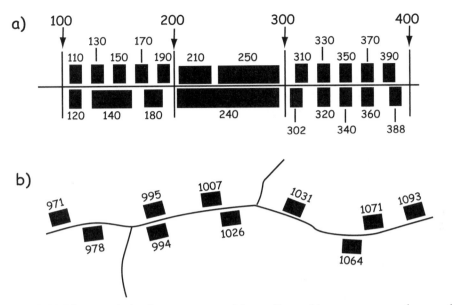

Figure 9-51: Idiosyncratic development may result in non-linear address sequences on the ground, so errors may result when geocoding is applied. Geocoding is typically applied with an assumption of a linear distribution of addresses across a range. When this is not true, as illustrated with address 250 in part a, above, or 1026 in part b, geocoded locations will be in error.

Summary

Spatial analysis, along with map production, is one of the most important uses of GIS. Spatial analytical capabilities are often the reason we obtain GIS and invest substantial time and money to develop a working system. Any analytical operation we perform on our spatial or associated attribute data may be considered as spatial analysis.

Spatial operations are applied to input data and generate output data. Inputs may be one to many layers of spatial data, as well as non-spatial data. Outputs may also number from one to many. Operations also have a spatial scope, the area of the input data that contributes to output values. Scopes are commonly local, neighborhood, or global.

Selection and classification are among the most oft-used spatial data operations. A selection identifies a subset of the features in a spatial database. The selection may be based on attribute data, spatial data, or some combination of the two. Selection may apply set or Boolean algebra, and may combine these with analyses of adjacency, connectivity, or containment. A selected set may be classified in that variables may be changed or new variables added that reflect membership in the selected set.

Classifications may be assigned automatically, but the user should be careful in choosing the assignment. Equal-area, equal-interval, and natural-breaks classifications are often used. The resulting classifications may depend substantially on the frequency histogram of the input data layer, particularly when outliers are present.

A dissolve operation is often used in spatial analysis. Dissolves are routinely applied after a classification, as they remove redundant boundaries that may slow processing.

Proximity functions and buffers are also commonly applied spatial data operations. These functions answer questions regarding distance and separation among features in the same or different data layers.

Buffering may be applied to raster or vector data, and may be simple, with a uniform buffer distance, or complex, with multiple nested buffers or variable buffer distances.

Overlay involves the vertical combination of data from two or more layers. Both geometry (coordinates) and attributes are combined. Any combination of points, lines, and area features is possible, although overlays involving at least one layer of area features are most common. The results of an overlay usually take the lowest geometric dimension of the input layers.

Overlay sometimes creates gaps and slivers. These occur most often when a common feature occurs in two or more layers. These gaps and slivers may be removed by several techniques.

Network models may be temporally dynamic or static, but they are constrained to model the flow of resources through a connected set of linear and point features. Traffic flow, oil and gas delivery, or electrical networks are examples of features analyzed and managed with network models. Route finding, allocation, and flow are commonly modeled in networks.

Geocoding, or linear referencing, is used to calculate approximate locations along a linear segment when the end-point addresses are known. Often used in census and delivery applications, geocoding works best when addresses are uniformly spaced across the segment. Because it is an approximation, geocoded locations are expected to sometimes be in error, and these errors are often more frequent in rural or sparsely addressed segments. Linear referencing may also be used to locate changes in linear feature characteristics, for example, road surface or accident locations.

Suggested Reading

Ahuja, R.K., Magnanti, T.L., & Orlin, J.B. (1993). *Network Flows: Theory, Algorithms, and Applications.* Prentice Hall: Englewood Cliffs.

Aronoff, S. (1989). *Geographic Information Systems, A Management Perspective.* WDL Publications: Ottawa.

Batty, M & Xie, Y. (1994). Model structures, exploratory spatial data analysis, and aggregation. *International Journal of Geographical Information Systems*, 8:291-307.

Bonham-Carter, G. F. (1996). *Geographic Information Systems for Geoscientists: Modelling with GIS.* Pergamon: Ottawa.

Carver, S. J. (1991). Integrating multi-criteria evaluation with geographical information systems. *International Journal of Geographical Information Systems*, 5:321-340.

Chou, Y. H. (1997). *Exploring Spatial Analysis in Geographic Information Systems.* Onword Press: Albuquerque.

Cliff, A.D & Ord, J.K. (1981). *Spatial Processes: Models and Applications.* Pion: London.

Cooper, L. (1963). Location-allocation problems. *Operations Research*, 11:331-342.

Dale, P. (2005). *Introduction to Mathematical Techniques Used in GIS.* CRC Press: Boca Raton.

Daskins, M.S. (1995). *Network and Discrete Location - Models, Algorithms, and Applications.* Wiley: New York.

DeMers, M. (2000). *Fundamentals of Geographic Information Systems* (2nd ed.). Wiley: New York.

Heuvelink, G.B.M. & Burrough, P.A. (1993). Error propagation in cartographic modelling using Boolean logic and continuous classification. *International Journal of Geographical Information Systems*, 7:231-246.

Laurini, R. & Thompson, D. (1992). *Fundamentals of Spatial Information Systems.* Academic Press: London.

Malczewski, J. (1999). *GIS and Multicriteria Decision Analysis.* Wiley: New York.

Martin, D. (1996). *Geographical Information Systems and their Socio-economic Applications* (2nd ed.). Routledge: London.

McMaster, S., & McMaster, R.B. (2002). Biophysical and human-social applications. In J.D. Bossler (Ed.), *Manual of Geospatial Science and Technology.* Taylor and Francis: London.

Monmonier, M. (1993). *How To Lie With Maps*. University of Chicago Press: Chicago.

National Research Council of the National Academies (2006). *Beyond Mapping: Meeting National Needs Through Enhanced Geographic Information Science*. The National Academies Press: Washington D.C.

Openshaw, S. & Taylor, P. (1979). A million or so correlation coefficients: three experiments on the modifiable areal unit problem. In N. Wrigley (Ed.) *Statistical Applications in the Spatial Sciences*. Pion: London.

Smith, M.J.de, Goodchild, M.F., & Longley, P.A. (2007). Geospatial Analysis: A Comprehensive Guide to Principles, Techniques and Software Tools. Winchelsea Press: Leicester.

Stillwell, J.A., & Clarke, G. (2004). *Applied GIS and Spatial Analysis*. Wiley: New York.

Steinitz, C P. & Jordan, L. (1976). Hand-drawn overlays: their history and prospective uses. *Landscape Architecture*, 56:146-157.

Worboys, M.F., & Duckham, M. (2004). *GIS: A Computing Perspective* (2nd ed.). CRC Press: Boca Raton.

Study Problems and Questions

9.1 - Can you define and give examples of local, neighborhood, and global spatial operations?

9.2 - Describe selection operations.

9.3 - Can you describe set and Boolean algebra?

9.4 - Write the simplest Boolean expression that results in the grey area selections:

a)

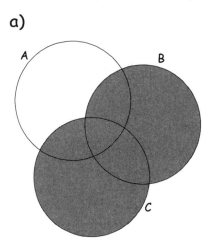

b)

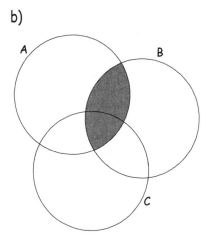

c)

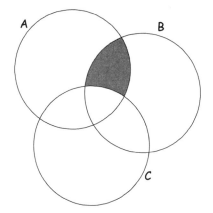

d)

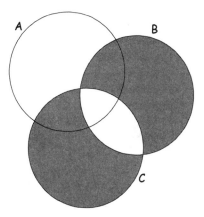

9.5 - Perform the following reclassification:

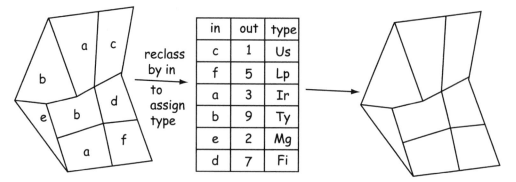

in	out	type
c	1	Us
f	5	Lp
a	3	Ir
b	9	Ty
e	2	Mg
d	7	Fi

9.6 - Reclassify the following polygons, according to the column Area, into small (<18000), medium (18000 to 45000), and large (> 45000).

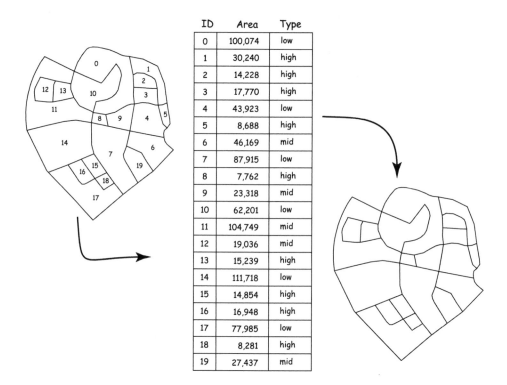

ID	Area	Type
0	100,074	low
1	30,240	high
2	14,228	high
3	17,770	high
4	43,923	low
5	8,688	high
6	46,169	mid
7	87,915	low
8	7,762	high
9	23,318	mid
10	62,201	low
11	104,749	mid
12	19,036	mid
13	15,239	high
14	111,718	low
15	14,854	high
16	16,948	high
17	77,985	low
18	8,281	high
19	27,437	mid

9.7 - List and describe three different classification methods.

9.8 - What is the modifiable area unit problem (MAUP)? Why is it important? What is the zone effect, and what is the area effect?

9.9 - What is a dissolve operation? What are they typically used for?

9.10 - Perform a dissolve operation on the variable Type, for the layer depicted below:

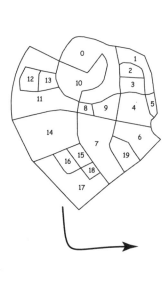

ID	Area	Type
0	100,074	low
1	30,240	high
2	14,228	high
3	17,770	high
4	43,923	low
5	8,688	high
6	46,169	mid
7	87,915	low
8	7,762	high
9	23,318	mid
10	62,201	low
11	104,749	mid
12	19,036	mid
13	15,239	high
14	111,718	low
15	14,854	high
16	16,948	high
17	77,985	low
18	8,281	high
19	27,437	mid

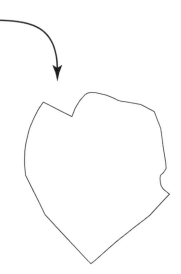

9.11 - Select the most appropriate characteristics for the buffer below.

Is it simple, multi-distance, or variable distance?

Does it retain or dissolve intersections?

Is it interior or exterior?

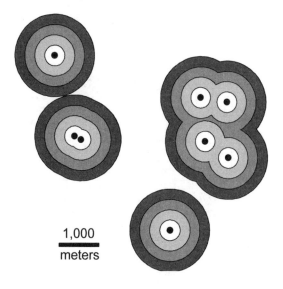

1,000
meters

9.12 - Sketch out the output from a variable distance buffer applied to the set of points shown below. Draw output buffers that dissolve the boundaries between areas that fall within multiple buffers.

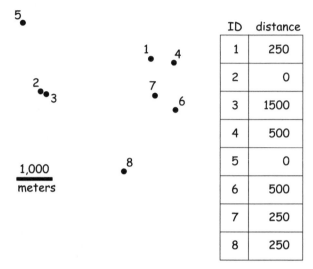

ID	distance
1	250
2	0
3	1500
4	500
5	0
6	500
7	250
8	250

9.13 - How are raster proximity functions different from vector proximity functions?

9.14 - Describe the basic concept behind layer overlay.

9.15 - Diagram and contrast raster and vector overlay.

9.16 - Why are output features in vector overlay typically set to the minimum dimensional order (point, line or polygon) of the input features?

9.17 - Sketch both the output polygons and the resultant attribute table from the overlay shown below:

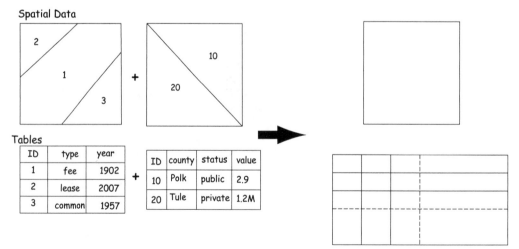

9.18 - What is the sliver problem in vector layer overlay? How might this problem be resolved?

9.19 - Sketch the output of the raster overlay, below, providing both cell values and the output table with ID and count variables.

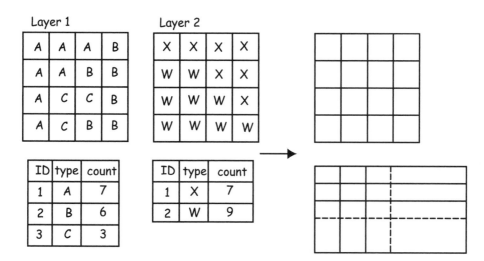

9.20 - Describe/define network models. What distinguishes them from other spatial or temporal models?

9.21 - What are the common uses for network models? Why are these models so important?

10 Topics in Raster Analysis

Introduction

Raster analyses range from the simple to the complex, largely due to the early invention, simplicity, and flexibility of the raster data model. Rasters are based on two-dimensional arrays, data structures supported by many of the earliest programming languages. The raster row and column format is among the easiest to specify in computer code and is easily understood, thereby encouraging modification, experimentation, and the creation of new raster operations. Raster cells can store nominal, ordinal, or interval/ratio data, so a wide range of variables may be represented. Complex constructs may be built from raster data, including networks of connected cells, or groups of cells to form areas.

The flexibility of raster analyses has been amply demonstrated by the wide range of problems they have helped solve. Raster analyses are routinely used to predict the fate of pollutants in the atmosphere, the spread of disease, animal migration, and crop yields. Time varying and wide-area phenomena are often analyzed using raster data, particularly when remotely-sensed inputs are available. Raster analyses are applied to a range of scales, from fine-grained problems, for example by the U.S. Environmental Protection Agency in hazard analysis of urban superfund sites, to global-scale estimates of forest growth. Local, state, and regional organizations have used raster analyses at many scales in between.

Numerous research projects have expanded and embellished the basic raster data structure, and developed a general set of raster tools for spatial data analyses. Yale University, the Ohio State University, the Idrisi Project of Clark University, and the Harvard School of Design are among the public institutions that have developed raster analysis packages for research over the past four decades. Commercial raster GIS software has been created by a number of companies.

The long history of raster analyses has resulted in a set of tools that should be understood by every GIS user. Many of these tools have a common conceptual basis and they may be adapted to several types of problems. In addition, specialized raster analysis methods have been developed for less frequently encountered problems. The GIS user may more effectively apply raster data analysis if she understands the underlying concepts and has become acquainted with a range of raster analysis methods.

Map Algebra

Map algebra is the cell-by-cell combination of raster data layers. The combination entails applying a set of local and neighborhood functions, and to a lesser extent global functions, to raster data.

The concept of map algebra is based on the simple yet flexible and useful data structure of numbers stored in a raster grid. Each number represents a value at a raster cell location. Simple operations may be applied to each number in a raster. Further, raster layers may be combined through operations such as layer addition, subtraction, and multiplication.

Map algebra entails operations applied to one or more raster data layers. *Unary* operations apply to one data layer. *Binary* operations apply to two data layers, and higher-order operations may involve many data layers.

A simple unary operation applies a function to each cell in an input raster layer, and records a calculated value to the corresponding cell in an output raster. Figure 10-1a illustrates the multiplication of a raster by a scalar (a single number). In this example the raster is multiplied by two. This might be denoted by the equation:

$$Outlayer = Inlayer * 2 \qquad (10.1)$$

Each cell value of In_layer is multiplied by two, and the result placed in the corresponding cell in Outlayer. Other unary functions are applied in a similar manner, for example, each cell may be raised to an expo-

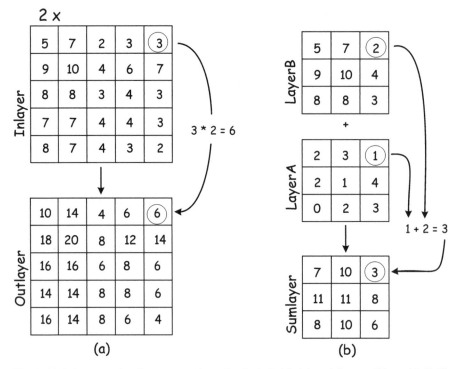

Figure 10-1: An example of raster operations. On the left side (a) each input cell is multiplied by the value 2, and the result stored in the corresponding output location. The right side (b) of the figure illustrates layer addition.

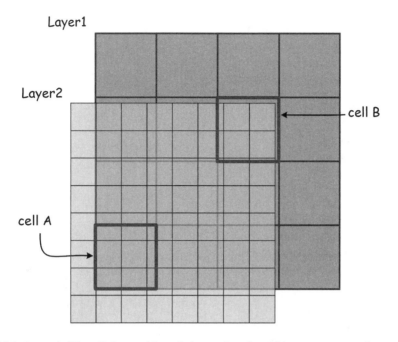

Figure 10-2: Incompatible cell sizes and boundaries confound multi-layer raster operations. This figure illustrates ambiguities in selecting input cell values from Layer2 in combination with Layer1. Multiple full and partial cells may contribute values for an operation applied to cell A. A portion of cell B is undefined in Layer2. These ambiguities are best resolved by resampling prior to layer combination.

nent, divided by a fixed number, or converted to an absolute value.

Binary operations are similar to unary operations in that they involve cell-by-cell application of operations or functions, but they combine data from two raster layers. Addition of two layers might be specified by:

$$\text{Sumlayer} = \text{LayerA} + \text{LayerB} \qquad (10.2)$$

Figure 10-1b illustrates this raster addition operation. Each value in LayerA is added to the value found in the corresponding cell in LayerB. These values are then placed in the appropriate raster cell of Sumlayer. The cell-by-cell addition is applied for the area covered by both LayerA and LayerB, and the results are placed in Sumlayer.

Note that in our example LayerA and LayerB have the same extent – they cover the same area. This may not always be true.

When layer extents differ, most GIS software will either restrict the operation to the area where input layers overlap. A *null* or a "missing data" number is usually assigned to cells where input data are lacking. This number acts as a flag, indicating there are no results. It is often a number, such as –9999, that will not occur from an operation, but any placeholder may be used as long as the software and users understand the placeholder indicates no valid data are present.

Incompatible raster cell sizes cause ambiguities when raster layers are combined (Figure 10-2). This problem was described briefly in the previous chapter, and is illustrated with an additional example here. Consider cell A in Figure 10-2 when Layer1 and Layer2 are combined in a raster operation. Several cells in Layer2 correspond to cell A in Layer1. If these two layers are added, there are likely to be several different input values for Layer2 corresponding to one input value for Layer1. The problem is compounded in cell B, because a portion of the cell is not defined for Layer2. It falls outside the layer boundary. Which Layer2 value should be used in a raster operation? Is it best to choose only the values in Layer2 from the cells with complete overlap, or to use the median number, the average cell number, or some weighted average? This ambiguity will arise whenever raster data sets are not aligned or have incompatible cell sizes. While the GIS software may have a default method for choosing the "best" input when cells are different sizes or do not align, these decisions are best controlled by the human analyst prior to the application of the raster operation. The analyst may resample the data into a compatible coordinate system, using transformation and resampling methods described in Chapter 4.

As with vector operations, raster operations may be categorized as local, neighborhood, or global (Figure 10-3). Local operations use only the data in a single cell to calculate an output value. Neighborhood operations use data from a set of cells, and global operations use all data from a raster data layer.

The concepts of local and neighborhood operations are more uniformly specified with raster data than with vector data. Cells within a layer have uniform size, so a local operation has a uniform input area. In contrast, vector areas represented by polygons may have vastly different areas. Irregular polygonal boundaries cover differing areas and have differing footprints, for example, the local area defined by Alaska is different than the local area defined by Rhode Island. A local operation in a given raster is uniform in that it specifies a particular cell size and dimension.

Neighborhood operations in raster data sets are also more uniformly defined than in vector data sets. A neighborhood may be defined by a fixed number of cells in a specific arrangement, for example, the neighborhood might be defined as a cell plus the eight surrounding cells. This neighborhood has a uniform area and dimension for most of the raster, with some minor adjustments needed near the edges of the raster data layer. Vector neighborhoods may depend not only on the shape and size of the target feature, but the shape and sizes of adjacent vector features.

Global operations in map algebra may produce uniform output, or they may produce different values for each raster cell. Global operations that return a uniform value are in effect returning a single number that is placed in every cell of the output layer. For example, the global maximum function for a layer might be specified as:

$$Out\_num = globalmax(In\_layer) \qquad (10.3)$$

This would assign a single value to Out_num. The value would be the largest number found when searching all the cells of In_layer. This "collapsing" of data from a two-dimensional raster may reduce the map algebra to scalar algebra. Many other functions return a single global value placed in every cell for a layer, for example, the global mean, maximum, or minimum.

Global operations are at times quite useful. Consider an analysis of regional temperature. We may wish to identify the areas where daily maximum temperatures were warmer this year than the highest regional temperature ever recorded. This analysis might help us to identify the extent of a warming trend. We would first apply a maximum function to all previous yearly weather records. This would provide a scalar value, a single number representing the regional maximum temperature. We would then compare a raster data set of maximum temperature for each day in the current year against the "highest ever" scalar. If the value for a day were higher than the regional maximum, we would output a flag to a cell location. If it were not, we would output a different value. The final output raster would provide a map of the cells exceeding the previous regional maximum. Here we use a global operation first to create our single scalar value (highest regional temperature). This scalar is then used in subsequent operations that output raster data layers.

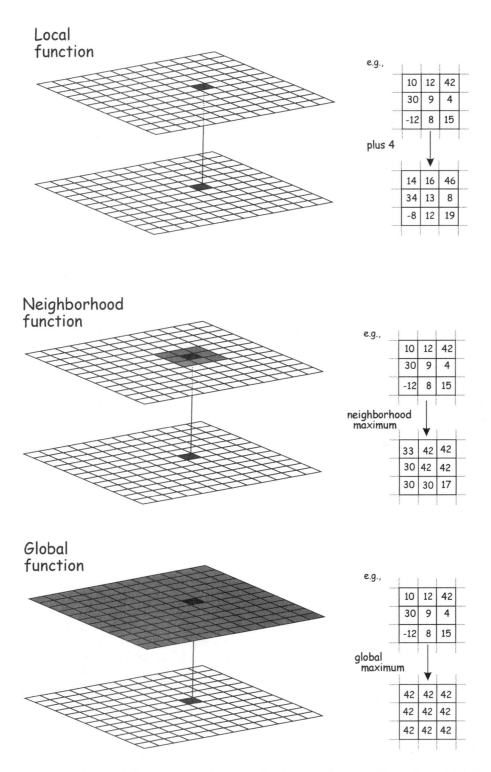

Figure 10-3: Raster operations may be local, neighborhood, or global. Target cells are shown in black, and source cells contributing to the target cell value in the output are shown in gray. Local operations show a cell-to-cell correspondence between input and output. Neighborhood operations include a group of nearby cells as input. Global operations use all cells in an input layer to determine values in an output layer.

Local Functions

There is a broad number of local functions (or operations) that can be conveniently placed in one of four classes: mathematical functions, Boolean or logical operations, reclassification, and multi-layer overlay.

Mathematical Functions

We may generate a new data layer by applying mathematical functions on a cell-by-cell basis to input layers (Figure 10-1, Table 10-1). Most functions use one input layer and create one output layer, but any number of inputs and outputs may be supported, depending on the function.

A broad array of mathematical functions may be used, limited only by the constraints of the raster data model. Raster data value and type are perhaps the most common constraints. Most raster models store one data value in a cell. Each raster data set has a data type and maximum size that applies to each cell, for example, a two-byte signed integer may be stored. Mathematical operations that create non-integer values, or values larger than 32,768 (the capacity of a two-byte integer), may not be stored accurately in a two-byte integer output layer.

Most systems will do some form of automatic type conversion, but there are often limits on the largest values that can be stored, even with automatic conversion. Most raster GIS packages support a basic suite of mathematical functions (Table 10-1).

Although the set of functions and function names differ among software packages, nearly all packages support the basic arithmetic operations of addition through division, and most provide the trigonometric functions and their inverses (e.g., sin, asin). Truncation, power, and modulus functions are also commonly supported, and vendors often include additional functions they perceive to be of special interest.

Table 10-1: Common local mathematical functions

Function	Description
Add, subtract, multiply, and divide	cell-by-cell combination with the arithmetic operation
ABS	Absolute value of each cell
EXP, EXP10, LN, LN10	Applies base e and base 10 exponentiation and logarithms
SIN, COS, TAN, ASIN, ACOS, ATAN	Apply trigonometric functions on a cell-by-cell basis
INT, TRUNC	Truncate cell values, output integer portion
MODULUS	Assigns the decimal portion of each cell
ROUND	Rounds a cell value up or down to nearest integer value
SQRT, ROOT	Calculates the square root or specifies other root of each cell value
POWER	Raises each cell to a defined power

Examples of addition and multiplication functions are shown in Figure 10-1. We will not show additional examples because most differ only in the function that is applied. These mathematical functions are often applied in raster analysis, e.g., when multiplying each cell by 3.28 to convert height values from units in meters to units measured in feet.

Note that although many systems will let you perform these operations on any type of raster data, they often only make sense for interval/ratio data, and may return erroneous results when applied to nominal or ordinal data. Numbers may be assigned to indicate population density by high, medium, and low, and while the sin function may be applied to these data, the results will usually have little meaning.

Logical Operations

There are a large number of local functions that apply logical (also known as Boolean) operations to raster data. A logical operation typically involves the comparison of a cell to a scalar value or set of values and outputs a "true" or a "false" value. True is often represented by an output value of 1, and false by an output value of 0.

There are three basic logical operations, AND, OR, and NOT (Figure 10-4). The AND and OR operations require two input layers. These layers serve as a basis for comparison and assignment. AND requires both input values be true for the assignment of a true output. Typically, any non-zero value is considered to be true, and zeros false. Note in Figure 10-4a that output values are typically restricted to 1 and 0, even though there may be a range of input values. Also note that there may be cells where no data are recorded. How these are assigned depends on the specific GIS system. Most systems assign null output when any input is null; others assign false values when any input is null.

Figure 10-4b shows an example of the OR operation. This cell-by-cell comparison assigns true to the output if either of the corresponding input cells is true. Note that the

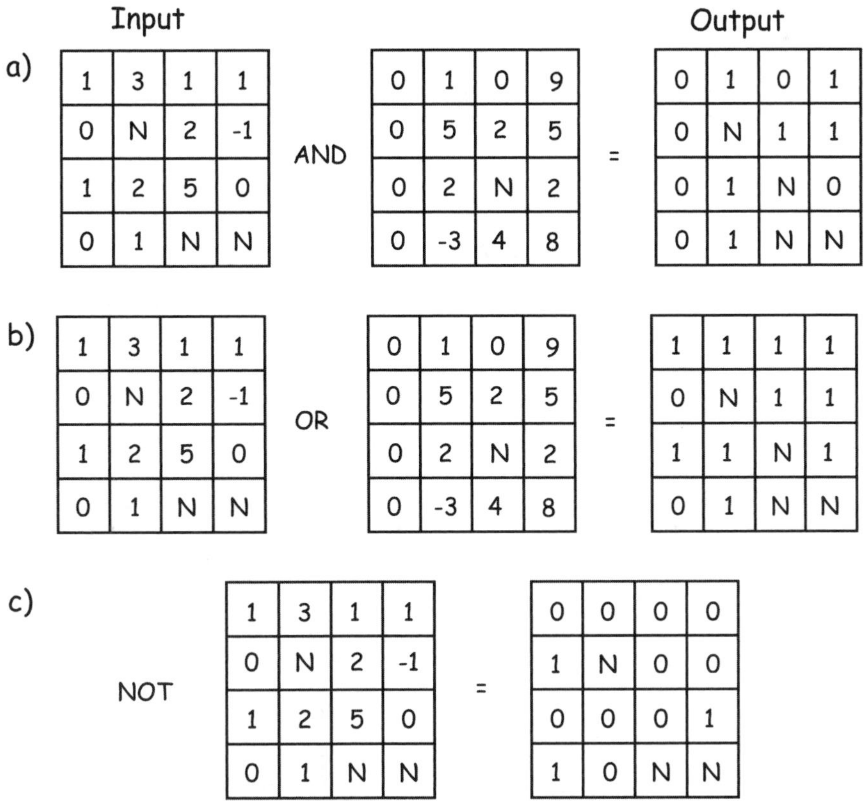

Figure 10-4: Examples of logical operations applied to raster data. Operations place true (non-zero) or false values (0) depending on the input values. AND requires both corresponding input cells be true for a true output, OR assigns true if either input is true, and NOT simply reverses the true and false values. Null or unassigned cells are denoted with an N.

cells in either layer or both layers may be true for a true assignment, and that in this example, null values (N) are assigned when either of the inputs is null. Some implementations assign a true value to the output cell if any of the inputs is non-null and non-zero; the reader should consult the manual for the specific software tool they use.

Figure 10-4c shows an example of the NOT operation. This operation switches true for false, and false for true. Note that null input assigns null output.

Finally, note that many systems provide an XOR operation, known as an eXclusive OR (not illustrated in our examples). This is similar to an OR operation, except that true values are assigned to the output when only one or the other of the inputs is true, but not when both inputs are true. This is a more restrictive case than the general OR, and

may be used in instances when we wish to distinguish among origins for a true assignment.

Logical operations may be provided that perform ordinal or equality comparisons, or that test if cell values are null (Figure 10-5). Ordinal comparisons include less than, greater than, less than or equal to, greater than or equal to, equal, and not equal. Examples of these logical comparisons are shown in Figure 10-5a and b, respectively. These operations are applied cell-by-cell, and the corresponding true or false output assigned. As shown in Figure 10-5a, the upper left cell of the first input layer is not less than the upper left cell of the second input layer, so a 0 (false) is assigned to the upper left cell in the output layer. The upper right cell in the first layer is less than the corresponding cell in the second input layer,

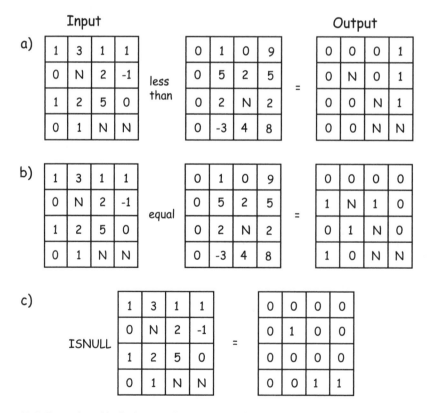

Figure 10-5: Examples of logical comparison operators for raster data sets. Output values record the ordinal comparisons for inputs, with a 1 signifying true and a 0 false. The ISNULL operation assigns true values to the output whenever there are missing input data.

resulting in the assignment of 1 (true) in the output layer.

We often need to test for missing or unassigned values in a raster data layer. The operation has no standard name, and may be variously called via `ISMISSING`, `ISNULL`, or some other descriptive name. The operation tests each cell for a null value, shown as N in Figure 10-5c. A 0 is assigned to the corresponding output cell if a non-null value is found, otherwise a 1 is assigned. These tests for missing values are helpful when identifying and replacing missing data, or when determining the adequacy of a data set and identifying areas in need of additional sampling.

Figure 10-6 shows an example of a logical comparison among two data layers. The left and central panels show landcover for an agricultural area, with three categories: corn (1), soybeans (2), and all others (0). We may be interested in identifying acres that were rotated between these two crops, or from these two crops to other crops over the 2009-2010 time period. The logical *equal* comparison between these layers reveals areas that have changed. If the cell values are not equal across the years, the logical *equal* comparison will return a value of 0, while areas that remain the same will maintain the value of 1.

Further logical comparisons, using class values, could identify how much each of the component crop types had changed.

Note that logical operators may be applied to interval/ratio, ordinal, and categorical data, although ordinal comparisons should be carefully applied to categorical data. For example, in our crop types example in Figure 10-6, soybeans are assigned a value of 2, and are "larger" than corn, but this distinction does not imply that soybeans are somehow 2 times larger, more valuable or anything other than just different from corn.

Reclassification

Raster reclassification assigns output values that depend on the specific set of input values. Assignment is most often defined by a table, ranges of values, or a conditional test.

Raster reclassification by a table is based on matching input cell values to a reclassification table (Figure 10-7a). The reclassification table specifies the mapping between input values and output values. Each input cell value is compared to entries for an "in" column in the table. When a match is found, the corresponding "out"

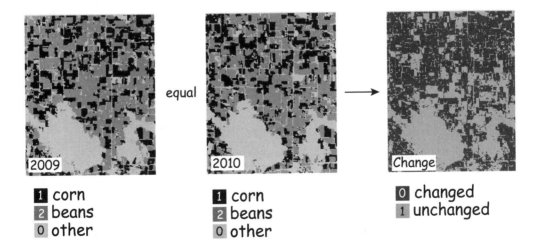

Figure 10-6: An example of a logical operation applied to categorical data, here landuse classes of corn (value = 1), soybeans (2), and other crops (0) over two time periods. The equal operation returns 0 where land use has changed, and 1 where it has remained constant over these two periods.

value is assigned to the output raster layer. Unmatched input values can be handled in one of several ways. The most logically consistent manner is to assign a null value, as shown in Figure 10-7a for the input value of -1. Some software simply assigns the input cell value when there is no match. As with all spatial processing tools, the specifics of the implementation must be documented and understood.

Figure 10-7b illustrates a reclassification by a range of values. This process is similar to a reclassification by a table, except that a range of values appears for each entry in the reclassification table. Each range corresponds to an output value. This allows a more compact representation of the reclassification. A reclassification over a range is also a simple way to apply the automated reclassification rules discussed at length in Chapter 8 -- the equal interval, equal area, natural breaks, or other automated class-creation methods. These automated assignment methods are often used for raster data sets

because of the large number of values they contain.

Data can also be reclassified to select the input source based on a condition. These "conditional" functions have varying syntax, but typically require a condition that results in a true or false outcome. The value or source layer assigned for a true outcome is specified, as is the value or source layer assigned for a false outcome. An example of one conditional function may be:

Output = CON (test, out if true, out if false)

where CON is the conditional function, test is the condition to be tested, out if true defines the value assigned if the condition tests true, and out if false defines the value assigned if the condition tests false (Figure 10-8). Note that the value that is output may be a scalar value, for example, the number 2, or the value output may come from the corresponding location in a specified raster

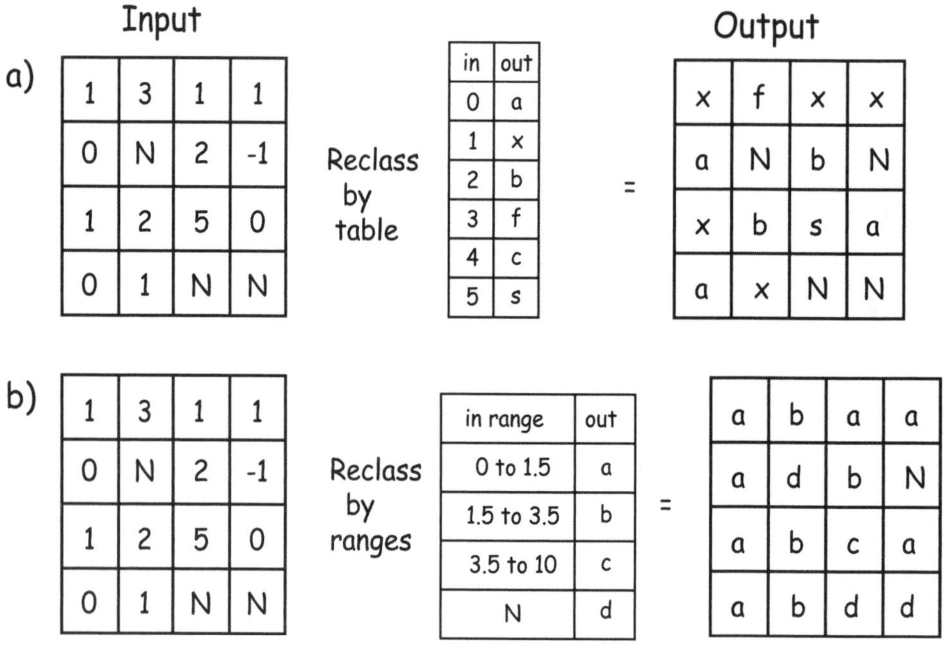

Figure 10-7: Raster reclassification by table matching (a) and by table range (b). In both cases, input cell values are compared to the "in" column of the table. A match is found and the corresponding "out" values assigned.

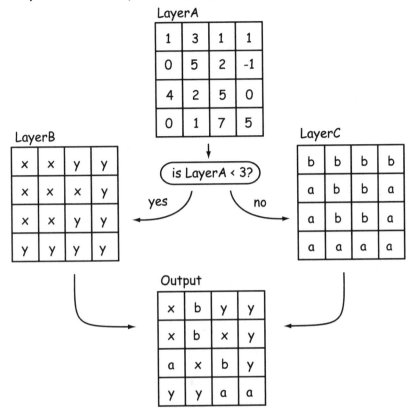

Output = CON (LayerA < 3, LayerB, LayerC)

Figure 10-8: Reclassification by condition assigns an output based on a conditional test. In this example, each cell in **LayerA** is compared to the number 3. For **Layer A** cells with values less than **3**, the condition evaluates to true, and the output cell value is assigned from **LayerB**. If the **LayerA** cell value is equal to or greater than **3**, then the output cell value is assigned from **LayerC**.

layer. The condition is applied on a cell-by-cell basis, and the output value assigned based on the results of the conditional test.

Nested Functions

Local functions may be nested in analyses. Functions are nested when a function is used as the argument of another function. For example, we may wish to take the natural logarithm (LN) of all the cells in a layer. The mathematical LN function is only defined for positive values. When inputs are negative we need to either accept null values in the output data layer, or process these input cells in a different manner. We could do this by applying the absolute value function (ABS) to create an intermediate output.

We could then apply the LN function to this output for our final result. This could be described as the equations:

InitOutput = ABS (Input_Layer)

FinalOutput = LN (InitOutput).

We could do the same thing by nesting the functions, if allowed by the GIS software:

FinalOutput = LN (ABS (Input_Layer))

Output = CON (ISNULL(LayerA), LayerB, LayerC)

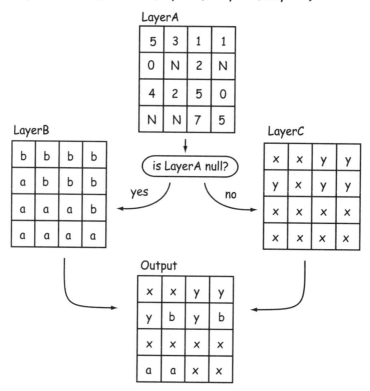

Figure 10-9: Local logical operations may be combined. This example shows the ISNULL function embedded in the conditional function, **CON**. As described in Figure 10-7, the first argument in this **CON** function defines the Boolean test. Here, if the cell value in **LayerA** is null (N), then the conditional test returns a true value. This executes the following entry, an assignment to the output from **LayerB**. If the cell is not null, **ISNULL** returns a false, and the **Output** cell value assigned from the corresponding location in **LayerC**. Note the **Output** values are **a** or **b**, from **LayerC**, for the cells in **LayerA** that have null (N) values.

Figure 10-9 shows another example of nested function. Output values are assigned from two different input layers. Cell values are assigned from **LayerB** when **LayerA** values are null, and from **LayerC** when **LayerA** values are not null. This might be desirable if we have an incomplete but otherwise high quality data set, and we wish to fill missing values from the next best available data. Map algebraic expressions with nested functions can become quite complex, but also may be quite effective and efficient in solving complex problems.

Overlay

Raster overlay combines features from two or more data layers, and is among the most useful of spatial functions. The features in raster data correspond to cells, or perhaps groups of cells with the same values, but as with vector overlay, great utility is often gained from combining data from different layers.

There are some differences between raster and vector overlay due to the differences in the data model. Raster overlay is often restricted to nominal data. The cell values do not typically represent continuous variables such as temperature, but rather categorical variables such as type or township name.

Although raster overlay may be implemented so that it admits continuous numbers, this typically results in too many unique cell combinations to be of much value. If continuous data are used, they are often converted to categories first, for example, rainfall may be assigned to low, medium, or high classes.

Raster overlay involves the cell-by-cell comparison of values in two or more layers (Figure 10-10). The values in each input data layer are associated with a specific combination of additional variables, and these additional variables may be recorded in an attribute table. Each unique combination of

cells from the two layers is identified, and assigned a new identifier (Out-ID) in the Output layer. Note the two input attribute tables are combined in a corresponding fashion. In Figure 10-10 you can see the upper left corner of the Output layer has the corresponding type and name attribute values from Input layer 1, and the ID and cost attribute values from Input layer 2.

Recall that in many implementations of the raster data model there is a many-to-one relationship between the raster cells and the attribute rows. This occurs because multiple cells correspond to each row. Also note that the cells may form disjunct regions of the

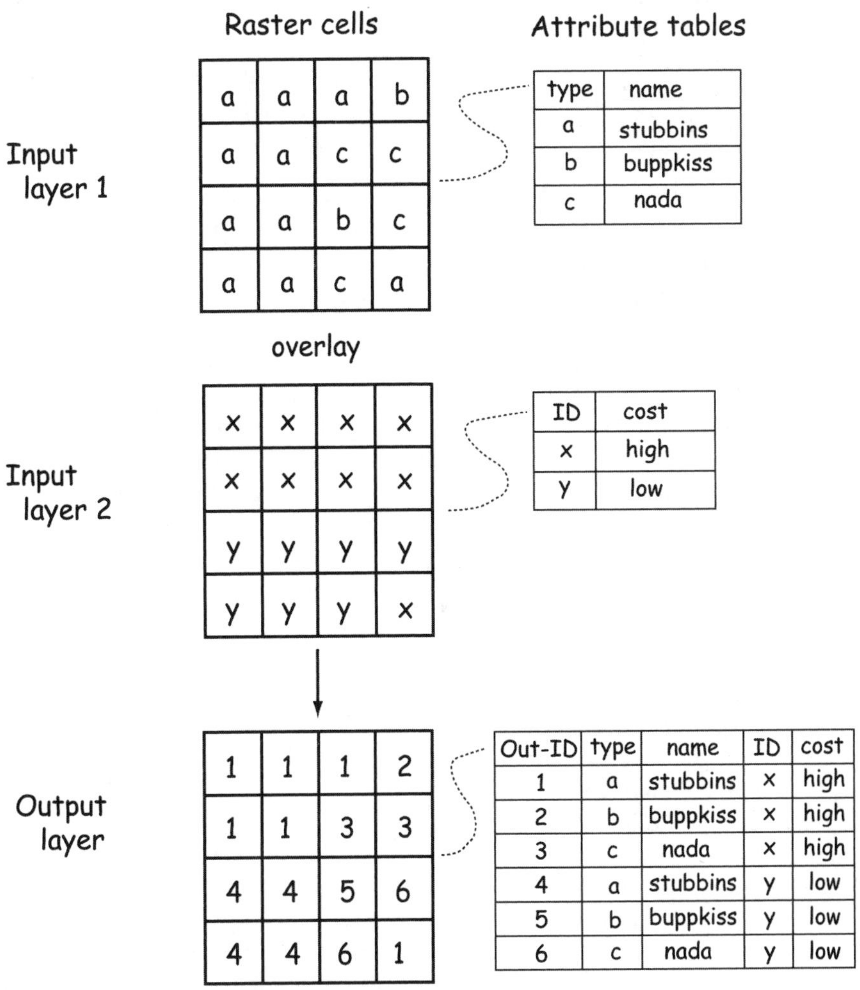

Figure 10-10: Raster overlay involves the cell-by-cell combination of data in multiple layers. New output values are generated for each unique combination of input values.

same type, for example, Figure 10-10 shows a cell of type a in the lower left corner of Input layer 1 that is not contiguous with the rest of the a cells in Layer 1. This combination carries through to the output, where there are disjunct groups of cells with Out-ID values of 6.

Clip (or extraction) is another common type of local raster function (Figure 10-11). Source and template data layers are specified and an output data layer created. This output layer contains only the values of the source that are indicated by the template layer. The nature of the template and output data layer values depend on the specific implementation of the raster extraction. Template values are typically assigned a value of 1 for those cells that are to pass through to the output, and a 0 or null value for those that are to be ignored. Output values for the clipped area are copied from the source, while output values for the area outside the clipped region are typically assigned a null value, or the value 0.

Care must be taken to ensure there are no ambiguous cells created by this convention. For example, if there are null values in both the source data layer and the area outside the clip, one cannot be certain if the nulls come from the source or indicate a region outside the clip area. Special coding or other provisions can be used to avoid these ambiguities.

Overlay functions in map algebra can be created through addition and multiplication functions. Union operations can be performed with layer addition, and clip operations through multiplication. These two raster algebra operations may be used to combine raster data layers in a number of ways, even if the specific software implementation of the raster GIS does not provide an explicit or specialized overlay function.

The union of two raster data layers may be performed through addition. When two layers are added, values are added on a cell-by-cell basis and the results placed in corresponding cell locations for an output data layer. Each output value may be used to identify the combinations of input values.

Figure 10-12 shows the overlay of two data layers through raster addition. Cells in Layer A have values 10 through 40, and cells in Layer B have values 1 through 7. These might correspond to four different species types in Layer A and six different landuse types in Layer B. Data layers are combined on a cell-by-cell basis, so each cell value in Layer A is added to the corresponding cell value in Layer B. In the upper right corner cell, the value 20 from Layer A is added to 2 in Layer B, and the resultant value 22 is placed in the Output layer. Cell addition of these two layers will result in a set of numbers between 12 and 47. These numbers correspond to the various combinations of species and landuse.

Source

1	3	4	7
6	3	2	-1
1	2	5	0
0	1	3	2

Clip

Template

0	0	0	1
0	0	1	1
0	1	1	1
0	1	1	0

=

Output

N	N	N	7
N	N	2	-1
N	2	5	0
N	1	3	N

Figure 10-11: An illustration of a raster clip (or extraction) operation. Values from a source layer are extracted based on values in an extraction template. Output cells are often assigned a null value, N, in the "outside" area.

Attribute data may be matched to the corresponding attribute tables. Each unique value in the input data layer is identified and associated with the appropriate combinations in the output data layer. Attribute assignment is illustrated in Figure 10-12. Cells in Layer A with an ID value of 10 are associated with Type = A and Spec = 22b. Every cell in the Output layer that exhibit an ID = 10 for input Layer A have a Type = A and Spec = 22b in the corresponding output cells.

Note that identical values may be discontinuous in the output data layer. The ID values of 12 are found in two disjunct sets in Figure 10-12, one in the upper-left corner and one in the lower-right corner of the output table. These cells with ID=12 are all referred to by the same entry in the attribute table. This many-to-one relationship may occur quite often, for example, it also exists for ID=22 in the Output layer.

Raster data sets often do not uniquely identify disjunct but otherwise identical

areas because of limits on attribute table size. Raster data sets often have thousands to millions of unique cell locations. The average area of contiguous, identical cells typically decreases when data layers are combined through raster overlay. There are often many small areas that have the same combination of input values, but are separated from other cells with the same value. If each group of cells is assigned a unique identifier, the attribute table may grow to be quite large. Large data sets are becoming less of a handicap as computing power and space increases, but many software packages by default implement a one-to-many relationship. This contrasts with the common approach applied by vector overlay software packages, which typically identify each polygon uniquely, whether or not there are other polygons with an identical set of attribute values. The GIS user needs to understand the output convention for the specific software so that output from overlay

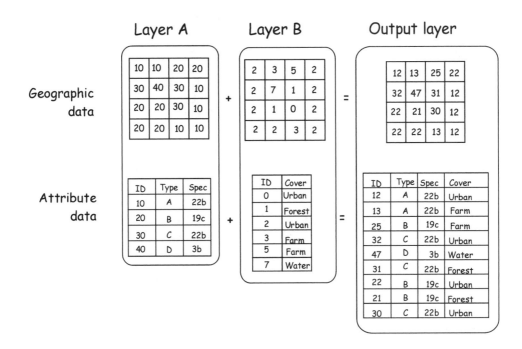

Figure 10-12: Overlay through raster addition. Cell values in **Layer A** and **Layer B** are added to yield values in the **Output layer**. Attribute values associated with each input layer may be combined in an associated table.

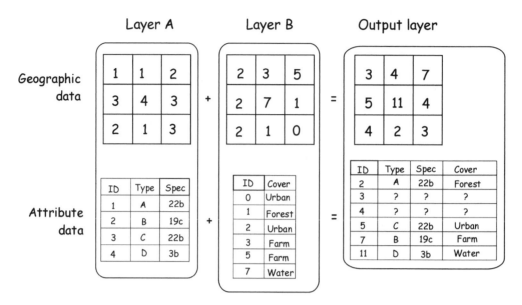

Figure 10-13: Raster addition will lead to ambiguous output when different combinations of inputs lead to identical output combinations (unknown type, spec, and cover in the output layer). Input layers may be classified or renumbered to ensure unique output combinations.

operations may be properly interpreted and applied.

Raster addition can be used to mimic raster overlay. However, care must be taken to avoid ambiguous combinations when using raster layer addition for overlay. Output numbers derived from two different combinations must be avoided, because it represents an ambiguous result. Consider the example shown in Figure 10-13. Two data layers are overlain through addition. A value of 4 may occur in the output layer from multiple combinations: 2 in Layer A and 2 in Layer B, or 1 in Layer A and 3 in Layer B. There are similar problems for output values equal to 3, so our results are ambiguous. We must ensure this cannot occur. We typically do this by reclassifying a data layer. For example, we could multiply Layer A by 10, thus giving values of 10, 20, 30, and 40. The output values will then uniquely identify the combination of inputs.

A clip using raster data may be implemented as a reclassification and then a multiplication. Note that in Chapter 9 we

described the vector clip functions as a special case of overlay in which the attributes and interior geometry were saved based on the boundaries in a clipping layer. This clipping layer serves as an outline or area template for which data are retained. In a raster clip, the clipping layer may be represented as a set of cells with a value of 1 embedded in non-clipping cells with values of 0.

Figure 10-14 illustrates a raster clip operation that is a combination of cell reclassification and multiplication. The first step is to identify the set of values that defines the clip area. This is the portion of the input data layer to be transferred to the output data layer. Individual cell values or cell values over an interval or range may be defined. These may come from a selection based on raster values, from a list of values, or from a previous spatial operation such as a buffer.

A clip template is created that defines a *binary mask*, a set of cells that "mask" out a portion of an input layer. Cells to be passed through to the output layer are set to the value 1 (Figure 10-14). Cells to be "clipped

Input raster

2	2	2	8	8	2	2	2
2	2	2	8	8	8	2	2
2	3	3	3	8	8	8	7
2	3	3	3	8	8	8	7
3	3	3	6	6	6	7	7
3	3	3	3	6	6	6	7
3	6	3	6	6	6	6	6
3	6	6	6	6	6	6	6

×

Clip raster

0	0	0	0	1	1	1	1
0	0	0	0	1	1	1	1
0	0	1	1	1	1	1	0
0	0	1	1	1	1	0	0
0	0	1	1	1	0	0	0
0	0	1	1	0	0	0	0
0	1	0	0	0	0	0	0
0	0	0	0	0	0	0	0

Output raster

0	0	0	0	8	2	2	2
0	0	0	0	8	8	2	2
0	0	3	3	8	8	8	0
0	0	3	3	3	8	0	0
0	0	3	6	6	0	0	0
0	0	3	3	0	0	0	0
0	6	0	0	0	0	0	0
0	0	0	0	0	0	0	0

Figure 10-14: A raster clip operation may be performed via multiplication of a binary mask. An input raster (left) has been multiplied by a binary raster. Cell-by-cell multiplication passes through each input cell corresponding to a 1 in the clip raster (center), and passes through a 0 for all other cells to the output raster (right). Note that zero values are not allowed in the input raster because they have the same effect as zeros in the clip raster.

away" are set to the value 0. The clip template or layer is then multiplied by the input raster, yielding an output raster. Cell-by-cell multiplication by 1 passes values through to the output layer. Multiplication by 0 discards values for the cell, resulting in a clipped raster to the area of interest.

Neighborhood, Zonal, Distance, and Global Functions

Neighborhood functions (or operations) in raster analyses deserve an extended discussion because they offer substantial analytical power and flexibility. Neighborhood operations are applied in many analyses across a broad range of topics, including the calculation of slope, aspect, and spatial correlation.

Neighborhood operations most often depend on the concept of a *moving window*. A "window" is a configuration of raster cells used to specify the input values for an operation (Figure 10-15). The window is positioned on a given location over the input raster, and an operation applied that involves the cells contained in the window. The result of the operation is usually associated with

the cell at the center of the window position. The result of the operation is saved to an output layer at the center cell location. The window is then "moved" to be centered over the adjacent cell and the computation repeated (Figure 10-15). The window is swept across a raster data layer, usually from left to right in successive rows from top to bottom. At each window location the moving window function is calculated and the result output to the new data layer.

Moving windows are defined in part by their dimensions. For example, a 3 by 3 moving window has an edge length of three cells in the x and y directions, for a total area of nine cells. Moving windows may be any size and shape, but they are typically odd-

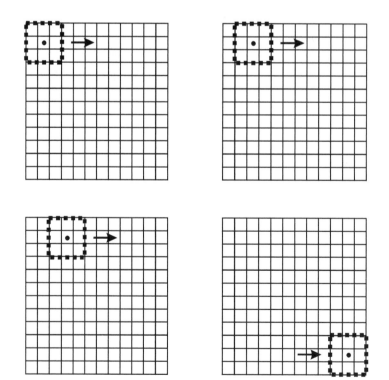

Figure 10-15: The concept of a moving window in raster neighborhood operations. Here a 3 by 3 window is swept from left to right and from top to bottom across a raster layer. The window at each location defines the input cells used in a raster operation.

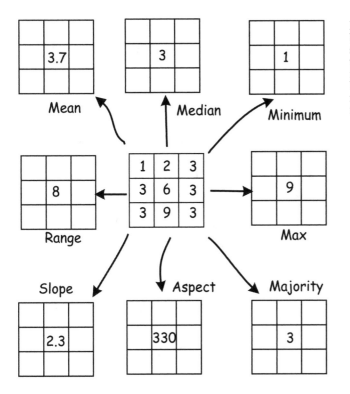

Figure 10-16: A given raster neighborhood may define the input for several raster neighborhood operations. Here a 3 by 3 neighborhood is specified. These nine cells may be used as input for mean, median, minimum, or a number of other functions.

numbered in both the x and y directions to provide a natural center cell, and they are typically square. A 3 by 3 cell window is the most common size, although windows may also be rectangular. Windows may also have irregular shapes, for example, L-shaped, circular, or wedge-shaped moving windows are sometimes specified.

There are many neighborhood functions that use a moving window. These include simple functions such as mean, maximum, minimum, or range (Figure 10-16). Neighborhood functions may be complicated, e.g., the statistical standard deviation, and they may be non-arithmetic, for example, a count of the number of unique values, the mode, or a Boolean occurrence. Any function that combines information from a consistently shaped group of raster cells may be implemented with a moving window.

Moving window functions may be arithmetic, adding, subtracting, averaging, or otherwise mathematically combining the values

around a central cell, or they may be comparative or otherwise extract values from a set of cells. Common statistical operations include calculating the largest value, the mode (peak of a histogram), median (middle value), the range (largest minus smallest), or diversity (number of different values). These neighborhood operations are useful for many kinds of processing.

Consider the *majority* operation, also known as a *majority filter*. You might wonder why one would want to calculate a majority filter for a data layer. Data smoothing is a common application. We described in Chapter 6 how multi-band satellite data are often converted from raw image data to landcover classification maps.These classifiers often assign values on a pixel basis, and often result in many single pixels of one landcover type embedded within another landcover type. These single pixels are often smaller than the minimum mapping unit, the smallest uniform area that we care to map. A majority filter is often used to remove this classification "noise."

A majority filter is illustrated in Figure 10-17. It illustrates NASS crop data for an area in central Indiana, based on classified satellite images. There are over 40 common landcover types in the area, but these have been reclassified into the dominant types of developed (road), corn, beans, and other crops. Each pixel is 30 meters across, and the image on the left is NASS data as delivered. Note that corn and beans dominate, but there are many "stray" pixels of a dissonant vegetation type embedded or on the edge of a dominant type in an area, e.g., bean pixels in a corn field, or corn pixels in a bean field. These stray pixels usually do not represent reality, in that although there may be the isolated plant or two from previously deposited seed in these annual crop rotations, they almost never approach 30 meters in size. The embedded cells are most often mis-classifications due to canopy thinning or perhaps

weeds below the crop, and in any event, are often below the minimum mapping unit.

The illustrated majority filter counts the values of the four cells sharing an edge with any given cell. If a majority, meaning three or more cells, are of a type, then the cell is output as this majority type (top, Figure 10-17). If a majority is not reached, e.g., only one or two cells add up to the most frequent type in the four bordering cells, then the center cell value is unchanged in the output (Figure 10-17). The removal of most of the single pixel "noise" by the majority filter can be observed in the classified image on the right side of Figure 10-17.

There may be many variants for a given operation. The majority filter just discussed may assign an output value if only two of the four adjacent cell values are most frequent, or use the 8 or 24 nearest cells to calculate a majority. The dependence of output on algorithm specifics should always be recognized when applying any raster operation.

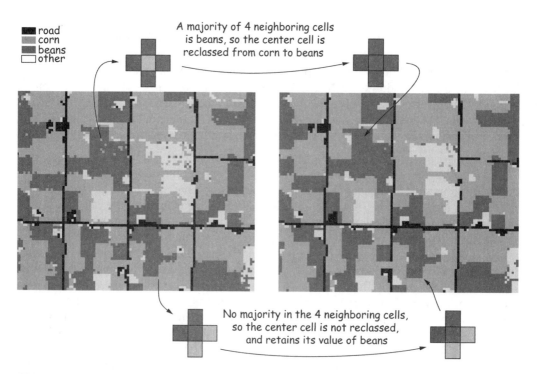

Figure 10-17: An example of a majority filter applied to a raster data layer from a classified satellite image. Many isolated cells are converted to the category of the dominant surrounding class.

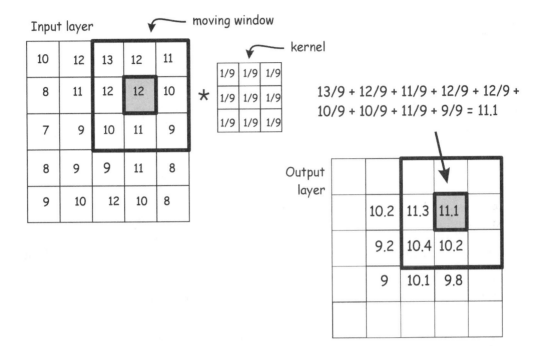

Figure 10-18: An example of a mean function applied via a moving window. Input layer cell values from a 3 by 3 moving window (upper left) are multiplied by a kernel to specify corresponding output cell values. This process is repeated for each cell in the input layer.

Figure 10-18 shows an example of a mean calculation using a moving window. The function scans all nine values in the window. It sums them and divides the sum by the number of cells in the window, thus calculating the mean cell value for the input window. The multiplication may be represented by a 3 by 3 grid containing the value one-ninth (1/9). The mean value is then stored in an output data layer in the location corresponding to the center cell of the moving window. The window is then shifted to the right and the process repeated. When the end of a row is reached the window is returned to the left-most columns, shifted down one row, and the process repeated until all rows have been included.

The moving window for many simple mathematical functions may be defined by a *kernel*. A kernel for a moving window function is the set of cell constants for a given window size and shape. These constants are used in a function at every moving window

location. The kernel in Figure 10-18 specifies a mean. As the figure shows, each cell value for the Input layer at a given window position is multiplied by the corresponding kernel constant. The result is placed in the Output layer.

Note that when the edge of the moving window is placed on the margin of the original raster grid, we are left with at least one border row or column for which output values are undefined. This is illustrated in Figure 10-18. The moving window is shown in the upper right corner of the input raster. The window is as near the top and to the right side of the raster as can be without placing input cell locations in the undefined region, outside the boundaries of the raster layer. The center cell for the window is one cell to the left and one cell down from the corner of the input raster. Output values are not defined for the cells along the top, bottom, and side margins of the output raster when using a 3 by 3 window, because a portion of

Mean function kernels

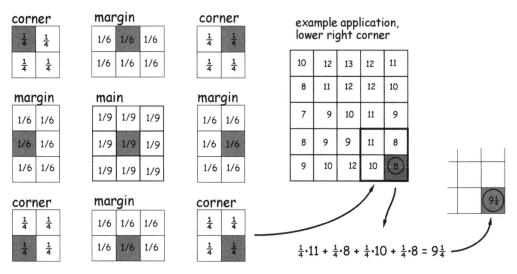

Figure 10-19: Kernels may be modified to fill values for a moving window function at the margin of a raster data set. Here the margin and corner kernels are defined differently than the "main" kernel. Output values are placed in the shaded cell for each kernel. The margin and corner kernels are similar to the main kernel, but are adjusted in shape and value to ignore areas "outside" the raster.

the moving window would lay outside the raster data set. Each operation applied to successive output layers may erode the margin further.

There are several common methods of addressing this margin erosion. One is to define a study area larger than the area of interest. Data may be lost at the margins, but these data are not important if they are outside the area of interest. A second common approach defines a different kernel for margin windows (Figure 10-19). Margin kernels are similar to the main kernels, but modified as needed by the change in shape and size. Figure 10-19 illustrates a 3 by 3 kernel for the bulk of the raster. Corner values may be determined using a 2 by 2 kernel, and edges can be determined with 2 by 3 kernels. Outputs from these kernels are placed in the appropriate edge cells.

Different moving windows and kernels may be specified to implement many differ-

ent neighborhood functions. For example, kernels may be used to detect edges within a raster layer. We might be interested in the difference in a variable across a landscape. For example, a railway accident may have caused a chemical spill and seepage through adjacent soils. We may wish to identify the boundary of the spill from a set of soil samples. Suppose there is a soil property, such as a chemical signature, that has a high concentration where the spill occurred, but a low concentration in other areas. We may apply kernels to identify where abrupt changes in the levels of this chemical create edges.

Edge detection is based on comparing differences across a kernel. The values on one side of the kernel are subtracted from the values on the other side. Large differences result in large output values, while small differences result in small output values. Edges are defined as those cells with output values larger than some threshold.

Figure 10-20 illustrates the application of an edge-detection operation. The kernel on the left side of Figure 10-20 amplifies differences in the x direction. The values in the left of three adjacent columns are subtracted from the value in the corresponding right-hand row of cells. This process is repeated for each cell in the kernel, and the values summed across all nine cells. Large differences result in large values, either positive or negative, saved in the center cell. Small differences between the left and right rows lead to a small number in the center cell. Thus, if there are large differences between values when moving in the x direction, this difference is highlighted. Spatial structure such as an abrupt change in elevation may be detected by this kernel. The kernel in the middle-right of Figure 10-20 may be used to detect differences in the y direction.

Neighborhood functions may also smooth the data. An averaging kernel, described above in Figure 10-18 and Figure 10-19, will reduce the difference between a cell and surrounding cells. This is because windows average across a group of cells, so there is much similarity in the output values calculated from adjacent window placements.

Raster data may contain "noise". Noise are values that are large or small relative to their spatial context. Noise may come from several sources, including measurement errors, mistakes in recording the original data, miscalculations, or data loss. There is often a need to correct these errors. If it is impossible or expensive to revisit the study area and collect new data, the noisy data may be smoothed using a kernel and moving window.

There are functions known as *high-pass filters* with kernels that accentuate differences between adjacent cells. These high-pass filter kernels may be useful in identifying the spikes or pits that are characteristic of noisy data. Cells identified as spikes or pits may then be evaluated and edited as appropriate, removing the erroneous values. High-pass kernels generally contain both negative and positive values in a pattern that accentuates local differences.

Figure 10-21 demonstrates the use of a high-pass kernel on a data set containing noise. The elevation data set shown in the top portion of the figure contains a number of anomalous cells. These cells have extremely high values (spikes, shown in black) or low values (pits, shown in white) relative to nearby cells. If uncorrected, pits and spikes will affect slope, aspect, and other terrain-based calculation. These locally extreme values should be identified and modified.

The high-pass kernel shown contains a value of 9 in the center and -1 in all other cells. Each value is divided by nine to reduce the range of the output variable. The kernel returns a value near the local average in smoothly-changing areas. The positive and negative values balance, returning small numbers in flat areas.

The high-pass kernel generates a large positive value when centered on a spike. The large differences between the center cell and adjacent cells are accentuated. Conversely, a large negative value is generated when a pit is encountered. An example shows the application of the high-pass filter for a cell near the upper left corner of the input data layer (Figure 10-21). Each cell value is multiplied by the corresponding kernel coefficient. These numbers are summed, and divided by 9, and the result placed in the corresponding output location. Calculation results are shown as real numbers, but cell values are shown here recorded as integers. Output values may be real numbers or integers, depending on the programming algorithm and perhaps the specifications set by the user.

The mean filter is representative of many moving window functions in that it increases the *spatial covariance* in the output data set. High spatial covariance means values are autocorrelated (discussed in greater depth when in the spatial prediction section of Chapter 13). A large positive spatial covariance means cells near each other

Input
layer

980	980	980	980	980	940	940	940	940	940
980	980	980	980	980	940	940	940	940	940
980	980	980	980	980	940	940	940	940	940
980	980	980	980	980	940	940	940	940	940
980	980	980	980	980	940	940	940	940	940
980	980	980	980	980	940	940	940	940	940
980	980	980	980	980	900	900	900	900	900
980	980	980	980	980	900	900	900	900	900
980	980	980	980	980	900	900	900	900	900
980	980	980	980	980	900	900	900	900	900

Kernels

0	0	0
1	0	-1
0	0	0

horizontal
difference

vertical
difference

0	1	0
0	0	0
0	-1	0

Output A

0	0	0	0	40	40	0	0	0	0
0	0	0	0	40	40	0	0	0	0
0	0	0	0	40	40	0	0	0	0
0	0	0	0	40	40	0	0	0	0
0	0	0	0	40	40	0	0	0	0
0	0	0	0	40	40	0	0	0	0
0	0	0	0	80	80	0	0	0	0
0	0	0	0	80	80	0	0	0	0
0	0	0	0	80	80	0	0	0	0
0	0	0	0	80	80	0	0	0	0

Output B

0	0	0	0	0	0	0	0	0	0
0	0	0	0	0	0	0	0	0	0
0	0	0	0	0	0	0	0	0	0
0	0	0	0	0	0	0	0	0	0
0	0	0	0	0	0	0	0	0	0
0	0	0	0	0	40	40	40	40	40
0	0	0	0	0	40	40	40	40	40
0	0	0	0	0	0	0	0	0	0
0	0	0	0	0	0	0	0	0	0
0	0	0	0	0	0	0	0	0	0

Figure 10-20: There are a large number of kernels used with moving windows. The kernel on the left amplifies differences in the x direction, while the kernel on the right amplifies differences in the y direction. These and other kernels may be used to detect specific features in a data layer.

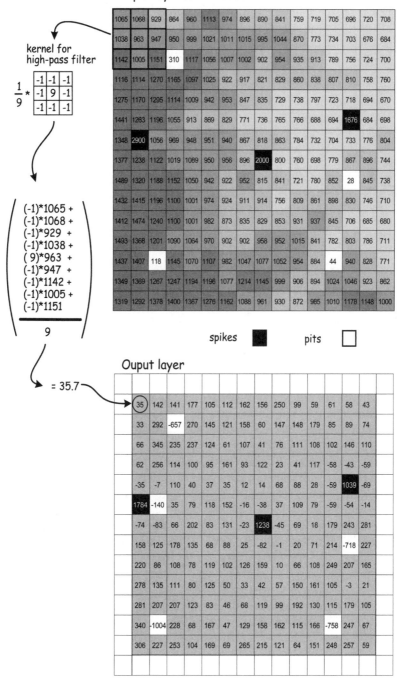

Figure 10-21: An example of a moving window function. Raster data often contain anomalous "noise" (dark and light cells). A high-pass filter and kernel, shown at top left, highlights "noisy" cells. Local differences are amplified so that anomalous cells are easily identified.

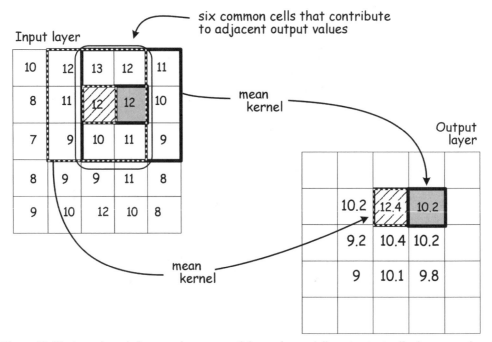

Figure 10-22: A moving window may increase spatial covariance. Adjacent output cells share many input cell values. In the mean function shown here, this results in similar output cell values.

are likely to have similar values. Where you find one cell with a large number, you are likely to find more cells with large numbers. If spatial data have high spatial covariance, then small numbers are also likely to be found near each other. Low spatial covariance means nearby values are unrelated – knowing the value at one cell does not provide much information about the values at nearby cells. High spatial covariance in the "real world" may be a good thing. If we are prospecting for minerals then a sample with a high value indicates we are probably near a larger area of ore-bearing deposits. However if the spatial autocorrelation is increased by the moving window function we may get an overly optimistic impression of our likelihood of striking it rich.

The spatial covariance increases with many moving window functions because these functions share cells in adjacent calculations. Note the average function in Figure 10-22. The left of Figure 10-22 shows

sequential positions of a 3 by 3 window. In the first window location the mean is calculated and placed in the output layer. The window center is then shifted one cell to the right, and the mean for this location calculated and placed in the next output cell to the right. Note that there are six cells in common for these two means. Adjacent cells in the output data layer share six of nine cells in the mean calculation. When a particularly low or high cell occurs, it affects the mean of many cells in the output data layer. This causes the outputs to be quite similar, and increases the spatial covariance.

Zonal Functions

Zonal functions apply operations based on defined regions, or zones, within an area. Typically the zones are recorded in a data layer, with a unique identifier for each zone. A function is then applied based on the zone.

There are many reasons for applying zonal functions. We often want to summarize data for defined units in a region, including total population in a county, average rainfall in a watershed, or number of impoverished families across neighborhoods. More complicated analyses may require different operations be applied to different zones, for example, we may be creating an elevation data set from many sources, and we may wish to use the highest quality data in zones where it exists, and use successively poorer data in other zones. Zonal functions give us these capabilities.

Figure 10-23 illustrates the application of a zonal function. In this example the function calculates the zonal average for In_Layer, based on zones defined by Zone_Layer. The syntax here is

Out_Layer = ZoneAvg(In_Layer, Zone_Layer)

There is no standard syntax across software, so the specific order and interpretation of operands depends on the software used.

Note that the output here is a raster, with identical values in all the cells of a given zone. This is typically how zonal functions are specified. Most systems can create a table with zonal identifiers and summary values to accompany the layer, and in some cases the operation only results in a table.

Zonal functions typically require compatible cell sizes. Generally, this means the cells in the input layers and zone-defining layer have the same cell size and orientation. The zone layer may have dimensions that are integer multiples of the input layers, but the reverse is generally not recommended. Input layer cell sizes larger than zone layer cell sizes may lead to ambiguous zone definition when more than one zone may correspond to an input cell.

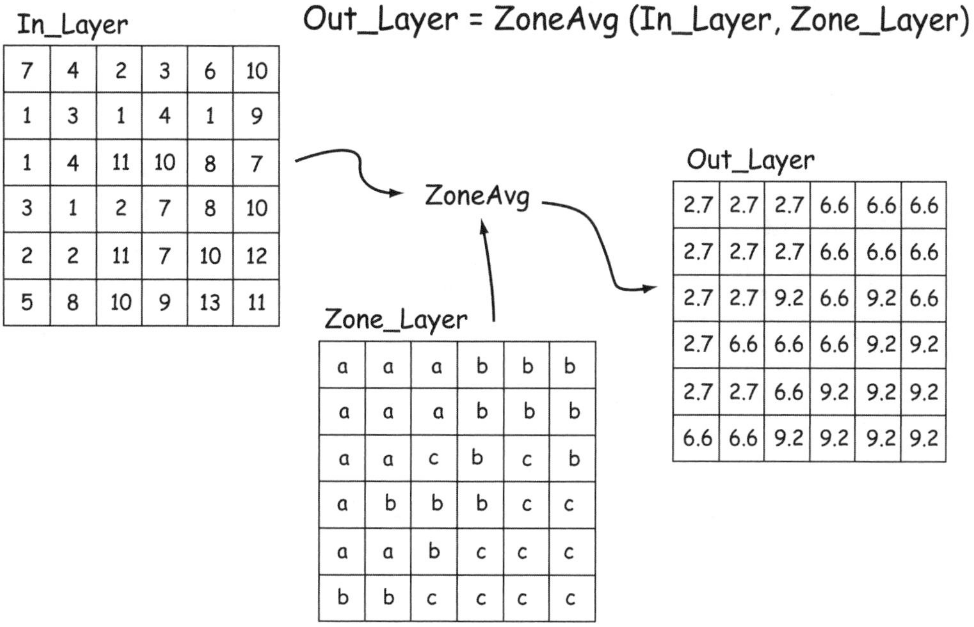

Figure 10-23: An example of a zonal function. Averages are calculated based on the zones stored in Zone_Layer.

Cost Surfaces

Many problems require an analysis of travel costs. These may be monetary costs of travel, such as the price one must charge to profitably deliver a package from the nearest distribution center to all points in a region. Travel costs might also be measured in other units, for instance, the time it takes to travel from a school to the nearest hospital, or as a likelihood, such as the chance of a noxious foreign weed spreading out from an introduction point. These analyses may be performed with the help of *cost surfaces*. A cost surface contains the minimum cost of reaching cells in a layer from one or more source cells (Figure 10-24).

The simplest cost surface is based on a uniform travel cost. Travel cost depends only on the distance covered, with a fixed cost applied per unit distance traveled. This cost per unit distance does not change from cell to cell. There are no barriers, so the straight line distance is converted to a cost. First, the distance is calculated from our source or starting location to each cell. As illustrated in Figure 10-24, the distance is calculated based on the Pythagorean formula. Distances to each cell in the x and y directions contribute to the total distance from a source cell or cells.

The distance from a source cell is combined with a fixed cost per unit distance to calculate travel cost. As shown in the right side of Figure 10-24, each distance value is multiplied by the fixed cost factor. This results in a cost surface, a raster layer containing the travel cost to each cell. If there are multiple source cells, travel costs are calculated from each source cell, and the lowest cost is typically placed in the output cell.

Note that distance is commonly measured at least two ways - a straight-line (Euclidian) distance, as shown in Figure 10-24, or as a row-column distance. A row-col-

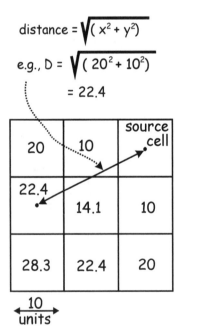

distance $= \sqrt{(x^2 + y^2)}$

e.g., $D = \sqrt{(20^2 + 10^2)}$
$= 22.4$

cost = distance * fixed cost factor
e.g.,
cost = distance * 2

Figure 10-24: A cost surface based on a fixed cost per unit distance. Minimum distance from a set of source cells is multiplied by a fixed cost factor to yield a cost surface.

cost = cell distance * friction

output cost surface

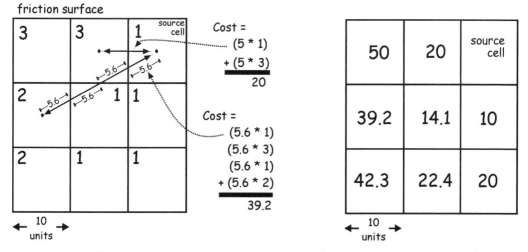

Figure 10-25: A cost surface based on spatially-variable travel costs. A friction surface specifies the spatially varying cost of traveling through raster cells. The distance traversed through each cell is multiplied by the cost in the friction surface. The values are summed for each path to yield a total cost.

umn distance is measured along the row and column axes, and is by definition longer than the straight-line distance. Straight-line distances are preferred in most applications, although they are more difficult to implement.

Travel costs may also be calculated using a *friction surface*. The cell values of a friction surface represent the cost per unit travel distance for crossing each cell. Friction surfaces are used to represent areas with a variable travel cost. Imagine a large military base. Part of the base may include flat, smooth areas such as drill fields, parking lots, or parade grounds. These areas are relatively easy to cross, with correspondingly low travel times per unit distance. Other parts of the base may be covered by open grasslands. While the surface may be a bit rougher, travel times are still moderate. Other parts may be comprised of forests. These areas would have correspondingly high travel times, as a vehicle would have to pick a path among the trees. Finally, there may be areas occupied by water, fences, or buildings. These areas would have effectively infinite travel times.

Each cell in the friction surface contains the cost required to traverse a portion of the cell. A value of 3 indicates it costs three units (of time, money, or other factor) per unit distance in the cell. If a cell is 10 wide and costs 3 units per unit distance, and the cell is crossed along the width, then the cost for traversing the cell is 10 times 3, or 30 units.

The actual cost for traversing the cell depends on the distance traveled through the cell. When a cell is traversed parallel to the row or column edge then the distance is simply the cell dimension. When a cell is traversed at any other angle the distance will vary. It may be greater or less than the cell dimension, depending on the angle and location of the path.

The travel cost required to reach each cell is the minimum accumulated total of the cost times the distance to a source cell. We specify a minimum accumulated cost because if there is more than one source cell, there is a large number of potential paths to each of these source cells. Distance across each cell is multiplied by the friction surface cost for that cell and summed for a path to accumulate the total travel cost. The lowest cost path from a source location to a cell is usually assigned as the travel cost to that cell.

Figure 10-25 shows an example of calculations for the friction cost along a set of paths. These are straight-line paths that travel either parallel to the cell boundaries (purely in an x or y direction) or at some angle across cells.

Sample calculation of the friction costs for a path parallel to the x axis is shown at the top middle and on the left side of Figure 10-25. Note that when travelling parallel to a cell boundary, one half-cell width is traversed in the starting and ending cells. Intermediate cells are crossed at a full cell width. When moving from the starting cell to the adjacent left cell, a friction surface value of 1 is encountered, then a friction surface value of 3. One-half the distance, 5 units, is through the top-right cell at a per-unit friction cost of 1. One-half the distance is through the adjacent cell to the left, at a per-unit friction cost of 3. The total cost is then the distance traveled in each cell multiplied by the per-unit friction cost of the cell:

$$5 * 1 + 5 * 3 = 20 \tag{10.4}$$

The friction cost when traversing cells at an angle is illustrated at the bottom left and bottom center of Figure 10-25. The friction cost is the sum of the cell cost per unit distance multiplied by the distance traveled in each cell. The path begins at the source cell and ends two cells to the left and one cell down. Each intervening cell is traversed for a distance of 5.6 cell units. The distance traversed in each cell is multiplied by the friction value for each cell. The total cost for this leg is:

$$5.6*1 + 5.6*3 + 5.6*1 + 5.6*2 = 39.2 \tag{10.5}$$

In general, the cost of any path is expressed as:

$$Totalcost = d1*c1 + d2*c2 + ...dn*cn \tag{10.6}$$

where d_i is the distance and c_i is the cost across each cell of a path.

Many softwares calculate the cumulative cost for the most direct path using a slightly different approach, called the *row-*

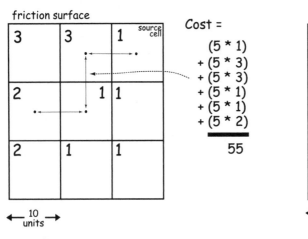

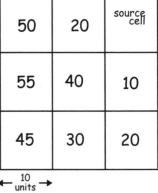

Figure 10-26: Calculations of the travel cost from a source cell to each other cell using row/column distance calculations.

column distance. Rather than travel along a straight-line path, the row-column distance travels from cell-center to cell center (Figure 10-26). Calculations are much easier because the length of the path within each cell is constant with row column distance, and for square cells this distance equals the cell width (or height). The distance in each cell varies when using the straight line distance, and so the time required to calculate the accumulated distance is substantially increased. The row/column distance gives the same relative costs for travel from a source cell to each target cell, but the absolute costs change (compare the costs on the right of Figure 10-25 to the right of Figure 10-26).

Many implementations of a friction surface or cost function allow you to search for the minimum cost to travel to a cell from a set of source cells. The straight line distance may not be the "least costly," and so alternatives may be examined. There are many routes from any source cell to any destination cell, thousands of distinct routes in most instances. Software typically implements some optimization algorithm to eliminate routes early on and reduce search time, thereby arriving at the cost surface in some acceptable time period.

Note that barriers may be placed on a cost surface to preclude travel across portions of the surface. These barriers may be specified by setting the cost so high that no path will include them. Any circuitous route will be less expensive than traveling over the barriers. Some software allows the specification of a unique code to identify barriers, and this code precludes movement across the cell.

Summary

Raster analyses are essential tools in GIS, and should be understood by all users. Raster analyses are widespread and well developed for many reasons, in part due to the simplicity of the data structure, the ease with which continuous variables may be rep-

resented, and the long history of raster analyses.

Map algebra is a concept in which raster data layers are combined via summation and multiplication. Values are combined on a cell-by-cell basis, and may be added, subtracted, multiplied, or divided. Care must be taken to avoid ambiguous combinations in the output that originate from distinct input combinations.

Raster analyses can be local, neighborhood, or global, and general analyses such as buffering and overlay may be applied using raster data sets. Neighborhood operations are particularly common in raster analyses, and may be applied with a moving window approach. A moving window is swept across all cells in a data layer, typically multiplying kernel values by data found around a center cell. Window size and shape may be modified at the edges of the data layers. Moving windows may be used to specify a wide range of combinatorial, terrain, and statistical functions.

Cost or friction surfaces are an important subset of proximity analyses that may be easily applied in raster analyses. A cost surface identifies the travel costs required for movement from a specified set of locations.

Suggested Reading

Berry, J.K. (1987). Fundamental operations in computer-assisted mapping. *International Journal of Geographic Information Systems*, 1:119-136.

Berry, J.K. (1986). A mathematical structure for analyzing maps. *Environmental Management*, 11:317-325.

Bonhame-Carter, G.F. (1996). *Geographic Information Systems for Geoscientists: Modelling with GIS*. Pergamon: Ottawa.

Burrough, P.A. & McDonnell, R.A. (1998). *Principles of Geographical Information Systems* (2nd ed.). Oxford University Press: New York.

Cliff, A.D., & Ord, J.K. (1987). *Spatial Autocorrelation*. Methuen: New York.

DeMers, M.N. (2002). *GIS Modeling in Raster*. Wiley: New York.

de Smith, M.J., Goodchild, M.F., & Longley, P.A. (2007). *Geospatial Analysis, a Comprehensive Guide to Principles, Techniques, and Software Tools*, Matador: Leicester.

Eastman, J.R., Jin, W., Keym, P.A.K., & Toledano, J. (1995). Raster procedures for multi-criteria/multi-objective decisions. *Photogrammetric Engineering and Remote Sensing*, 61:539-547.

Eastman, J.R. (1997). *Idrisi for Windows*. Clark University: Worcester.

Hengl, T. (2006). Finding the right pixel size, *Computers and Geosciences*, 32:1283-1298.

Morain, S., & Baros, S.L. (1996). *Raster Imagery in Geographic Information Systems*. OnWord Press: Santa Fe.

Mitchell, A. (1999). *The ESRI Guide to GIS Analysis: Geographic Patterns and Relationships*. ESRI Press: Redlands.

Tomlin, C. D. (1990). *Geographic Information Systems and Cartographic Modeling*. Prentice-Hall: Upper Saddle River.

Study Questions

10.1 - What is map algebra?

10.2 - Why must raster layers have compatible cell sizes and orientations for most raster combination operations?

10.3 - What is a null value in a raster data set? How is this null value typically treated in a raster operation?

10.4 - Perform the listed local raster operations.

3	2	4	11	9	1	3
1	⑥	5	20	14	8	7
7	13	2	1	4	△9	11
12	11	10	8	5	6	10
3	2	☐1	17	12	☆11	9
⑧	5	6	8	3	13	16
19	17	9	11	⬠12	7	15

Perform the following operations with a 3x3 window, centered on the noted cells

-average, on the circle,
-standard deviation, on the circle,
-maximum, on the triangle
-value range, on the square
-average, on the ellipse,
-median, on the star,

10.5 - What are the values in cells C1, C2, C3, and C12 in the output layer?

$$Con(Layer1 < 2, 0, 1)$$

Layer1

2	N	1	2
1	N	1	3
4	1	2	0
N	2	N	1

Output

C1	C2	C3	C4
C5	C6	C7	C8
C9	C10	C11	C12
C13	C14	C15	C16

10.6 - What are the cell values for cells C1, C3, C4, and C10 in the output layer, below?

Output = CON((layerA==N), 1, layerA)

layerA

N	N	1	0
1	N	2	N
N	4	N	N
0	1	N	1

Output

C1	C2	C3	C4
C5	C6	C7	C8
C9	C10	C11	C12
C13	C14	C15	C16

10.7 - Give an example of a nested operation.

10.8 - What are the values in output cells C9, C10, C11, and C12?

Output = CON(ISNULL(layerA), 1, N)

layerA

1	N	N	0
0	N	2	1
N	1	5	0
N	1	N	1

Output

C1	C2	C3	C4
C5	C6	C7	C8
C9	C10	C11	C12
C13	C14	C15	C16

10.9 - What is the scope of a raster operation?

10.10 - Does a NOT operation applied to a raster cell value containing a NULL value return a NULL value, a zero value, a 1, or some other non-null value?

10.11 - Diagram AND, OR, and NOT operations on a raster data set.

10.12 - Provide the answer for the following logical operations:

1	1	0	0
0	0	0	1
1	1	0	0
1	0	0	1

and

0	1	0	1
1	0	0	1
1	1	1	0
1	0	1	1

=

1	1	0	0
0	0	0	1
1	1	0	0
1	0	0	1

or

0	1	0	1
1	0	0	1
1	1	1	0
1	0	1	1

=

10.13 - Describe how local arithmetic functions can be used to apply overlay or clip functions in a raster environment. What are the primary restrictions on values for the raster cells in the input data layers?

10.14 - What is a kernel in a moving window operation? Does the kernel size or shape change for different portions of the raster data set? Why or why not?

10.15 - What is the kernel below likely used for in a moving window operation?

-2	-1	-2	-1	-2
-1	0	0	0	-1
-2	0	25	0	-2
-1	0	0	0	-1
-2	-1	-2	-1	-2

10.16 - What is the kernel below likely used for in a moving window operation?

1	2	1
0	0	0
-1	-2	-1

10.17 - What is meant by high or low spatial covariance in a raster data layer?

10.18 - Calculate the cost of travel between A and B, and A and C, over the cost surface below, both by straight line, and by row-column paths.

Source/target cells

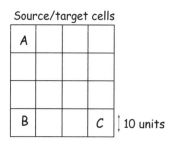

B   C ┆ 10 units

Cost surface

3	5	6	8
4	1	7	5
2	5	1	6
2	4	1	1

remember, the
diagonal of a square
is 1.41 x the edge

11 Terrain Analysis

Introduction

Elevation and related terrain variables are important at some point in almost everyone's life. Terrain determines the availability and location of surface water, and therefore soil moisture and drainage. Terrain in large part affects water quality through sediment generation and transport. Elevation and slope define flood zones, watershed boundaries, and hydrologic networks. Terrain also strongly influences the location and nature of transportation networks, and the cost and methods of house and road construction. Terrain data are used in shading hardcopy and digital maps, thereby depicting three-dimensions on two-dimensional media (Figure 11-1). Terrain variables are frequently applied in a broad range of spatial analyses (Table 11-1).

Figure 11-1: An example of a terrain-based image of the western United States. Shading, based on local elevation, emphasizes terrain shape. Topographic features are clearly identified, including the Central Valley of California at the left of the image, and to the center and right the parallel mountains and valleys of the Basin and Range region. (courtesy USGS)

Given the importance of elevation and other terrain variables in resource management, and the difficulties of manual terrain analysis, it is not surprising that terrain analysis is well-developed in GIS. Indeed, it is often impractical to perform consistent terrain analyses without a GIS. For example, slope is of vital importance in many resource management problems. Slope calculations over large areas based on hardcopy maps are slow, error prone, and inconsistent. Elevation differences over a horizontal distance are difficult to measure. Further, these measurements are slow and likely to vary among human analysts. In contrast, digital slope calculations are easy to program, consistent, and have proven to be as accurate as field measurements.

Both data and methods exist to extract important terrain variables via a GIS. Digital elevation models (DEMs), described in

Table 11-1: A subset of commonly used terrain variables. (adapted from Moore *et al.*, 1993).

Variable	Description	Importance
Height	Elevation above base	Temperature, vegetation, visibility
Slope	Rise relative to horizontal distance	Water flow, flooding, erosion, travel cost, construction suitability, geology, insolation, soil depth
Aspect	Downhill direction of steepest slope	Insolation, temperature, vegetation, soil characteristics and moisture, visibility
Upslope area	Watershed area above a point	Soil moisture, water runoff volume and timing, pollution or erosion hazards
Flow length	Longest upstream flow path to a point	Sediment and erosion rates
Upslope length	Mean or total upstream flow path length from a point	Sediment and erosion rates
Profile curvature	Curvature parallel to slope direction	Erosion, water flow acceleration
Plan curvature	Curvature perpendicular to slope direction	Water flow convergence, soil water, erosion
Visibility	Site obstruction from given viewpoints	Utility location, viewshed preservation

Chapters 2 and 7, have been developed for most of the world using methods described in Chapters 5 and 6. DEM data have been produced at considerable cost, and there is a continuing process of DEM renewal and improvement.

Most terrain analyses are performed using a raster data model. While the TIN (Chapter 2) and other data models have been developed to store and facilitate terrain analyses, raster data structures are more commonly used. Raster data sets facilitate the easy and uniform calculation of slope, aspect, and other important terrain variables.

Calculations are based on cell values assigned to a regular grid. We use the concept of Z values, the height stored in the raster arrays, to extract information about terrain, using the magnitudes and patterns of changes in Z across the grid (Figure 11-2). For example, the height differences between adjacent cells or in a neighborhood of cells is used to calculate a local slope (Slope in Figure 11-2), and the angle and orientation of lines defined by X, Y, and Z values around a point may be used to calculate the direction of the normal vector, at right angles to the local surface (Figure 11-2). Local curvature and slope direction are also calculated by differences in Z values in a neighborhood.

Many terrain analysis functions can be specified by a mathematical operation applied to an appropriate moving window. The results from these mathematical operations in turn provide important information about terrain characteristics that are helpful in spatial analysis and problem solving.

Slope and Aspect

Slope and aspect are two commonly used terrain variables. They are required in many studies of hydrology, conservation, site planning, and infrastructure development, and are the basis for many other terrain analysis functions. Road construction costs and safety are sensitive to slope, as are most other construction activities. Watershed boundaries, flowpaths and direction, erosion modeling, and viewshed determination (discussed later in this chapter) all use slope and/ or aspect data as input. Slope or aspect may be useful in mapping both vegetation and soil resources, thereby increasing the accu-

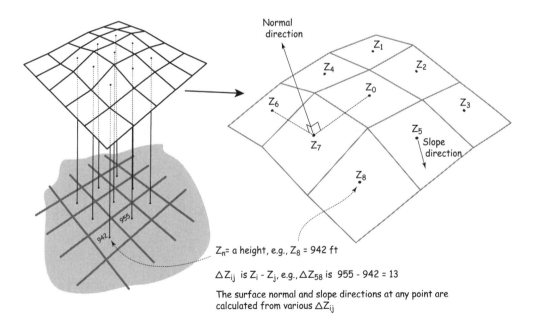

Z_n= a height, e.g., Z_8 = 942 ft

ΔZ_{ij} is $Z_i - Z_j$, e.g., ΔZ_{58} is 955 - 942 = 13

The surface normal and slope directions at any point are calculated from various ΔZ_{ij}

Figure 11-2: A depiction of a surface represented by a raster DEM, and changes in Z values for cells used for calculating various terrain attributes.

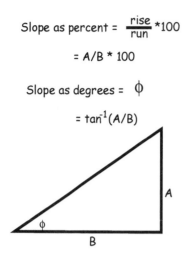

$$\text{Slope as percent} = \frac{\text{rise}}{\text{run}} *100$$

$$= A/B * 100$$

$$\text{Slope as degrees} = \phi$$

$$= \tan^{-1}(A/B)$$

To convert from percent slope to degrees, apply formula,

e.g. 3% = how many degrees?

$A/B * 100 = 3$, then $A/B = 3/100 = 0.03$

$= \tan^{-1}(0.03) = 1.72$ degrees

Figure 11-3: Slope formula, showing the rise (**A**), run (**B**), and slope angle (ϕ).

racy and specificity of spatial data, and at times reducing costs.

Slope is defined as the change in elevation (a rise) with a change in horizontal position (a run). Seen in cross section, the slope is related to the rise in elevation over the run in horizontal position (Figure 11-3). Slope is often reported in degrees, between zero (flat), and 90 (vertical). The slope is equal to 45 degrees when the rise is equal to the run. The slope in degrees is calculated from the rise and run through the tangent trigonometric function. By definition, the tangent of the slope angle (ϕ) is the ratio of the rise over the run, as shown in (Figure 11-3). The inverse tangent of a measured rise over a run gives the slope angle. A steeper rise or shorter run lead to a higher ϕ and hence steeper slope.

Slope may also be expressed as a percent, calculated by 100 times the rise over the run (Figure 11-3). Slopes expressed as a percent range from zero (flat) to infinite

(vertical). A slope of 100% occurs when the rise equals the run.

Calculating slope from a raster data layer is more complicated than in the cross-section view shown in Figure 11-3. The raster cells occur at regular intervals across an irregular terrain surface. Slope direction at a point in the landscape is typically measured in the steepest direction of elevation change (Figure 11-4). Slope changes in a complex way across many landscapes, and calculations of slope must factor in the relative changes in elevations around a central cell.

As demonstrated in Figure 11-4, the slope direction often does not point parallel to the raster rows or columns. Consider the cells depicted in Figure 11-5. Higher elevations occur at the lower right corner, and lower elevations occur towards the upper left. The direction of steepest slope trends from one corner towards the other, but does not pass directly through the center of any cell. How do we obtain values for the rise and run? Which elevations should be used to calculate slope? Intuitively we should use some combination of a number of cells in the vicinity of the center cell, perhaps all of them. A number of researchers have investigated methods for calculating slope, but

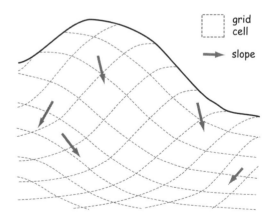

Figure 11-4: Slope direction, shown as gray arrows for some example locations above, often changes substantially among cells on a raster surface. Slope calculations in three dimensions require the consideration of all values surrounding a cell.

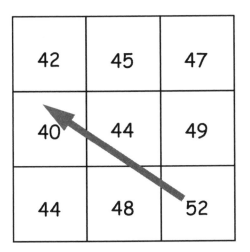

Figure 11-5: Slope direction on a raster surface usually does not point from cell center to cell center. Therefore, formulae that accurately represent slope on a surface integrate several cells surrounding the center cell.

and other odd-numbered windows are also used. Each cell in the window is assigned a subscript, and the elevation values found at window locations referenced by subscripted Z values.

Figure 11-6 shows an example of a 3 by 3 cell window. The central cell has a value of 44, and is referred to as cell Z_0. The upper left cell is referred to as Z_1, the upper center cell as Z_2, and so on through cell Z_8 in the lower right corner.

Slope at each center cell is most commonly calculated from the formula:

$$s = atan \sqrt{\left(\frac{dZ}{dx}\right)^2 + \left(\frac{dZ}{dy}\right)^2} \qquad (11.1)$$

before we discuss these methods we should describe some general characteristics of slope calculation using raster data sets.

Elevation is often represented by the letter Z in terrain functions. These terrain functions are usually calculated with a symmetrical moving window. A 3 by 3 cell window is most common, although 5 by 5

where s is slope, $atan$ is the inverse tangent function, Z is elevation, x and y are the respective coordinate axes, and dZ/dx and dZ/dy are calculated for each cell based on elevation values surrounding a given cell. The symbol dZ/dx represents the rise (change in Z) over the run in the x direction, and dZ/dy represents the rise over the run in the y direction. These formula are combined to calculate the slope for each cell based on

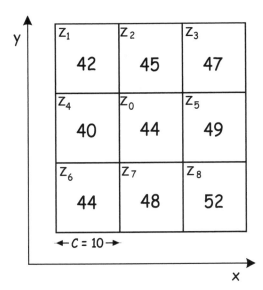

for Z_0:

$dZ/dx = (49 - 40)/20 = 0.45$

$dZ/dy = (45 - 48)/20 = -0.15$

slope $= atan\ [(0.45)^2 + (-0.15)^2\ ]^{0.5}$

$= 25.3^\circ$

Figure 11-6: Slope calculation based on cells adjacent to the center cell.

the combined change in elevation in the x and y directions.

Many different formulas and methods have been proposed for calculating dZ/dx and dZ/dy. The simplest, shown in Figure 11-6 and at the top of Figure 11-7, use cells adjacent to Z_0:

$$dZ/dx = (Z_5 - Z_4)/(2C) \qquad (11.2)$$

$$dZ/dy = (Z_2 - Z_7)/(2C) \qquad (11.3)$$

where C is the cell dimension and the Zs are defined as in Figure 11-6. This method uses the "four nearest" cells, Z_4, Z_5, Z_2, and Z_7, in calculating dZ/dx and dZ/dy. These four cells share the largest common border with the center. This "four nearest" method is perhaps the most obvious and provides reasonable slope values under many circumstances.

A common alternate method is known as a *3rd-order finite difference* approach (Figure 11-7, bottom). This method for calculating dZ/dx and dZ/dy differs mainly in the number and weighting it gives to cells in the vicinity of the center cell. The four nearest cells are given a higher weight than the "corner" cells, but data from all eight nearest cells are used.

Several other methods have been developed that are better for calculating slope under certain conditions. A method may be judged as better when, on average, it produces more accurate slope estimates when compared to carefully collected field measurements of slope. However, no method has proved best under all terrain conditions. Literature on the methods, their derivation, and application are listed in the suggested reading section at the end of this chapter.

Comparative studies have shown the two methods described here to be among the best for calculating slope and aspect when applied over a wide range of conditions. The method using the four nearest cells was among the best for smooth terrain, and the

3rd-order finite difference approach is often among the best when applied to rough terrain. Alternative methods typically perform no better than the two methods described above.

Aspect is also an important terrain variable that is commonly derived from digital elevation data. Aspect is used to define the direction water will flow, the amount of sunlight a site may receive, and to determine what portion of the landscape is visible from any viewing point.

The aspect at a point is the steepest downhill direction. The direction is typically reported as an azimuth angle, with zero in the direction of grid North, and the azimuth angle increasing in a clockwise direction. Aspects defined this way take values between 0 and 360 degrees. Flat areas by definition have no aspect because there is no downhill direction.

Aspect (α) is most often calculated using dZ/dx and dZ/dy:

$$\alpha = 180 - \arctan\left(\frac{\left(\frac{dZ}{dy}\right)}{\left(\frac{dZ}{dx}\right)}\right) + 90\left(\frac{\left(\frac{dZ}{dx}\right)}{\left|\frac{dZ}{dx}\right|}\right) \qquad (11.4)$$

where $atan$ is the inverse tangent function that returns degrees, and dZ/dy and dZ/dx are defined as above.

As with slope calculations, estimated aspect varies with the methods used to determine dZ/dx and dZ/dy. Tests have shown the four nearest cell and 3rd-order finite difference methods again yield among the most accurate results, with the 3rd-order method among the best under a wide range of terrain conditions.

Profile curvature and *plan curvature* are two other local topographic indices that are important in terrain analysis and may be derived from gridded elevation data. Profile and plan curvature are helpful in measuring and predicting soil water content, overland flow, rainfall-runoff response in small catchments, and the distribution of vegetation.

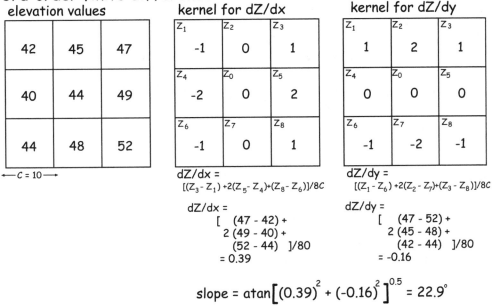

Four nearest cells
elevation values

42	45	47
40	44	49
44	48	52

←— $C = 10$ —→

kernel for dZ/dx

Z_1	Z_2	Z_3
0	0	0
Z_4 -1	Z_0 0	Z_5 1
Z_6 0	Z_7 0	Z_8 0

$dZ/dx = (Z^5 - Z^4)/2C$
$dZ/dx = (49 - 40)/20 = 0.45$

kernel for dZ/dy

Z_1	Z_2	Z_3
0	1	0
Z_4 0	Z_0 0	Z_5 0
Z_6 0	Z_7 -1	Z_8 0

$dZ/dy = (Z^2 - Z^1)/2C$
$dZ/dy = (45 - 48)/20 = -0.15$

$$slope = atan\left[(0.45)^2 + (-0.15)^2\right]^{0.5} = 25.3°$$

3rd-order finite difference
elevation values

42	45	47
40	44	49
44	48	52

←— $C = 10$ —→

kernel for dZ/dx

Z_1	Z_2	Z_3
-1	0	1
Z_4 -2	Z_0 0	Z_5 2
Z_6 -1	Z_7 0	Z_8 1

$dZ/dx =$
$[(Z_3 - Z_1) + 2(Z_5 - Z_4) + (Z_8 - Z_6)]/8C$

$dZ/dx =$
$[\quad(47 - 42) +$
$2(49 - 40) +$
$(52 - 44)\quad]/80$
$= 0.39$

kernel for dZ/dy

Z_1	Z_2	Z_3
1	2	1
Z_4 0	Z_0 0	Z_5 0
Z_6 -1	Z_7 -2	Z_8 -1

$dZ/dy =$
$[(Z_1 - Z_6) + 2(Z_2 - Z_7) + (Z_3 - Z_8)]/8C$

$dZ/dy =$
$[\quad(47 - 52) +$
$2(45 - 48) +$
$(42 - 44)\quad]/80$
$= -0.16$

$$slope = atan\left[(0.39)^2 + (-0.16)^2\right]^{0.5} = 22.9°$$

Figure 11-7: Four nearest cells (top) and 3rd-order finite difference (bottom) methods used in calculating slope. C is cell size and dZ/dx and dZ/dy are the changes in elevation (rise) with changes in horizontal position (run). Note that different slope values are produced by the different methods.

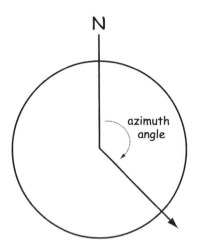

Figure 11-8: Aspect may be reported as an azimuth angle, measured clockwise in degrees from north.

Profile curvature is an index of the surface shape in the steepest downhill direction (Figure 11-9). The profile curvature may be envisioned by imagining a vertical plane, slicing downward into the Earth surface, with the plane containing the line of steepest descent (aspect direction). The surface traces a path along the face of this plane, and the curvature is defined by the shape of this path. Smaller values of profile curvature

indicate a concave (bowl shaped) path in the downhill direction, and larger values of profile curvature indicate a convex (peaked) shape in the downhill direction.

Plan curvature is the profile shape in the local direction of level, at right angle to the steepest direction. This means plan curvature is measured at a right angle to profile curvature (Figure 11-9). Plan curvature may also be envisioned as a vertical plane slicing into the surface, but with plan curvature the plane intersects at a 90° angle to the direction of steepest slope. The surface traces a path on the face of the plane, and the plan curvature is a measure of the shape of that path. Concave plan curvature values are small or negative for sloping valleys or clefts, while convex plan curvature values at ridge and peak sites are large or positive.

These concepts of directional terrain shape may be developed further to identify *terrain* or *morphometric features*. These are characteristic terrain elements including planes, peaks, passes, saddles, channels, ridges, and pits. Each of these shapes has particular terrain attributes that often affect important spatial variables. For example, soil is thinner and water scarcer on ridges and peaks because they are convex, while materials accumulate in pits and channels.

Z_1	Z_2	Z_3
Z_4	Z_0	Z_5
Z_6	Z_7	Z_8

$\leftarrow c \rightarrow$

$D = [(Z_4 + Z_5)/2 - Z_0] / c^2$

$E = [(Z_2 + Z_7)/2 - Z_0] / c^2$

$F = (Z_3 - Z_1 + Z_6 - Z_8) / 4c^2$

$G = (Z_5 - Z_4) / 2c$

$H = (Z_2 - Z_7) / 2c$

Figure 11-9: Profile curvature and plan curvature measure the local terrain shape. They are calculated by combining the values surrounding a center cell

plan curvature
$$\frac{2 (DH^2 + EG^2 - FGH)}{G^2 + H^2}$$

profile curvature
$$\frac{-2 (DG^2 + EH^2 + FGH)}{G^2 + H^2}$$

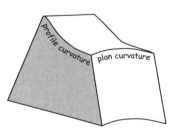

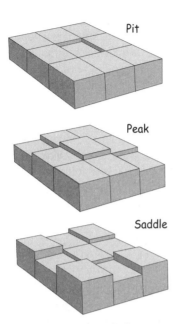

Figure 11-10: Morphometric feature types may be defined by the relative heights, and hence directional convexity, of adjacent cells (adapted from Jo Wood, 1996).

Terrain features may be identified by observing the convexity in orthogonal, or right-angle, directions. For example, peaks are characterized by convex shapes in both the x and y directions, a ridge is convex in one direction but relatively flat in another, while a pit is concave in orthogonal directions (Figure 11-10). Formulas similar to those in Figure 11-9 have been developed to measure the convexity and concavity in specified orthogonal directions. The various combinations may then be applied to identify terrain features (Figure 11-11).

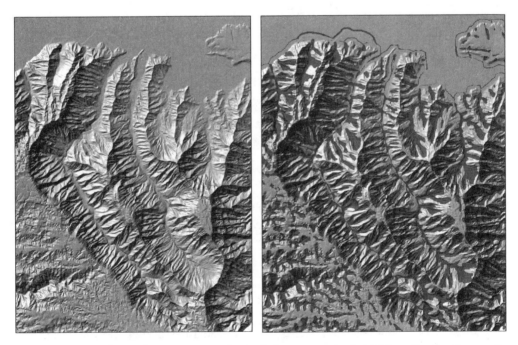

Figure 11-11: Terrain morphometery, or morphometric features may be derived from directional convexity measures. A shaded map of a mountainous area shows valleys and channels (above left), which are identified via morphometric terrain analyses and shown as dark areas, above right.

Hydrologic Functions

Digital elevation models are used extensively in hydrologic analyses. Water is basic to life, commerce, and comfort, and there is a substantial investment in water resource monitoring, gathering, protection, and management. Spatial functions are applied to DEMs to yield important information on hydrology.

A *watershed* is an area that contributes flow to a point on the landscape (Figure 11-12). Watersheds may also be named basins, contributing areas, catchments, drainages, and sub-basins or sub-catchments. The uphill area that drains to any point on a landscape is the watershed for that point. Water falling anywhere in the upstream area of a watershed will pass through that point. Watersheds may be quite small. For example, the watershed may cover only a few square meters on a ridge or high slope. Local high points have watersheds of zero area because all water drains away. Watersheds may also be quite large, including continental areas that drain large rivers such as the Amazon or Mississippi Rivers. Any point in the main channel of a large river has a large upstream watershed.

The drainage network, or set of streams and rivers in a watershed, is completely con-tained within the watershed. As shown in Figure 11-12, the stream network shows a dendritic pattern, with smaller watercourses branching off from larger segments as one moves upstream. The base of the drainage network is often called a *pour point* or *outlet*.

Flow direction is used in many hydrologic analyses. The true surface flow direction is the path water would take, if dumped in sufficient excess on a point so as to generate surface flow. This excess water flows in the steepest downhill direction, usually set equivalent to the local aspect.

The use of aspect to assign flow direction may cause errors in flow estimation, particularly in nearly flat and in built environments. Water flows both above and below the surface; if subsurface flow is large, ignoring it may cause errors. If soils have different permeabilities, or resistance to flow, then subsurface flow direction may be different than surface flow direction. In steep, undeveloped terrain, there is a strong downslope gravitational gradient that often dominates, and surface and subsurface flow directions are often similar, so aspect provides a reasonable approximations of overall flow direction. In flat or nearly flat terrain, soil permeability may dominate, causing different subsurface and surface flow directions. Ditches, culverts, buried stormsewers, and other built features alter flow directions in ways that aren't represented by terrain. However, subsurface drainage and built features are often addressed by modifications of flow directions first based on surface shape.

Flow directions may be stored as compass angles in a raster data layer (Figure 11-13). Acceptable values are from 0 to 360 if the angle is expressed in degrees azimuth. Alternately, flow direction can be stored as a number indicating the adjacent cell to which water flows, taking a value from 1 to 8 or some other unique identifier for each adjacent cell.

The use of a single flow direction is an incomplete representation on peaks, some ridges, and in flat areas. Peaks and ridges

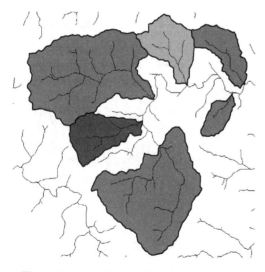

Figure 11-12: Drainage network and watersheds derived from a DEM.

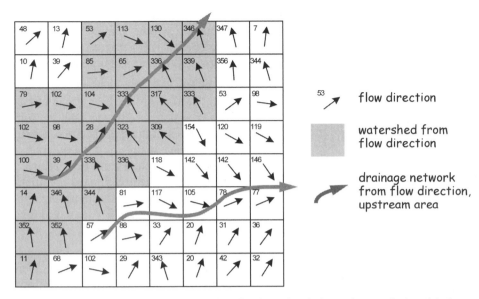

Figure 11-13: Flow direction (arrow, and number reflecting azimuth degrees), watershed, and drainage network shown for a raster grid. Elevation data are used to define the flow direction for each cell. These flow directions are then used to determine a number of important hydrologic functions.

show divergent flow, flow in multiple directions out of a cell (Figure 11-14), and convergent flow, into a cell, can occur. Most flow direction methods provide a single direction for each cell, so divergent and some convergent flows are not represented. One solution involves recording sub-cell flow directions for peak or ridge-spanning cells, but this leads to complicated raster

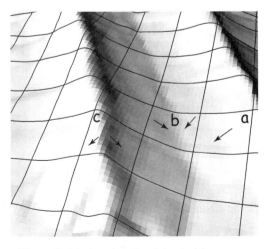

Figure 11-14: An example of simple (a), convergent (b), and divergent (c) flow.

structures and calculations. A more common option is to reduce raster cell size such that divergent and convergent cells encompass an insignificant area relative to study objectives.

Flow direction in flat areas may also cause problems. Aspect is undefined in a truly flat region, because there is zero gradient. Flow directions in these cases are either manually specified, or the aspect calculated using a larger cell size or neighborhood. The neighborhood may be successively expanded until an unambiguous flow direction is defined.

There are various ways to distribute flows among adjacent cells. The *D8* method is common, and assigns all flow to the cell with the steepest downhill gradient (Figure 11-15). The D8 is simple to understand, program, and store, but is particularly poor at representing divergent flow. This can cause large errors in derived measures such as upslope contributing area, or soil moisture indices. Output flow direction rasters derived from the D8 method may be represented with only 8 codes, allowing a simple and compact data layer.

Alternative flow direction methods may assign flow to multiple cells, and hence represent some forms of divergent flow. One common method, known as *D-infinity*, distributes flow to one downslope cell when the flow direction is exactly toward the center of the cell, and otherwise assigns a portion of the flow to each of the two adjacent cells in the downslope direction (Figure 11-15). The split is proportional to the angles between the steepest downslope direction and the respective cell centers. This reduces the main shortcoming of the D8 method, while slightly increasing complexity.

Flow accumulation area, contributing area, or *upslope area* is another important hydrologic characteristic. A flow accumulation area function is based on a flow direction surface, and typically returns a value to each cell that is equal to the area draining through that cell. In effect, the watershed area for each cell is recorded as the cell value.

Upslope areas may be calculated in several ways. Perhaps the simplest to understand is based on identifying all the local high points, or elevation maximas, choosing one to start, then summing area as one moves downslope until the boundary of the study area is reached. A second maxima is

selected and area accumulated as one moves downslope, adding to the area again accumulated from the previous cell. This process is repeated for all local high points.

Watersheds may be identified once a flow direction surface has been determined. Flow direction is followed "uphill" from a point, until a peak is reached. Each uphill cell may have many contributing cells, and the flow into each of these cells is also followed uphill. The uphill list is accumulated recursively until all cells contributing to a point have been identified, and thus the watershed is defined.

A *drainage network* is the set of cells through which surface water flows. Drainage networks are also based on the flow direction surface. Streams, creeks, and rivers occur where flow directions converge. Thus, flow direction may be used to produce a map of likely stream location, prior to field mapping a stream (Figure 11-13). A drainage network may be simply defined as any cell that has a contributing watershed larger than some locally-defined threshold area. These drainage networks are only approximations, because a drainage network defined this way does not incorporate subsurface properties such as soil texture, depth, porosity, or subsoil water movement. Nonetheless a drain-

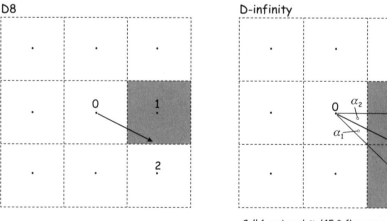

Cell 1 assigned $\alpha_1/45$ * flow area for cell 0
Cell 2 assigned $\alpha_2/45$ * flow area for cell 0

Figure 11-15: The D8 flow direction method (above left) assigns all flow to the cell center closest to the flow direction (cell 1), while the D-infinity method partitions the flow to the two cells nearest the flow direction, proportional to the flow direction angles (cells 1 and 2, above right).

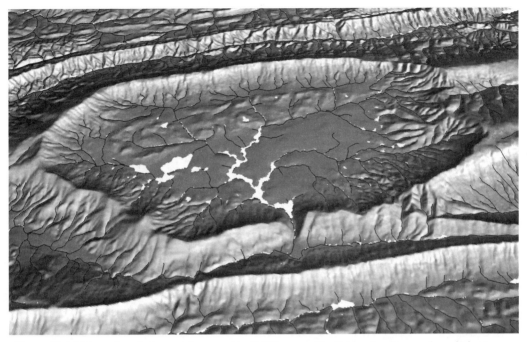

Figure 11-16: Examples of pits or sinks caused by DEM errors. The light-colored areas along drainageways show local depressions that are artifacts of data errors, and don't exist on the landscape. Drainage networks and watersheds based on unfilled DEMs will be in error, because the flow directions based on the DEMs will be inward at all pits.

age network derived from terrain data alone is often a useful first approximation. The watershed for each cell may be calculated, and the area compared to the threshold area. The cell is marked as part of the drainage network if the area surpasses the threshold.

Random errors in DEM elevation values often create spurious *pits* (also known as *sinks)*. Pits are cells that are lower than all surrounding cells, and they often cause problems when determining flow direction (Figure 11-16). Pit cells have no outward direction of steepest descent, so water drains in, but not out. Pits may exist in the real terrain surface, e.g., in *karst* regions where sinkholes occur on the surface due to collapsed subterranean caverns. In these cases watersheds may have no outlets, and streams may regularly disappear underground. Pits are also common in areas of deranged topography, for example, in the relatively flat, recently glaciated terrain of Minnesota, Wisconsin, and much of central Canada, but in most other areas, pits are data artifacts, and do not represent real geography.

Spurious pits are found in most DEMs due to small random errors. For example, DEM data collected from LiDAR mapping often have a small ground footprint, and may sample small features that are above the surrounding ground level. A laser image over a recently ploughed field may return spot heights for local mounds and furrows, incompletely harvested crops, and farm machinery. A log or dense shrubs in a steepsided ravine may be mis-identified as the ground surface, creating a barrier in the data that doesn't represent true conditions. Pits can be artifacts of interpolation methods that are used to fill in the grid values in unsampled locations. Post-processing attempts to remove these spurious readings, but they are common nonetheless.

Pits may cause problems over locally flat or valley bottom surfaces, or along drainage ways (Figure 11-16). Many hydrologic functions give erroneous results near pits, so pits often must be filled prior to further analyses. Flow direction and flow accumulation functions, described earlier, often

Figure 11-17: Watersheds and stream networks delineated from unfilled DEMs (left) often result in missing stream segments, shown at a, b, d, and e, and incomplete watersheds (above a and b, and at c).

return erroneous results, particularly when pits fall along watercourses. Drainage networks are incomplete, flow accumulation values too low, and watersheds may be improperly identified when pits are encountered (Figure 11-17).

Pit removal involves pit identification, followed by either filling the pits or downcutting adjacent downstream cells to remove the pit. A threshold is often specified below which a pit is not removed. This threshold is typically larger than common vertical errors in the data but also less than any true, "on the ground," pit depth. Alternately, or in addition, known pits may be identified prior to the filling process, and these are then left unfilled. Once spurious pits are filled, further processing to identify watersheds and drainage networks may proceed.

To review, the steps for identifying a watershed from a DEM is shown in Figure 11-18 and Figure 11-19. DEMs are filled, as needed, and then the flow direction, accumulation, stream threshold, and watershed boundaries calculated. Different fill thresholds and flow accumulation methods may

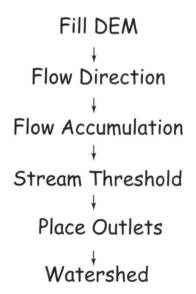

Figure 11-19: Typical steps in a watershed and stream delineation.

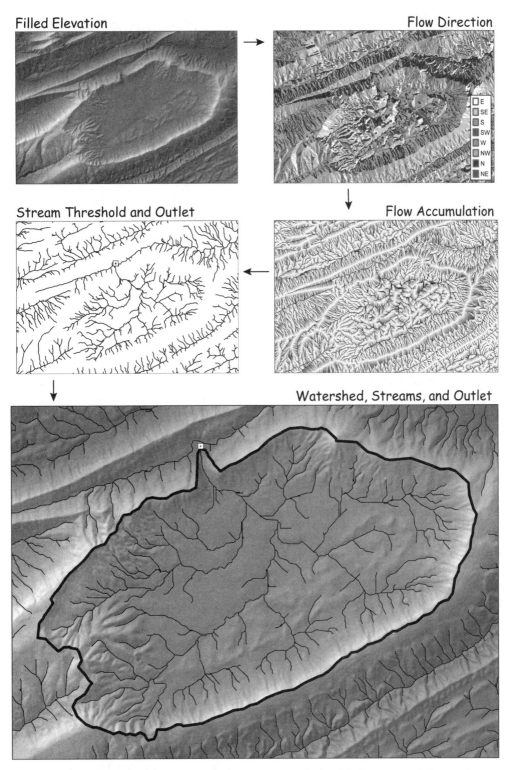

Figure 11-18: An example of the steps required to create watershed and drainage network features from a digital elevation model.

result in slightly different stream locations, and in some cases, watershed boundaries.

Several other hydrologic indices have been developed to identify locally convergent or divergent terrain positions, or terrain morphometery related to hydrography. These indices are used in many subsequent topographic and hydrologic analyses, such as predicting plant community composition or growth, erosion modeling, or estimating the rainfall required to saturate an area and predict the likelihood and intensity of flooding.

The *specific catchment area* (SCA) is defined as the total area draining to a point relative to drainage width, in raster data sets calculated as

$$SCA = AREA/C \qquad (11.5)$$

where AREA is the accumulated surface area upstream from a point, and C is the raster cell dimension. Stream power index (SPI) is defined as:

$$SPI = SCA * \tan(\beta) \qquad (11.6)$$

where b is the slope at a point, and SCA is as defined above. SPI is used to identify the potential erosion at a point, which depends both on the upstream area and hence ability to accumulate water, and the local slope, which drives the erosive energy in water flow.

Perhaps the most commonly applied wetness index is calculated by:

$$w = \ln\left(\frac{SCA}{\tan\beta}\right) \qquad (11.7)$$

where w is the wetness index at a cell, SCA is the specific catchment area, and β is the slope at the cell. This index has been shown

to effectively represent the increased soil wetness due to large upslope areas and low slopes, particularly when combined with plan curvature and profile curvature measurements. These factors sort terrain along ridge-to-stream and convex-to-concave gradients.

There are many other, more specialized topographic indices for estimating topoclimatic variables such as total solar radiation or surface air drainage. There are also indices for surface roughness, variability, or directional change. These and others are described in the references at the end of this chapter.

Viewsheds

The *viewshed* for a point is the collection of areas visible from that point. Views from any non-flat location are blocked by terrain. Elevations will hide points if the elevations are higher than the line of sight between the viewing point and target point (Figure 11-20).

Viewsheds and visibility analyses are quite important in many instances. High-voltage power lines or cell towers are often placed after careful consideration of their visibility, because most people are averse to viewing them. Communications antennas, large industrial complexes, and roads are often located at least partly based on their visibility, and viewsheds are specifically managed for many parks and scenic areas.

A viewshed is calculated based on cell-to-cell intervisibility. A line of sight is drawn between the view cell and a potentially visible target cell (Figure 11-20). The elevation of this line of sight is calculated for every intervening cell. If the slope to a target cell is less than the slope to a cell closer to the viewpoint along the line of sight, then the target cell is not visible from the viewpoint. Specialized algorithms have been developed to substantially reduce the time required to calculate viewsheds, but in concept lines of sight are drawn from each viewpoint to each cell in the digital elevation data. If there is

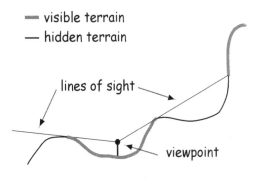

Figure 11-20: Mechanics of defining a viewshed.

no intervening terrain, the cell is classified as visible. The classification identifies areas that are visible and areas that are hidden (Figure 11-21). Viewsheds for line or area features are the accumulated viewsheds from all the cells in those features.

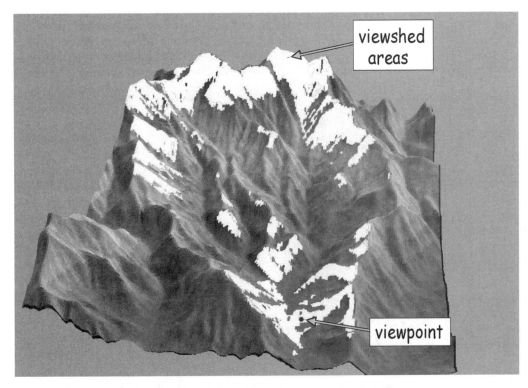

Figure 11-21: An example of a viewpoint, and corresponding viewshed.

Profile Plots

Profile plots are another common derivative of elevation data. These plots sample elevation along a linear *profile path*, and display elevation against distance in a graph (Figure 11-22). Elevation is typically plotted on the y-axis, and horizontal distance on the x-axis These profile plots are helpful in visualizing elevation change, slope, and cumulative travel distance along the specific profile path. Profile plots are common on the edges of maps, particularly maps of off-road, bicycle, or cross-country routes.

Profile plots often have some level of vertical exaggeration. Vertical exaggeration is a scaling factor applied to the elevation data when shown on the graph. For example, Figure 11-22 shows a square graph that depicts approximately 31 km across the Earth's surface. The vertical elevation axis spans approximately 2.5 km over the same dimensions on the graph. This is a vertical exaggeration of approximately 12 (from 31/2.5). Exaggeration factors are required because vertical elevation gain is generally small relative to the horizontal distance travelled.

Contour Lines

Contour lines, or topographic contours, are connected lines of uniform elevation that run at right angles to the local slope (Figure 11-23). Contour lines are a common feature on many map series, for example, they are depicted on the USGS 1:24,000 scale nationwide series, and Britain's 1:50,000 Ordinance Survey maps. The shape and density of contour lines provide detailed information on terrain height and shape in a two-dimensional map, without the need for continuous tone shading. Both color and continuous tone printing were important limitations for past cartographers. Contour lines could be easily drawn with simple drafting tools. Although continuous tone printing is much less expensive today, contours will remain common as they have entered the culture of map making and map reading.

Several rapid, efficient methods have been developed for calculating contours, either from points or from grid data (Figure 11-24). Early contour maps and DEMs were developed from height measurements at a set of points. While useful, these points did not provide clear depictions of elevation. Contour lines of fixed values were interpolated

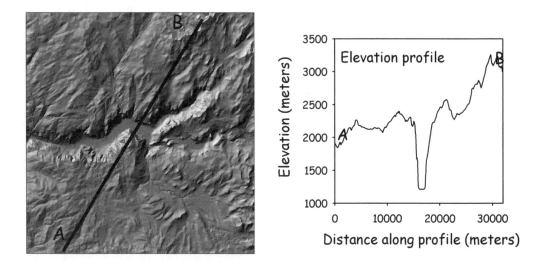

Figure 11-22: An example of a profile plot. The profile path is shown on the shaded relief image (left), with the starting point A and ending point B. The profile plot is shown on the right, with corresponding starting and ending points. The plot shows the change in elevation along the path. Note that the vertical exaggeration here is approximately 9 to 1.

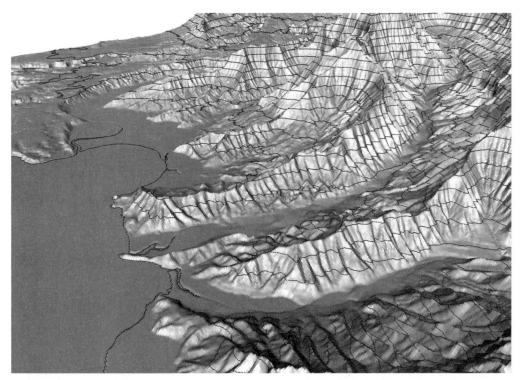

Figure 11-23: Contours represent lines of uniform elevation. This rendering shows contour lines starting at sea level, at the left, and continuing upward at a contour interval of 100 meters.

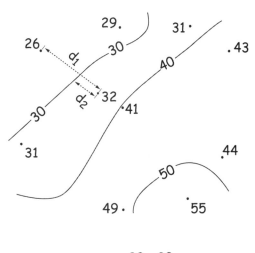

$$d_2 = d_1 \cdot \frac{32 - 30}{32 - 26}$$

Figure 11-24: Contour line locations are often estimated from point height locations, as a linear proportion of the height and distance differences between points.

linearly between nearest measurement points, as shown in Figure 11-24. Later measurement methods either identified contour lines directly from stereopairs (see Chapter 6), or derived them from mechanically or electronically-produced rasters. Raster to contour generation also typically follows a linear interpolation. For a raster, appropriate adjacent cell centers are selected, and contour values interpolated as illustrated in Figure 11-24.

Contour lines are typically created at fixed height intervals, for example every 30 meters (100 feet) from a base height (Figure 11-25). Because each line represents a fixed elevation above or below adjacent lines, the density of contour lines indicates terrain steepness. Point *A* in Figure 11-25 falls in a flat area (the foreground of the photo, at bottom), where elevation does not change much, and there are few contour lines. Steep areas are depicted by an increase in contour

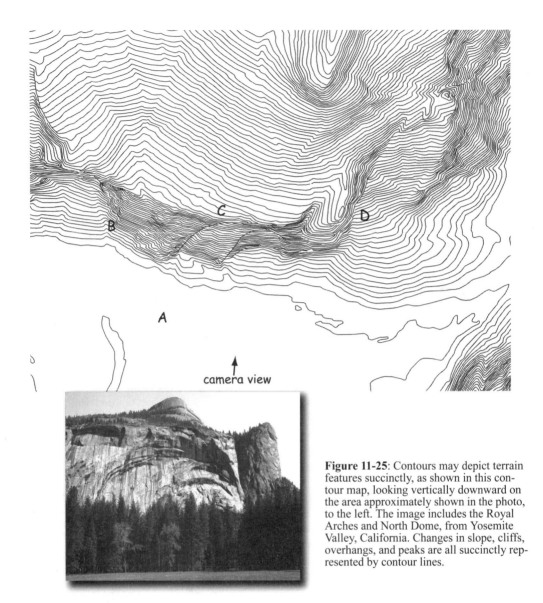

Figure 11-25: Contours may depict terrain features succinctly, as shown in this contour map, looking vertically downward on the area approximately shown in the photo, to the left. The image includes the Royal Arches and North Dome, from Yosemite Valley, California. Changes in slope, cliffs, overhangs, and peaks are all succinctly represented by contour lines.

density, as shown at point **B**, with changes in steepness depicted by changes in density (above and below point **C**). Note that contours may succinctly represent complex terrain structures, such as the curving arches in the center of the photograph, and shown below point **C**, and the overhanging cliff, to the left of point **D**.

Until recently contour lines were most commonly created by direct interpretation from airphoto stereo-pairs. Many contour maps are now developed through estimation from grid-based DEM data, or from spot height measurements by field surveying or LiDAR.

Shaded Relief Maps

A *shaded relief map*, also often referred to as a *hillshade map*, is a depiction of the brightness of terrain reflections given a terrain surface and sun location. Although shaded relief maps are rarely used in analyses, they are among the most effective ways to communicate the shape and structure of terrain features, and many maps include relief shading (Figure 11-26).

Shaded relief maps are developed from digital elevation data and models of light reflectance. An artificial sun is "positioned" at a location in the sky and light rays projected onto the surface depicted by the elevation data. Light is modeled that strikes a surface either as a direct beam, from the sun to the surface, or from background "diffuse" sunlight. Diffuse light is scattered by the atmosphere, and illuminates "shaded" areas,

although the illumination is typically much less than that from direct beam.

The brightness of a cell depends on the local incidence angle, the angle between the incoming light ray and the surface normal, shown as θ in Figure 11-27. The surface normal is defined as a line perpendicular to the local surface. Direct beam sunlight striking the surface at a right angle (θ = 0) provides the brightest return, and hence appears light. As θ increases, the angle between the direct beam and the ground surface deviates from perpendicular, and the brightness decreases. Diffuse sunlight alone provides a relatively weak return, and hence appears dark. Combinations of direct and diffuse light result in a range of gray shades, and this range depends on the terrain slope and angle relative to the sun's location. Hence, subtle variations in terrain are visible on shaded relief maps.

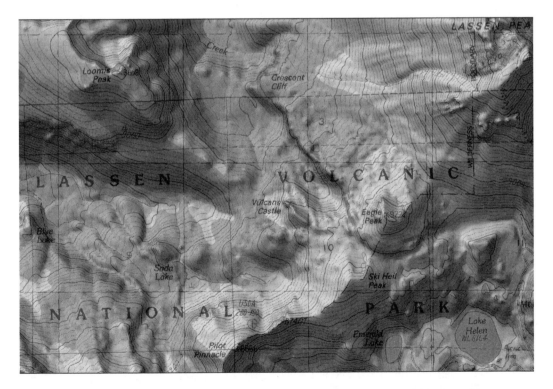

Figure 11-26: Relief shading is often added as a background "under" other mapped data to provide a sense of terrain shape and steepness. This shading provides a three-dimensional perspective for a mapped area, as demonstrated in this relief shading of a U.S. Geological Survey 1:24,000 scale quandrangle map.

Calculating a shaded relief surface requires specifying the sun's position, usually via the solar zenith angle, measured from vertical down to the sun's location, and the solar azimuth angle, measured from north clockwise to the sun's position (Figure 11-28). Local slope and surface azimuth define a surface normal direction. An angle may be defined between the solar direction and the surface normal direction, shown as θ in Figure 11-28. As noted earlier, the amount of reflected energy decreases as θ increases, and this may be shown as various shades of grey in a hillshade surface.

A shaded relief map also requires a calculation of visibility, often prior to calculating the reflectances. Visibility to the sun is determined; if a cell is visible from the sun, the slope and aspect values are used to assign the cell brightness.

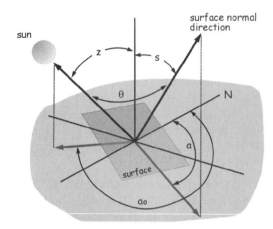

incidence angle θ is equal to:
$$\cos^{-1}[\ \cos(z)\cos(s) + \sin(z)\sin(s)\cos(a_o - a)\ ]$$
where: z is the solar zenith angle
          a_o is the solar azimuth angle
          s is the surface normal slope angle
          a is the surface normal azimuth angle

Figure 11-28: Direct beam reflectance may be calculated as shown above from the incidence angle, θ, between the incoming sun beam and the local surface "normal". The surface normal is defined by a line perpendicular to the local surface plane.

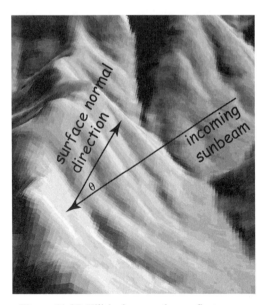

Figure 11-27: Hillshade maps show reflectance as a function of the angle, θ, between sunbeams and surface normals.

Terrain Analysis Software

Terrain analysis and DEM data management and analysis are important enough to be included in most general-purpose GIS packages, including ArcGIS, GRASS, ERDAS, Idrisi, and Manifold. While they support the most common set of terrain and hydrologic analyses, none of these packages includes the broadest range of terrain processing and analysis functions, and so specialized analyses are often pursued via other software with a specific focus on terrain data processing and analysis. These tools are often experimental, developed by universities, and designed to address a broad range of terrain analysis functions, e.g., the Terrain Analysis System (TAS) from the University of Guelph, or commercial, written by a private organization with a very narrow focus, e.g., the Watershed Modeling System (WMS) by the Scientific Software Group.

TAS, since succeeded by Whitebox GAT, contains what is likely the most comprehensive set of terrain analysis functions in a freely available package. Support is particularly strong for hydrologic surface and stream link processing and analysis, with functions for calculating various flow direction, accumulation, and watershed delineation methods typically not supported by other packages. Basic terrain modification, LiDAR data input and processing, and general raster GIS functions are also supported.

Landserf is a package with particularly strong support for terrain shape and geomorphological analysis, in addition to a strong focus on surface visualization. Multiple methods of calculating and combining first and second order terrain gradients are supported, as well as basic elevation data conversion and processing. Landserf is written in Java, and hence available across the widest range of operating systems.

ArcHydro is a set of hydrologic analysis tools written as an extension to ArcGIS. It supports a fairly complete set of hydrologic and watershed delineation functions.

There are many other packages available, including RiverTools, TAUDEM, Surfer, TAPES, and MicroDEM, which provide various specialized capabilities, and may be worth investigating for users interested in terrain and hydrologic analysis.

Summary

Terrain analyses are commonly performed within the framework of a GIS. These analyses are important because terrain governs where and how much water will accumulate on the landscape, how much sunlight a site receives, and the visibility of human activities.

Slope and aspect are two of the most used terrain variables. Both are commonly calculated via trigonometric functions applied in a moving window to a raster DEM. Several kernels have been developed to calculate changes of elevation in x and y

directions, and these component gradients are combined to calculate slope and aspect.

Profile curvature and plan curvature are two other important terrain analysis functions. These functions measure the relative convexity or concavity in the terrain, relative to the downslope direction for profile curvature and the cross-slope direction for plan curvature.

Terrain analyses are also used to develop and apply hydrologic functions and models. Watershed boundaries, flow directions, flow paths, and drainage networks may all be defined from digital elevation data.

Viewsheds are another commonly applied terrain analysis function. Intervisibility may be computed from any location on a DEM. A line of sight may be drawn from any point to any other point, and if there is no intervening terrain, then the two points are intervisible. Viewsheds are often used to analyze the visibility of landscape alterations or additions, for example, when siting new roads, powerlines, or large buildings.

Finally, relief shading is another common use of terrain data. A shaded relief map is among the most effective ways to depict terrain. Terrain shading is often derived from DEMs and depicted on maps.

Suggested Reading

Ayeni, O. O. (1982). Optimum sampling for digital terrain models, *Photogrammetric Engineering and Remote Sensing*, 48:1687-1694.

Baral, D.J., & Gupta, R.P. (1997). Integration of satellite sensor data with DEM for the study of snow cover distribution and depletion patterns. *International Journal of Remote Sensing*, 18:3889-3894.

Band, L.E. (1986). Topographic partition of watersheds with digital elevation models. *Water Resources Research*, 22:15-24.

Berry, J.K. (1987). Fundamental operations in computer-assisted mapping. *International Journal of Geographic Information Systems*, 1:119-136.

Berry, J. K. (1986). A mathematical structure for analyzing maps. *Environmental Management*, 11:317-325.

Beven, K.J., & Kirby, M.J., A physically-based variable contributing area model of basin hydrology, *Hydrological Sciences Bulletin*, 1979, 24:43-69.

Bolstad, P.V., & T. Stowe. (1994). An evaluation of DEM accuracy: elevation, slope and aspect. Photogrammetric Engineering and Remote Sensing, 60:1327–1332.

Bonham-Carter, G. F., (1996). *Geographic Information Systems for Geoscientists: Modelling with GIS*. Pergamon: Ottawa.

Burrough, P.A., & McDonnell, R.A. (1998). *Principles of Geographical Information Systems* (2nd ed.). Oxford University Press: New York.

Collins, S.H., & Moon, G.C. (1981). Algorithms for dense digital terrain models. *Photogrammetric Engineering and Remote Sensing*, 47:71-76.

DeFloriani, L. & Magillo, P. (1994). Visibility algorithms on triangulated digital terrain models. *International Journal of Geographical Information Systems*, 8:13-41.

Dozier, J. & Frew, J. (1990). Rapid calculation of terrain parameters for radiation modeling from digital elevation data. *IEEE Transactions on Geoscience and Remote Sensing*, 28:963-969.

Dubayah, R. & Rich, P.M. (1995). Topographic solar radiation models for GIS. *International Journal of Geographical Information Systems*, 9:405-419.

Fisher, P. F. (1996). Reconsideration of the viewshed function in terrain modelling. *Geographical Systems*, 3:33-58.

Flint, A.L., & Childs, S.W. (1987). Calculation of solar radiation in mountainous terrain. *Agricultural and Forest Meteorology*, 40:233-249.

Hengl, T., & Reuter, H.I., Eds. (2009). *Geomorphometry: Concepts, Software, Applications*, Elsevier: Amsterdam.

Hodgson, M.E. (1995). What cell size does the computed slope/aspect angle represent? *Photogrammetric Engineering and Remote Sensing*, 61:513-517.

Horn, B.K.(1981). Hill shading and the reflectance map. *IEEE Proceedings on Geosciences*, 69:14-47.

Hutchinson, M.F. (1989). A new procedure for gridding elevation and stream line data with automatic removal of spurious pits. *Journal of Hydrology*, 106:211-232.

Hutchinson, M.F.(1993). Development of a continent-wide DEM with applications to terrain and climate analysis, In: M. F. Goodchild et al. (Eds.), *Environmental Modeling with GIS* (pp. 392-399). Oxford University Press: New York.

Jain, M.K., Kothyari, U.C., & Ranga, R.K.G. (2004). A GIS based distributed rainfall-runoff model, *Journal of Hydrology*, 299:105-122.

Jenson, S.K. (1991). Applications of hydrologic information automatically extracted from digital elevation models. *Hydrologic Processes*, 5:31-44.

Jenson, S.K., & Domingue, J.O. (1988). Extracting topographic structure from digital elevation data for geographic information system analysis. *Photogrammetric Engineering and Remote Sensing*, 54:1593-1600.

Jones, N. L., Wright, S.G., & Maidment, D R. (1990). Watershed delineation with triangle-based terrain models. *Journal of Hydraulic Engineering*, 116:1232-1251.

Lindsay, J.B. (2005). The terrain analysis system: a tool for hydro-geomorphic applications. *Hydrological Processes*, 19:1123-1130.

Louhaichi, M., Borman, M.M., Johnson, A.L., & Johnson, D.E. (2003). Creating low-cost high-resolution digital elevation models. *Journal of Range Management*, 56:92-96.

Martz, L.W., & Garbrecht, J. (1998). The treatment of flat areas and depressions in automated drainage analysis of raster digital elevation models. *Hydrological Processes*, 12:843-856.

Maune, D.F. (Ed.). (2007). *Digital Elevation Model Technologies and Applications: The DEM User's Manual* (2nd ed.). American Society of Photogrammetry and Remote Sensing: Bethesda.

Moore, I.D., & Grayson, R.B. (1991). Terrain-based catchment partitioning and runoff prediction using vector elevation data. *Water Resources Research*, 27:1177-1191.

Moore, I.D., Turner, A, Jenson, S., & Band, L. (1993). GIS and land surface-subsurface process modelling, In M.F. Goodchild et al., (Eds.), *Environmental Modeling with GIS*, Oxford University Press: New York.

Skidmore, A. K. (1989). A comparison of techniques for calculating gradient and aspect from a gridded digital elevation model. *International Journal of Geographical Information Systems*, 3:323-334.

Strahler, A.N. (1957). Quantitative analysis of watershed geomorphology, *Transactions of the American Geophysical Union*, 8:913-920.

Tarboton, D.G., Bras, R.L., & Rodriquez-Iturbe, I. (1992). A new method for the determination of flow directions and upslope areas in grid digital elevation models. *Water Resources Research*, 33:309-319.

Tomlin, C.D. (1990). *Geographic Information Systems and Cartographic Modeling*, Prentice-Hall: Upper Saddle River.

Wilson, J., & Gallant, J. (Eds.). (2000). *Terrain Analysis: Principles and Applications*, Wiley: New York.

Wood, J. (1996). The geomorphological characterization of digital elevation models. Ph.D. thesis, University of Leicester, UK.

Wood, R., Sivapalan, M., & Robinson, J. (1997). Modeling the spatial variability of surface runoff using a topographic index. *Water Resources Research*, 33:1061-1073.

Zevenbergen, L.W., and Thorne, C.R. (1987). Quantative analysis of land surface topography, *Earth Surface Processes and Landforms*, 12:47-56.

Ziadat, F.M. (2005). Analyzing digital terrain attributes to predict soil attributes for a relatively large area. Soil Science Society of America Journal, 69:1590-1599.

Zhou, Q., & Liu, X. (2004). Analysis of errors of derived slope and aspect related to DEM data properties. *Computers & Geosciences,* 30:369-378.

Study Questions

11.1 - What are digital elevation models, and why are they used so often in spatial analyses?

11.2 - How are digital elevation data created?

11.3 - Write the definition of slope and aspect, and the mathematical formulas used to derive them from digital elevation data?

11.4 - Calculate dz/dx and dz/dy for the following 3 x 3 windows. Elevations and the cell dimension are in meters.

	windows	4-nearest neighbor	3rd-order finite difference

a)

110	113	118
112	114	119
111	117	121

↤10↦

4-nearest neighbor
dz/dx =

dz/dy=

3rd-order finite difference
dz/dx =

dz/dy =

b)

67	63	62
65	64	64
70	68	66

dz/dx =

dz/dy=

dz/dx =

dz/dy =

c)

18	23	17
21	24	19
20	22	18

dz/dx =

dz/dy=

dz/dx =

dz/dy =

11.5 - Calculate the slope and aspect for the underlined cell values, using the four nearest cell method.

712	709	707	703	704
710	<u>706</u>	704	700	702
708	705	705	<u>697</u>	700
711	<u>709</u>	705	696	694
714	712	708	703	698

11.6 - Calculate the slope and aspect for the underlined cell values above, using the third-order finite difference method.

11.7 - Which usually has a larger value, slope expressed as degrees, or slope expressed as percent?

11.8 - Plot a graph of slope in degrees (on x-axis) against slope in percent (y-axis).

11.9 - What is a contour, and how is it calculated from DEM data?

11.10 - Draw the contours for the following set of points, starting at the 960 elevation value, and placing contours every 30 units.

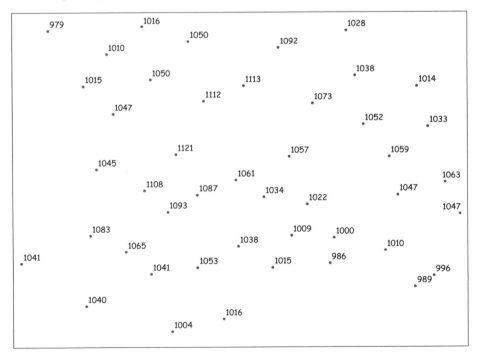

11.11 - What are plan curvature and profile curvature, and how do they differ?

11.12 - Define the watershed boundaries and possible stream locations in the digital elevation data depicted below:

162	108	67	103	56	66	130	214	153	122	70		56	91	165
169	160	101	120	95	115	119	202	212	121	55	43	101	158	261
254	224	182	158	214	142	208	249	225	129	58	121	137	253	344
323	312	204	191	214	228	300	345	195	126	58	105	188	298	381
338	334	267	307	231	194	200	190	176	114	63	141	199	277	278
438	471	405	344	228	242	194	137	103	81	111	103	198	262	195
550	550	387	304	301	330	245	257	175	110	163	204	225	206	144
669	557	502	414	451	378	396	329	180	148	242	349	293	191	148
604	639	490	442	433	425	406	264	169	169	278	401	297	241	167
742	666	536	443	340	294	265	202	221	227	339	342	260	260	245
799	630	509	438	456	414	304	344	337	322	359	377	387	375	308
767	685	608	578	457	426	318	442	371	421	430	330	275	292	226
734	789	721	578	512	421	443	512	506	503	378	315	227	213	173
668	765	826	728	579	558	489	534	513	366	330	244	266	190	170
705	767	784	785	761	675	607	545	440	275	226	202	165	104	55

11.13 - Try to identify the following: Solar zenith angle, solar azimuth angle, and solar incidence angle.

11.14 - Draw a diagram illustrating the solar incidence angle, and identify what site/terrain factors affect the solar incidence angle?

11.15 - What are viewsheds, when are they used, and how are they calculated?

11.16 - What is a shaded relief map, also known as a hillshade surface? How are the values for each cell of the hillshade surface calculated?

12 Spatial Estimation: Interpolation, Prediction, and Core Area Delineation

Introduction

Spatial prediction methods are used to estimate values at unsampled locations (Figure 12-1). An obvious question is why estimate? Why not just measure the value at all locations? Predictions are required because time and money are limiting. There is an infinite number of potential sampling locations for any continuous variable in any study area, and it is impossible to measure at all locations. While there is a finite number of discrete objects in all studies, there are usually too many to measure them all. Practical constraints usually limit samples to a subset of the possible lines, polygons, points, or raster cell locations.

Spatial prediction may be required for other reasons. Besides cost, some areas may be difficult or impossible to visit. A parcel owner may prohibit entry, or it may be unsafe to collect data in desired location. For example, it may be too dangerous to collect soil samples in part of a park because lions may eat the sampling crew, or elephants trample them.

Spatial prediction may be required due to missing or otherwise unsuitable samples. If it is difficult, expensive, or the wrong season for sampling, it may be impossible to recover lost samples. Samples may be discovered to be unreliable or suspect once the measuring crew has returned to the office.

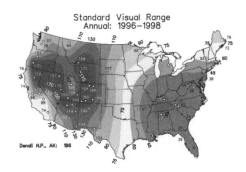

Figure 12-1: Air quality varies across space and time, and is only measured in a few locations. Spatial estimation methods are commonly used to predict air quality at unsampled locations. (courtesy U.S. NPS)

These suspect, "outlier," points should be dropped from the data set. These now missing points may be crucial to the analysis, and if so, the missing values estimated. Replacing the samples may be impossible, because too much time has passed, or because the collection sites are no longer accessible. Finally, estimates may be required when changing to a smaller cell size in a raster data set. The "sampling" frequency is set by the original raster, and values must be estimated for the new, smaller cells.

Spatial interpolation is the prediction of variables at unmeasured locations based on a sampling of the same variables at known locations. Most interpolation methods rely on the nearest points to estimate missing values, and use some measure of distance from known to unknown values. We might have measured air pollution at a set of towers across a region, but need estimates for all locations in that region. Interpolation is routinely used to estimate air and water temperature, soil moisture, elevation, ocean productivity, population density, and a host of additional variables.

Spatial prediction also involves the estimation of variables at unsampled locations, but differs from interpolation in that estimates are based at least in part on other variables, and often on a total set of measurements. We may use elevation to help estimate temperature because it is often cooler at higher locations. A map of elevations may be combined with a set of measured temperatures to estimate temperatures at unknown locations.

A *core area* is characterized by high use, density, intensity, or probability of occurrence for a variable or event. Core areas are defined from a set of samples, and are used to predict the frequency or likelihood of occurrence of an object or event. Home ranges for individual animals, concentra-tions of business activity, or centers of criminal activity are all examples of core areas. There are several methods that may be used in identifying these core areas. These methods typically draw from a set of sample points that constitute events, such as an observation of an animal, a business location, or a crime that has been committed.

Spatial prediction typically translates from lower spatial dimensions to the same or higher dimensions. This means we typically generate points or lines from point data, or areas from point, line, or area data. Prediction methods allow us to extend the information we have collected, most often to "fill in" between sampled locations, but also to improve the quality of the data we have collected.

Spatial prediction methods may also be used to translate information from a higher order to a lower order, that is, to estimate point values from data collected or aggregated to area or lines. We may have population data reported for an area, and we may wish to estimate population for a specific point within this area. This may be affected by the *modifiable areal unit problem*, a common hazard in spatial estimation methods that was described in Chapter 9.

Whatever the methods used, spatial estimation is based on a set of samples. An individual sample consists at least of the coordinates of the sample location and a measurement of the variable of interest at the sample location. We may also measure additional, "related" variables at the sample location. Coordinates should be measured to the highest accuracy and precision practical, given cost and time constraints and the intended use of the data. Sample variables should be measured using accurate, standardized, repeatable methods.

Sampling

Estimation is based on a sample of known points. The aim is to estimate the values for a variable at unknown locations based on values measured at sampled locations. Planning will improve the quality of the samples, and usually leads to a more efficient and accurate interpolation.

We control two main aspects of the sampling process. First, we may control the location of the samples. Samples must be spread across our working area. However, we may choose among different patterns in dispersing our samples. The pattern we choose will in turn affect the quality of our interpolation. A poor distribution of sample points may increase errors or may be inefficient, resulting in unnecessary costs.

Sample number is the second main aspect of the sampling process we can control. One might believe the correct number is "as many as you can afford," however this is not always the case. A law of diminishing returns may be reached, and further samples may add relatively little information for substantially increased costs. Unfortunately, in most practical applications the available funds are the main limiting factor. Most surfaces are undersampled, and additional funds and samples would almost always increase the quality of the interpolated surface. To date there have been relatively few studies or well established guidelines for determining the optimum sample number for most interpolation methods.

There are times when we control neither the distribution nor number of sampling points. This often occurs when we are working with "found" variables, for example, the distribution of illness in a population. We may identify the households where a family member has contracted a given illness. Although we can control neither the number nor the distribution of samples, we may wish to use these "samples" in an interpolation procedure.

Sampling Patterns

There are a number of commonly applied sampling patterns. A *systematic sampling pattern* is the simplest (Figure 12-2a), because samples are spaced uniformly at fixed x and y intervals. The intervals may not be the same in both directions, and the x and y axes are not required to align with the northing and easting grid directions. The sampling pattern often appears as points placed systematically along parallel lines.

Systematic sampling has an advantage over other sampling patterns by way of ease in planning and description. Field crews quickly understand how to lay out the sample pattern, and there is little subjective judgement required.

However, systematic sampling may suffer from a number of disadvantages. It is usually not the most statistically efficient sampling pattern because all areas receive the same sampling intensity. If there is more interest or variation in certain portions of the study area, this preference is not addressed by systematic sampling. The difficulty and cost of traveling to the sample points is another potential disadvantage. It may be difficult or impossible to stay on line between sampling points. Rough terrain, physical barriers, or lack of legal access may preclude sampling at prescribed locations.

In addition, systematic sampling may introduce a bias, particularly if there are patterns in the measured variable that coincide with the sampling interval. For example, there may be a regular succession of ridges and valleys associated with underlying geologic conditions. If the systematic sampling interval coincides with this pattern, there may be a bias in sample values. Another bias might result from oversampling a type of terrain, for example, a preponderance of samples might come from valley locations. This bias might in turn result in inaccurate interpolations for values on ridge locations.

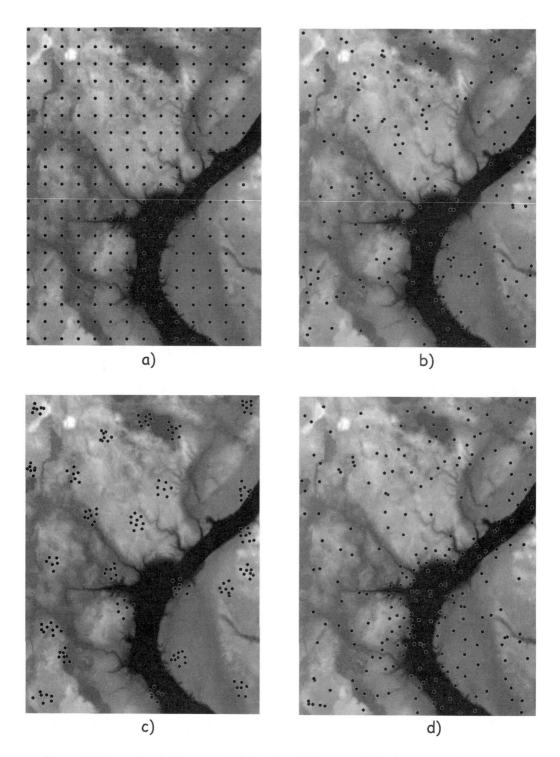

Figure 12-2: Examples of (a) systematic, (b) random, (c) cluster, and (d) adaptive sampling patterns. Sample points are shown as solid circles. Contours for the surface are shown as lines. Sampling methods differ in the distribution of sample points.

Random sampling (Figure 12-2b) may avoid some, but not all, of the problems that affect systematic sampling. Random sampling entails selecting point locations based on random numbers. Typically, both the easting and northing coordinates are chosen by independent random processes. These may be plotted on a map and/or listed, and then visited with the aid of a GPS or other positioning technology to collect the sample. The points do not have to be visited in the order in which they were selected, so in some instances travel distances between points will be quite small. On average, the distances will be no shorter than with a systematic sample, so travel costs are likely to be at least no worse than with systematic sampling.

Random samples have an advantage over systematic samples in that they are unlikely to match any pattern in the landscape. Hence, the chances for biased sampling and inaccurate predictions are less likely.

However, like systematic sampling, random sampling does nothing to distribute samples in areas of high variation. More samples than necessary may be collected in uniform areas, and fewer samples than needed may be collected in variable areas. In addition, random sampling is more complicated and hence more difficult to understand than systematic sampling. More training may be required for sampling crews when implementing random sampling. Random sampling is seldom chosen when sampling over large areas, due to these disadvantages and relatively few advantages over alternative sampling strategies.

Cluster sampling is a technique that groups samples (Figure 12-2c). Cluster centers are chosen by some random or systematic method, with a cluster of samples arranged around each center. The distances between samples within a cluster are generally much smaller than the distances between cluster centers.

Reduced travel time is the primary advantage of cluster samples. Because groups of sample points are found in relatively close proximity, the travel times within a cluster are generally quite small. A sampling crew may travel several hours to reach a cluster center, but they may spend only a few minutes between each sample within a cluster. Cluster sampling is often used in natural resource surveys that entail significant off-road travel because of the reduction in travel times.

There are several variants of cluster sampling. Cluster centers may be located randomly or systematically. Samples within a cluster may also be placed at random or systematically around the cluster center. Both approaches have merit, although it is more common to locate cluster centers at random and distribute samples within a cluster according to some systematic pattern. This is a common approach used by the U.S. Forest Service to conduct national surveys of forest production and forest conditions, and by prospectors during mineral exploration.

Adaptive sampling is a final method we will describe. Adaptive sampling is characterized by higher sampling densities where the feature of interest is more variable (Figure 12-2d). Samples are more frequent in these areas, and less frequent in "flatter", less variable areas. Adaptive sampling greatly increases sampling efficiency because small-scale variation is better sampled. Large, relatively homogenous areas are well represented by a few samples, reserving more samples for areas with higher spatial variation.

Adaptive sampling requires some method for estimating feature variation while in the field, or it requires repeat visits to the sampling areas. Sample density is adaptively increased in areas of high variation. In some instances it is quite obvious where the variation is greatest while in the field. For example, when measuring elevation it is obvious where the terrain is more variable. Sample density may be increased based on field observations of steepness.

If there is no method of identifying where the features are most variable while in the field, then sample density cannot be increased "on the spot." Samples may be returned to the office or lab for analysis and a preliminary map produced. Sample locations are then selected based on local variation. The list or map of coordinate locations may be taken to the field and used as a guide in collecting samples.

There are many other aspects of sampling designs that may be important for specific variables or problems. Sampling may occur with or without replacement. "Without replacement" for spatial samples is generally interpreted to mean that if a sample location is selected once, for example, by a random selection process, the sample location may not be used again, even if it is selected a second time in the random process. This is typically not a problem when point locations are sampled. However, the issue of replacement may arise when linear or area features are used as sampling units. For example, we may be sampling county populations, and wish to estimate populations for all counties. If a county is selected for our sample set, sampling without replacement means we may not select it again, and unintentionally enter it twice into our set. These and other important considerations in sampling design and execution are treated in the substantial literature on general and spatial sampling.

Spatial Interpolation Methods

There are many different interpolation methods. While methods vary, all combine the sampled values and positions to estimate values at unmeasured locations. Often, mathematical functions are used that incorporate both distance from interpolation points to sample points and the values at those sample points. Methods differ in the mathematical functions used to weight each observation, and the number of observations used. Some interpolators use every observation when estimating values at unsampled locations, while other interpolators use a subset of samples, for example, the three points nearest an unmeasured location.

Different interpolation methods will often produce different results, even when using the same input data. This is due to the differences in the mathematical functions and number of data points used when estimating values for the unsampled locations. Each method may have unique characteristics, and the overall accuracy of an interpolation will often depend on the method and samples used.

Accuracy is often judged by the difference between the measured and interpolated values at a number of withheld sample points. These withheld points are not used when performing the interpolation, but are checked against the interpolated surface. However, no single interpolation method has been shown to be more accurate than all others for every application. Each individual or organization should test several sampling regimes and interpolation methods before adopting an interpolation method.

Interpolation methods may produce one or more of a number of different output types. Interpolation is usually used to estimate values for a raster data layer. Other methods produce *contour lines*, more generally known as *isolines*, lines of uniform value. Contour lines are less frequently produced by interpolation methods, but are a common way of depicting a continuous surface. At least one interpolation method defines polygon boundaries.

Interpolation to a raster surface requires estimates of unmeasured values at the center of each grid cell. Raster layer boundaries and cell dimensions are specified, in turn defining the location of each raster cell. The interpolation method uses the sample values to estimate values for each cell in the raster data layer.

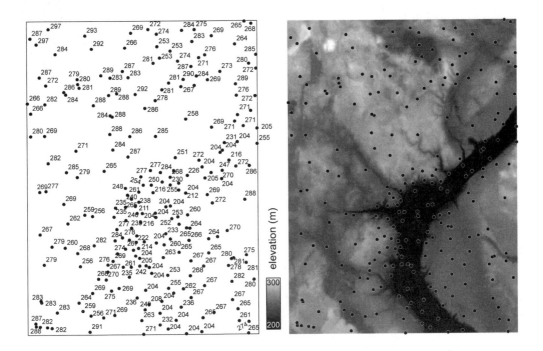

Figure 12-3: Contour lines (left) and sample points (right) for an elevation surface. These points will be used later in this chapter to demonstrate interpolation methods.

Contour line generation is more involved, and may require an iterative process. The location of a set of known levels is determined. For example, a set of points where temperature is exactly 10 degree centigrade may be estimated. These points are connected to form a line. Other sets of points may be estimated for 12, 14, 16, and other temperatures. Points for any given temperature are joined with the restrictions that lines of different temperatures do not cross. These contour lines depict the changes in temperature (or any other plotted variable) across the landscape.

We will describe the most common interpolation methods and apply them all to a single data set to facilitate comparisons. The left side of Figure 12-3 shows sample points and elevation values for an elevation surface, and the right side shows the set of sample points on a shaded elevation map of the same area. These sample points will be used to demonstrate the application of various interpolation and spatial prediction methods in the following sections of this chapter. Estimated elevation surfaces for each method will be shown.

Note that the comparisons and figures are only to illustrate different interpolation methods. They are not to establish the relative merit or accuracy of the various methods. The best interpolation method for any given application depends on the characteristics of the variable to be estimated, the cost of sampling, available resources, and the accuracy requirements of the users. The relative performance of interpolators has been determined for some variables in some locations. Enough comparisons have been conducted to establish that no interpolation method is superior for all data sets or conditions. Future studies may perhaps establish the likely best method for each application.

Nearest Neighbor Interpolation

Nearest neighbor interpolation, also known as Thiessen polygon interpolation, assigns a value for any unsampled location that is equal to the value found at the nearest sample location. This is conceptually the simplest interpolation method, in the sense that the mathematical function used is the simple equality function, and only one point, the nearest point, is used to assign a value to an unknown location.

The nearest neighbor interpolator defines a set of polygons, known as Thiessen polygons. All locations within a given Thiessen polygon have an identical value for the Z variable (in this and other chapters, Z will be used to denote the value of a variable of interest at an x and y sample location). Z may be elevation, size, production in pounds per acre, or any other variable we may measure at a point. Thiessen polygons define a region around each sampled point that have

a value equal to the value at the sampled point. The transition between polygon edges is abrupt, that is, the variable jumps from one value to the next across the Thiessen polygon boundary.

Figure 12-4 shows an elevation surface, sample points, and Thiessen polygons based on the sample points. The left side of Figure 12-4 shows the shaded surface from a 7.5 minute DEM in central Minnesota. The land is gently undulating, with an abrupt decrease in elevation near the right edge. A river runs from the lower edge up to the right edge of the DEM. Cliffs are indicated by abrupt changes in tone along the river's edge. Sample points are indicated by filled circles, and sampling is sparse. In most cases there are more than 500 meters between samples, however sampling is denser in portions of the map near the river. Thiessen polygons are shown on the right of Figure 12-4, with values assigned based on the nearest point.

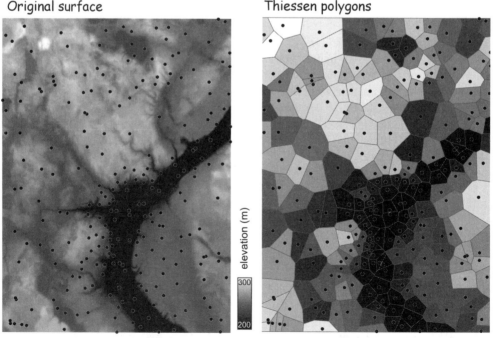

Original surface   Thiessen polygons

Figure 12-4: Original data and sample points for a surface (left) and sample points and Thiessen polygons (right).

Polygons are smaller where sampling density is highest.

Thiessen polygons provide an *exact interpolator*. This means the interpolated surface equals the sampled values at each sample point. The value for each sample location is preserved, so there is no difference between the true and interpolated values at the sample points. Exact interpolators have this admirable quality, but often are not the best interpolators at unsampled points; for example, the Thiessen polygon method is usually in error at non-sampled locations, often moreso than other inexact interpolators.

An independent error measure is required if we are to obtain a good estimate of the interpolation accuracy. Accuracy estimates may be obtained with a withheld samples technique, where the surface is fit to the data withholding one data point. The error is estimated at the withheld point as the observed minus the interpolated values. The sample is replaced, a new sample selected and withheld, and the surface fit and error again determined. This is repeated for each data point. A less efficient testing method

entails collecting an independent set of sample points that are withheld from the interpolation process. Their measured values are then compared to the interpolated values, and the mean error, maximum error, and perhaps other error statistics identified.

Fixed Radius – Local Averaging

Fixed radius interpolation is more complex than nearest neighbor interpolation, but less complex than most other interpolation methods. In a fixed radius interpolation, a raster grid is specified in a region of interest. Cell values are estimated based on the average of nearby samples.

The samples used to calculate a cell value depend on a *search radius*. The search radius defines the size of a circle that is centered on each cell. Any sample points found inside the circle are used to interpolate the value for that cell (Figure 12-5). Points that fall within the circle are averaged, those outside the circle ignored.

Figure 12-6 shows a perspective view of fixed radius sampling. Note that there is a sample data layer, shown at the top of Figure

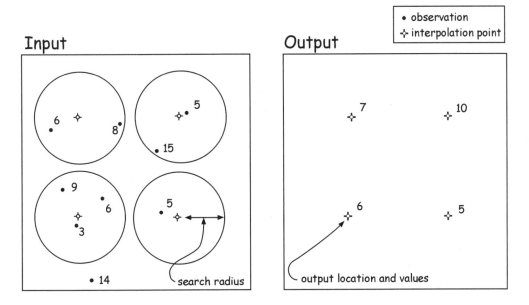

Figure 12-5: A diagram and example of a fixed radius interpolation. Values within each sampling circle are averaged to estimate an output value for the corresponding point.

12-6, vertically aligned with the interpolated surface. This surface is a raster data layer with interpolated values in each raster cell. A fixed-radius circle is centered over a raster cell. The average is calculated for all samples contained within the sample circle, and this average is placed in the appropriate output raster cell. The process is repeated for each raster cell in the surface. The fixed-radius circles are shown corresponding to three raster cells, containing three, zero, and one sample points, respectively. Circles may contain no points, in which case a zero or no data value must be placed in the raster cell. The radius for the circle is typically much larger than the raster cell's width. This means circles overlap for adjoining cells, which causes neighboring cell values to be similar.

Fixed radius interpolators are often used to create a moving average of the samples. Each sample point may correspond to a sum or density value, and points may be averaged spatially to interpolate the values for nearby cells. For example, an agronomist may measure corn production in bushels per acre at several points in a county. These may be converted to a raster surface by averaging the measurements of bushels per acre that fall within the circles centered on each cell.

The fixed radius interpolator tends to smooth the sample data (Figure 12-7). Large or small values sampled at a given point are maintained when only that one sample point falls within a search radius for a cell. Large or small values are brought toward the overall sample mean when they occur within a search radius with other sample points.

The search radius affects the values and shape of the interpolated surface. Too small a search radius results in many empty cells. The cell value is typically set to a value that indicates no data are present. Too large a search radius may smooth the data too much. In the extreme case, a search radius may be defined that includes all sample points for all cells. This would result in a single interpo-

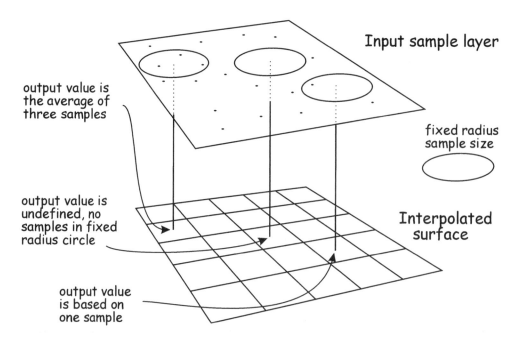

Figure 12-6: A perspective diagram of fixed radius sampling. A circle is centered on each raster cell location. Samples within the circle contribute to the value assigned to each corresponding raster cell. (adapted from Mitchell, 1999)

Original surface   Fixed radius

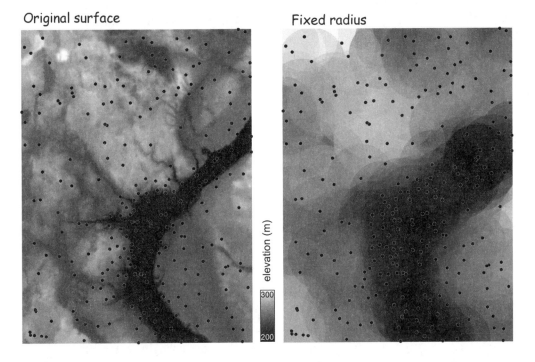

Figure 12-7: Original data and sample points (left), and a fixed radius interpolation (right).

lated value repeated for all cells. Some intermediate search radius should be chosen. If many cell values change with a small change in the search radius, this may be an indication that the samples are too sparse, and more sample points may be required

Fixed radius interpolators are not exact interpolators because they may average several points in the vicinity of a sample, and so they are unlikely to place the measured value at sample points in the interpolated surface.

Inverse Distance Weighted Interpolation

The inverse distance weighted (IDW) interpolator estimates the value at unknown points using the sampled values and distance to nearby known points. The weight of each sample point is an inverse proportion to the distance, thus the name. The farther away the point, the less weight the point has in

helping define the value at an unsampled location. Values are estimated by:

$$Z_j = \frac{\sum_i \dfrac{Z_i}{d^n_{ij}}}{\sum_i \dfrac{1}{d^n_{ij}}} \qquad (12.1)$$

where Z_j is the estimated value for the unknown point at location j, d_{ij} is the distance from known point i to unknown point j, Z_i is the value for the known point i, and n is a user-defined exponent. Any number of points greater than two may be used, up to all points in the sample. Typically some fixed number of close points is used, for example, the three nearest sampled points will be used to estimate values at unknown locations. Note that n controls how fast a

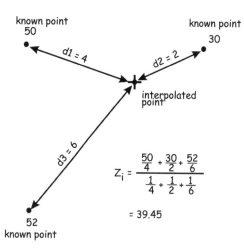

Figure 12-8: An example calculation for a linear inverse distance weighted interpolator. The values at each known point (50, 52, 30) are averaged, with weights based on the distances (d1, d2, d3) from the interpolated point.

point's influence wanes with distance. The larger the n, the smaller the weight ($1/d^n_{ij}$), so the less influence a point has on the estimate of the unknown point.

Figure 12-8 illustrates an IDW interpolation calculation. The three nearest samples are used. Each measured sample value is weighted by the inverse of the distance from the unknown, interpolated. These weighted values are added. The result is divided by the sum of the weights to "scale" the weights to the measurement units. This produces an estimate for the unsampled location.

IDW is an exact interpolator. Interpolated values are equal to the sampled values at each sampled point. As a d_{ij} becomes very small (sample points near the interpolated location), the $1/d_{ij}$ becomes very large. The contribution from the nearby sample point dwarfs the contributions from all other points. The values $1/d_{ij}$ are very near zero for all i values except the one very near the sampled point, so the values at all other points are effectively multiplied by zero in the numerator of the IDW equation. The sum in the denominator reduces to the weight $1/d_{ij}$. The weights on the top and the bottom of the IDW equation become more similar, and

the fraction approaches 1. Thus, at a sampled point the IDW interpolation formula reduces to:

$$\frac{\dfrac{Z_i}{d^n_{ij}}}{\dfrac{1}{d^n_{ij}}} \qquad (12.2)$$

By simple division this is reduced mathematically to Z_i, the value measured at the sampling location.

Inverse distance weighting results in smooth interpolated surfaces. The values do not jump discontinuously at edges, as occurs with Thiessen polygons. While IDW is easily and widely applied, care must be taken in evaluating the particular n and i selected. The effects of changing n and i should be tested in an oversampled case, where adequate withheld points can be compared to interpolated points. The IDW, and all other interpolators, should be applied only after the user is convinced the method provides estimates with sufficient accuracy. In the case of IDW, this may mean testing the interpolator over a range of n and i values, and selecting the combination that most often gives acceptable results.

The size of the user-defined exponent, n, affects the shape of the interpolated surface (Figure 12-9). When a larger n is specified, the closer points become more influential. Higher exponents result in surfaces with higher peaks, lower valleys, and steeper gradients near the sampled points. Contours become much more concentrated near sample points when $n = 2$ (Figure 12-9c) than when $n = 1$ (Figure 12-9b). These changing shades reflect steeper gradients near the known data points.

The number of points, i, used to estimate an interpolated point, j, also affects the estimated surface (Figure 12-9). Both bottom panels (c and d) use an exponent of $n = 2$ but differ in the number of nearby sample points used in each interpolation. Panel c, on the

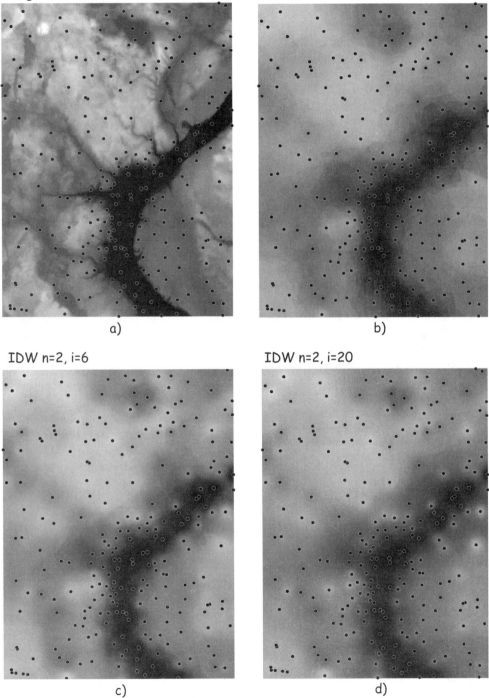

Original surface

IDW n=1, i=6

a)

b)

IDW n=2, i=6

IDW n=2, i=20

c)

d)

Figure 12-9: The effects of exponent order, n, and sample size, i, on the interpolated values for an inverse distance weighted interpolator. Local influences are stronger as the exponent increases and the number of sample points decreases.

bottom left, interpolates using the nearest six points, while panel **d**, on the bottom right of Figure 12-9, uses the twenty nearest points. These panels show complex patterns and no distinct trends. In some regions the gradients are steeper, in others shallower. A larger number of sample points tends to result in a smoother interpolated surface. However the effects depend also on the distribution and values of known data points, and the impacts of changing sample number are difficult to generalize.

Splines

A *spline* is a flexible ruler that was commonly used by draftsmen to create smooth curves through a set of points. Mathematical spline functions, also referred to as splines, are used to interpolate along a smooth curve. These functions serve the same purpose as the flexible ruler in that they force a smooth line to pass through a desired set of points. Spline functions are more flexible because they may be used for lines or surfaces and they may be estimated and changed rapidly. The sample points are analogous to the drafted points in that these points serve as the "guides" through which the spline passes.

Spline functions are constructed from a set of joined polynomial functions. Line functions will be described here, but the principles also apply to surface splines. Polynomial functions are fit to short segments. An exact or a least-squares method may be used to fit the lines through the points found in the segment. For example, a third-order polynomial may be fit to a line segment (Figure 12-10). A different third-order polynomial will be fit to the next line segment. These polynomials are by their nature smooth curves within a given segment.

Splines are typically first, second, or third order, corresponding to the maximum exponent in the equation used to fit each segment (for example, 2nd order for x^2, 3rd order for x^3 or x^2y). Segments meet at *knots*,

or *join points*. These join points may fall on a sampled point, or they may fall between sampled points.

Constraints are set on spline functions to ensure the entire line remains smooth at the join points. These constraints are incorporated into the mathematical form of the function for each segment. These constraints require that the slope of the lines and the change in slope of the lines be equal across segments on either side of the join point. Typically, spline functions give exact interpolation (the splines pass through the sample points) and show a smooth transition (Figure 12-11). Strictly enforcing exact interpolation can sometimes lead to artifacts at the knots or between points. Large loops or deviations may occur. The spline functions are often modified to allow some error in the fit, particularly when fitting surfaces rather than lines. This usually removes the artifacts of spline fits, while maintaining the smooth and continuous interpolated lines or surfaces.

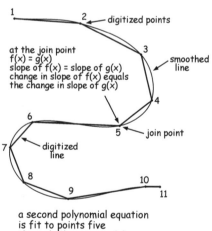

Figure 12-10: Diagram of a two-dimensional (line) spline. Segments are fit to portions of a line. Segments are constrained to join smoothly at knots, where they meet.

Original surface   Spline interpolation

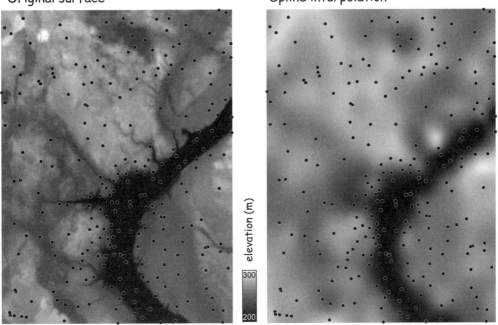

elevation (m)

300

200

Figure 12-11: Samples and an original elevation surface (left) and a spline-fit surface (right).

Spatial Prediction

Spatial predictions are based on mathematical models, often built via a statistical process. These statistically-based models use coordinate location and measured or observed "independent" variables to predict values for important but unknown "dependent" variables. Spatial prediction is different than interpolation because it uses a statistical fitting process rather than a set algorithm, and because spatial prediction uses independent variables as well as coordinate locations to estimate unknown variables. We admit that our distinction between spatial prediction and interpolation is artificial, but it is useful in organizing our discussion, and highlights an important distinction between our data-driven models and our fixed interpolation methods.

Spatial predictions are a special case of general predictive modeling, the focus of applied statistics. There is a rich, broad, and deep literature on statistical modeling. There is a substantial and growing body of work devoted to spatial statistics in general, and spatial predictive modeling in particular. We will only scratch the surface of this field; the reader is referred to the introductory spatial statistics texts listed at the end of this chapter. Isaaks and Srivistava (1989), Fotheringham et al. (2000), and O'Sullivan and Unwin (2003) provide good introductions to the issues and methods of statistically-based spatial modeling, and cite the literature that may carry one further.

While there are many types of spatial prediction, our discussions will be restricted to predicting continuous spatial variables. These variables are conceptualized as *spatial fields* that may be observed at any point across a spatial domain, are measured on an interval/ratio scale, and typically have values that vary in concert -- that is, they are spatially correlated. This is in contrast to discrete objects, such as point, line, or polygon features. While the occurrence and properties of discrete features may be predicted using spatial models, this is done less commonly, and most discrete object predictions use a different set of tools that will not be discussed here.

Spatial prediction may be considered more general than interpolation. Both methods are used to estimate values of a target variable at unknown locations. Interpolation methods use only the measured target variable and sample coordinates to estimate the target variables at unknown locations.

Spatial prediction may also be viewed as different from spatial interpolation because prediction methods often address the presence of *spatial autocorrelation*. Spatial autocorrelation is the tendency of nearby objects to vary in concert. High values are found near high values, and low values found near low values. Explanations of this common condition often refer to the observation of Waldo Tobler, that "...everything in the universe is related to everything else, but closer things are more related." However, the nature of the correlation may change from one variable to the next, or it may change in space. Correlations may be strong in one region but poor in another, or positive in one area and negative in another. We have an opportunity to improve our predictions if we study the correlation and incorporate this knowledge of the correlation structure into the predictive process.

In addition to spatial autocorrelation, variables may show *cross-correlation* between different variables: the tendency for two variables to change in concert. This means two different variables at the same or nearby locations may be high or low together. Spatial prediction methods may incorporate neither, one, or both of these types of correlation in predictions

Surfaces with low and high spatial autocorrelation and with strong cross-correlation are shown in (Figure 12-12). Figure 12-12a shows two surfaces, **Layer 1**, with a high autocorrelation, and **Layer 2**, with a low autocorrelation. Scatter diagrams of sample pairs separated by a uniform, short lag distance are shown to the right of each corresponding layer. Higher autocorrelation, as shown in **Layer 1**, indicates that points near each other are alike. A sample from a surface with high autocorrelation provides substantial information about the values at nearby locations (Figure 12-12a, top). Samples from a surface with low autocorrelation do not provide much information at values in the vicinity of the sample point (Figure 12-12a, bottom).

Two cross-correlated raster layers are shown in (Figure 12-12b). Positive cross-correlated layers have values that tend to both be high in some regions and both be low in other regions. Many features are positively correlated, such as, housing prices and average income, or donut shop density and number of security guards. Negative cross-correlation occurs when variables change in the opposite sense -- areas with high values for one variable are low for the other, for example, low temperatures at higher elevations.

a) spatial autocorrelation

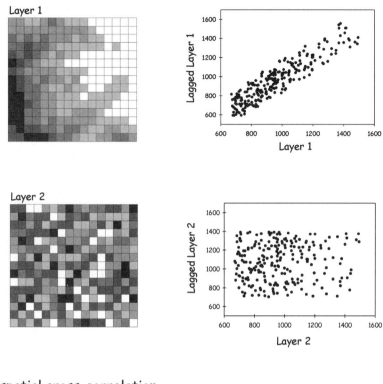

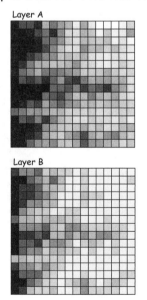

b) spatial cross-correlation

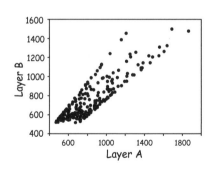

Figure 12-12: Part **a** shows spatially autocorrelated (**Layer** 1) and spatially uncorrelated (**Layer** 1) data layers. Plots of sample pairs with a lag distance **h** = 1 show similar values for the autocorrelated **Layer** 1, and unrelated values for uncorrelated **Layer** 2. Panel **b** shows two cross-correlated layers. **Layer** A has higher values on average than **Layer** B, but the two layers vary in concert. Both reach high and low values in similar areas.

The Moran's I statistic is an established measure of spatial correlation:

(12.3)

$$I = \frac{n \sum\limits_{i=1}^{n} \sum\limits_{j=1}^{n} w_{ij}(Z_i - \bar{Z})(Z_j - \bar{Z})}{\sum\limits_{i=1}^{n}(Z_i - \bar{Z})^2 \sum\limits_{i=1}^{n} \sum\limits_{j=1}^{n} w_{ij}}$$

Where Z_i and Z_j are the variable values at points i and j, respectively, $\bar{Z}$ is the variable mean, and w_{ij} are weight values that take the value 1 if Z_i and Z_j are adjacent, and 0 if the values are not. An example of Moran's I calculations is shown in Figure 12-13.

Moran's I values approach a value of +1 in areas of positive spatial correlation, meaning large values tend to be clumped together, and small values clumped together. Values near zero occur in areas of low spatial correlation, and indicate knowing a value at a location does not provide much information about values in adjacent locations -- they are just as likely to be different or similar to the observed value. Moran's I approaches -1 when values are anti-correlated -- a large value is more likely to be next to small values than next to other large values.

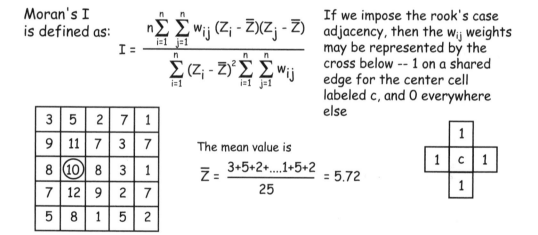

Moran's I is defined as:

$$I = \frac{n \sum\limits_{i=1}^{n} \sum\limits_{j=1}^{n} w_{ij} (Z_i - \bar{Z})(Z_j - \bar{Z})}{\sum\limits_{i=1}^{n}(Z_i - \bar{Z})^2 \sum\limits_{i=1}^{n} \sum\limits_{j=1}^{n} w_{ij}}$$

If we impose the rook's case adjacency, then the w_{ij} weights may be represented by the cross below -- 1 on a shared edge for the center cell labeled c, and 0 everywhere else

3	5	2	7	1
9	11	7	3	7
8	(10)	8	3	1
7	12	9	2	7
5	8	1	5	2

The mean value is

$$\bar{Z} = \frac{3+5+2+....1+5+2}{25} = 5.72$$

The calculation of Moran's I for the circled cell is then:

$$= \frac{4[1 \cdot (10 - 5.72)(11 - 5.72) + 1 \cdot (10 - 5.72)(8 - 5.72) + 1 \cdot (10 - 5.72)(12 - 5.72) + 1 \cdot (10 - 5.72)(8 - 5.72)]}{(10 - 5.72)^2 \cdot 4}$$

$$= \frac{68.9}{73.3} = 0.94$$

Figure 12-13: Moran's I is a measure of the correlation among nearby observations. This example shows the calculation of Moran's I for a cell in a raster data set. The formula is a weighted combination of the value at a location and neighboring locations. Positive Moran's I values indicate positive spatial correlation, and negative values indicate anti-correlation.

Moran's I may be calculated for both raster and vector data sets. Moran's I values for raster data sets are usually based on the cells immediately adjacent to the focal (center) cell. Weights are typically 1 for cells that share an edge with the focal cell, and 0 otherwise (Figure 12-13). The neighborhood may be "rook's case" and include only cells that share a full edge, as in Figure 12-13, or they may be "king's case" and include all of the eight neighbors in the calculation.

Moran's I may also be calculated for vector polygon data sets, but this requires a definition of adjacency. Polygon adjacency may be defined in various ways, most commonly when two polygons share an edge. However, two polygons may be considered adjacent if they share a node or vertex, or if their edges are separated by less than a specified distance. As with raster calculations of Moran's I, weights are typically set to one for adjacent polygon pairs and set to zero for separated polygon pairs.

There are many other local indices of spatial autocorrelation, or LISA, including Geary's C, or the Gi of Getis and Ord (1992), and they perform in a manner similar to Moran's I (Anselin, 1995). The indices vary slightly in how they estimate the correlation and in the specific calculations of relatedness and separation. These and a number of additional topics are quite well covered by Anselin (1995), Fotheringham et al., (2000), and O'Sullivan and Unwin (2003), listed in suggested reading at the end of this chapter.

Spatial Regression

Spatial regression and other statistically-based prediction methods typically use observations of dependent variables, other "independent" variables, and sample coordinates to develop the prediction equations. For example, we might wish to estimate temperature across a region, and have access to a network of temperature stations. We may use the interpolation techniques described in the previous section to estimate temperature, and these techniques use only the station coordinates and the temperature values measured at those coordinates. However, we may also note that there are strong relationships between temperature and elevation (generally cooler at higher elevations). We may combine measurements from our temperature stations with data on elevation, latitude, and longitude in a statistical process to develop a mathematical relationship between temperature, latitude, longitude, and elevation. We would then use this relationship to estimate the temperature field in the region.

Spatial predictions are often described mathematically by a general function, such as:

$$Z_i = f(x_i, y_i, \alpha_i, \beta_j) \qquad (12.4)$$

where Z_i is the estimated output value, at the coordinates X_i, Y_i at point i, α_i are variables measured at point i, and β_j are variables measured at other locations.

Trend Surface and Simple Spatial Regression

Trend surface prediction is a type of spatial regression that involves fitting a statistical model, or trend surface, through the measured points. The surface is typically a polynomial in the X and Y coordinate system. For example, a second-order polynomial model would be:

$$Z = a_0 + a_1x + a_2y + a_3x^2 + a_4y^2 + a_5xy \quad (12.5)$$

where Z is the value at any point x and y, and each a_p is a coefficient estimated in a regression model. Least-squares methods, described in most introductory statistical textbooks, are used to estimate the best set of a_p values. The a_p values are chosen to minimize the average difference between the measured Z values and the prediction surface.

There must be at least one more sample point than the number of estimated a_p coefficients due to statistical constraints. This does not pose a practical problem for most applications, because the best polynomial models are often second or third order and have fewer than 10 coefficients. More than 10 sample points are typically collected to ensure adequate coverage of a study region.

Trend surfaces are not exact predictors in that the surface typically does not pass through the measured points. This means that there is an error at each sample location, measured as the difference between the interpolated surface and the measurement. Trend surfaces are often among the most accurate methods when fitting smoothly varying surfaces, such as mean daily temperature over large areas Trend surfaces typically do not have the "bulls-eye" artifact that

may appear with excessive local influence in many interpolators.

Trend surface methods typically do not perform well when there is a highly convoluted surface (Figure 12-14). Elevation in a dissected mountain region, precipitation from a single summer thunderstorm, or population density in a mixed-use neighborhood are often poorly estimated with a trend surface. Even high-order polynomials may not be sufficiently flexible to fit these complex, convoluted surfaces.

Trend surfaces may be extended to include independent variables that provide some help in predicting the variable of interest:

$$Z = a_0 + a_1x + a_2y + a_3Q + a_4W \qquad (12.6)$$

where x and y are the coordinate locations, and Q and W are independent variables

Original surface

Trend surface

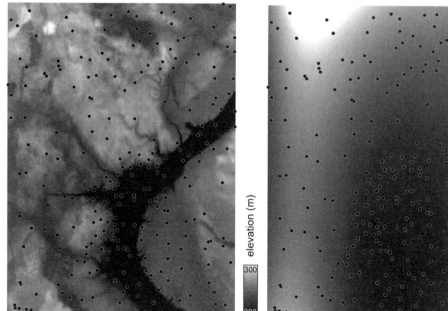

Figure 12-14: Sample points and original surface (left), and sample points and a third-order trend surface fit to the sample points (right).

measured at the point (x, y), and Z is the dependent variable to be predicted at the point (x,y). The a_p values are coefficients for the predictive equation, usually estimated through a least-squares statistical process. The value Z may be predicted at any location we have values for x, y, Q, and W.

Kriging and Co-Kriging

Kriging is a statistically-based estimator of spatial variables. It differs from the trend-surface approach in that predictions are based on regionalized variable theory, which includes three main components. The first component in the kriging model is the spatial trend, an increase or decrease in a variable that depends on direction; for example, temperature may decrease toward the northwest.

The second component describes the local spatial autocorrelation, that is, the tendency for points near each other to have similar values. Kriging is unique and powerful because we use the observed change in spatial autocorrelation with distance to estimate values at our unknown locations.

The third component in the prediction is random, stochastic variation. These three components are combined in a mathematical model to develop an estimation function. The function is then applied to the measured data to estimate values across the study area.

Much like IDW interpolators, weights in kriging are used with measured sample variables to estimate values at unknown locations. With kriging, the weights are chosen in a statistically optimal fashion, given a specific kriging model and assumptions about the trend, autocorrelation, and stochastic variation in the predicted variable.

Kriging methods are the centerpiece of geostatistics. Geostatistics was initially developed in the early 1900s by D.G. Krige and Georges Matheron for use in mining. Prospecting samples may be expensive to obtain or process, and accurate ore density or occurrence predictions may be quite difficult, but valuable. Krige and Matheron

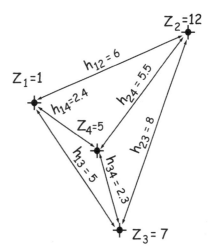

Figure 12-15: Lag distances, used in calculating semivariances for kriging.

sought to develop estimators that would incorporate trends, autocorrelation, and stochastic variation and also provide some estimate of the local variance in the predicted variable.

Kriging uses the concept of a *lag distance,* often symbolized by the letter h. Consider the sample set shown in Figure 12-15. Each value for the variable Z is shown plotted over a region. Individual points may be listed as Z_1, Z_2, Z_3, etc., to Z_k, when there are k sample points. The lag distance for a pair of points is the distance between them, and by convention is denoted by h. The lag distance is calculated from the x and y coordinate values for the sample points, based on the Pythagorean formula. In our example in Figure 12-15, the lag (horizontal) distance between the locations of sample points Z_1 and Z_2 is approximately 6 units. The difference in values measured at those points, $Z_1 - Z_2$, is equal to 11. Each pair of sample points is separated by a distance, and also has a difference in the values measured at the points. For example, Z_1 is 2.4 units from Z_4, and Z_1 is 5 units from Z_3. Each pair has a given difference in the Z values, for example, Z_1 minus Z_4 is 4. Every possible set of pairs Z_a, Z_b, defines a distance h_{ab}, and is different by the amount $Z_a - Z_b$. The distance h_{ab} is known as the lag distance between points

+/- lag tolerance

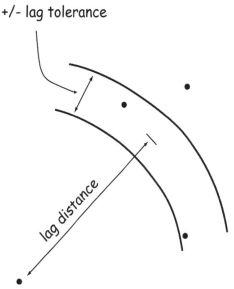

lag distance

Figure 12-16: A lag tolerance defines a range for grouping samples. The grouping aids estimation of spatial covariance.

a and b, and in general there is a subset of points in a sample set that are a given lag distance apart.

Lag distances often are applied with an associated *lag tolerance*. A lag tolerance defines a small range that is "close enough" to a lag distance (Figure 12-16). A lag tolerance is required because the individual lag distances typically are not repeated in the sample data. Most or all distances between sample points are unique, so there is little or no replication with which to calculate the variability at each lag. Some distances may be quite similar, but distances usually will differ in the smallest decimal places. A lag tolerance circumvents this problem.

The lag tolerance defines when distances are similar enough to be grouped in spatial covariance calculations. For example, we may wish to calculate the semivariance for points that are 112 meters apart. If we are inflexible and only use point pairs that are exactly 112 meters apart (within the precision of our measurement system), we may have only a few, or perhaps even no points

that meet this strict criterion. Our lag tolerance expands the number of points available for a spatial variance calculation. By allowing a tolerance, distances that are plus or minus that tolerance from the given lag distance can be used to calculate a spatial variability. For example, we might set a tolerance for h of 10 units. Any pair of points between 102 and 122 units apart are used to calculate an index of spatial covariance for the lag distance h = 112.

Geostatistical prediction uses the key concept of a *semivariance* to represent spatial covariance. A semivariance is the variance based on nearby samples, and it is defined mathematically as:

$$\gamma(h) = 1/2n * \sum (Z_a - Z_b)^2 \qquad (12.7)$$

Z_a is the variable measured at one point, Z_b is the variable measured at another point h distance away, and n is the number of pairs that are approximately the distance h apart.

The semivariance at a given lag distance is a measure of spatial autocorrelation at that distance. Note that when nearby points (small h) are similar, the difference ($Z_a - Z_b$) is small, and so the semivariance is small. High spatial autocorrelation means points near each other have similar Z values.

The semivariance may be calculated for any h. For example, when $h=1$, the semivariance, $\gamma(h)$ may be equal to 0.3; when $h=2$, then $\gamma(h)$ may be 0.5; when $h=3$, then $\gamma(h)$ may be 0.8. We may calculate a semivariance provided there are sufficient point pairs that are h distance apart to give a good estimate.

We may plot the semivariance over a range of lag distances (Figure 12-17), and this plot is known as a *variogram* or *semivariogram*. A variogram summarizes the spatial autocorrelation of a variable. Note that the semivariance is usually small at small lag distances, and increases to a plateau as the lag distance h increases. This is the typical form of a variogram. The *nugget*

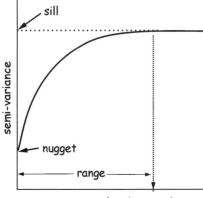

Figure 12-17: An idealized variogram, with the nugget, sill, and range identified.

is the initial semivariance when the autocorrelation typically is highest. The nugget is shown at the left of the diagram in Figure 12-17, the semivariance at a lag distance of zero. This is the intercept of the variogram. The *sill* is the point at which the variogram levels off. This is the "background" variance, and may be thought of as the inherent variation when there is little autocorrelation. The *range* is the lag distance at which the sill is reached. The nugget, sill, and range will differ among spatial variables.

A set of sample points is used to estimate the shape of the variogram. First, a set of lag distances h_1, h_2, h_3, etc., are defined; each distance signifies a given lag distance, plus or minus the lag tolerance. The semivariance is then calculated for each lag distance. An example is shown in Figure 12-18. Remember, each of these points is calculated from Equation 12.7 for a given lag distance apart. A line may then be fit through the set of semivariance points, and the variogram estimated. This line is sometimes called the variogram model.

Spatial prediction is among the most important applications of the variogram model (Figure 12-19). There are many variations and types of kriging models, but the simplest and most commonly applied rely on the variogram to estimate "optimal" weights

for prediction. These weights are used to estimate values at unknown locations by:

$$Q = \sum_{j=1}^{n} w_j \cdot v \qquad (12.8)$$

where Q is the estimated value at an unmeasured point, w_j are weights for each sample j, and v is the known value at sample point j.

Weights are optimal in the sense that they minimize the error in a prediction, and they are unbiased, given a specific data set and model. The calculation of optimal weights requires some rather involved mathematics, beyond our present scope, but is described in great detail in references listed at the end of this chapter.

Estimating each w_j involves a constrained minimization process. A set of equations may be written that expresses the errors as the differences between our measured values and the predicted values by a function of a set of unknown weights. This set of equations is solved under the constraints that the weights sum to zero and the error variance is minimized. The solution involves calculating the expected values of covariances between points according to a variogram model, for example, by fitting a smooth relationship between the observed semivariogram points, as shown in Figure 12-18. The covariances are a function of the specific lag distances observed in the sam-

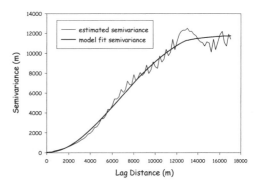

Figure 12-18: A variogram, a plot of calculated and fit semivariance vs. lag distance.

Original surface   Kriged surface

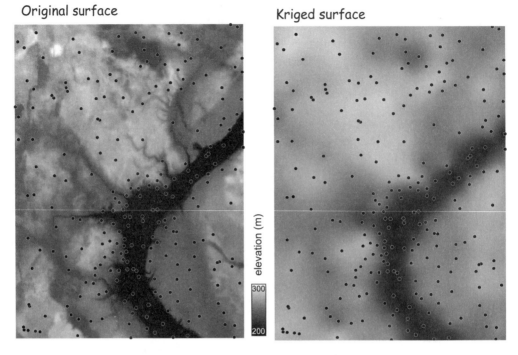

Figure 12-19: Contours and sample points from an elevation surface (left) and contours from a kriged surface, interpolated using the sample points as input (right).

ple, and are used to solve for the optimal set of weights in equation 12.8.

As stated earlier, kriging is similar to IDW interpolation in that a weighted average is calculated. However kriging uses the minimum variance method to calculate the weights, rather than applying some arbitrary and perhaps more imprecise weighting scheme as with IDW.

Co-kriging is an extension of kriging that includes the measurement of a separate, correlated variable at the sample locations in addition to the variable of interest. There may be an easily measured secondary variable that is to some extent related to the primary variable, but that is easier or less expensive to measure. In many analyses, temperature might be a primary variable and elevation a secondary variable. Co-kriging exploits the covariance between the primary and secondary variables to improve our estimate of the primary variable. Co-kriging is similar in motivation to kriging in that a set

of optimal weights is estimated, but with co-kriging there are weights for both the primary and secondary variables.

Spatial prediction with kriging, co-kriging, and other geostatistical methods can be a complex and nuanced process. There is a wide range of models that may be fit, and these in part depend on the characteristics of the data. Different data characteristics indicate particular modeling methods or model forms, for example, if there are trends in the data, or directional differences in the variance. These considerations are beyond the scope of our present discussion, and the interested reader is referred to more complete treatments, such as Isaaks and Srivastava (1989), and Burrough and McDonnell (1998) listed under suggested reading at the end of this chapter.

Interpolation Accuracy

The preceding pages describe tools for estimating important variables at unknown locations. This raises important questions: How accurately does an interpolated surface match the real surface, and which interpolation method should be used? Although the answer is often the most accurate model, some times it is not, as when there are nearly as accurate models that require fewer samples, are more robust, have fewer parameters, or are easier to create. Model assessment is a well-developed field, and will not be thoroughly reviewed here, but a few main concepts and techniques will be introduced.

We will focus here on determining the accuracy of a given estimated surface. When repeatedly interpolating the same type of problem, we would also seek to identify a consistently superior interpolation method, but this is a complex process we leave to more advanced texts; however, it depends on the accuracy measures we introduce here.

Accuracy is measured at assessment points, locations where we know both the true value and the estimated values for a variable. We often describe a sample set with n points, with estimated or interpolated values at any ith point denoted by P_i, and the true or observed value at the point denoted by O_i. Each assessment point provides an error estimate:

$$e_i = P_i - O_i \qquad (12.9)$$

There are several metrics that are commonly used to characterize aggregate error, perhaps chief among them the root mean squared error:

$$RMSE = \left[\left(\sum_{i=1}^{n} e_i^2 \right) / n \right]^{0.5} \qquad (12.10)$$

Error values are squared to remove the sign effect, and then the square root taken on the sum to return to the measured unit scale,

instead of a squared unit scale. Predictions either above or below the observed values are generally considered to be considered equally bad, and the error is averaged over all samples. However, squaring the errors magnifies the influence of outliers, extremely large positive or negative errors, so some argue that this is an overly pessimistic estimate of error, or at least when there are large outliers.

The *mean absolute error* is an alternative error metric, less often used but less sensitive to outliers than the RMSE. The MAE is defined as:

$$MAE = \left| \left(\sum_{i=1}^{n} |e_i| \right) / n \right| \qquad (12.11)$$

It substitutes the absolute value operation for the squaring/square-root operations and so is less sensitive to outliers, but otherwise is quite similar to the RMSE.

Another accuracy metric is the *mean bias error*:

$$MBE = \left| \left(\sum_{i=1}^{n} e_i \right) / n \right| \qquad (12.12)$$

MBE measures the average bias in the predictions, the amount by which, on average, an estimated surface over or under-predicts the true values. MBE conveys useful information overall, but provides little information on the magnitude of individual errors and should be used in conjunction with RMSE, or preferably, MAE.

Overall measures of agreement between an estimated and true surface have been proposed, including Willmott's index of agreement:

$$d = 1 - \frac{\sum\limits_{i=1}^{n} (P_i - O_i)^2}{\sum\limits_{i=1} (|P_i'| + |O_i'|)^2} \qquad (12.13)$$

with

$$P_i' = P_i - \bar{O} \qquad (12.14)$$

and

$$O_i' = O_i - \bar{O} \qquad (12.15)$$

Primary citations of these and other accuracy metrics are provided at the end of this chapter, and in the considerable literature on interpolation spatial estimation.

Assessing the accuracy of an interpolated surface requires we collect both observed and predicted values at a set of points. In an ideal assessment these would be independent of the samples we use to estimate the surface, but this is rarely possible. Samples are often expensive, difficult to collect, and sparse, and most interpolated surfaces would benefit from additional sampling. If each new sample can materially improve our interpolation, we are hard-pressed to hold them in reserve for an accuracy assessment. We are tempted to use most or all of our samples while interpolating, and leave few or none for an accuracy assessment.

Exact interpolators are particularly vexing. As you might recall, Thiessen polygons, inverse-distance weighted, and some spline interpolators have zero error at all sample points by definition, because they are formulated to exactly return the observed values at the fitted points. One might think that we must hold a set of points in reserve in order to get a true estimate of the interpolator accuracy.

There is a technique, known variously as leave-one-out, bootstrapping, or cross-validation, which addresses both the undersampling and robust accuracy estimation requirements. Bootstrapping involves fitting the surface as many times as there are sample points, each time withholding one of the points. We fit the surface the first time, withholding the first point. We can then subtract the withheld measured value (O_1) to the interpolated value (P_1), and obtain one estimate of the error. We then repeat this process for the rest of the sample points. For n samples, we fit the surface n times. We can then compare the withheld point's true value, O_i, to the fit value P_i, giving us n error values, e_i. We can then apply equations 12.10 through 12.15 to characterize the accuracy of our fits.

Unfortunately, most surface interpolation tools in GIS do not support bootstrapping or similar validation methods. This is unfortunate, doubly so because they typically provide only the RMSE value, and then only without bootstrapping or other cross-validation, and for inexact interpolators. RMSE estimated from the fit points without bootstrapping may well give an optimistic estimate of accuracy, particularly when sample size is small, and should not be accepted in lieu of a bootstrap or similar validation. This should be remedied, if not within the specific GIS software used in fitting, then by exporting the sample data to a statistically-oriented surface fitting system, e.g., the open source statistical package R.

Core Area Mapping

Core area mapping is another common and useful spatial analysis tool. A *core area* is a primary area of influence or activity for an organism, object, or resource of interest. Detectives may wish to map a series of burglaries to uncover clustering or patterns in occurrence. Wildlife managers may wish to map the home range of an endangered organism, or a business owner the home locations of her customers.

Core area mapping typically involves identifying area features (polygons, raster areas, or volumes) from a set of point or line observations. Individual burglaries, for example, are recorded as point locations, perhaps tagged to the address or building where they occurred. These points may be used to define a polygon by one of several core area mapping techniques. In this way, the core area is a higher dimensional spatial object (area) that is defined from a set of lower dimensional objects (points or lines). This core area represents some central or important region where features occur frequently, in this example, burglaries. Additional resources may be focused on this core area, such as increased patrols or surviellence.

Core area mapping is commonly used. Perhaps the most frequent applications to date have involved analysis of patterns of human activity, particularly crime occurrence, as illustrated in the previous example. In addition, plant and animal species densities are often analyzed and summarized using these methods, particularly when the organism is highly valued or endangered. Resource managers record organism occurrences in the field, perhaps using GPS or other spatial positioning technologies. These observations may be combined and abundance patterns are analyzed after a sufficient number of observations have been gathered. Core areas may be identified, and key habitat conditions or requirements inferred. These may guide management actions such as the protection of areas with a high concentration of endangered species and the enhancement of other areas by adding key habitat requirements.

Mean Center and Mean Circle

The *mean center* and associated *mean circle* are perhaps the simplest and most obvious measure of a central location and a core area. The mean center is simply the average x and y coordinates of the sample points. Each sample point has an associated pair of coordinates. These may be summed and the average calculated, and this mean point identified as the center of the core area.

Mean circles may be associated with the mean center to define a core area (Figure 12-20). The mean circles are defined by a radius measured from the mean center. The mean circle radius is commonly the distance to the farthest sample point, the average distance from the mean center to the set of sample points, or some other statistical measures based on the variance of the distance to sample points. These distances may be calculated easily from the sample x and y coordinates, first by calculating the mean, and then by applying the general formulas to

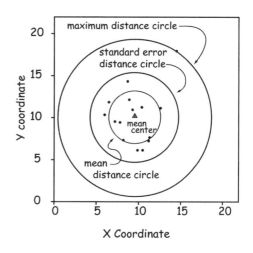

Figure 12-20: An example of a mean center and corresponding mean circles for a set of sample points.

calculate distance from sample points to the mean center. The largest distance, average distance, or the standard deviation of the distance from points to the center then may be determined.

Mean circles have the advantages of simplicity and ease of construction, but they assume a uniformly circular shape for the core area. Some measures of mean center may be biased by extreme points, for example the maximum distance circle in Figure 12-20. Note that the outlier near x = 15 and y = 17.5 results in a large maximum distance circle. This circle contains substantial area with no points nearby, and it is probably an overestimation of the core area. It is not clear that the mean distance or standard error distance circles are better at defining a core area. The core areas defined by these measures may be appropriate for some applications, but they are often too small in others. Some multiple of the mean distance or standard error may be chosen based on statistical assumptions, or past experience. For example, if we assume the samples follow a random normal distribution, then a core area defined by a circle approximately 1.8 times

the standard error distance should contain 68% of the data. Previous experience may help, for example one might know that in a particular region 90% or more of a wolf pack core area is within 10.8 kilometers of a mean center.

In many cases circular core areas are sub-optimal because many variables are known to exhibit non-regular shapes and a circular core area is identified when using the mean center / mean circle methods. While mean circle methods are often used in exploratory data analyses, other methods have been developed to more effectively identify irregularly shaped core areas.

Convex Hulls

Convex hulls, also known as minimum convex polygons, are perhaps the simplest way to identify core areas with irregular shapes. A *convex hull* is the smallest polygon created by edges (lines) that completely enclose a set of points and for which all exterior angles between edges are greater than or equal to 180 degrees (Figure 12-21). An exterior angle is measured from

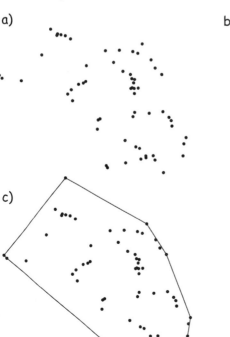

Figure 12-21: A set of points (a) may be enclosed by connecting the "outermost" points in the set. This "hull" defines a polygon. Hulls may be characterized as concave (b), when some exterior angles are less than 180 degrees, or convex (c), when all exterior angles are greater than or equal to 180 degrees.

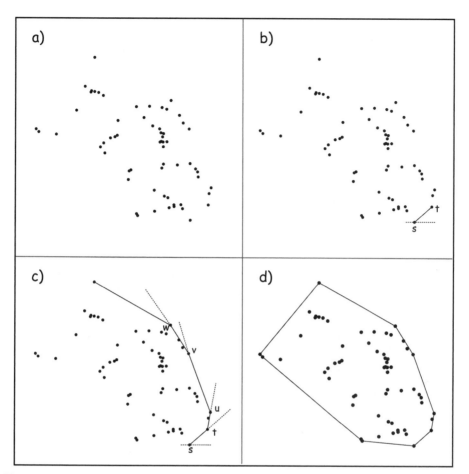

Figure 12-22: The convex hull for a set of points (a) may be calculated from a sweep algorithm. Starting with an extreme point such as **s** (b), successive minimum deflection angles are selected (c) until the starting point is reached (d).

one edge or side to another through the region "outside" of a polygon. Squares, triangles, and regular pentagons are all examples of convex hulls, while stars and crosses are examples of non-convex hulls. While these geometric figures have regular shapes, most convex hulls derived from sampled points will not.

Convex hulls are often considered a natural bounding area for a set of points. This assertion is accepted by most analysts when there are no outlying data points, far removed from the rest. When outliers are present, the convex hull will often be unreasonably large.

Convex hulls are widely used because they are simple to develop and interpret, and there is little or no subjectivity in their application. The shape of the convex polygon is determined solely by the arrangement of the sample points, and not by controlling parameters that must be specified by the human applying the method. They represent the irregular shapes common to most sampling.

A convex hull may be easily created with a "sweep" algorithm applied to a set of sample points (Figure 12-22a). These are the locations of the events of interest, for example observations of a rare animal or crime locations. An extreme point is identified from the set, usually the sample with the largest or smallest x or y coordinate (point **s** in Figure 12-22b). The angles of deflection from the current point to all other points are calculated, and the smallest positive clockwise or counter-clockwise angle and corresponding point are identified (point t in Figure 12-22b). This point is the next in the convex hull, and becomes the starting point for the next calculation. This process is repeated until the starting point is reached (Figure 12-22c and d).

Convex hulls are often considered a natural bounding area for a set of points. However, convex hulls often ignore clustering in the data. A dense cluster of points in an interior region does not influence the shape of the core area. We lose much of the information on density or frequency of occurrence in the interior region of the bounding polygon. Algorithms defining optimum concave polygons have been developed, generally fitting convex hulls to successive subsets of bounding points, and discarding outlier points, or areas defined by the outlying points. One such method is described next.

Characteristic Hull Polygons

An alternative to convex hulls has been developed, known as characteristic hull polygons (CHP). A Delaunay triangulation is created among the sampled points, the same method described in Chapter 2 when developing a triangulated irregular network. A set of minimum spanning triangles is created, and this set of triangles winnowed to remove a largest area or longest perimeter subset. Figure 12-23a shows a set of sample points and the resulting convex hull, while Figure 12-23b shows the Delaunay triangulation for the same set of points. In this example, the top five percent of polygons with the longest perimeter have been discarded, and the remaining shaded to represent a core area. This reduces the influence of distant points and allows for "holes" embedded within a core area, two advantages over convex hulls. One must choose whether to use area, perimeter, or some other metric of size, so the resultant CHP size and shape depend on the threshold value chosen, e.g., 5 vs. 10

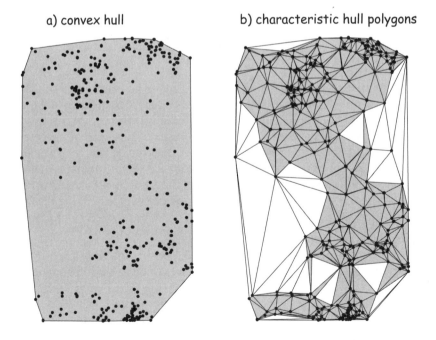

a) convex hull   b) characteristic hull polygons

Figure 12-23: An example of a core area defined by a convex hull (a), and characteristic hull polygons (CHP, in b). The shaded area is offered as a core area, with a smaller, higher sample density, and arguably more accurate area identified by the CHP.

largest polygons; however, the method is easy to apply and arguably provides a better estimate of core areas when compared to a convex hull, particularly when outliers are frequent.

Kernel Mapping

Kernel mapping uses a set of sample locations to estimate a continuous density surface. Kernel mapping is widely applied because it is mathematically flexible, relatively easy to implement, may be robust to outliers, readily incorporates clustered samples, can represent irregular shaped core areas, and are often statistically-based.

Kernel mapping is based on a density distribution that is assumed for each sample point. These density distributions are placed over the sample plane, one for each observation point, and vertically added to determine the composite density from the sample. This composite density may be used to identify a core area, selecting the densest areas first.

An example will help illustrate these ideas and the process of kernel mapping. Consider samples to detect the density of defects in a tile floor. Each tile is 0.5 inches across. We count the number of defects per tile, beginning at one edge of the tile mosaic. We will show the samples collected along a line, but the process and principles are similar in two dimensions.

Figure 12-24 shows the results of a sampling along a line segment. One defect, or fault, is found on a tile located 2 inches from the start, and it is represented by a rectangle two units tall. Each fault represents a density of two units, because each tile is 0.5 inches across -- hence $1/0.5 = 2$ faults/inch. We observe two faults at 2.5 inches (four faults/inch), one at 3.5 inches, and additional observations until our last fault observed at 12.5 inches. Note that the density is in the form of rectangles that are "stacked" two units high for each fault observed for a tile.

Note two things about the density estimates. First, we assume a characteristic shape for the density derived from each observation. In Figure 12-24 we assume the shape of a rectangle for each observa-

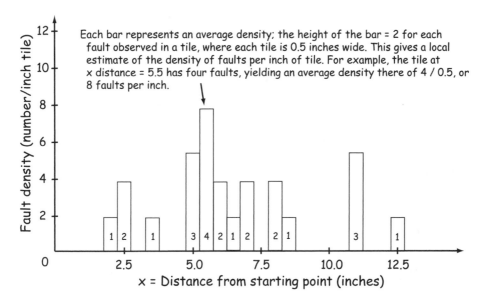

Figure 12-24: Kernel mapping is based on the concept of a distribution of observations in space. Each observation contributes information about our cumulative distribution and the observations are combined to approximate our cumulative distribution.

tion, with a uniform density across the tile. This may not be true, but in our case we are using a discrete sample, and so it is a valid approximation. In general this shape is called a density distribution. This characteristic shape (density distribution) is then placed for each observed sample, for example, note that there is a rectangle placed for each defect we observe at a distance from the starting point in Figure 12-24.

Second, note that the shapes (density distributions) are added vertically in areas where they coincide, as shown in Figure 12-24. In our example, rectangles are stacked. With more complex, mathematical-defined density distributions, the values are added over each point. The cumulative density distribution is the sum of the distributions associated with each sample.

Density distributions typically are not squares or other geometric figures, but rather symmetric shapes such as parabolas, Gaussian curves, or otherwise smoothly varying surfaces about a center point. These shapes can be mathematically defined and specified for each sample point. For example, a general Gaussian curve for one variable has the form:

$$f(x) = \frac{1}{\sqrt{2\pi\sigma^2}} \cdot e^{\frac{-(x-x_o)^2}{2\sigma^2}} \quad (12.16)$$

where x_o is the sample location and σ is a scaling constant. This is a symmetric function about x_o, meaning the function is a mirror image reflected across both sides of the point x_o (Figure 12-25). Note that the density distribution in the figure reaches a peak at x_o, and the area under the curve is typically equal to one. The formula is often written with $\sigma^2 = 1$, or may be scaled by dividing by a value h, so that it appears as:

$$f(x)_h = \frac{1}{h\sqrt{2\pi}} \cdot e^{\frac{-(x-x_o)^2}{2h^2}} \quad (12.17)$$

The value h is also known as a *bandwidth parameter*, and is described in the next few paragraphs.

Many functional forms can be used to represent the kernel densities. Typically these shapes are "bumps", in that they smoothly rise to a peak and then descend to near zero. Different forms of the kernel density function may have characteristic shapes -- how fast they reach the peak, how pointed the peak becomes, and how quickly they return to values near zero at points more distant from the peak.

The composite density distribution is created by "stacking" our individual density distributions from the set of observations (Figure 12-26a and b). Density distributions may be plotted for each observation, for example, two of many observations are shown in Figure 12-26a. Each point yields a smooth "bump" centered on the observation. When all observed points are plotted, there is a commensurately large number of small, overlapping bumps, as shown by the thin lines in Figure 12-26b. These may then be summed vertically to create the cumulative density distribution, shown by the thick line in Figure 12-26b.

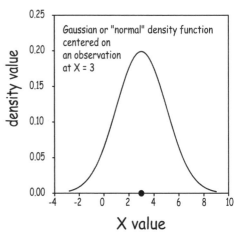

Figure 12-25: A density distribution is assumed, and plotted for each observation. Here an observation at $x = 3$ (plotted dot) generates a bell-shaped curve centered on the observation.

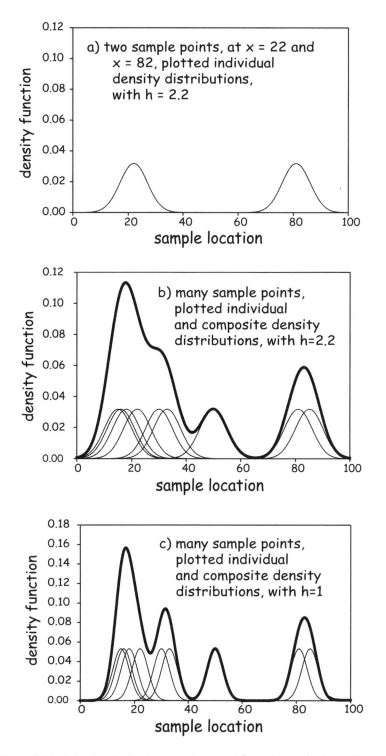

Figure 12-26: Individual density distributions may be plotted for each sample observation, as shown for two points in a one-dimensional sampling (**a**, above). Distributions for the entire sample set are plotted, and added to create a combined estimate of the density distribution (**b**). We usually choose a bandwidth parameter, h, that controls the shape of the individual and hence composite density distributions. Narrower bandwidths result in higher and narrower peaks (**c**).

We often choose bandwidth parameters, symbolized by h, that define the "spread" or width of the individual density distributions (Figure 12-26c and Figure 12-27). Perhaps the simplest way to understand the bandwidth is to think of the binning interval in our example in Figure 12-24. There, we counted tile defects for each 0.5 inch tile, and plotted a rectangle corresponding to the resultant fault density. Our bandwidth was set at 0.5 inches. We just as well could use a bandwidth of 1 inch, counting the number of defects per two tiles (one inch), along our sampling line. This would give a related, but slightly different estimate of the density distribution of defects along our sampling line. As shown in the right panel of Figure 12-27, the estimated density distribution for the first seven inches of our sampled line is less "peaked" or "spikey." Although the same sample set is used to estimate both density distributions, each observation is spread across a broader interval when we choose a larger bandwidth.

We observe the same change in the peakedness when we change the bandwidth for continuous density distributions, such as the Gaussian distribution shown in Figure 12-26 and Equation 12.9. A sample is plotted using a Gaussian density function for each observation and a bandwidth of h = 2.2 in Figure 12-26b. Reducing the bandwidth to h = 1 narrows the shape for each individual sample and results in higher, narrower, more peaked shapes in the cumulative distribution shown in Figure 12-26c.

Kernel mapping is generally a three-step process, as may be surmised from the preceding discussion. First we collect samples and the concomitant coordinate locations. Second we choose a kernel density function. Finally, we choose a bandwidth, h, apply the kernel density distribution, and sum across each sample area to achieve our composite estimate of density.

Mathematically, this process is summarized by the equation:

$$\lambda(x, y) = \frac{1}{nh^2} \cdot \sum_{i=1}^{n} \frac{K(x_i, y_i)}{h} \quad (12.18)$$

where $\lambda(x,y)$ is the composite density distribution, n is the number of samples, h is the

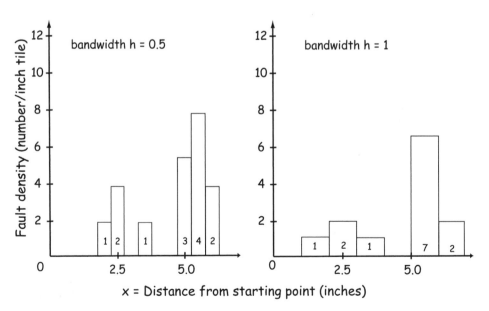

Figure 12-27: We choose both the general form of the density distribution, and a bandwidth parameter that affects the shape of the distribution. Here, the rectangular shape of the fault density becomes broader and shorter as the bandwidth changes from 0. 5 to 1.

bandwidth, and $K(x_i,y_i)$ is the individual density distribution applied at each sample point i.

An example of kernel density estimation is shown in Figure 12-28. An individual sample point is shown in Figure 12-28a, with a single peak corresponding to the Gaussian density distribution chosen. A more complex shape with multiple peaks occurs when all sample points are plotted, as shown in Figure 12-28b. Individual distributions are summed vertically, resulting in an undulating, complex surface. This surface represents the density or probability of occurrence of the underlying variable, for example, the density of defects in a tile floor, the crime density mapped across a city, or the utilization density for a wolf pack in their home range.

While the choice of bandwidth affects our results, there is no uniformly best method to select the appropriate value for h. One commonly applied method is to plot several density surfaces, one for each of a given h value, and select the h that most closely approximates your perception of the best density. Insights in the distribution and behavior of the data set are often gained by analyzing densities across a range of bandwidth values.

Formulae exist for optimum bandwidths under various conditions. One method for calculating optimum bandwidth has been proposed by Fotheringham *et al.* (2000) for a Gaussian kernel:

$$h_{opt} = \left[\frac{2}{3n}\right]^{\frac{1}{4}} \sigma \qquad (12.19)$$

Where h_{opt} is the optimum bandwidth, n is the number of samples, and σ is the standard deviation parameter, unknown, but estimated from the sample.

Numerous formulas exist defining optimum bandwidths, and one is faced with a rather different choice of selecting the correct optimum. The motivations behind various optimum bandwidths are described in the books by Silverman (1986) and by Fotheringham *et al.* (2000), listed at the end of this chapter.

Core area delineation is a primary use for estimated density distributions. As expected, the identified core areas are dependant on the selected bandwidth. Figure 12-29 shows vertical views of two-dimensional density distributions for optimum (a),

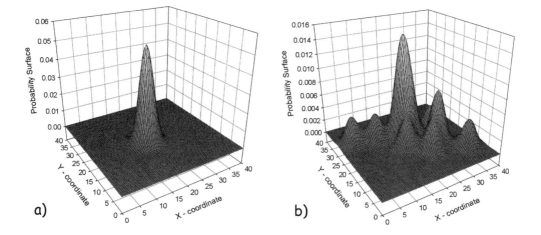

Figure 12-28: Kernels may be used to map density distributions across a two-dimensional surface. Density distributions from individual samples (a) are summed to create a composite estimate of the density surface (b).

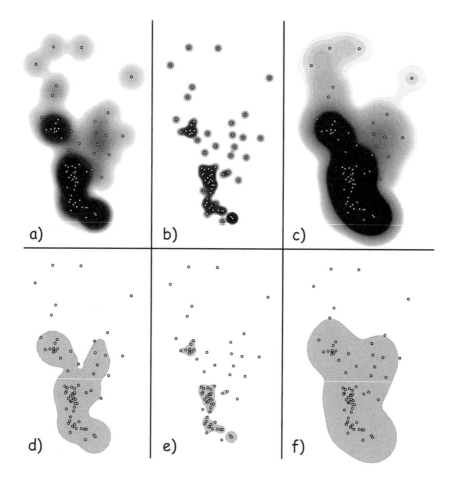

Figure 12-29: Kernel mapping may be used to identify core areas, although the areas that will be defined depend on the method used. Panels a through c show Gaussian density distributions for a sample set under varying bandwidths, while d through f show corresponding core areas encompassing 90% of the density distributions.

below-optimum (b), and above-optimum (c) bandwidth. Darker shades of gray show higher densities, and note the narrower, more concentrated distributions at the lowest bandwidth (b) relative to the largest bandwidth (c). These different bandwidths result in different core area polygons (d through f). Empirical tests and experience guide the choice of best bandwidth.

Time-Geographic Density Estimation

Density estimators have been developed for space utilization by moving objects, typically animals for home range analysis, although sometimes other objects. An object may be observed periodically through space, e.g., when a GNSS is attached to a migrating penguin, and the position relayed to a base station. These positions are often called control points, because they establish the location of the tracked object at a fixed point in time. This sequence of control points defines a path (Figure 12-30).While locations between observations cannot be precisely determined, the control points constrain where the penguin might have been, because there is an upper limit on how fast the bird can travel. We may establish a maximum velocity, v, either from previous observations, from the current tracking effort, or from theoretical limits. Time-geographic density estimation (TDGE) combines a sequential set of control points with knowledge about maximum velocity to estimate spatial occurrence probabilities.

TDGE depends on the concept of a geo-ellipse between two points. If P_i is the control point at time i and P_j the control point at time j, then the geoellipse g_{ij} may be defined as:

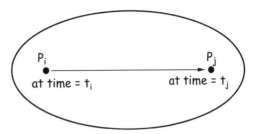

Figure 12-31: An object is estimated to have been contained within an ellipse, given two subsequent control points. The ellipse describes the furthest an object could have reached in any path traveling from P_i and P_j over the time interval t_i to t_j.

$$g_{ij} = \{P|[D(P, P_i) + D((P, P_j) \le ML)]\} \quad (12.20)$$

where $D(P, P_i)$ is a distance between any point P and the control point P_i, and ML is the maximum distance the object could possibly travel between the successive control points P_i and P_j. ML may be estimated by:

$$ML = (t_j - t_i) \cdot v \quad (12.21)$$

where t_j is the time of observation of control point j, and v is the maximum velocity for the object.

Figure 12-31 illustrates a geoellipse for two control points, P_i and P_j. Note that the distance function need not be Euclidian distance, but it usually is. The tracked object is restricted to have been within the drawn ellipse, provided our estimate of v is valid The size and shape of the ellipse depend on the distance between the successive control points, the time interval between the observations, and the maximum velocity possible. Successive points near each other relative to the maximum distance, given the time difference and maximum velocity, will be enclosed in a nearly circular ellipse, while successive points very near the maximum possible distance will be joined by a long, very narrow ellipse.

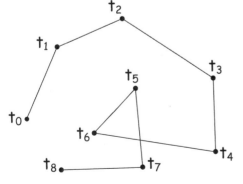

Figure 12-30: A sequence of observations on a moving object, from an initial time (t_0) to a final time (t_8).

Much as when using kernel density functions for estimating a core area, a time-geographic estimate of space use is a composite of many observations. Here, each pair of observations may be considered a density volume, proportional to the probability that the object occupied a location during the time interval (Figure 12-32). A uniform density function is the simplest to understand, implying the object was moving at maximum velocity between the two controlling observations, but along an unknown path within the ellipse. A uniform density function should have a volume equal to the likelihood of occupancy, and as with a standard kernel density estimator. For simple shapes such as a 3-dimensional uniform probability distribution, the volume is equal to the area times height.

Figure 12-32 shows two points, P_1 observed at time t_1, and P_2 observed at time t_2. Our task is to calculate the area of the elliptic volume which represents the occupancy probability, given our observations. Two paths are shown between the points, one traveling distance d_m and another d_n. These two paths have the same length, by the definition of the bounding ellipsoid, and they are also each equal to the long-axis length of the ellipsoid, $2a$. Geometric relationships between the inter-point distances and the dimensions of an ellipse allow us to calculate a and b, two characteristic dimensions, which in turn allow us to calculate the area, πab. This may then be scaled by height to assign an occupation probability (Figure 12-32, lower half).

The process is repeated for overlapping point pairs across the set of observed control

P_1 at (x_1, y_1), and P_2 at (x_2, y_2)

$d_m = d_n = ML = (t_2 - t_1) \cdot v$

$a = ML/2$

$c^2 = (x_2 - x_1)^2 + (y_2 - y_1)^2$

$b^2 = a^2 - c^2$

Area $= \pi ab$

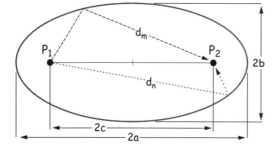

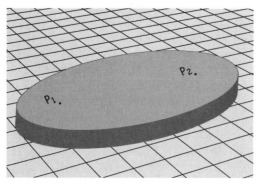

Figure 12-32: The process of calculating a geoellipse density between two sequential points, representing the likelihood of occupation. Here a uniform probability is assumed across the observations.

- Two points, P_1 and P_2, are measured at time t_1 and t_2, for an object with a maximum velocity v. The point locations, time interval, and v define an ellipse.

- The ellipse area can be calculated, with the ellipse height scaled to a density volume proportional to the likelihood of occupation.

- The subsequent pair of points (P_2 and P_3, not shown) are processed, and a new volume added to the occurrence surface, similar to kernel mapping

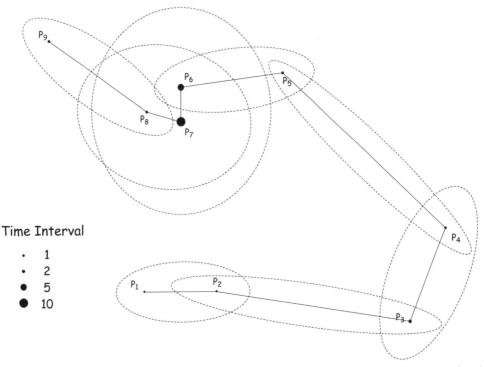

Figure 12-33: An example of the overlapping set of geoellipses used to create a composite density function from a time-geographic data set. The sequence begins with P_1 in the lower left, through P_9 in the upper left. Note that the control point symbol size denotes the time interval since the last control point observation (adapted from Downs et al., 2011).

points. The next two points in the sequence, P_2 and P_3, are paired, and the density ellipse calculated, summing the densities where geoellipse overlap. The process is repeated for points P_i, P_{i+1} until the last point is reached. Figure 12-33 illustrates the overlapping set of geoellipses from a sample set.

Ellipses may vary in shape, depending on time interval, distance between features, and the maximum velocity (Figure 12-33). Longer time intervals between observations result in larger ellipses, irrespective of the distance between subsequent points. As the inter-point distance approaches the maximum set by the maximum velocity, v, the ellipses become longer and narrower, and reduce to a line when the points are spaced at the maximum possible distance. Conversely,

the ellipses approach circles when the time interval between points is long but the observed distance between points is small. This occurs when the object has not moved much, relative to how far it might have moved in the time interval between observations.

The composite time geographic density function is shown in equation (12.22), where f(x) is the density at any point across a surface, n is the number of observed control points, t_s and t_e are start and end times, respectively, t_i and t_j are consecutive point pairs, v is the maximum velocity, and D(P, P_i) is the distance function, as described in equation (12.20). This equation is used for a set of points to estimate the density across space. The numerator sums the weighted

$$f(x) = \frac{1}{(n-1)[(t_e - t_s) \cdot v]^2} \sum_{i=1}^{n-1} H\left[\frac{D(P,P_i) + D(P,P_j)}{(t_i - t_j) \cdot v}\right] \qquad (12.22)$$

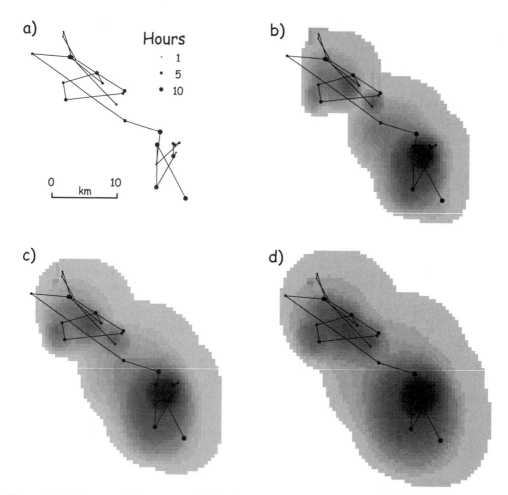

Figure 12-34: An example time geographic density function for a sequential set of control points. The path (**a**) shows both the trajectory and interval between observations. Successive figures (**b** through **d**) show different maximum velocity values, with velocities of 2 (**b**), 2.5 (**c**), and 3 (**d**) kilometers/hour (courtesy J. Downs).

distance ellipse functions for each pair of sampling points, and the denominator scales this by the maximum distance that may have been travelled during that time interval.

The composite of individual ellipses may result in complex aggregate density volumes. Densities will be highest where points are clustered or near where paths intersect frequently. Sharp edges and sampling artifacts may occur when using a uniform density function, at least until sample size becomes large.

Although Figure 12-32 illustrates a TDGE using a uniform distribution function, other functions may be used. One form

assumes the likelihood of occupation decreases linearly with the distance from the line connecting two subsequent control points. This is often called a linear decay function, because the occupation likelihood is assumed to decrease linearly with distance. A more rapid or less rapid decrease with distance may be represented by other functions.

The composite time geographic density estimate in Figure 12-34 illustrates a space-time path and a linear decay function applied to estimate an occupancy or home range estimate for the path. Panel **a** shows the control points for a path, with the time between successive observations shown by point size. Panels b through d show TGDE calculated using different maximum velocities. Note that the highest densities (darkest shades) show where the control points are clustered and where the distance between observations is short relative to time period between observations, which in turn is dependent on maximum velocities. This implies the net object movement was small between control points, although there is a denser area of likelihood there.

While a maximum velocity may be established from observations or theoretical values, the shape of the distance function often is not. A uniform function may be more defensible if the object is moving at near the maximum speed for most of the duration. However, a linear decay function may make more sense when the sampling interval varies in frequency, and the object is often moving much more slowly than the maximum velocity. TGDE is a developing field, and the interested reader should refer to the papers by J. Downs and colleagues listed at the end of this chapter.

Summary

Interpolation and spatial prediction allow us to estimate values at locations where they have not been measured. These methods are commonly used because our budgets are limited, samples may be lost or found wanting, or because time has passed since data collection. We may also interpolate when converting between data models, for example, when calculating a raster grid from a set of contour lines, or when resampling a raster grid to a finer resolution.

Spatial prediction involves collecting samples at known locations and using rules and equations to assign values at unsampled locations. There are many ways to distribute a sample, including a random selection of sample locations, a systematic pattern, clustering samples, adaptive sampling, or a combination of these. The sampling regime should consider the cost of travel and collecting samples, as well as the nature of the spatial variability of the target feature and the intended use of the interpolated surface.

Sample values are combined with sample locations to estimate or predict values at unsampled locations. There are many spatial prediction methods but the most common are Thiessen (nearest neighbor) polygon, local averaging (fixed radius), inverse distance weighted, trend surface, and kriging interpolation. Each of these methods has advantages and disadvantages relative to each other, and there is no method that is uniformly best. Each method should be tested for the variables of interest, and under conditions in the study area of interest. The best tests involve comparisons of interpolator estimates against withheld sample points.

Measures of core area are commonly identified from spatially-distributed observations. This form of prediction identifies regions of high probability for an object or event. Mean center or mean circle are simple measures. A convex hull, defined as the minimum area polygon encompassing all points and with convex exterior angles, is commonly applied. More sophisticated measures include kernel mapping, based on centering scaled distribution functions over each observation, and vertically summing the distribution functions.

Suggested Reading

Anderson, D.J. (1982). The home range: a new nonparamteric estimation technique. *Ecology*, 6:103-112.

Anselin, L. (1988). *Spatial Econometrics: Methods and Models*. Dordrecht: Kluwer Academic.

Anselin, L. (1995). Local indicators of spatial association - LISA. *Geographical Analysis*, 27:93-115.

Anselin, L. (2002). Under the hood: issues in the specification and interpretation of spatial regression models. *Agricultural Economics*, 17:247-267.

Anselin, L., Syabri, I., & Kho Y. (2006). GeoDa: An introduction to spatial data analysis.*Geographical Analysis*, 38:5-22.

Angulo-Martínez, M., López-Vicente, M., Vicente-Serrano, S.M., & Beguería, S. (2009). Mapping rainfall erosivity at a regional scale: a comparison of interpolation methods in the Ebro Basin (NE Spain). *Hydrology and Earth Systems Science*, 13:1910-1920.

Ayeni, O. O. (1982). Optimum sampling for digital terrain models. *Photogrammetric Engineering and Remote Sensing*, 48:1687-1694.

Besag, J. (1974). Spatial interaction and the statistical analysis of lattice systems. *Journal of the Royal Statistical Society*, 43B:192-225.

Besag, J., & Kooperberg, C.L. (1995). On conditional and intrinsic autoregressions. *Biometrika*, 82:733-746.

Bowman, A.W., & Azzalini, A. (1997). *Applied Smoothing Techniques for Data Analysis: The Kernel Approach with S-Plus Illustrations*. Oxford: Oxford University Press.

Burgess, T.M. & Webster, R. (1984). Optimal sampling strategies for mapping soil types. I. Distribution of boundary spacing. *Journal of Soil Science*, 35:641-654.

Burrough, P.A.& McDonnell, R.A.(1998). *Principles of Geographical Information Systems*, New York: Oxford University Press.

Cressie, N. (1989). Geostatistics. *American Statistician*, 43:197-202.

Cressie, N. (1991). *Statistics for Spatial Data*. New York: Wiley.

DeGruijter, J.J. & Ter Braak, C.J.F. (1990). Model-free estimation from spatial samples: a reappraisal of classical sampling theory. *Mathematical Geology*, 22:407-415.

Downs, J.A. & Horner, M.W. (2009). A characteristic-hull based method for home range estimation. *Transactions in GIS* 13:527-537.

Downs, J.A, Horner, M.W., & Tucker, A.D. (2011). Time-geographic density estimation for home-range analysis. *Annals of GIS*, 17:163-171.

Dubrule, O. (1994). Comparing splines and kriging. *Computers and Geosciences*, 10:327-338.

Fotheringham, A., Brunsdon, C., & Charlton, M. (2000). *Quantitative Geography: Perspectives on Spatial Data Analysis*. London: Sage Publications.

Getis, A. & Ord, J.K. (1992). The analysis of spatial association by use of distance statistics, *Geographical Analysis*, 24: 189-206.

Goovaerts, P. (1997). *Geostatistics for Natural Resource Evaluation*. New York: Oxford University Press.

Griffith, D.A., & L.J. Layne, A. (1999). *Casebook for Spatial Statistical Data Analysis*. Oxford: Oxford University Press.

Hutchinson, M. F. (1995). Interpolating mean rainfall with thin plate smoothing splines. *International Journal of Geographical Information Systems*, 9:385-404.

Isaaks, E.H. & Srivastava, R.M. (1989). *An Introduction to Applied Geostatistics*. New York: Oxford University Press.

Lam, N.S. (1983). Spatial interpolation methods: a review. *American Cartographer*, 10:129-149.

Laurini, R. & Thompson, D. (1992). *Fundamentals of Spatial Information Systems*. London: Academic Press.

Legendre, P. (1993). Spatial autocorrelation: Trouble or new paradigm? *Ecology*, 74:1659-1673.

Mark, D.M. (1987). Recursive algorithm for determination of proximal (Thiessen) polygons in any metric space. *Geographical Analysis*, 19:264-272.

Mitasova, H. & Hofierka, J. (1993). Interpolation by regularized spline with tension: application to terrain modeling and surface geometry analysis. *Mathematical Geology*, 25:657-669.

Mitchell, A. (1999). *The ESRI Guide to GIS Analysis*. Redlands: ESRI Press.

O'Sullivan, D., & Unwin, D.J. (2003). *Geographic Information Analysis*. New York: John Wiley and Sons.

Silverman, B.W. (1986). *Density Estimation*. London: Chapman and Hall.

Varekamp, C., Skidmore, A.K., & Burrough, P.A. (1996). Using public domain geostatistical and GIS software for spatial interpolation. *Photogrammetric Engineering and Remote Sensing*, 62:845-854.

Willmott, C.J. (1981). On the validation of models. *Physical Geography*, 2:184-191.

Willmott, C.J., & Matsuura, K. (2005). Advantages of the mean absolute error (MAE) over the root mean square error (RMSE) in assessing average model performance. *Climate Research*, 30:79-82.

Worton, B.J. (1987). A review of models of home range for animal movement. *Ecological Modelling*, 38:277-298.

Study Questions

12.1 - Why perform a spatial interpolation?

12.2 - Describe four different sampling patterns, and provide the relative advantages or disadvantages of each? Which do you think is used most in practice, and why?

12.3 - Draw a systematic sampling pattern on the area below, left, and an adaptive sampling pattern on the area below, right. Use the same number of sample points, e.g., approximately 50, on both. Which do you think will give a better estimate of terrain locations at unknown points? Why? Would increasing the sample number change which sampling design you would think is best?

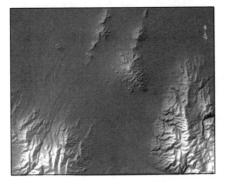 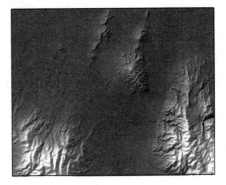

12.4 - Draw the Thiessen polygons (nearest neighbor interpolation) for the set of points below.

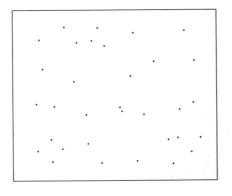

12.5 - Calculate the cell values indicated at the crosses, below, using fixed radius sampling with the shown circle.

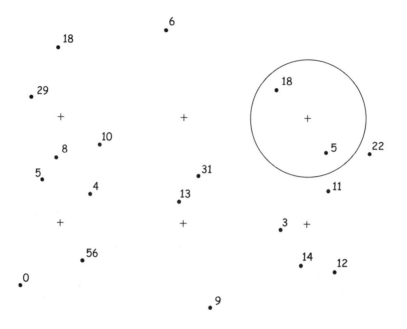

12.6 - Calculate the z values for the unknown points, listed below, using an inverse distance weighted approach. Use the three nearest known points (use i = 3. Known points are shown in map as filled circles and corresponding coordinate and z values in the table at right), and a distance exponent (n) of 1.

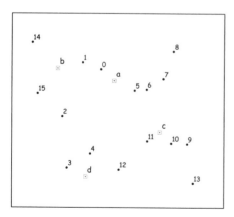

ID	X	Y	Z
0	153,951.9	4,478,714.6	2040.6
1	151,280.9	4,479,647.3	1863.0
2	148,228.5	4,472,143.4	1992.1
3	148,906.8	4,464,978.6	2540.1
4	152,383.2	4,466,928.8	2106.3
5	158,827.3	4,475,746.9	2283.2
6	160,607.9	4,475,874.2	1933.5
7	163,024.4	4,477,357.9	1836.4
8	164,465.8	4,481,173.5	1838.3
9	166,416.0	4,468,285.4	2523.9
10	164,169.1	4,468,370.4	2138.8
11	160,692.7	4,468,709.3	1854.2
12	156,537.9	4,464,724.2	1866.9
13	167,306.3	4,462,816.5	2453.8
14	143,946.6	4,482,445.4	1837.9
15	144,709.7	4,475,323.0	1912.8

Unknown points:

ID	X	Y	Z
a	155,859.6	4,477,146.0	
b	147,592.6	4,478,884.2	
c	162,515.7	4,469,981.2	
d	151,704.9	4,463,749.2	

12.7 - What is a primary difference between a spline interpolation method and a trend surface interpolation?

12.8 - What is the primary difference between a trend surface interpolation and a kriged interpolation?

12.9 - Describe the variogram. What does it represent on the x and y axes, and what are the important regions/points of the plot?

12.10 - Draw the approximate mean center, standard deviation circle, and maximum circle for the following data:

12.11 - What is the convex hull? How is it calculated/determined?

12.12 - Draw the convex hull for the points depicted below:

12.13 - Can you describe/define a kernel density map? How are the values based on samples?

12.14 - Can you write down and describe at least one function or equation used to generate a density surface?

12.15 - What is a "bandwidth" in a kernel mapping approach? Can you write down the equation of a generic spatial optimum?

13 Spatial Models and Modeling

Introduction

A model is a description of reality. Here, our interest is restricted to computer-based models of spatial phenomena. These models describe the basic properties or processes for a set of spatial features, and help us understand their form and behavior.

Many computer-based models use spatial data, and are developed and run using some combination of GIS, general and specialized computer programming languages, and spatial and nonspatial analytical tools. Spatially explicit models are a primary benefit of GIS technologies, and many spatial models are based on data in a GIS. These models may be run in the GIS, or the spatial data may be prepared in a GIS, and exported to a model that is developed and run outside a GIS.

While there may be as many classes of models as there are modelers, here we split spatial models into three broad and overlapping classes: *cartographic models*, *simple spatial models*, and *spatio-temporal models*. Joseph Berry, an early and well-known developer and proponent of spatial modeling, described cartographic models as automating manual map analysis and processing, while spatial models focus on applying mathematical relationships. Cartographic models are most often applied to rank areas in support of decision making, while simple spatial models often apply sets of equations to predict a specific con-

tinuous variable across space. Cartographic model outputs are often nominal (suitable or unsuitable, Figure 13-1) or ordinal (low, medium, or high suitability), while the outputs from simple spatial models are often interval/ratio (e.g., population density, accident frequency, or soil erosion rates).

Cartographic models solve problems via spatial layer combination in overlay, buffers, reclassification, and other spatial operations. These models often employ the concepts of map algebra, described in Chapter 10, but may include a much broader range of operations. Suitability analyses, defined here as the classification of land according to their utility for specific uses, are among the most common cartographic models.

Most cartographic models are temporally static because they represent spatial features at a fixed point in time. Data in base layers are mapped for given periods. These data are the basis for spatial operations that may create new data layers. For example, we may be interested in identifying the land that is currently most valuable for agriculture. Costs of production may depend on the slope (steeper is costlier), soil type (some soils require more fertilizer), current land cover (built-up is unsuitable, forests more expensive to clear), or distance to roads or markets. Agricultural production may also depend on soil types, topography (neither flooded nor drought-prone), and the ability to irrigate. Spatial data on elevation, soil properties, current land use, roads, market location, and irrigation potential may be combined to rank sites by production value. We may use a mathematical relationship for specific calculations of average costs and revenues, for example, agronomists may have developed the rela-

tionships between soil types and average corn production in the region, and we may use a cost-per-mile for transport based on local rates. These spatial data are combined in a cartographic model to assign a land value for each parcel in a study region. The model is temporally static in that the values for the spatial variables, such as soil fertility

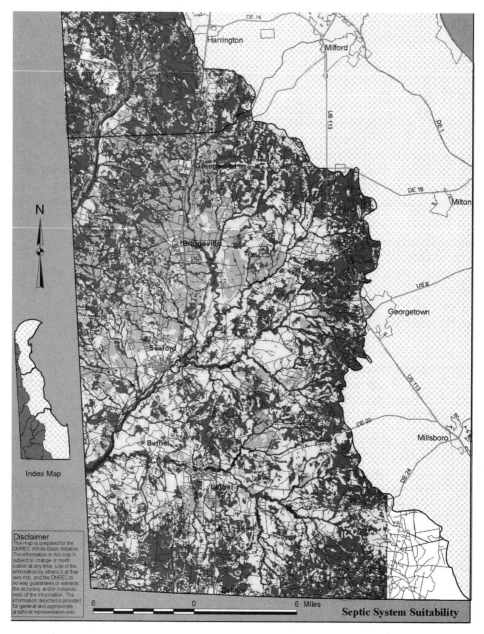

Figure 13-1: An example of a suitability map, produced by combining soils, elevation, wetland, and watercourse data. These analyses are often automated via a cartographic model to produce suitability or other nominal or ordinal rankings over large areas (courtesy State of Delaware).

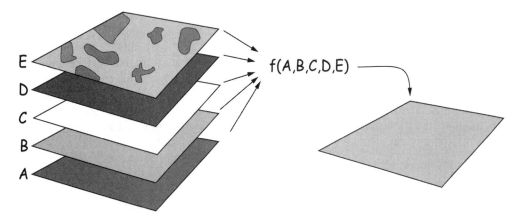

Figure 13-2: An example of a simple spatial model, where a function is applied to spatial inputs to estimate an important spatial output.

or distance to roads, do not change during the analyses.

Cartographic models are generally not temporally dynamic, even though they may be used to analyze change. For example, we may wish to analyze vegetation change over a 10-year period, based largely on vegetation maps produced at the start and end of the period. Each data layer represents the vegetation boundaries at a fixed point in time. The model is static in that the polygon boundaries for a given layer do not change. There may be two vegetation data layers, each corresponding to a different point in time, and the vegetation boundaries are mapped as found at each time interval. Our cartographic model includes a temporal component in that it compares vegetation change through time, but the cartographic model does not generate new boundaries of polygons or any other characteristics of spatial features. Boundaries may be a composite of those lines that exist in the input data layers, but new lines at new coordinate locations are not generated. Most spatial modeling or models conducted in the framework of GIS have been cartographic models that are temporally static in this manner.

Simple spatial models typically apply a set of equations to spatially-resolved variables. They often rest on equations developed from data at a set of observations at points or sub-areas, and then applied across broader geographic areas.

An example may help understand simple spatial models. William Cooke and colleagues reported on a model of West Nile virus infection among birds, and the risk of transmission to humans. West Nile virus is a sometimes fatal introduced disease, and varies in prevalence through space and time. Cooke and his associates compiled data on the frequency of bird and human infections within each zip code in Mississippi over several years. Human and bird cases were clustered, with outbreaks concentrated in rural areas. Road density was used as a surrogate for rural/urban land use.

Spatial variables related to mosquito habitat quality were compiled statewide, including stream density, vegetation type, temperature, and precipitation surplus. These were combined with virus infection frequency at specific locations to fit a predictive statistical model. Mapped spatial variables were then applied in the model to predict outbreak risk across the state.

Simple spatial models are common, with hundreds of examples found across a range of disciplines. They typically include a model derived through sampling and a statistical fitting process, a model that is subsequently applied across space to estimate important events, densities, or other characteristics.

Spatio-temporal models are dynamic in both space and time. They differ from cartographic or predictive spatial models in that time passes explicitly within the running of the model, and changes in time-driven processes within the model cause changes in spatial variables. Spatio-temporal models often attempt to explicitly represent processes within the model.

The dispersion of oil after a spill is an example of a process that might be analyzed via a spatio-temporal model. Currents, winds, wave action, and the physics of oil separation and evaporation on exposure to air might be combined in a model to predict the changing location of an oil slick. The actions of objects as they move across an environment may also be represented in a spatio-temporal model.

Spatio-temporal models include time-driven processes within the framework of the model. These processes are typically quite detailed and include substantial computer code to represent important sub-processes. Our oil evaporation example demonstrates the subprocesses represented in a dynamic spatial model. Oil evaporation rates depend on many factors, including oil viscosity, component oil fractions, wind speed, temperature, wave height and action, and sunlight intensity. These processes may be modeled by suitable functions applied to spatially defined patches of oil. The sub-model may estimate evaporation of various components of the oil in the patch, and update the characteristics of oil in that patch. Oil chemistry and viscosity may change due to more rapid evaporation of lighter components, in turn affecting future evaporation calculations. Spatial features may change through time due to the represented dynamic process, for example, the boundary defining an oil spill may vary as the model progresses.

Spatio-temporal models are typically more limited than other modeling approaches in the range and number of spatial themes analyzed, but they provide a more mechanistic representation of dynamic processes. Substantial effort goes into developing sub-models of important processes. Model components and structures focus on one or a few key output spatial variables, and input data themes are included only as they are needed by these sub-process models. These temporally dynamic models explicitly calculate the changes in the output spatial variables through time. Feature boundaries, point feature locations, and attribute variables that reflect the spatial and aspatial characteristics of key output variables may change within the model run, typically multiple times, and with an explicit temporal frequency.

Simple spatial models and spatial statistical analyses are often used as precursors to spatio-temporal models. By uncovering key processes or rates, they can guide other analysis. For example, in our oil spill example, the specific relationship between wave height or frequency and oil separation may be represented by an equation, but the specific parameters that define the shape of the relationship may be estimated via a statistical process. Experiments or observations on separation rates at various wave heights may be collected, and the specific model parameters estimated. These may then be included as a component of the larger spatio-temporal model.

Cartographic Modeling

A *cartographic model* provides information through a combination of spatial data sets, functions, and operations. These functions and operations often include reclassification, overlay, interpolation, terrain analyses, buffering and other functions. Multiple data layers are combined via these operations, and the information is typically in the form of a spatial data layer. Map algebra, described in Chapter 10, is often used to specify cartographic models for raster data sets.

Suitability analyses are perhaps the most common examples of cartographic models. These analyses rank land according to its utility for various purposes. Suitability analyses often involve the overlay, weighting, and rating of multiple data layers to categorize lands into various classes. Relevant data layers are combined and the resultant polygons are classified based on the combination of attributes. Figure 13-3 illustrates a simplistic cartographic model for the identification of potential park sites. Suitable sites are those that are near lakes, near roads, and not

wetlands. The model uses three input data layers, containing lakes, roads, and hydric status for a common study area. Spatial operations are applied to the spatial data layers, including reclassification, buffering, and overlay. These result in a suitability layer. This suitability layer can then be used to narrow sites for further evaluation, identify owners, or otherwise aid in park site selection.

Cartographic models have been used for a variety of applications. These include land use planning, transportation route and corridor studies, the design and development of water distribution systems, modeling the spread of human disease or introduced plant and animal species, building and business site selection, pollution response planning, and endangered species preservation. Cartographic models are so extensively used because they provide information useful to managers, the public, and policy makers, and help guide decisions requiring the consideration of spatial location across multiple themes.

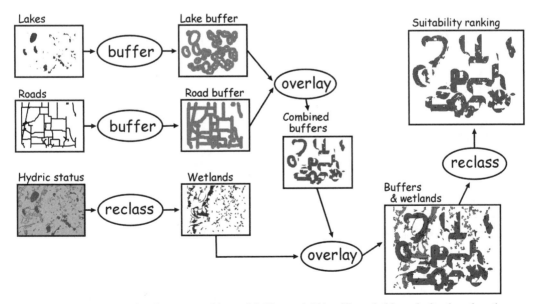

Figure 13-3: An example of a cartographic model. The model identifies suitable park sites based on the proximity to roads and lakes, and the absence of wetlands.

Cartographic models are often succinctly represented by *flowcharts*. A flowchart is a graphic representation of the spatial data, operations, and their sequence of use in a cartographic model. Figure 13-4 illustrates a flowchart of the cartographic model illustrated in Figure 13-3. Suitable sites need to be identified that are near roads, near lakes, and are not wetlands. Data layers are represented by rectangles, operations by ellipses, and the sequence of operations by arrows. Operations are denoted inside each ellipse. Flowcharts are often required by an agency or organization to document a completed spatial analysis. Because a consistent set of symbols aids in the effective communication of the cartographic model, a standard set of symbols and flowcharting methods may help in understanding the data and operations used in an analysis.

Flowcharts are useful during the development and application of a cartographic model. Flowcharts aid in the conceptualization and comparison of various competing approaches and may aid in the selection of the final model. A sketch of the flowchart is often a useful and efficient framework for

documenting the application of the cartographic model. File locations, work dates, and intermediate observations can be noted with reference to the flowchart, or directly onto a copy of the flowchart.

Cartographic modeling often produces a large number of "intermediate" or temporary data layers that are not required in the final output or decision-making. Our example in Figure 13-4 illustrates this. The needed information is contained entirely within the suitability ranking data layer. This layer summarizes the ranking of lands based on the provided criteria. Five other data layers were produced within the illustrated cartographic model. Buffered, recoded, and overlay layers were necessary intermediate steps, but in this analysis their utility was temporary. Once the layers were included in subsequent operations, they were no longer needed. This proliferation of data layers is common in cartographic modeling, and it can cause problems as the new layers and other files accumulate in the computer workspace. Frequent removal of uneeded files is often helpful.

Much of the power of cartographic modeling comes from the flexibility of spatial analysis functions. Spatial functions and operations are a set of tools that may be mixed and matched in cartographic models. Overlay, proximity, reclassification, and most other spatial analysis tools are quite general. These tools may be combined in an astoundingly large number of ways, by selecting different tools and by changing their sequence of application. For example, differences in distances, thresholds, and reclassification tables may be specified. These variations will result in different output data layers, even when using the same input data layers. With a small set of tools and data layers, we can create a huge number of cartographic models. Designing the best cartographic model to solve a problem -- the selection of the appropriate spatial tools and the specification of their sequence -- is perhaps the most important and often the most difficult process in cartographic modeling.

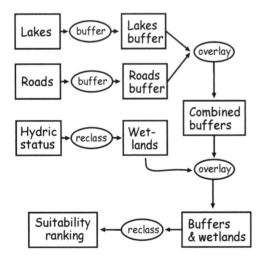

Figure 13-4: A flowchart depicting the cartographic model in Figure 13-3. The flowchart is a rapidly-produced shorthand method of representing a sequence of spatial operations.

Designing a Cartographic Model

Most cartographic models are based on a set of criteria. Unfortunately, these criteria are often initially specified in qualitative terms, such as "the slopes must not be steep." A substantial amount of interpretation may be required in translating the criteria in a suitability analysis into a specific sequence of spatial operations. In our present example we must quantify what is meant by "too steep." General or qualitative criteria may be provided and these must be converted to specific, quantitative measures. The conversion from a qualitative to quantitative specification is often an iterative process, with repeated interaction between the analyst developing and applying the cartographic model and the manager or decision-maker who will act on the resultant information.

We will use a home-site selection exercise to demonstrate this process. The problem consists of ranking sites by suitability for home construction. The area to be analyzed has steep terrain and is in a seasonally cold climate. There are four criteria:

a) Slopes should not be too steep. Steep slopes may substantially increase costs or may preclude construction.

b) A southern aspect is preferred, to enhance solar warming.

c) Soils suitable for on-site septic systems are required. There is a range of soil types in the study area, with a range of suitabilities for septic system installation.

d) Sites should be far enough from a main road to offer some privacy, but not so far as to be isolated.

These criteria must be converted to more specific restrictions prior to the development and application in a cartographic model. The decision-maker must specify what sort of classification is required. Is a simple binary classification needed, with suitable and unsuitable classes, or is a broader range of classes needed? If a range of classes is specified, is an ordinal ranking acceptable, or is an interval/ratio scale preferred? These questions are typically answered via discussions between the analyst and the decision-makers. Each criterion can then be defined once the type and measurement scale of the results are specified. It may be fairly simple to establish the local slope limit that prohibits construction. For example, conversations with local building experts may identify 30 degrees as a threshold beyond which construction is infeasible. Further work is required to quantify how slopes affect construction costs. Similar refinements must be made for each criterion. We must quantify the range and any relative preferences for southern aspects, relative soil suitabilities, what defines a main road, and what constitutes short and long distances.

A second key consideration involves the availability and quality of data. Do the required data layers exist for the study area? Are the spatial accuracies, spatial resolution, and attributes appropriate for the intended analysis? How will map generalizations affect the analysis, for example, will inclusions of different soil types in a soil polygon lead to inappropriate results? Is the minimum mapping unit appropriate? If not, then the requisite data must be obtained or developed, or the goals and cartographic model modified.

Weightings and Rankings

While some cartographic models are simple and restrictive, many more cartographic models require the combination of criteria that vary across a range of values, and require an explicit ranking of the relative importance of different classes or types of criteria. A simple, restrictive example might require us to identify parcels greater than a certain size and within a certain distance of water. We may clearly identify areas that meet these desired conditions.

A much more common class of problems requires us to integrate multiple criteria that are qualitatively different. For example, site suitability for hazardous waste storage depends on a number of factors, including distance to population centers, transportation, geology, and aquifer depth and type. We must rate sites across a range of values for all of these variables. Once criteria are precisely defined we must obtain appropriate data, develop a flowchart or plan for our analysis, and address the more difficult problem of assigning rankings within each criteria, and assigning the relative weightings among criteria. Note that in the following discussions we use the word "rankings" when describing the assignment of relative values within the same layer, such as how we rank a sandy soil vs. a silty soil in a soils layer, and the word "weightings" when assigning the relative values of different layers, for example, how we weight the values in an elevation layer vs. the values in a land use layer.

Rankings Within Criteria

Each criterion in our cartographic model is usually expressed by a data layer, or "criterion layer." Each criterion layer is a spatial representation of some constraint or selection condition, for example, the criterion we build outside a floodplain may consist of a set of numbers in a layer identifying floodplain locations. Floodplain sites may be assigned a value of 0, and upland sites a value of 1.

Before we can assign a value to any site, we must first obtain floodplain maps and interpret the codes in the maps to delineate the most flood-prone areas. Floodplain maps may exist, or we may have to generate them from other sources. This allows us to rank areas based on the likelihood of flooding.

We must explicitly formalize our ranking for each layer used to represent a criterion. One early decision is whether ranks should be discrete or continuous (Figure 13-5). Rankings are discrete when input data are interpreted such that the criterion data layer is a map of discrete values. Soils are either good or bad for construction, slopes either too steep or acceptable, and the final map

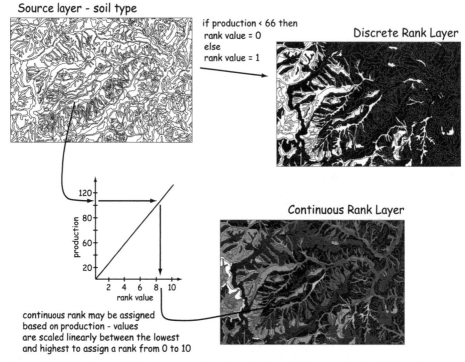

Figure 13-5: Rankings within a layer are often discrete (top right) or continuous (bottom).

defines two or a few discrete classes, for example, sites are categorized as either suitable or unsuitable. Ranks are continuous when they vary along a scale, for example, soils may be rated from 1 to 100 for construction suitability.

Figure 13-5, top right, shows the assignment of discrete ranking of land productivity based on values in a soil layer. The source layer in the top left of the figure is analyzed. If the expected production for a given soil polygon is less than 66, then the output ranking is set to 0. If the production is greater than or equal to 66, then the output rank is set to 1. A range of input values has been placed into two discrete classes, illustrated as the discrete rank layer in the top-right part of Figure 13-5.

Discrete rankings are most often used when there are clear, discrete classes to be represented in criteria. A disease may be present or absent, a country an ally or enemy, or a block inventoried or not. The values to be represented are discrete categories.

In contrast, we may apply criteria as continuous rankings within a cartographic model. These continuous rankings provide a range of values to characterize a suitability or restriction, and they result in a set of incrementally varying ranks. Ranks (or scores) typically range over a real or large integer interval, for example, from 0 to 1 or 0 to 1000. Highest suitability is usually assigned to the highest rank, and lowest to the bottom.

The bottom right of Figure 13-5 shows a continuous ranking over a range of 0 to 10. A high value of 10 is specified for the most productive soils, and a low value of 0 for the least productive. We may use production data gathered over a set of soil types, and a map of soil types to assign the relative value of each soil. We could scale production from the lowest to the highest observed over the range of 0 to 10, and in so doing create a layer that represents a soil productivity criteria.

We are not constrained to linear or always increasing or decreasing relationships between our input layers and our criteria layers. There may be complex relationships between an input value and our output ranking scores or values. Any curve or relationship we can create with a combination of mathematical and logical functions may be represented, to reflect increasing, decreasing, or complex relationships.

We should have some justification for adopting a specific curve when establishing relative ranks within a layer. For example, we may wish to represent the mercury hazard based on methyl mercury concentrations in water supplies across a state. There may be a broad range of low mercury concentrations for which there are no or few negative health impacts. However, as a threshold concentration is reached, there may be a rapid upturn into a very steep curve, where the risk of severe damage is great (Figure 13-6). The shapes of these curves should be established through sets of epidemiological studies, in

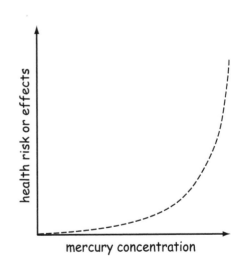

Figure 13-6: The ranking within a criterion layer should be based on a defensible relationship, whose shape has been established through sufficient study or experience. Here, risk for mercury exposure via concentration in drinking water has been related to negative health impacts. Suitability or hazard rankings in cartographic models should be well supported by measurements.

which mercury concentration in human blood or tissue was related to drinking water, and health impacts were recorded for thousands of people at various levels of mercury exposure.

Figure 13-7 illustrates two examples of continuous criteria scores. Figure 13-7a shows the representation of a complex road criterion for a cartographic model. This criterion specifies that desirable sites are greater than 300 but less than 2000 meters from a road. The top left graphic of Figure 13-7a shows the original roads layer. Following the arrows counter-clockwise, you find the distance layer, a raster with the distance from the nearest road recorded in each cell value. In this example the distances range from 0 to 6000 meters. The graphic in the lower right of Figure 13-7a shows a suitability assignment function. Distance values are recorded along the horizontal axis, and are used to assign suitability for building, shown on the vertical axis. This function assigns suitability scores of 0 for distances less than 300 meters. Suitabilities increase and distance increases, in a linear fashion, to a score of 1 at a distance of 1150 meters, half way between 300 and 2000. Scores then decline linearly to a value of 0 at 2000 meters, and remain 0 for all distances greater than 2000 meters.

Figure 13-7b illustrates a continuous ranking of suitability scores, in this instance for slope. Slopes are calculated from the elevation layer (Figure 13-7b, left), ranging from 0 to 49.6 degrees for this data set. Slope values are transformed to continuous slope suitability values using a smoothly decaying function (lower right, Figure 13-7b). These values are assigned to each cell location in an output slope suitability data layer (top right, Figure 13-7).

Note that these continuous rankings may be combined, often through a weighted addition process, to generate a combined suitability score. The various suitability layers sum vertically to give a total composite score for each cell. This score may be used to rank areas on relative suitability. Discrete and continuous suitability layers may be combined using a mix of Boolean and addition operations to provide a final ranking. This combination often requires that we define the relative importance of each criteria layer, a process known as weighting among criteria.

Weighting Among Criteria

Distinct criteria must be combined in many spatial analyses, usually in some overlay or addition process (Figure 13-8). We must choose how to weight one layer relative to another. How important is slope relative to aspect? Will an optimum aspect offset a moderately steep site? How important is isolation relative to other factors? Because the criteria will be combined in a suitability data layer, the relative weightings given each criterion will influence the results. Different relative weights are likely to result in different suitability rankings. It is often difficult to assign these relative weights in an objective fashion, particularly when suitability depends on non-quantifiable measures.

The assignment of relative weightings is easiest when the importance of the various criteria may be expressed on a common scale, such as money. In our example, we may be able to assign a monetary value to slope effects due to increased construction costs. Soil types may be categorized based on their septic capacity. Different septic systems may be required for different types of soils, either through larger drain fields or the specification of mound vs. field systems. Costs could be estimated based on the variable requirements set by soil type. Nuisance cost for noise and distance cost in lost time or travel might also be quantified monetarily. Reducing all criteria to a common scale removes differential weighting among criteria.

There are many instances where a common measurement scale does not exist and it is not possible to develop one. Many rankings are based on variables that are difficult to quantify. Personal values may define the distances from a road that constitute "isolated" versus "private," or what is the rela-

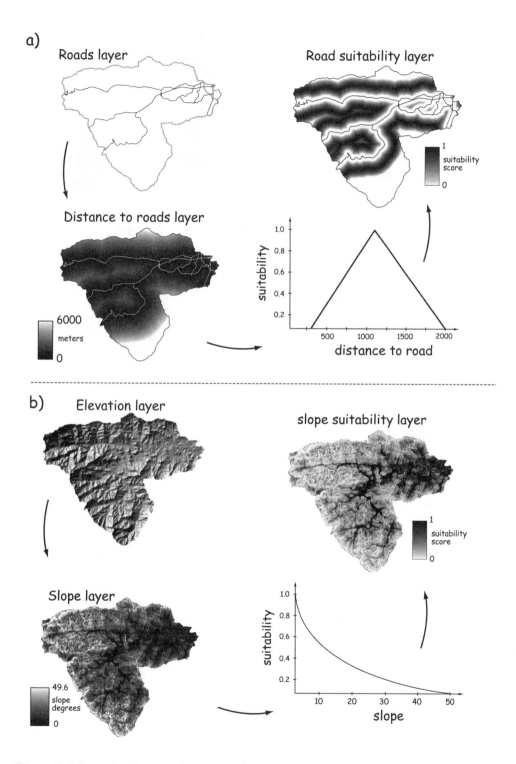

Figure 13-7: Examples of two continuous score layers to represent criteria in a cartographic model. Panel a, top, shows the dual distance to road criteria applied using a suitability function. This results in a continuous range of suitability scores between 0 and 1. In panel b, a different suitability function is applied to the slope layer, resulting in a continuous suitability score for the slope criterion.

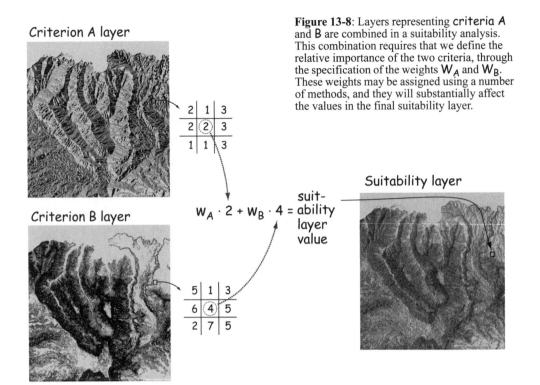

Criterion A layer

Criterion B layer

Figure 13-8: Layers representing criteria A and B are combined in a suitability analysis. This combination requires that we define the relative importance of the two criteria, through the specification of the weights W_A and W_B. These weights may be assigned using a number of methods, and they will substantially affect the values in the final suitability layer.

Suitability layer

$W_A \cdot 2 + W_B \cdot 4 = $ suit-ability layer value

tive importance of slope vs. construction cost. In such cases, the scales and weights may be defined in conference with the decision-maker. Expert opinion, group interviews, or stakeholder meetings may be required when there are multiple or competing parties. Multiple model runs may be required, each run with a different set of relative weights within and among criteria. These multiple runs may reveal criteria that are important or unimportant under all viable weightings, as well as those that are influential under a limited range of weightings.

One method of assigning weights is based on their "importance ranking." The factors (criteria) used to decide the quality of a site may be ranked in their importance, from most important to least. We may then calculate the relative weights according to:

$$w_i = \frac{n - r_i + 1}{\sum\limits_{k=1}^{n} (n - r_k + 1)} \qquad (13.1)$$

where w_i is the weighting for criterion i, n is the number of criteria, and k is a counter for summing across all criteria.

Suppose we wish to rank sites for store placement based on four factors: distance to nearest competitor, distance to nearest major road, parking density, and parcel cost. Figure 13-9 shows an example calculation of criteria weights based on importance ranking. Each criterion is listed in the leftmost column. Ranks are assigned to criteria by the planner, client, decision maker, or interested group. The numerator of Equation 13.1 is calculated for each criterion, giving the most important criterion the highest value and the least important the lowest value. The denominator is calculated by summation, and then the individual weights calculated, as shown in the right-most column. These weights may then be used to combine the data from the various criteria.

Note that there are several assumptions in this example. First, we assume that the values in each layer associated with each criterion have appropriate ranges, or at least are on similar scales. In Figure 13-8, the values for criterion layer A and criterion layer B vary over an approximately equal range. If one layer had a range from 1000 to 5000 and the other had values of 1 to 5, then this would affect the combination and final suitability ranking.

Second, we may be implicitly assuming that the scales are approximately linear in our ranking within and across the criteria. We often combine the values within a criterion layer using an arithmetic operation, for example, by summing values with weights (Figure 13-9). The relative weights among and within each layer are mixed, which is often a logical course of action under an assumption of linearity. Strongly non-linear relationships in the ratings and weightings scales often lead to counter-intuitive and unwanted suitabilities.

There are a many other methods for defining the values for each criterion layer and the relative weightings among layers. These include methods that attempt to ensure consistency among weights, but they are beyond the scope of this introductory text. You may find more detailed descrip-

tions in the excellent book by Malczewski (1999) listed in the suggested readings section at the end of this chapter.

Cartographic Models: A Detailed Example

Here we provide a detailed description of the steps involved in specifying and applying a cartographic model. We use a refinement of the general criteria for homesite selection described in the previous section. These general criteria are listed on the left side and the refined criteria are shown on the right side of Table 13-1. The refined criteria may have been defined after further discussion with the decision-makers, local area experts, and a review of available data and methods.

Note that we adopt the simplest weighting and ranking scheme in applying the criteria in Table 13-1. All criteria are equally weighted, and all criteria are binary -- land is categorized as unsuitable or suitable based on each criterion. A location must pass all criteria to be suitable, and the final rating is suitable or unsuitable.

Criterion	Rank	Numerator $(n - r_i + 1)$	Weight $(n - r_i + 1) \Big/ \sum_{k=1}^{n} (n - r_k + 1)$
distance to nearest competitor	2	4-2+1 = 3	3/10 = 0.3
distance to major road	3	4-3+1 = 2	2/10 = 0.2
parking density	4	4-4+1 = 1	1/10 = 0.1
parcel cost	1	4-1+1 = 4	4/10 = 0.4

$$\sum_{k=1}^{n} (n - r_k + 1) = 10$$

Figure 13-9: An example of one method for calculating relative weights for each of four criteria according to Equation (13.1).

In our example we will apply the cartographic model described by the flowchart in Figure 13-4 to a small watershed in a mountainous study area. Application of the refined criteria require three base data layers -- elevation, soils, and roads. For this example we assume the three data layers are available at the required positional and attribute accuracy, clipped to the study area of interest. The need for new data layers often becomes apparent during the process of translating the initial, general criteria to specific, refined criteria, or during the development of the flowchart describing the cartographic model. Once data availability and quality have been assured, we can complete the final flowchart.

Figure 13-10 contains a flowchart of a cartographic model that may identify suitable sites. Spatial data layers are shown as rectangles, and a descriptive data layer name is included within the rectangle. Spatial operations or functions are contained in ellipses, and arrows define the sequence of data layers and spatial operations. The three base data layers, elevation, soils, and roads, are shown at the top of the flowchart.

There are three main branches in the flowchart in Figure 13-10. The left-most branch addresses the terrain-related criteria, the center branch addresses the soils criteria, and the right branch applies the road distance criteria. All three branches join in the cartographic model, producing a final suitability classification.

The left branch of the cartographic model is shown in detail in Figure 13-11. This and subsequent detailed figures show a thumbnail of the spatial data layers at each step in the process. Data layer names are adjacent to the spatial data layer. The first two criteria involve terrain-related constraints. Suitable sites are required to possess a restricted set of slopes and aspects. These criteria require slope and aspect data layers, to be calculated and then classified into areas that do and do not meet the respective criteria. The elevation data layer is shown at the top of Figure 13-11; low elevations in black through higher elevations in lighter shades. There are two main river systems in the study area, one running from west to east in the northern portion of the study area, and one running from south to north. Highland areas are found along the north, west, and east margins of the study area.

Slope and aspect are derived from the elevation data layer (Figure 13-11). Lower

Table 13-1 Original and refined criteria for cartographic model example.

General Criteria	Refined Criteria
Slopes not too steep	Slopes < 30 degrees
Southern aspect preferred	90 < Aspect < 270
Soils suitable for septic system	Specified list of septic-suitable soil units
Far enough from road to provide privacy, but not isolated	300 meter < distance to road < 2000 meters

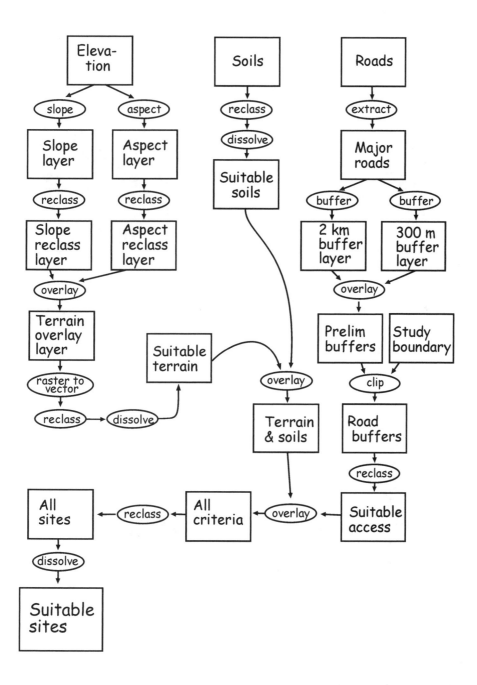

Figure 13-10: Flowchart for the homesite suitability cartographic model. Three basic data layers are entered. A sequence of spatial operations is used to apply criteria and produce a map of suitable sites.

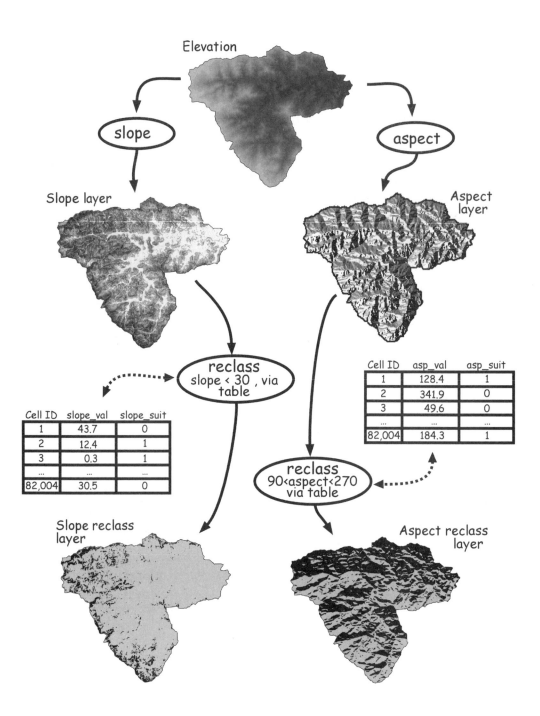

Figure 13-11: A detailed depiction of the left-most branches of the cartographic model shown in Figure 13-10. Slope and aspect are derived from an elevation data layer for the study region. Both layers are then reclassified using a table assignment. Slope values < 30 are reclassified as suitable (gray), all other slopes as unsuitable (black). Aspect values between 90 and 270 are reclassified as suitable (gray), all others as unsuitable (black).

slope values are shown in light shades, higher slope values are shown in dark shades, and aspects are shown in a range of light to dark shades from 0 to 360 degrees. Slope and aspect layers are reclassified based on the threshold values specified in the criteria listed in Table 13-1. A reclassification table is used to assign values to the **slope_suit** variable based on the slope layer. Cells with a **slope_val** less than 30 are assigned a **slope_suit** of 1, while cells with a **slope_val** of 30 or higher are given a value **slope_suit** of 0. Aspect values are also reclassified using a table.

Slope and aspect layers are combined in an overlay, converted from raster to vector, and reclassified to produce a suitable terrain layer (Figure 13-12). Raster to vector conversion is chosen because two of the three base data layers are in a vector format, and because future complex selections might be better supported by the attribute data structure used for vector data sets. This conversion creates polygons that have the attributes of the input raster data layer. Note this conversion takes place after the raster layers have been reclassified into a small number of classes, and after the data have been combined to a single layer in an overlay. Raster-to-vector conversion proceeds more quickly after the number of raster classes has been reduced and the data combined in a single terrain-suitability layer.

The terrain overlay must then be reclassified to identify those areas that meet both the slope and the aspect criteria (see the terrain suitability coding in Figure 13-12). Those polygons with a 1 for both **slope_suit** and **asp_suit** are assigned a value of 1 for **terrain_suit**. All others are given a value of 0, indicating they are unsuitable home sites based on the slope and/or aspect criteria.

Because we wish to reduce the number of redundant polygons where possible, a dissolve is applied after the reclassification. This substantially reduces the size of the output data set, and speeds future processing. Reclassified, dissolved terrain data are saved in a layer labeled **Suitable Terrain** (Figure 13-12).

The central branch of the cartographic model is shown in Figure 13-13. Digital soil surveys are available that depict homogenous soil units as polygons. Attribute data are attached to each polygon, including soil type and soil suitability for septic systems. Soils data for the study area may be reclassified based on these septic suitability attributes. A reclassification table assigns a value of 1 to the variable **soil_suit** if the soil type is suitable for septic systems, 0 if the soil type is not (Figure 13-13).

After reclassification there may be many adjacent soil polygons with the same **soil_suit** value. These are grouped using a dissolve operation (data between reclass and dissolve are not shown in the figure; see Chapter 9 for an example). The dissolve removes boundaries between like polygons, thereby substantially reducing the number of polygons and hence the number of entries in the attribute table. This may be particularly important with complex data sets such as soils data, or with converted raster data, as these often have thousands of entries, many of which will be combined after the dissolve.

The right branch of the cartographic model is presented in Figure 13-14. The **Roads** data layer is obtained and **Major roads** extracted. This has the effect of removing all minor roads from consideration in further analyses. What constitutes a major road has been defined prior to this step. In this case, all divided and multi-lane roads in the study area were selected. Two buffers are applied, one at a 300-meter distance and one at a 2-kilometer distance from major roads. These buffers are then overlain. Because the buffer regions extend outside the study area, the buffers must be clipped to the boundary of the study area. These data are then reclassified into suitable and unsuitable areas, resulting in the **Road buffers** layer (lower left, Figure 13-14).

All data layers are combined in a final set of overlays and reclassifications (Figure 13-15). The **Suitable access** layer, derived from the roads data and criteria, is combined with the **Terrain & soils** layer. The **All criteria** layer contains the required spatial data to identify suitable vs. unsuitable sites. This overlay layer must be reclassified based on the road, soil, and terrain suitability variables, classifying all potential sites into a suitable or unsuitable class. A final dissolve yields the final digital data layer, `Suitable sites`.

This example analysis, while simple and limited in scope, illustrates both the flexibility and complexity of spatial data analysis using cartographic models. In some respects the cartographic model was simple because only three input spatial data layers were required, and a small set of spatial data operations were used. Reclass, overlay, and dissolve operations were used repeatedly, with

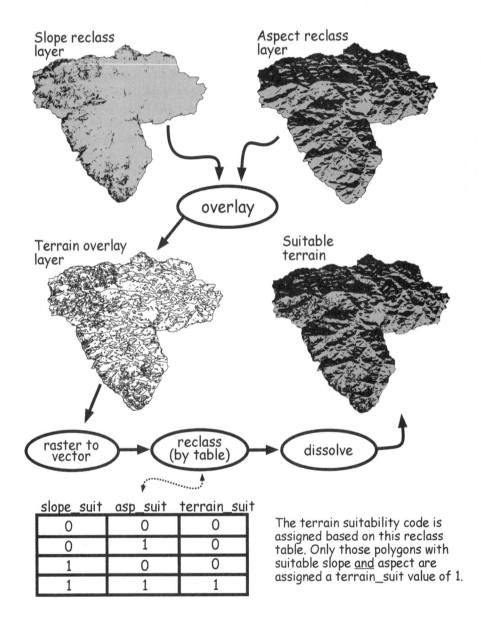

slope_suit	asp_suit	terrain_suit
0	0	0
0	1	0
1	0	0
1	1	1

The terrain suitability code is assigned based on this reclass table. Only those polygons with suitable slope <u>and</u> aspect are assigned a terrain_suit value of 1.

Figure 13-12: The recoded slope and aspect data layers are combined in an overlay operation, and the result reclassified. Suitable terrain is shown in gray, unsuitable in black.

buffer, slope, aspect, and raster-to-vector conversion also applied. The modeling is flexible in that the spatial operations may be selected for the particular problem, for example, each recode may be tailored in each application. Finally, this example illustrates the complexity that can be obtained with cartographic modeling, in that over 20 different instances of a spatial operation were applied, in a defined sequence, resulting in at least 15 intermediate data layers as well as the final result layer.

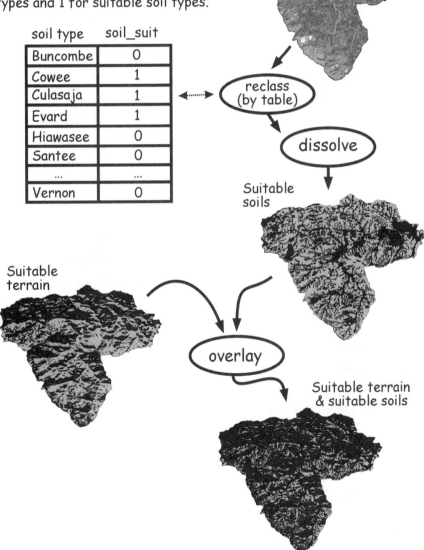

Soils polygons are recoded based on the soil type. The variable soil_suit is assigned a value 0 for unsuitable soil types and 1 for suitable soil types.

soil type	soil_suit
Buncombe	0
Cowee	1
Culasaja	1
Evard	1
Hiawasee	0
Santee	0
...	...
Vernon	0

Soils

reclass (by table)

dissolve

Suitable soils

Suitable terrain

overlay

Suitable terrain & suitable soils

Figure 13-13: A detailed prediction of the center branch of the cartographic model. Soils data are reclassified into those suitable for septic systems and those not, and then combined with the suitable terrain data layer to identify sites acceptable based on both criteria.

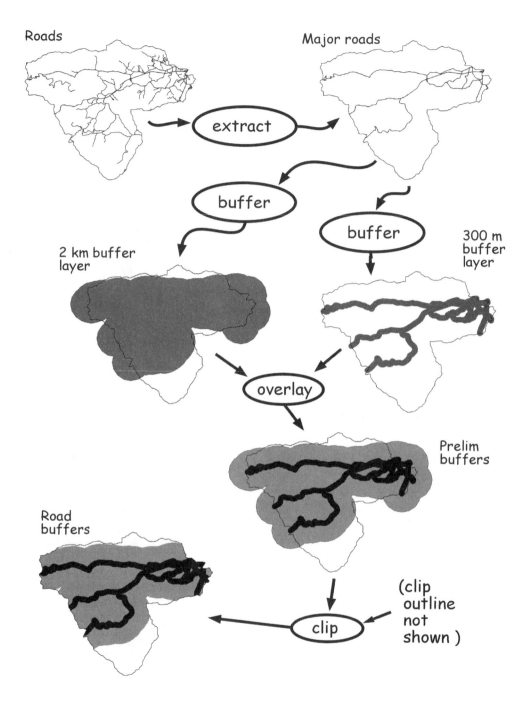

Figure 13-14: A detailed representation of the right branch of the cartographic model shown in Figure 13-10. Roads are buffered at 300 meters and 2 kilometers, and these buffers overlain. The buffers are clipped to the study region, and suitable areas more than 300 meters and less than 2 kilometers from roads identified.

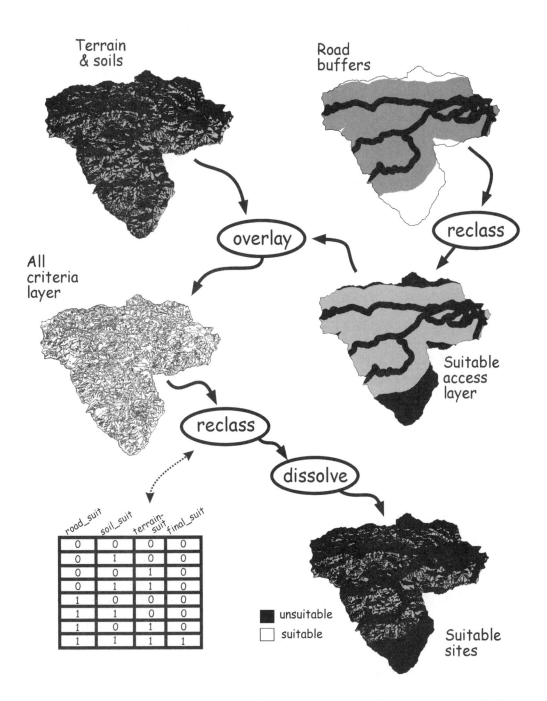

Figure 13-15: The overlay and reclassification of the combined data layers. Terrain, soils, and road buffer data are combined in an overlay. These data are reclassed based on the suitability criteria. A final dissolve is applied to reduce the number of polygons, resulting in a final layer of suitable sites.

Simple Spatial Models

Predictive spatial models are commonly applied, particularly when there is a well-established model based on point or small-scale observations and analysis, and when the output is a continuous variable, e.g., temperature, housing value, soil erosion rates, or cancer frequency.

As noted earlier, our simple spatial models typically are based on one or a few equations, described as:

$$O = f(A, B, C, D, \beta_1, \beta_2, ...) \quad \text{13.2}$$

where O is the spatially-reference output, $f()$ is an equation, A, B, C, D, are variables, and β_i's are equation parameters. For example, NASA has sponsored the development of global models of gross primary productivity (GPP), the total biomass produced globally by plants in any given year. One common model takes the form:

$$GPP = \varepsilon \cdot NDVI \cdot PAR \quad \text{13.3}$$

Where NDVI is a satellite-based measure of plant abundance, PAR is the amount of sunlight available for photosynthesis, and ε is a conversion efficiency, which may be fixed, or may depend on additional factors, such as vegetation type or soil dryness or type. In this example our equation is simple multiplication of the components, and ε is the unique parameter in the simplest case of a fixed ε. In more complicated forms, there is a different ε for each vegetation type, applied accordingly.

Simple spatial models require spatial fields of all variables, and appropriate parameters for all conditions in the modeled area. In our GPP example above, we must have estimates of NDVI and PAR over our prediction region. In this specific case, robust measurements of NDVI have been developed based on repeat satellite measurements, as have methods to estimate PAR from the available meteorology networks

and measurement systems. Values of ε have been estimated for dominant vegetation types, and how these parameters vary with other environmental factors like temperature and available moisture. Model estimates have been compared to measurements across a broad range of conditions.

While we call these simple spatial models, as the previous and subsequent examples will show, it is often time consuming and difficult to develop the spatial data and estimate the parameters to apply these models across space. The models are often based on observed relationships and measurements at points or small plots, for example, crop growth on sunny vs. cloudy days, or the change in GPP across nearby forest stands with different NDVI values. These may suffice to estimate ε for the specific types, but differences among vegetation types may require repeat measurements over a broad range of conditions. A network of field stations, perhaps in combination with remotely-sensed data, may be required to estimate the input variables, e.g., PAR at the required intervals across the landscape.

The Revised Universal Soil Loss Equation (RUSLE) and its precursor the Universal Soil Loss Equation (USLE) is one of the most widely-used, simple spatial models:

$$E = R \cdot K \cdot C \cdot P \cdot L \cdot S \quad \text{13.4}$$

where E is average annual erosion, R is a rainfall factor, K reflects soil erodibility, C integrates crop effects, P accounts for management practices, L reflects slope length, and S represents steepness.

The USLE/RUSLE predict soil erosion on farm fields, and have been under development since the 1930s. Rainfall intensity, soil properties, crop type, slope steepness, and slope length factors have been measured in tens of thousands of plots. Supporting information has been developed for the entire country by the U.S. Natural Resource Conservation Service, including soil and climate factors for the U.S., and the impacts of common crop types and management regimes. The USLE and RUSLE has been widely applied in other countries.

The RUSLE has been widely applied within a GIS framework for erosion estimates on a catchment or larger scales. The model is relatively simple, much of the input data have been developed and are publicly available, and the output of broad interest. Methods for applying the model have varied, in part because the model was developed for individual fields, but spatial data are often not available on a per-field basis. While the rainfall factor, R, is generally similar across county-sized areas spanning tens of kilometers, other factors often change on a field or sub-field basis. Applications often differ in the methods for estimating the management and crop factors, and in particular a combination of slope length and steepness factors.

Estimating driving variables across space often presents choices, as illustrated in the calculation of the RUSLE slope factors, L and S (Figure 13-16). Simple spatial models are often based on small area studies for which all variables may be easily measured. This is often not true when applying the models to larger areas. Slope steepness (S factor) is easily estimated within a raster framework, but slope length (L) is considered uniform at a fixed length of 22.1 m (72.6 ft) in the standard RUSLE. Application of the RUSLE to convergent or divergent slopes or to lengths or cell sizes different than the standard may result in prediction errors. This challenge has been the focus of

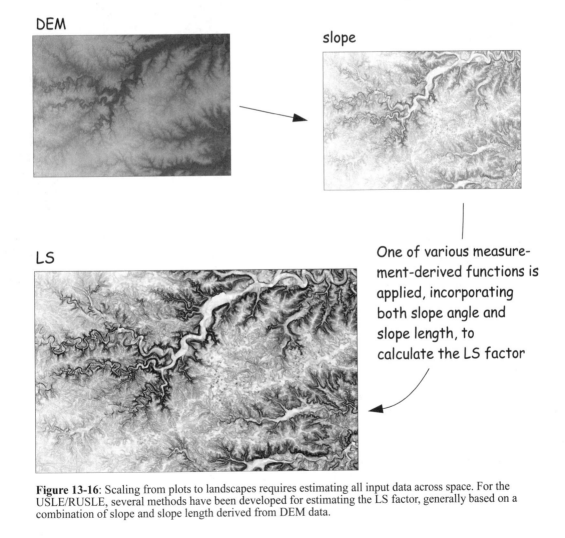

DEM

slope

One of various measurement-derived functions is applied, incorporating both slope angle and slope length, to calculate the LS factor

LS

Figure 13-16: Scaling from plots to landscapes requires estimating all input data across space. For the USLE/RUSLE, several methods have been developed for estimating the LS factor, generally based on a combination of slope and slope length derived from DEM data.

many studies, and the book chapter by Wilson and Lorang, listed in the references, describes some of the methods used to effectively estimate a combination of L and S.

Remaining K, C, and P factors may be derived from standard spatial data sets, e.g., NASS or NLCD data for landcover/crop type and treatment, and K factors from SSURGO data (Figure 13-17). Application of the model to the spatial data, here in a cell-by-cell multiplication, yields estimates of erosion across a region.

Erosion = R K C P L S

1) R is constant for the study area, within Filmore County, and set at 155, from NRCS literature

2) P is assumed constant across all types, with a value of 0.5

3) K is derived from SSURGO soils data, contained in the horizon table. K values are extracted for the surface horizon:

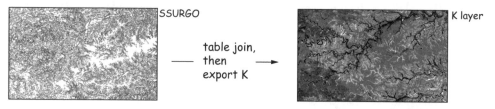

table join, then export K

4) C values are assigned via a reclassification, based on NRCS tables per crop type

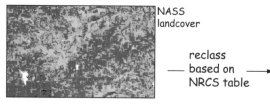

reclass based on NRCS table

5) LS values are calculated together, according to McCool et al., 1987, 1989:

$$LS = \left[\frac{\lambda}{a}\right](b\ sin\beta + c)$$

6) RULSE model applied on a cell by cell basis:

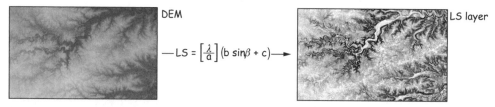

K
C
LS
P · R

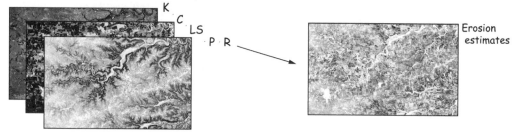

Erosion estimates

Figure 13-17: USLE/RUSLE erosion estimates may be calculated from appropriately developed base layers.

Spatio-temporal Models

Spatio-temporal models have been developed and applied in a number of disciplines. This is an active area of research, as there are many fields of study and management that require analysis and predictions of spatially and time varying phenomena. We will briefly discuss some basic characteristics of spatio-temporal models. We will then describe their differences from other models, some basic analysis approaches, and describe two examples of spatio-temporal models.

Spatio-temporal models use spatially-explicit inputs to calculate or predict spatially-explicit outputs (Figure 13-18). Rules, functions, or some other processes are applied using spatial and often non-spatial data. Input variables such as elevation, vegetation type, human population density, or rainfall may be used as inputs to one or more mathematical equations. These equations are then used to calculate a value for one or more spatial locations. The values are often saved in a spatial data format, such as a layer in a GIS.

Spatio-temporal models involve at least a three dimensional representation of one or more key attributes – variation in planar (x-y) space and through time. A fourth dimension may be added if the vertical (z) direction is also modeled. We arbitrarily treat spatially variable network analyses separately, because networks are constrained to a subset of two-dimensional space. Spatio-temporal models may also be classified by a number of other criteria: whether they treat continuous fields or discontinuous objects, if they are process-based or rely on purely-fit models, and if they are stochastic or deterministic. Combinations of these model characteristics lead to a broad array of spatio-temporal model types.

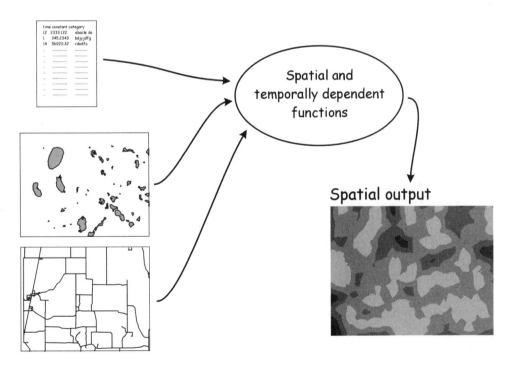

Figure 13-18: Spatio-temporal models combine spatial and aspatial data with time-variant functions to produce spatial output.

Models of continuous phenomena predict values that vary smoothly across time or space. Air temperature, precipitation, soil moisture, and atmospheric pollutants are examples of continuous variables that are predicted using spatio-temporal models. Soil moisture this month may depend on soil moisture last month and the temperature, precipitation, and sunshine duration in the intervening period. All these factors may be entered in spatial data layers, and the soil moisture predicted for a set of points.

Models of discrete phenomena predict spatial or attribute characteristics for discontinuous features. Boundaries for vegetation types are an example of features that are often considered discrete. We use a line to identify the separation between two types, for example, between a grassland and a forest. A spatial model may consider the current position of the forest and grassland as well as soil type, fire prevention, and climatic data to predict the encroachment of forest on grassland sites. The boundaries between new forests and grasslands are always discrete, although their positions shift through time.

Models are considered process-based if their workings in some way represent a theoretical or mechanistic understanding of the processes underlying the observed changes, and models are purely-fit models when they do not. We may predict the amount of water flowing in a stream by a detailed spatial representation of the hydrologic cycle. Many processes may be explicitly represented by equations or subroutines in a spatial model. For example, rainfall location and intensity may be modeled through time for each raster cell in a study area. We can then follow the rainwater as it infiltrates into the soils and joins the stream system through overland flow, subsurface flow, and routing through stream channels. Calculations for these processes may be based on slope, topography, and channel characteristics. These processes are tied together in space. Calculations are performed at each point on the landscape; these calculations increase or decrease water

flow or other conditions at adjacent, downslope locations.

Rainfall might be modeled differently using a purely fit, statistical approach. A purely fit model might simply measure precipitation in the previous hour and average the precipitation for the previous week and previous month, and predict stream flow at a point. Processes such as evaporation or subsurface flow are not explicitly represented, and the output may be a statistical function of the inputs. The model may be more accurate than a process-based approach, in that the predicted outflow at any point in the stream may be closer to measured values than those derived from a process-based model. Conversely, the output may be poorer, in that the measurements may be farther from predictions. Process modelers argue that by incorporating the structure and function of the system into a process model we may better predict under new conditions, for example, for extreme drought or rainfall events never experienced before. They also argue that process models aid in our understanding a system and in generating new hypotheses about system function.

Besides being continuous or discrete and process or fit, models may be stochastic or deterministic. A deterministic model provides the same outputs every time it is given exactly the same inputs. If we enter a set of variables into a model without modifying the model, it will always produce exactly the same results. A stochastic model will not. Stochastic models often have random generation or some other variability generation procedures that change model results from run-to-run, even when using exactly the same inputs.

A disease spread process is a good example of a phenomenon that might be modeled with a stochastic process. Disease may occur at a set of locations, and may be spread through the atmosphere, water, or carried by animals or humans to initiate new disease centers. A doctor might model disease infection and growth stochastically. A random number might be generated, and the new center started at a location based on this number. The doctor might use another totally or partially random process to control how the new infection center grows or "dies" in the spatial model. Thus, the map of disease locations after different model runs may differ, even though the runs were initiated with identical input conditions.

With most spatial models, the target location of the model output is usually, but not always, the location of the inputs. For example, a demographics model may use a combination of current population in a census tract, housing availability and cost, job opportunities and location, general migration statistics, and age and marital status of those currently in the census tract to predict future population for the census tract. This model has a target location, the census tract, that is the same as the location for most of the input data.

In contrast, the target location of the model outputs may be different than the location of the inputs. Consider a fire behavior model. This model might predict the location of a wildfire based on current fire

location, wind speed, topography (fires burn faster upslope than down), and vegetation type and condition. Fire models often incorporate mechanisms to predict fire spread beyond the current burn front of a fire. Embers often are lifted above a fire by the upwelling heated air. These embers may be blown well in advance of a fire front, starting spot fires at some distance away from the main fire. In this case, the target location for a calculation in the spatial model is not the same as the input locations.

Cell-Based Models

Spatial-temporal models often are implemented as *cell-based models*. A cell-based model invokes a set of functions and logic, driven by cell values, to update these or other cell values through time. Input values at a starting time, t_0, may be derived from multiple layers. These input values are entered into functions that calculate the new values for the target layer or layers at the next time step, t_1. The process is then

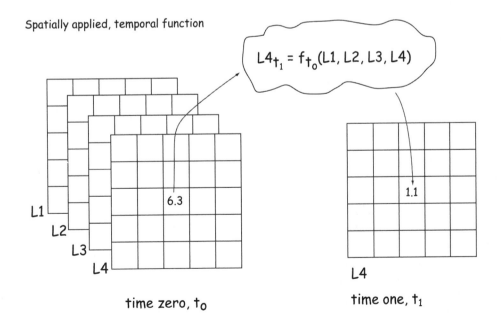

Spatially applied, temporal function

$$L4_{t_1} = f_{t_0}(L1, L2, L3, L4)$$

L1
L2
L3
L4

6.3

1.1

L4

time zero, t_0

time one, t_1

Figure 13-19: Time varying cell-based models use one to multiple input cell values from a starting time, t_0, and update cell values for some layer(s) at time t_1. The process is repeated for a specified length of time or until some specific output condition is reached.

repeated, and the values in the target layer(s) evolve through time (Figure 13-19).

Erosion due to surface runoff provides a good example of how cell-based modeling can be used. Erosion is the displacement and movement of soil from a surface. Although there are many erosive forces, water is the primary cause of erosion over most of the globe. The amount of soil erosion depends on many factors, including the rainfall rate, how fast the rainwater is absorbed by the soil (permeability), the type of soil, the slope at the site, and how much water is flowing from uphill cells. Some of these properties do not change with time, for example, slope or soil type, while other features do, such as rainfall rate and downflowing water. All of these factors may be provided as cell-based layers, some that change with time, and some that are static. These layers are then included in an equation to calculate erosion at each cell location for a grid. Rainfall and flow rates may be updated at each step, and the resultant erosion calculated and placed in an output layer, as shown in Figure 13-19.

Cellular automata (CA) are one of the simplest forms of cell-based models. Cellular automata operate on a single cell-based layer, using simple rules to derive the next state from the current state. Cells typically have two states, on and off or dark and light. The cell is on or off in the next step depending on the arrangement of on/off cells in the vicinity of the target cells. Cell values at each time step completely determine the cell pattern in the next time step, as shown in Figure 13-20. The four simple transition rules shown at the top of the figure govern cell assignment from one time step to the next. Starting with a single cell switched on in the black state at t_0, cells are turned on and off as the layer transitions to time steps t_1 through t_3.

While simple to construct and implement, cellular automata can produce quite sophisticated patterns and behavior. Convoluted, coral-like structures may evolve from simple beginnings. Patterns may grow and contract, appear to "consume" other patterns, or move across the field of cells.

Cellular automata have until recently been rarely applied for modeling outside of academic and research environments. While CA are useful learning tools and capable of providing some insights into how simple systems can exhibit complex patterns, there are few problems for which the simple space-transition rules may be clearly specified. Recent work has advanced the application of cellular automata, particularly in modeling urban development and land use.

Most temporally dynamic cell-based models are more complex than cellular automata. These models represent the levels of a variable or category with a number that can have more than the two possible values, and they use a function-based approach as shown in Figure 13-19.

Agent-based Modeling

Agent-based models are a third type of spatio-temporal model, often applied in cell-based environments. Also known as individ-

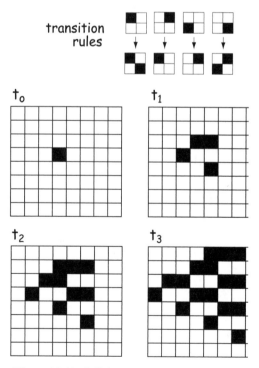

Figure 13-20: Cellular automata use simple transition rules to determine the time-step evolution of cell values.

ual-based models, these systems contain "agents" that may move across digital landscapes. The landscape is often described by a set of variables stored in cells. Agents may range across this grid, reacting to the environment or to other agents according to a small set of pre-specified rules. Complex behaviors may be exhibited from relatively simple sets of few rules.

Earliest agent-based models were designed as demonstrations, and addressed the seemingly orchestrated behavior of group action in animals, such as the wheeling movements of flocks of birds or schools of fish. This work is well described in the paper by Craig Reynolds (1987), listed in the references at the end of this chapter. These animal actions may be simulated rather closely using three rules of individual behavior, resulting in apparent coordination and complex aggregate behavior from a small set of instructions. Each agent, called a "boid," was governed by a desire to head in the same direction as its neighbors, move toward the center of its neighbors when the were "too far" from it, and move away from an individual neighbor when it was "too close." The artificial flocks were "released" in a virtual world, and the agents began to move in concert, turning with ever changing "leaders" as the flock moved about its environment. These early programs form the basis of much computer animation of group movement found in present-day films and video games.

Agent-based models are currently active areas of research and application, primarily in pedestrian and motorized transportation analyses, and in conservation biology. These analyses may be used in turn to help design better cities, roads or other infrastructure. For example, a network of sidewalks and pathways through a new facility may be represented in a raster dataset. Individual agents may be created to represent pedestrians, and programed with a set of simple behaviors, for example, minimum proximity to other pedestrians, pauses at exits/entryways or when following other pedestrians. Groups of agents may be placed in the envi-

ronment, and allowed to traverse according to their guiding behavior. Capacity limits, bottlenecks, and speed or flow may be monitored, and the effects of design changes considered.

While there are perhaps as many modeling frameworks and methods as there are models, we have attempted to describe a few of the most commonly applied approaches. We will now describe two examples of spatially-based modeling.

Example 1: Process-based Hydrologic Models

Water flows downhill. This simple knowledge was perhaps sufficient until humans began to build houses and roads, and populations grew to dominate most of the Earth's surface. Land scarcity has led humans to build in low-lying areas, and farming, wetland drainage, and upstream development have all contributed to more frequent and severe flooding.

Humans have been improving their understanding of water movement since the beginning of civilization, but the need for quantitative, spatially-explicit water flow models has increased substantially over the past few decades. Water models are needed because demands for water resources are exceeding the natural supply in many parts of the world. Water models are also needed because population pressures have driven farms, cities, and other human developments into flood-prone areas; these same developments have increased the speed and amount of rainfall runoff, thereby increasing flood frequency and severity. These factors are spurring the development of spatio-temporal hydrologic models. The models are often used to estimate stream water levels, such that we may better manage water resources and avoid loss of property or life due to flooding.

Many spatio-temporal hydrologic models predict the temporal fluctuations in soil moisture, lake or stream water levels, and discharge in hydrologic networks. The network typically consists of a set of connected rivers and streams, including impoundments such as lakes, ponds, and reservoirs (Figure 13-21). This network typically has a branching pattern. As you move upstream from the main discharge point for the network, streams are smaller and carry less water. Water level or discharge may be important at fixed points in the hydrologic network, at fixed points on land near the network, or at all points in the landscape. The hydrologic network is often embedded in a watershed, defined as the area that contributes downslope flow to the network.

Spatially-explicit hydrologic models are almost universally dependent on digital elevation data. DEMs are used to define watershed boundaries, water flow paths, the speed of downslope movement, and stream location (Chapter 12). Slope, aspect, and other factors that effect hydrologic systems may

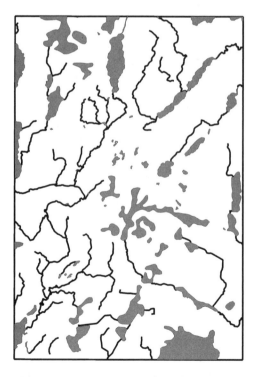

Figure 13-21: An example of a hydrographic network. Lakes and/or rivers form an interconnected network. Water may be routed from upland areas to and through this network.

be derived from DEMs. For example, evaporation of surface water and transpiration of soil water depend on the amount of solar radiation. Site solar radiation depends on the slope and aspect at each point, and in mountainous terrain it may also depend on surrounding elevations – sunrise is later and sunset earlier in valley locations, and north facing slopes in the northern hemisphere receive significantly less solar radiation than south facing slopes. Site-specific variables representing slope and aspect are used when estimating evaporation or plant use of water.

Slope and aspect are often used to define an important spatial data layer in hydrologic modeling – flow direction. This layer defines the direction of water flow at important points on the surface. If a raster data structure is used, flow direction is calculated for every cell. If a vector data structure is used, flow direction is defined between adjacent or connected vector elements.

Many spatio-temporal hydrologic models adopt a raster data structure. Raster data structures preserve variation in surface elevation that drives water movements. As described in Chapter 2, raster data sets have a relatively simple structure and so are easily integrated into hydrologic models. The connection between adjacent cells is explicitly recorded in raster data sets, so flow between cells is easily represented.

Most raster-based hydrologic models represent water flow through each grid cell (Figure 13-22). Water falls on each cell via precipitation. Precipitation either infiltrates into the soil or flows across the surface, depending on the surface permeability at the cell. For example, little water infiltrates for most human-made surfaces, such as parking lots or buildings. These sites have low permeability, so most precipitation becomes surface flow. Conversely, nearly all precipitation infiltrates into most forest soils.

Downslope water flow is also calculated in the model, depending on a number of factors at each cell. Slope and flow direction determine the rate at which water flows downhill. Downslope flow eventually reaches the hydrologic network and is routed via the network to the outlet. Mathematical functions describing cell-specific precipitation, flows, and discharge may be combined to predict the flow quantity and water level at points in the watershed and through the network.

Spatio-temporal hydrologic models often require substantial data development. Elevation, surface and subsurface permeability, vegetation, and stream network location must be developed prior to the application of many hydrologic models. DEM data may require substantial extra editing because terrain largely drives water movement. For example, local sinks occur much more frequently in DEMs than in real surfaces. Sinks may occur during data collection or during processing. Sinks are particularly troublesome when they occur at the bottom of a larger accumulation area. Modeled water may flow into the sink but may not flow out, depending on how water accumulation is modeled, while on the real earth surface the water may flow freely downhill. Local spikes in the model may push water incorrectly to surrounding cells, although they typically cause fewer problems than sinks. Both sinks and spikes must be removed prior to application of some hydrologic models.

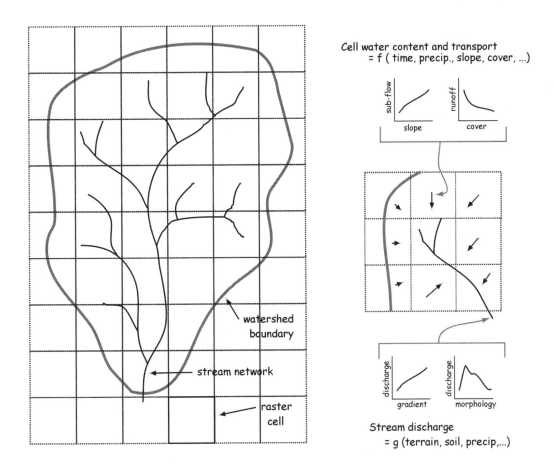

Figure 13-22: Watershed and stream network hydrology may be modeled in a raster environment. Cell characteristics for a watershed are modeled, and water accumulation and flow driven through the system. Soil water, stream levels, and stream discharge depend on spatially- and time-dependent functions.

Example 2: LANDIS, a Stochastic Model of Forest Change

Many human or natural phenomena are analyzed through spatially-explicit stochastic models. Disease spread, the development of past societies, animal movement, fire spread, and a host of other important spatial phenomena have been modeled. All these phenomena have a random element that substantially affects their behavior. Events too obscure or complex to predict may cause large changes in the system action or function. For example, wind speed or dryness on a given day dramatically affects fire spread, yet wind speed is notoriously difficult to predict. Spatially-explicit, stochastic models allow us to analyze the relative importance of component inputs and processes and the nature and variability of system response. Is it stochastic variation in wind, fuel amount, or fuel type that is most responsible for the variable nature of fire spread? We will discuss one spatial-stochastic model -- LANDIS -- that incorporates techniques used in a wide range of models.

Forest vegetation changes through time. Change may be caused by the natural aging and death of a group of trees, replacement by other species, or may be due to periodic disturbances such as fire, windstorms, logging, insects, or disease outbreaks. Because trees are long-lived organisms, the composition and structure of forests often change on temporal scales exceeding a normal human life span. Human actions today may substantially alter the trajectory of future change. We often need to analyze how past actions have led to current forest conditions, and how present actions will alter future conditions.

Forest disturbance and change are important spatial phenomena for many reasons. Humans are interested in producing wood and fiber, preserving rare species, protecting clean water supplies and fish spawning areas, protecting lives and property from wildfires, and enjoying forest-based recreation.

Forest change is inherently a spatial phenomenon. Fires, diseases, and other disturbances travel across space. The distribution of current forests largely affects the location and species composition of future forests. Seeds disperse through space, aided by wind and water or carried by organisms. Physical and biotic characteristics that largely determine seed and seedling survival and subsequent forest growth are variable in space. Some plants are better adapted to grow under existing forests, while others are aided by disturbances that open the canopy. Some species change soil or understory conditions in ways that prevent other species form growing beneath them. Plant succession, the replacement of one group of plants or species by another through time, is substantially affected by the current forest distribution and structure. It is not surprising that many process-based models of forest change incorporate spatial data.

Forests are extremely heterogeneous in space, and this complicates our understanding and predictions of forest change. Tree species, size, age, soils, water availability, and other factors change substantially over very short distances. Each forest stand is different, and we struggle to represent these differences. Given the long time scales, broad spatial scales, and inherent spatial variability of forests, many organizations have developed models based on spatial data, models that are in some way integrated into GIS.

LANDIS (LANdscape DISturbance) is an example of a spatially-explicit, process-based forest dynamics model. LANDIS has been developed by Dr. David Mladenoff and colleagues and has been applied to forest biomes across the globe. LANDIS incorporates natural and human disturbances with models of seed dispersal, plant establishment, and succession through time to predict forest composition over broad spatial scales and for its long temporal scales. LANDIS is notable for the broad areas it may treat at relatively high resolution, and long temporal scales. LANDIS has been used to model forest dynamics at a 30-meter resolution, over

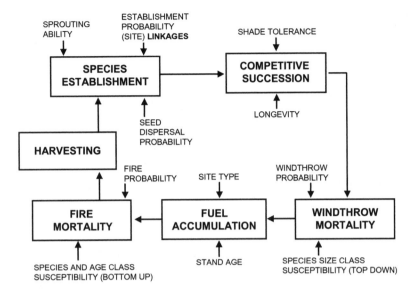

Figure 13-23: The major processes represented in LANDIS, a spatio-temporal forest succession model. (courtesy D. Mladenoff)

tens of thousands of hectares, and across five centuries.

LANDIS integrates information about forest disturbance and succession to predict changes in forest composition (Figure 13-23). Succession is the replacement of species through time. Succession is common in forests, for example when fast-growing, light-demanding tree species colonize a disturbed site, and are in turn replaced by more shade-tolerant, slower growing species. These shade-tolerant species may be self-replacing in that their seeds germinate and seedlings survive and grow, albeit slowly, in the dense shade. Small gaps from branch breakage or individual tree deaths allow small patches of light to reach these shade-tolerant seedlings, enabling them to eventually reach the upper canopy. This self replacement can result in a stable, same-species stand over long time periods. This cycle may be broken due to fire, windthrow, logging, or other disturbance event that opens up a stand to a broader range of species. LANDIS simulates large, heterogeneous landscapes, incorporates the interactions of dominant tree species, and includes spatially explicit

representations of ecological interactions. The model has been optimized to simulate millions of acres in reasonable run times, less than a day on desktop computers at the time of this writing.

LANDIS Design Elements

The design of LANDIS is driven by the overall objectives for the model, simulating forest disturbance and succession through time. LANDIS also satisfies a number of other requirements. LANDIS readily integrates satellite data sets and other appropriate spatial data, and it simulates the basic processes of disturbance, stand development, seed dispersal, and succession in a spatially explicit manner. These requirements led to the adoption of a number of specific design features in the model.

LANDIS is an object-oriented model. Specific features or processes are encapsulated in objects, and object-internal processes are isolated as much as possible from other portions of the model. As an example, there is a SPECIE object that encapsulates most of the important information and processes for each tree species included in the model. Each instance of a SPECIE has a name, for example, "Aspen," and other characteristics such as longevity, shade tolerance, or age to maturity, as well as methods for birth, death, and other actions or characteristics. Because these characteristics and processes are encapsulated in a SPECIE object, they may be easily changed as new data or a better understanding of forest succession processes become available. Many models are incorporating this object-oriented design, because it simplifies maintenance and modifications.

LANDIS uses a raster data model which eases the entry of classified satellite imagery, elevation, and other data sets reflecting short-range environmental and forest species variation. Interactions such as seed dispersal, competitions, and fire spread are explicitly modeled for each species occupying each grid cell.

LANDIS tracks the presence of age classes (cohorts) for a number of species in each cell and through time. The model begins with an initial condition: the distribution of species by age class across the landscape. Ten-year age classes are currently represented. The longevity, age of initial seed production, seed dispersal distance, shade tolerance, fire tolerance, and ability to sprout from damaged stumps or roots is recorded for each species. On undisturbed sites, cohorts pass through time until they reach their longevity. Older cohorts "die" and disappear from the cell. Younger cohorts may then appear, depending on the availability of seed.

The spatially explicit representation of seed sources and dispersal is an improvement of LANDIS over many earlier forest succession models. Previous models typically assumed constant or random seed availability. LANDIS is representative of spatially-explicit models, in that the specific locations of a process affect that process. Disturbed sites may be occupied by seedlings from a disturbed cell or nearby cells, or by sprouting from trees in a cell prior to disturbance. Cells cycle through the species establishment, succession, disturbance, and mortality processes (Figure 13-24).

The effects of site characteristics on species establishment and interactions are also

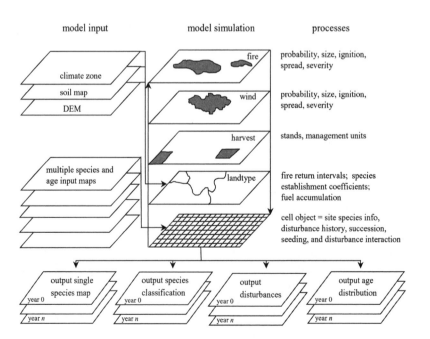

Figure 13-24: Basic spatial data and processes represented in LANDIS

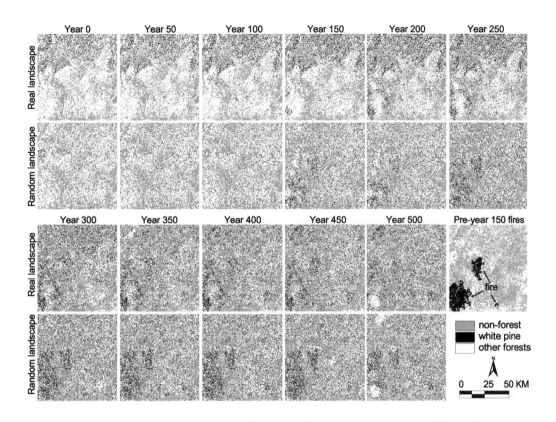

Figure 13-25: Changes in the spatial distribution of white pine, a forest tree species, through time as predicted by LANDIS. This graphic exemplifies the prediction of a feature of interest both spatially and temporally, and is representative of many analytical tools in use or under development.

represented in LANDIS. For example, establishment coefficients are used to represent the interaction between site characteristics and species establishment. Establishment coefficients vary by land type. Fire severity also varies by land type, as may seedling survival. Elevation, aspect, soils, and other spatial data are used as input to the spatial model.

Fire and wind disturbances are simulated based on historical records of disturbance sizes, frequencies, and severities. Disturbances vary in these properties across the landscape. For example, wind disturbances may be more frequent and severe on exposed ridges, and fires less frequent, less intense, and smaller in wetlands. Disturbances are stochastically generated, but the variability depends on landscape variables,

for example, fires are generated more frequently on dry upland sites.

LANDIS has been applied to a number of forest science and management problems, including the effects of climate change on forest composition and production, the impacts of changing harvesting regimes on landscape patterns, and regional forest assessments (Figure 13-25).

Hundreds of other spatially-explicit, temporally dynamic models have been developed, and many more are currently under development. As spatial data collection technologies improve and GIS systems become more powerful, spatio-temporal models are becoming standard tools in geographic science, planning, and in resource management.

Summary

Spatial analysis often involves the development of spatial models. These models can help us understand how phenomena or systems change through space and time, and they may be used to solve problems. In this chapter we described cartographic models, and spatio-temporal models.

Cartographic models often combine several data layers to satisfy a set of criteria. Data layers are combined through the application of a sequence of spatial operations, including overlay, reclassification, and buffering. The cartographic model may be specified with a flowchart, a diagram representing the data layers and sequence of spatial operations. Cartographic models are static in time relative to the other two model types.

Spatio-temporal models explicitly represent the changes in important phenomena through time within the model. These models are typically more detailed, and less flexible than cartographic models, in part because spatial-temporal models often include some representation of process. For example, many spatio-temporal models have been developed to model the flow of water through a region, and these models incorporate equations regarding the physics of water transport movement. Models may be stochastic or deterministic, process-based or statistical, or they may have a combination of these characteristics.

Suggested Reading

Anselin, L., Syabri, I, & Kho, Y. (2006). GeoDA: an introduction to spatial data analysis. *Geographical Analysis*, 38:5-22.

Burrough, P.A., & McDonnell, R.A. (1998). *Principles of Geographical Information Systems*. Oxford University Press: Oxford.

Brady, M., & Irwin, E. (2011). Accounting for spatial effects in economic models of land use: recent developments and challenges ahead. *Environmental and Resource Economics*, 48:487-509.

Brown, D., Riolog, R., Robinson, D.G., North, M., & Rand, W. (2005). Spatial processes and data models: Towards integration of agent-based models and GIS. *Journal of Geographical Systems*. 7:25-47.

Carlson, S. (2000). The amateur scientist: Boids of a feather flock together. *Scientific American*, 283:112-114.

Clarke, K.C., Hoppen, S., & Gaydos, L. (1997). A self-modifying cellular automaton model of historical urbanization in the San Francisco Bay area. *Environment and Planning*, 24:247-261.

Cliff, A.D. and Ord, J.K. (1981). *Spatial Processes: Models and Applications*. Pion: London.

Cooke, W.H.K., Katarzyna, G., & Wallis, R.C. (2006). Avian GIS models signal human risk for West Nile virus in Mississippi. *International Journal of Health Geography*, 5:36, doi:10.1186/1476-072X-5-36.

Fotheringham, S., & Wegener, M. (2000). *Spatial Models and GIS: New Potential and New Models*.Taylor and Francis: London.

Goodchild, M.F., Steyaert, L. T., & Parks, B. O. (1996). G*IS and Environmental Modeling: Progress and Research Issues*. GIS World Books: Fort Collins.

Griffith, D.A., & Layne, L.J. (1999). *A Casebook for Spatial Statistical Data Analysis*, Oxford University Press: Oxford.

He, H.S., Mladenoff, D. J., & Boeder, J. (1999). An object-oriented forest landscape model and its representation of tree species. *Ecological Modeling*, 119:1-19.

Horn, M.E.T. (2004). Modelling and assessment of demand-responsive passenger transport services. In J. Stillwell and G. Clarke (Ed.) *Applied GIS and Spatial Analysis*. Wiley: New York.

Huevelink, G.B.M. & Burrough, P.A. (1993). Error propagation in cartographic modelling using Boolean logic and continuous classification. *International Journal of Geographical Information Systems*, 7:231-246.

Jetten, V., Govers, G., & Hessel, R. (2003). Erosion models: quality of spatial predictions. Hydrologic Processes, 17:887-900.

Johnston, C. (1998). *GIS in Ecology*. Blackwell Scientific: Boston.

Kaufmann, A. (1975). *Introduction to the Theory of Fuzzy Subsets*. Academic Press: New York.

Klir, G.J., & Folger, T.A. (1988). *Fuzzy Sets, Uncertainty, and Information*. Prentice Hall: Englewood Cliffs.

Krzanowski, R. & Raper, J. (2001). *Spatial Evolutionary Modelling*. Oxford University Press: Oxford.

Malczewski, J.C. (1999). *GIS and Multicriteria Decision Analysis*. Wiley: New York.

McCool, D.K., Brown, L.C., Foster, G.R., Mutchler, C.K., & Meyer, LD. (1987). Revised slope steepness factor for the Universal Soil Loss Equation. *Transactions of the American Society of Agricultural Engineers*, 30:1387-1396.

McCool, D.K., Foster, G.R., Mutchler, C.K., & Meyer, LD. (1989). Revised slope length equation for the Universal Soil Loss Equation, *Transactions of the American Society of Agricultural Engineers*, 32:1571-1576.

Mladenoff, D.J., & He, H.S. (1999). Design, behavior and application of LANDIS, an object-oriented model of forest landscape disturbance and succession, In D.J. Mladenoff, & W.L. Baker (Eds.), *Advances in Spatial Modeling of Forest Landscape Change: Approaches and Applications*. Cambridge University Press: Cambridge.

Monmonnier, M. (1993). *How To Lie With Maps*. University of Chicago Press: Chicago.

Moore I.D., Gessler, P.E., Nielsen, G.A., & Peterson, G.A. (1993). Soil attribute prediction using terrain analysis. *Soil Science*, 57:443-452.

Parent, O., & LeSage, J.P. (2010). A spatial dynamic panel model with random effects applied to commuting times. *Transportation Research Part B: Methodological,* 44:633-645.

Pinske, J., & Slade, M.E. (2010). The future of spatial econometrics. *Journal of Regional Science*, 50:103-117.

Stillwell, J.A., & Clarke, G. (2004). *Applied GIS and Spatial Analysis*, Wiley: New York.

Reynolds, C.W. (1987). Flocks, herds, and schools: A distributed behavioral model. *Computer Graphics*, 21:25-34.

Rossiter, D.G. (1996). A theoretical framework for land evaluation. *Geoderma*, 72:165-190.

Running, S.W., Nemani, R.R., Heinsch, F.A., Zhao, M., Reeves, M., & Hashimoto, H. (2004). A continuous satellite-derived measured of global terrestrial primary production. *Bioscience*, 54:547-560.

Turner, M.G., & Gardener, R.H. (Eds.). (1991).*Quantitative Methods in Landscape Ecology*. Springer Verlag: New York.

Wagner, D.F. (1997). Cellular Automata and Geographic Information Systems. *Environment and Planning*, 24:219-234.

Wilson, J & Gallant, J. (Eds.) (2000). *Terrain Analysis: Principles and Applications.* Wiley: New York.

Wilson, J.P., & Lorang, M.S. (2000). Spatial models of soil erosion and GIS, in *Spatial Models and GIS*, Fotheringham, A.S., & Wegener, M. (eds.). Taylor & Francis: London.

Wolfram, S. (1994). *Cellular Automata and Complexity.* Addison-Wesley: Reading.

Study Questions

13.1 - Provide an example of a cartographic model, including the criteria and a flow-chart of the steps used to apply the model?

13.2 - Why must the criteria be refined in many cartographic modeling processes?

13.3 - What do we mean when we say that most cartographic models are temporally static?

13.4 - What is a discrete vs. continuous weighting in an input layer when combining layers in a cartographic overlay? How do you develop a reasonable continuous ranking function, that is, justify the shape of the curve vs. the level of the input variable?

13.5 - Define a cellular automata. How are they different from other cell-based models?

13.6 - What is an agent-based model? How are agents different from other elements in an agent-based model?

13.7 - Describe the main characteristics that distinguish spatio-temporal models from cartographic models.

14 Data Standards and Data Quality

Introduction

A standard is an established or sanctioned measure, form, or method. It is an agreed-upon way of doing, describing, or delivering something. Spatial data and analysis standards are important because of the range of organizations producing and using spatial data and because these data are often transferred among organizations. Data standards facilitate a common understanding of the components of a spatial data set, how data were developed, and the utility and limitations of these data.

GIS practitioners use several types of standards. *Data standards* are used to format, assess, document, and deliver spatial data. *Interoperability standards* identify how spatial data are served between heterogeneous networks of software and hardware systems, e.g., between wireless mobile devices and shared databases. *Analysis standards* ensure that the most appropriate methods are used, and that the spatial analyses provide the best information possible. *Professional* or *certification standards* establish the education, knowledge, or experience of the GIS analyst, thereby improving the likelihood that the technology will be used appropriately.

We have progressed farther in defining spatial data and interoperability standards than in defining analysis and professional standards. This is in part because of the newness of the technology, and in part

because GIS are used in such a wide range of disciplines. Urban planners, conservationists, civil and utility engineers, business people, and a number of other professions use GIS.

National and international standards organizations are important in defining and maintaining geospatial standards. The Federal Geographic Committee (FGDC) is the leading government organization in the United States in defining data standards. The FGDC focuses on the National Spatial Data Infrastructure (NSDI), a set of resources to aid the creation and sharing of digital geographic data. Standards are developed through a set of processes, from proposals through drafts to a FGDC adopted standard. Standards may be modified through an update process. Currently there are standards on methods (e.g., wetlands classification), content (Utilities Data Content Standard), metadata (data about data), and data transfer. Details are at www.fgdc.gov.

There are parallel initiatives in many countries, information on which can be found through the International Spatial Data Standards Commission. The Commission currently serves as a clearinghouse and gateway to national standards across the World.

Spatial data sharing, particularly in real time, is supported by system interoperability. Spatial data software runs on various computing platforms, and connects via physical or wireless networks. Interoperability standards are required because our methods of sharing have evolved through time. Until the late 1990s, most GIS analysis was on local data, meaning the data sets were stored on a hard disk in the same computer as the processing unit, or in a closely connected, often private network. Interoperability between systems was primarily through physical data sharing, via a tape or disk carried from one computer to another, and translators that converted file formats. Data standards focused on file formats and structures. Most computer systems are now connected via networks, most often the World Wide Web. Initially the web only facilitated delivery, and data were still transferred as whole files in older formats, and translated for use in local systems. However, a continuous, ubiquitous network allows the transfer of spatial data components, e.g., coordinates and attribute data, across computer systems, and data stored anywhere on the Web may appear no different than data in local storage.

The Open Geospatial Consortium (OGC) is an ad-hoc, self-selected group of companies, research institutions, government bodies, and individuals dedicated to developing interoperability standards. Interoperation problems are identified, such as general difficulties in accessing time-varying spatial location data through a distributed wireless network, and standards for access proposed. These are reviewed, discussed, amended, and adopted.

Web mapping services (WMS) standards are an example of OGC initiatives. Web mapping services allow GIS software to access data across the internet as if they were stored on the local hard disk. A GIS program or utility "maps" the WMS to the local computer, meaning it may access the data with the same protocols as if it were stored locally, without downloading a permanent copy to store on the local hard disk. WMS are particularly useful for high-resolution image data, which are often viewed for relatively small areas, but which may be collected and stored in large tiles. WMS may be an alternative to downloading large files, providing data on demand and only for an area of interest.

Web services such as WMS are important for the future of *cloud-based computing*, where data, programs, and processing is seamlessly distributed on computers connected across the web. Cloud-based geospatial computing is inherently dependent on robust, well-defined interoperability standards such as those being developed by the OGC. Standards identify data formats and content, parts and naming, metadata, how connections are made and data are passed between programs across distributed networks, and error checking in transfer. Standards allow data to be combined across different organizations, with local storage and access form and protocols, and a standard way of serving up data to others through a service.

It has proven more difficult to develop professional and analysis standards that are inclusive across all disciplines. Standard methods for one discipline may be inappropriate for another. For example, acceptable data collection methods for cadastral surveyors may be different than those for foresters. Field measurement techniques, data reduction, and positional reporting for cadastral surveys often require accuracies measured in centimeters (0.5 inches) or less, while relatively sparse attribute information is recorded. Conversely, forest inventory methods may allow relatively coarse-scale positional measurements, to the nearest few meters, but require standard methods for measuring a large set of attributes.

There has been recent progress on the development of a basic set of standards in the professional practice of GIS in the United States. Known as competency models, they define a set of skills considered essential for effective work in a field, and have been developed for a growing number of industries. All have a common foundation of basic personal and workplace competencies, with industry- and then occupancy-specific skills built on top.

The Geospatial Competency Model

The Geospatial Technology Competency Model (Figure 14-1) identifies a set of core and industry-sector geospatial abilities. The Competency Model identifies examples of over 40 "Critical Work Functions," geographic technology professionals are commonly expected to master and use in their careers, and the background knowledge on which these Critical Work Functions are based. The Geospatial Competency Model is based in part on the Geographic Information Science and Technology Body of Knowledge, first published by the Association of American Geographers in 2006. Critical work functions include operations in basic geodesy, data collection systems, data structures, GIS operation and programing, analyt-ical methods, cartography, the place of geographic information science and technology in society, and organization and institutions. A set of higher level requirements are noted for specific occupations.

We may reasonably expect this competency model to form the basis for professional certification, and to evolve into or form the basis of professional standards of knowledge in geospatial fields. One can envision professional or technical certification based on demonstrated knowledge in these areas, as in professional engineering exams, or perhaps in certification of completion of qualifying curricula. Currently there are no certification or testing mechanisms for this competency, but these may be developed in the future.

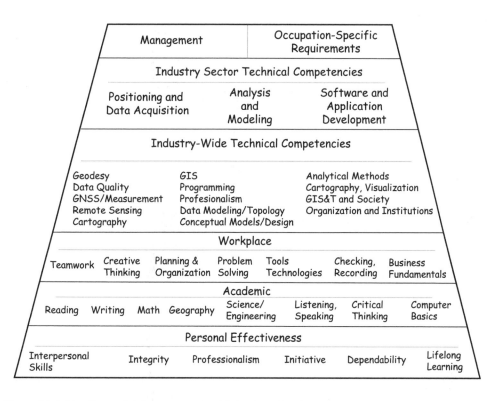

Figure 14-1: The Geospatial Competency Model developed by the U.S. Department of Labor. Please see: www.careeronestop.org/CompetencyModel/pyramid.aspx?GEO=Y for a complete description of the model and components.

Spatial Data Standards

Spatial data standards can be defined as methods for structuring, describing, and delivering spatially-referenced data. Spatial data standards may be categorized into four areas: media standards, format standards, accuracy standards, and documentation standards. All are important, although the last two are substantially more complex than the first two.

Media standards refer to the physical form in which data are transferred. They define specific formats for CD-ROM, magnetic tape, optical or solid-state storage, or some proprietary drive or other media type. Standardized formats are specified by the International Standards Organization (ISO).

Format standards specify data file components and structures. A format standard established the number of files used to store a spatial data set, as well as the basic components contained in each file. The order, size, and range of values for the data element contained in each file are defined. Information such as spacing, variable type, and file encoding may be included.

Format standards aid in the practical task of transferring data between computer systems, either within or between organizations. Producers and users may not use the same hardware or GIS software. The interchange between different software systems is aided by general, standard forms in which data may be delivered.

Many government or vendor formats have become widely supported because data are commonly delivered using the formats. For example, the U.S. Government supports the Spatial Data Transfer Standard (SDTS). This format specifies the logic, format, and encoding for raster, vector, and topological data transfer of spatial data. ESRI shapefiles (a cluster of files including .shp, .shx, and .dbf) are a commonly supported vector format, and many organizations transfer data using them. Image files in a .img format were first developed by ERDAS, an early remote sensing software package, and are widely supported, and ESRI geodatabases are becoming a common form of data delivery. These proprietary formats are not truly standards because the formats may be changed by the vendors that created them, their forms are not agreed upon by a standardizing body, and there is some interpretation on how they are applied, so there is often not complete transparency and hence interoperability.

Spatial data accuracy standards document the quality of the positional and attribute values stored in a spatial data set. Knowledge of data quality is crucial to the effective use of GIS, but we are often remiss in our assessment and reporting of spatial data quality. Many of us are accustomed to drawn maps. We accept erroneous or generalized data because the map features are logically consistent. We know the road widths are not plotted at map scale, buildings may be offset, and items grouped. We allow a certain fuzziness in the relative and absolute geographic position of objects in maps.

While we typically do not transfer this philosophy to our digital spatial data, we often pay less attention to documenting spatial data accuracy than we should. This is due in part to the cost of adequately estimating the errors in our spatial data sets. Field sampling is quite expensive, and we are tempted to spend the additional funds collecting additional data. Data production and analysis are often pushed to the limits of available time or monies, and the documentation of data accuracy may be given scant attention. Adherence to spatial data accuracy standards ensures we assess and communicate spatial data quality in a well-defined, established manner.

Documentation standards define how we describe spatial data. Data are derived from a set of original measurements taken by specific individuals or organizations at a specified time. Data may have been manipulated or somehow transformed, and data are stored in some format. Data documentation standards are an agreed-upon way of describing the source, development, and form of spatial data. When documentation standards are used they ensure a complete description of the data origin, methods of development, accuracy, and delivery formats. Standard documentation allows the data steward to maintain the data, and these standards allow any potential user to assess the appropriateness of these data for an intended task.

Data quality standards illustrate how our data have more value when we use standards. There are many ways to describe data quality, including the average distance error, the largest distance error, the percentage of points that are above an error threshold, or the total area that is misclassified. Attribute error may also be described in many different ways. The producer may describe the spatial data quality in one manner, but this may not allow a user to judge if the data are acceptable for an intended application. A data quality standard becomes familiar through use. We may know what levels of average error are likely to result in an unacceptable data. The standard allows us to compare two data sets in light of this past experience.

Data Accuracy

An accurate observation reflects the true shape, location, or characteristics of the phenomena represented in a GIS. When the concept of accuracy is applied to spatial variables, it is a measure of how often or by how much our data values are in error. Accuracy may be reported as a frequency, for example, when we report that 20% of the land cover class labelled as cropland is actually perennial grasses. Alternatively, accuracy may be expressed as an average error magnitude, for example, light poles may be displaced on average by 12.4 meters from their true locations.

Spatial data always contain some error. Errors can be caused by how we conceptualize the features, our methods of data collection and analysis, human error, or data may simply be out of date. Each of these causes a difference between reality and our spatial data representation.

Inadequacies in our spatial data model are a common cause of spatial data error. When we use a raster data set with a fixed cell size, we have set a limit on our positional accuracy. Using a raster model, we can locate objects to no better than one-half the cell resolution, on average. The raster model assumes a homogeneous pixel. If more than one category or value for a variable is found in the pixel, then the attribute value may be in error. Consider a raster data layer containing land use data and a 100 meter cell dimension. A small woodlot less than one-half hectare in size may not be represented if it is embedded in an urban or agricultural setting – it is too small and so may be "included" as part of the land use around it. This "generalization" or "inclusion" error may also occur in vector data sets. Any feature smaller than the minimum mapping unit may not be represented.

Vector data sets may poorly represent gradual changes, so there can be increased attribute error near vector boundaries. Digital soils data are often provided in a vector data model, yet the boundaries between soil types are often not discrete, but change over a zone of a few to several meters.

Errors are often introduced during spatial data collection. Many positional data are currently collected using GNSS technologies. The spatial uncertainty in GNSS positions described in Chapter 5 is incorporated into the positional data. Feature locations derived from digitized maps or aerial photographs also contain positional errors due to optical, mechanical, and human deficiencies. Lenses, cameras, or scanners may distort images, positional errors may be introduced during registration, or errors may be part of the digitization process. Blunders, fatigue, or differences among operators in abilities or attitudes may result in positional uncertainty.

Spatial data accuracy may be degraded during laboratory processing or data reduction. Mis-copies during the transcription of field notes, errors during keyboard entry, or mistakes during data manipulation may alter coordinate values used to represent a spatial data feature. Improper representation in the computer may cause problems, such as rounding errors when multiplying large numbers. These errors plus mistakes or improper laboratory techniques can also alter attribute values and introduce errors.

Data may also be in error due to changes through time (Figure 14-2). The world is dynamic, while our representation in a spatial data set often captures a snapshot at the time of data collection. Vegetation boundaries may be altered by fire, logging, construction, conversion to agriculture, or a host of other human or natural disturbances. Even in instances where positions are static, attributes may change through time. A two-lane gravel road may be paved or widened, causing attributes to be in error. Layers should have a recommended update interval which may vary by type. Elevation, geology, and soils may be updated rarely, and still maintain their accuracy. Vegetation, population, land use, or other factors change at faster rates, and should be updated more frequently if they are to remain accurate.

Documenting Spatial Data Accuracy

We must unambiguously identify true conditions if we are to document spatial data accuracy. For example, a road segment may be completely paved, or not. The data record for that road segment is accurate if it describes the surface correctly, and inaccurate if it does not. However, in many cases the truth is not completely known. The locations for the above roads may be precisely surveyed using the latest carrier-phase GNSS methods. Road centerlines and intersections may be known to the nearest 0.5 centimeters. While this is a very small error, this represents some ambiguity in what we deem to be the truth. Establishing the accuracy of a data set requires we know the accuracy of our measure of truth. In most cases, the truth is defined based on some independent, higher order measurements. In our roads example, we may desire that our data layer be accurate to 15 meters or better. Gauged on this scale, the 0.5 centimeter accuracy from our carrier-phase GNSS measurement may be considered true.

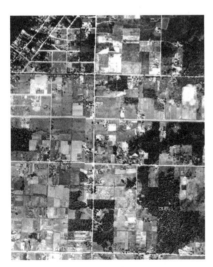 

Figure 14-2: Spatial data may be in error because of the passage of time. Road maps based on 1936 photographs (left) from the city of Bellevue, Washington, are likely to be in error in 1997 (right). (courtesy Washington Department of Natural Resources).

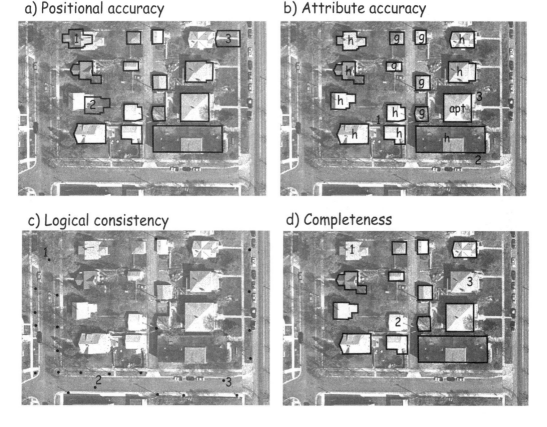

a) Positional accuracy

b) Attribute accuracy

c) Logical consistency

d) Completeness

Figure 14-3: Examples of errors of various types. This figure shows digitized features (lines or points) overlain on a source aerial photograph. Errors are labeled with numbers. In a the houses labeled 1, 2, and 3 suffer from positional inaccuracy, while b demonstrates attribute accuracy in that garages at 1 are labeled as houses, and 2 and 3 show apartments and houses mis-labeled. Panel c shows that data may not be logically consistent, with light poles at locations 1, 2, and 3 in a street, and d shows that data may lack completeness, with houses 1, 2, and 3 not digitized, as shown by the missing outline.

There are at four primary ways we describe spatial data accuracy: *positional accuracy, attribute accuracy, logical consistency*, and *completeness* (Figure 14-3). These four components may be complemented with information on the *lineage* of a data set to define the accuracy and quality of a data set. These components are described in turn below.

Positional accuracy describes how close the locations of objects represented in a digital data set correspond to the true locations for the real-world entities. In practice, truth is determined from some higher-order positioning technology.

Attribute accuracy summarizes how different the attributes are from the true values. Attribute accuracies are usually reported as a mean error or quantile above a threshold error for attributes measured on interval/ratio scales, and as percentages or proportions accurate for ordinal or categorical attributes.

Logical consistency reflects the presence, absence, or frequency of inconsistent data. Tests for logical consistency often require comparisons among themes, for example, all roads occur on dry land. This is different than positional accuracy in that both the road and the lake locations may contain positional error. However, these errors do not cause impossible or illogical juxtapositions. Logical consistency may also be applied to attributes, for example, wetland soils erroneously listed as suitable for construction, or lakes with zero depth.

Completeness describes how well the data set captures all the features it is intended to represent. A buildings data layer may omit certain structures, and the frequency of these omissions reflects an incomplete data set.

Data sets may be incomplete because of generalizations during map production or digitizing. For example, a minimum mapping unit may be set at 2 hectares when compiling a vegetation map. Isolated small pastures scattered through the forest may not be represented because they are only slightly larger than this minimum mapping unit, and erroneously they are not represented in the data layer. Completeness often refers to how well or often a data set contains those features it purports to represent. In our example above, completeness would define how often features greater than 2 hectares in size are included in the data set.

Lineage describes the sources, methods, timing, and persons responsible for the development of a data set. Lineage helps establish bounds on the other measures of accuracy described above, because knowledge about certain primary data sources helps define the accuracy of a data set.

Accuracy is most reliably determined by a comparison of true values to the values represented in a spatial data set. This requires we collect data at an adequate set of sample locations. "True" values are collected at these sample locations. Corresponding values are collected for the digital spatial data. The true and data values are compared, errors calculated, and summary statistics generated.

The source for our truth, the sampling method, our method for calculating error, and the summary statistics we chose will depend on the type of spatial data that are to be evaluated. Positional data will be assessed using different methods than attribute data. Nominal attribute data, for example, the type of land cover, will be assessed differently than a measurement recorded on a continuous range, e.g., the soil nitrogen content.

Positional Accuracy

Positional Accuracy measures how close a database representation of an object is to the true value. Accurate positions have small errors. Small is defined subjectively, but may at least be quantified.

Precision refers to the consistency of a measurement method. Precision is usually defined in terms of how dispersed a set of repeat measurements are from the average measurement. A precise measurement system provides tightly-packed results. Precise digitizing means we may repeatedly place a point in the same location.

Accuracy and precision are often confused, but they are two different characteristics, both desirable, that may change independently. A set of measurements may be precise but inaccurate. Repeat measurements may be well-clustered, meaning they are precise, but they may not be near the true value, meaning they are inaccurate. A *bias* may exist, defined as a systematic offset in coordinate values. A less precise process will result in a set of points that are more widely spread. However, their average error may be substantially less, therefore the set is more accurate.

Figure 14-4 illustrates the difference between accuracy and precision. Four digitizing sessions are shown. The goal is to place several points at the center of the cloverleaf intersection in Figure 14-4. The upper left panel shows a digitizing process that is both accurate and precise. Points, shown as white circles, are clustered tightly and accurately over the intended location.

The upper right panel of Figure 14-4 shows points that are precisely placed (tightly clustered), but not accurately located. This might be due to an equipment failure or some problem in registration; the operator may have made some blunder in photo registration and introduced a bias.

The lower left panel of Figure 14-4 shows points that are accurately but imprecisely digitized. The average location for these points is quite near the desired position, the center of the cloverleaf intersection, even though individual points are widely scattered. These points are not very close to the mean value and so precision is low, even though accuracy is high. Sensitive equip-

ment, an operator with unsteady hands, or equipment malfunction could all result in this situation.

The panel at the lower right of Figure 14-4 shows points with positions that are both imprecise and inaccurate. The mean value is not near the true location, nor are the values tightly clustered.

The threshold that constitutes high accuracy or precision is often subjectively defined. A duffer may consider as accurate

high average accuracy, high precision

low average accuracy, high precision

high average accuracy, low precision

low average accuracy, low precision

Figure 14-4: Accuracy and precision. Points (white circles) are digitized to represent the center of the cloverleaf intersection. Average accuracy is high when the average of the points falls near the true location, as in the panels on the left side of the figure. Precision is high when the points are all clustered near each other (top panels). A group of points may be accurate, but not precise (lower left), or precise, but not accurate (upper right). We typically strive for a process that provides both accuracy and precision (upper left), and avoid low accuracy and low precision (lower right).

any golf shot that lands on the green. This definition of accuracy may be based on thousands of previous attempts. However, for a professional golfer, anything farther than 2 meters from the hole may be an inaccurate shot. In a similar fashion, the high spatial accuracy requirements for a land surveyor may be different than those of a federal land manager. Cadastral surveys require the utmost in accuracy because people tend to get upset when there is material permanent trespass, as when a neighbor builds a garage on their land. Lower accuracy is acceptable in other applications, for example, a statewide map defining vegetation type may be acceptable even though boundaries are off by tens of meters.

The mean error and an error frequency threshold are the statistics most often used to document positional data accuracy. Consider a set of wells represented as point features in a spatial data layer. Suppose that after we have digitized our well locations we gain access to a GNSS system that effectively gives us the true coordinate locations for each well. We may then compare these well locations to the coordinate locations in our database. We begin by calculating the distance between our true and database coordinates for each well. This leaves us with a list of errors, one associated with each well location (Figure 14-5). Distance is measured using the Pythagorean formula with the true and database coordinates. Distances are always positive because of the form of the formula.

We may compute the mean error by summing the errors and dividing the sum by the number of observations. This gives us our average error, a useful statistic somewhere near the midpoint of our errors. We are often interested in the distribution of our errors, and so we also commonly use a frequency histogram to summarize our spatial error. The histogram is a graph of the number of error observations by a range of error values, for example, the number of error values between 0 and 1, between 1 and 2, between 2 and 3, and so on for all our observations. The graph will indicate the largest

and smallest errors, and also give some indication of the mean and most common errors.

Examples of error frequency histograms for two different data sets are shown in Figure 14-6. Each plot shows the frequency of errors across a range of error distances. For example the top graph shows that approximately 1 error in 100 has a value of near 4.5 meters, and the mean error is near 13 meters.

The mean error value does not indicate the distribution, or spread of the errors. Two data sets may have the same mean error but one may be inferior; the data set may have more large errors. The bottom graph in Figure 14-6 has the same mean error, 13 meters, as the top graph. Note that the errors have a narrower distribution, meaning the errors are clumped closer to the mean than in the top graph, and there are fewer large errors. Although the mean error is the same, many would consider the data represented in the bottom graph of Figure 14-6 to be more accurate.

Because the mean statistic alone does not provide information on the distribution of positional errors, an error frequency threshold can be reported. An error frequency threshold is a value above or below

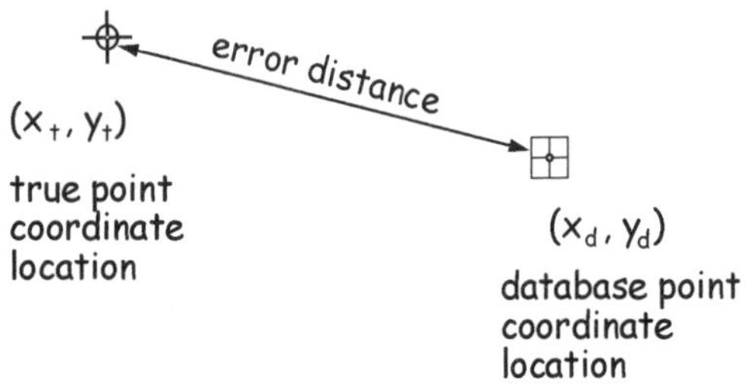

$$\text{error distance} = \sqrt{\left(x_t - x_d\right)^2 + \left(y_t - y_d\right)^2}$$

Figure 14-5: Positional errors are measured by the Pythagorean distance between a true and database coordinate for a location.

which a proportion of the error observations occurs. Figure 14-6 shows the 95% frequency threshold for two error distributions. The threshold is placed such that 95% of the errors are smaller than the threshold and approximately 5% are larger than the threshold. The top graph shows a 95% frequency threshold of approximately 21.8 meters. This indicates that approximately 95% of the positions tested from a sample of a spatial database are less than or equal to 21.8 meters from the true locations. The bottom panel in Figure 14-6 has a 95% frequency threshold at 17.6 meters. This means 5% of the errors in the second tested database are larger than 17.6 meters from their true location. If we are concerned with the frequency of large errors, this may be a better summary statistic than the mean error.

A Standard Method for Measuring Positional Accuracy

The Federal Geographic Data Committee of the United States (FGDC) has described a standard for measuring and reporting positional error. They have done so because positional accuracy is such an important characteristic of every digital data set, and there is a need for a standardized vocabulary and method for documenting spatial accuracy. This standard is known as the National Standard for Spatial Data Accuracy (NSSDA). The NSSDA specifies the number and distribution of sample points when performing an accuracy assessment, and prescribes the statistical methods used to summarize and report positional error. Separate methods are described for horizontal (x and y) accuracy assessment and vertical (z) accuracy assessment, although the methods differ primarily in the calculation of summary accuracy statistics. There are five steps in applying the NSSDA:

•Identify a set of test points from the digital data set under scrutiny.

•Identify a data set or method from which "true" values will be determined.

•Collect positional measurements from the test points as they are recorded in the test and "true" data sets.

•Calculate the positional error for each test point and summarize the positional accuracy for the test data set in a standard accuracy statistic.

•Record the accuracy statistic in a standardized form. Also include a description of the sample number, true data set, the accuracy of the true data set, and the methods used to develop and assess the accuracy of the true data set.

Test points must be clearly identifiable in both the test data set and in the "truth" data set. Points that are precisely, unambiguously defined are best. For example, we may wish to document the accuracy of roads data

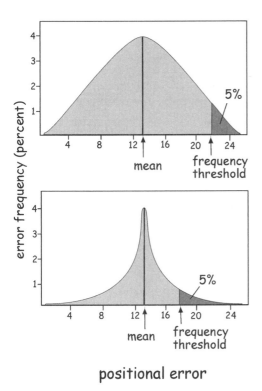

Figure 14-6: Mean error and frequency thresholds are often used to report positional error. The lightly shaded gray area represents 95% of the data.

compiled from medium-scale sources and represented by a single line in a digital layer. Right-angle road intersections are preferred over other features because the positions represented in the data base may be precisely determined. The coordinates for the precise center of the road intersections may also be determined from a higher accuracy data set, for example, from digital orthophotographs or field surveys. Other road features are less appropriate for test points, including road intersections at obtuse angles or acute curves, because there may be substantial uncertainty when matching the data layer to true coordinates. Obtuse road intersection points may be easily identified in the digital data layer, however the corresponding point on an orthophotograph may be difficult to define. Matching points on a curve is even more difficult.

The source of the true coordinate position should match our minimum accuracy specification, or at least an order of magnitude more accurate than the errors. GNSS are a common source of truth, as the accuracy may be set by collection equipment and methods, but any source of truth that matches our requirements is acceptable.

Figure 14-7 shows an example set of test points for road data layer, and an image backdrop. Prior knowledge lead us to expect average errors in excess of 20 meters. In this example, we have selected DOQ images (see Chapter 7) as our "true" data source. We know the DOQ-derived coordinates include error, an average of about 2 meters from metadata. DOQs were selected because they meet our accuracy requirements and are available for the entire work area.

-+- true point location

~ road data layer

Figure 14-7: A roads data layer displayed over a georeferenced DOQ. Test point locations are shown. The true coordinate values may be derived from the DOQ and the data coordinate values from the roads layer. Differences in these locations would be used to estimate the positional accuracy of the roads data layer.

The display of road locations on top of the DOQ shows there are substantial differences in true positions of features and their representations in the roads data layer. Any right-angle intersection is a prospective test point.

The inset in the lower left of Figure 14-7 shows the true point locations relative to the road intersections. Road centerlines were digitized. These true locations would be identified on the DOQs, perhaps by pointing a cursor at a georeferenced image displayed on a computer monitor. The data coordinates would then be extracted for the corresponding road intersection, and these two coordinate pairs, the true x, y and data x, y, would be one test point used in accuracy calculations.

The NSSDA specifies between 20 and 30 well-distributed test points (Figure 14-8). Test points should be distributed as evenly as possible throughout the data layer to be tested. Each quadrant of the tested data layer should contain at least 20% of the test points, and test points should be spaced no closer than one-tenth the longest spanning distance for the tested data layer (d, in Figure 14-8).

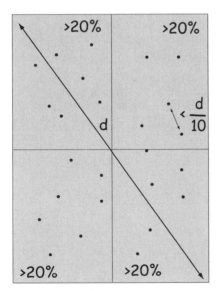

Figure 14-8: Recommended sampling for spatial data accuracy. Samples should be well distributed and well-spaced. (Adapted from LMIC, 1999).

Accuracy Calculations

The calculation of point accuracies and summary statistics are the next steps in accuracy assessment. First, the coordinates of both the true and data layer positions for a feature are recorded. These coordinates are used to calculate a positional difference, known as a positional error, based on the distance between the true coordinates and the data layer coordinates (Figure 14-9). This is a portion of the image inset shown in Figure 14-7. The true coordinates fall in a different location than the coordinates derived from the data layer. Each test point yields an error distance e, shown in Figure 14-9 and defined by the equation:

$$e = \sqrt{(x_t - x_d)^2 + (y_t - y_d)^2} \qquad (14.1)$$

where x_t, y_t are true coordinates and x_d, y_d are the data layer coordinates for a point.

The squared error differences are then calculated, and the sum, average, and root mean square error (RMSE) statistics determined for the data set. As previously defined in this book, the RMSE is:

$$RMSE = \sqrt{\frac{e_1^2 + e_2^2 + \ldots e_n^2}{n}} \qquad (14.2)$$

where e is defined as in Equation 14.1 above, and n is the number of test points used.

The RMSE is not the same as the average distance error, nor a "typical" distance error. The RMSE is a statistic that is useful in determining probability thresholds for error. The RMSE is related to the statistical variance of the positional error. If we assume the x and y errors follow a bell-shaped Gaussian curve commonly observed when sampling, then the RMSE tells us something about the distribution of distance errors. We can use knowledge about the RMSE that we get from our sample to determine what is the likelihood of a large or small error. A large RMSE means the errors are widely spread, and a small **RMSE** means the errors are packed tightly around the mean value.

Statistical theory allows us to establish fixed numbers that identify error thresholds. Because we have two variables, **x** and **y**, if we make appropriate assumptions, we can fix an error threshold at a given value. An error threshold is commonly set for 95%. When we fix a 95% error threshold, this means we identify the specific number such that 95% of our errors are expected to be less than or equal to the threshold. Statistical theory tells us that when we multiply the **RMSE** by the number 1.7308 and assume a Gaussian normal distribution, we obtain the 95% threshold. We will not cover the statistical assumptions and calculations used in deriving these constants, nor the theory behind them. A thorough treatment may be found in the references listed at the end of this chapter.

Accuracy calculations may be summarized in a standard table, shown in Table 14-1. The example shows a positional accuracy assessment based on a set of 22 points. Data for each point are organized in rows. The true and data layer coordinates are listed, as well as the difference and difference squared for both the **x** and **y** coordinate directions. The squared differences are summed, averaged, and the **RMSE** calculated, as shown in the summary boxes in the lower right portion of Table 14-1. The **RMSE** is multiplied by 1.7308 to estimate the 95% accuracy level, listed as the NSSDA. Ninety-five percent of the time the true horizontal errors are expected to be less than the estimated accuracy level of 12.9 meters listed in Table 14-1.

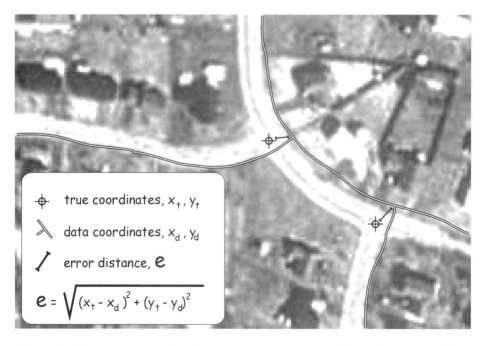

true coordinates, x_t, y_t

data coordinates, x_d, y_d

error distance, e

$$e = \sqrt{(x_t - x_d)^2 + (y_t - y_d)^2}$$

Figure 14-9: True and test point locations in an accuracy assessment. True and test point locations do not coincide. The differences in coordinates between true and test points are used to calculate an error distance.

Table 14-1: An accuracy assessment summary table.

ID	x (true)	x (data)	x difference	(x difference)²	y (true)	y (data)	y difference	(y difference)²	sum x diff² + y diff²
1	12	10	2	4	288	292	-4	16	20
2	18	22	-4	16	234	228	6	36	52
3	7	12	-5	25	265	266	-1	1	26
4	34	34	0	0	243	240	3	9	9
5	15	19	-4	16	291	287	4	16	32
6	33	24	9	81	211	215	-4	16	97
7	28	29	-1	1	267	271	-4	16	17
8	7	12	-5	25	273	268	5	25	50
9	45	44	1	1	245	244	1	1	2
10	110	99	11	121	221	225	-4	16	137
11	54	65	-11	121	212	208	4	16	137
12	87	93	-6	36	284	278	6	36	72
13	23	22	1	1	261	259	2	4	5
14	19	24	-5	25	230	235	-5	25	50
15	76	80	-4	16	255	260	-5	25	41
16	97	108	-11	121	201	204	-3	9	130
17	38	43	-5	25	290	288	2	4	29
18	65	72	-7	49	277	282	-5	25	74
19	85	78	7	49	205	201	4	16	65
20	39	44	-5	25	282	278	4	16	41
21	94	90	4	16	246	251	-5	25	41
22	64	56	8	64	233	227	6	36	100

Sum	1227
Average	55.8
RMSE	7.5
NSSDA	12.9

Errors in Linear or Area Features

The NSSDA as described above treats only the accuracies of point locations. It is based on a probabilistic view of point locations. We are not sure where each point is, however we can specify an error distance r for a set of features. A circle of radius r centered on a point feature in our spatial data layer will include the true point location 95% of the time. Unfortunately, there are no established standards for describing the accuracy or error of linear or area features.

Previous work in cartography and GIS used the concept of an *epsilon band* to characterize uncertainty in line position. An epsilon band may be defined as a region near a line that has a very high probability of containing the line (Figure 14-10). Within this epsilon band the line location is uncertain. The concept of an epsilon band is congruent with our model of point positional errors if we remember that most vector line data are recorded as a sequence of nodes and vertices. Lines are made up of point locations. Nodes and vertices defining line locations contain some error. The epsilon band may be thought of as having a high probability of

encompassing the true line segments. Larger epsilon bands are associated with poorer quality data, either because the node and vertex locations are poorly placed and far from the true line location, or because the nodes or vertices are too widely spaced.

In some instances we may assume the well-defined point features described in our accuracy test above may also represent the accuracy for nodes and vertices of lines in a data layer. Nodes or vertices may be used as test points, provided they are well defined and the true coordinates are known. However, the errors at intervening locations are not known, for example, midway along a line segment between two vertices. The error along a straight line segment may be at most equal to the largest error observed at the ends of the line segments. (Figure 14-11). If the data line segment is parallel to the true line segment, then the errors are uniform along the full length of the segment. Vertices that result in converging or crossing lines will lead to mid-point errors less than the larger of the two errors at the endpoints (Figure 14-11). These observations are not true if a straight line segment is used to approximate a substantially curved line. However, if

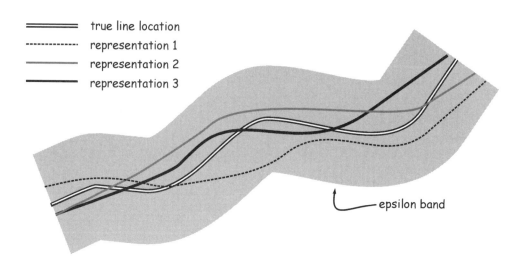

Figure 14-10: The concept of an epsilon band. A region to either side of a line is specified. This region has a high probability of containing the true line. Multiple digital representations may exist, e.g., multiple tries at digitizing a line. These representations will typically lie within the epsilon band.

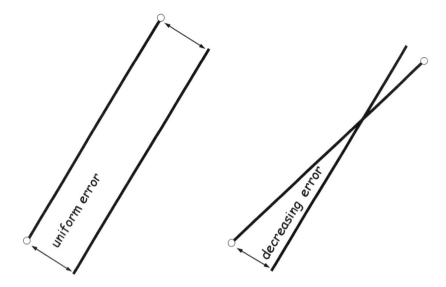

Figure 14-11: Errors for straight line segments are either the same (left) or less than (right) the maximum error observed at the end points. If nodes or vertices are sufficiently close such that the true line segments are approximately linear, then error assessments on nodes or vertices may provide an approximation of the line positional errors.

the line segments are sufficiently short (e.g., the interval along the line is small relative to the radius of a curve in the line), and the positional errors are distributed evenly on both sides of the line segments, then the NSSDA methods described above will provide an approximate upper limit on the linear error.

Attribute Accuracy

Unlike positional accuracy, there is no national standard for measuring and reporting attribute accuracy. Accuracy for continuous variables may be calculated in an analogous manner to positional accuracy. Accuracy for each observation is defined as the difference between the true and database values. A set of test data points may be identified, the true attribute value determined for each of those test data points, the difference calculated for each test point, and the accuracy summarized.

The accuracy of categorical attribute data may be summarized using an *error*

table and associated accuracy statistics. Points can be classified as correct, that is, the categorical variable matches the true category for a feature, or they may be incorrect. Incorrect observations occur when the true and layer category values are different. Error tables, also known as error matrices, confusion matrices, and accuracy tables, are a standard method of reporting error in classified remotely-sensed imagery. They have more rarely been used for categorical attribute accuracy assessment.

An error table summarizes a two-way classification for a set of test points (Figure 14-12). A categorical variable will have a fixed number of categories. These categories are listed across the columns and along the rows of the error table. Each test feature is tallied in the error table. The true category and the value in the data layer are known for each test feature. The test feature is tallied in the error table based on these values. The true values are entered via the appropriate column and the data layer values are entered via the appropriate row. The table is square, because there is the same number of categories in both the rows and columns. Correctly classified features are tallied on the diagonal – the true value and data layer value are identical, so they are noted at the intersection of the categories. Incorrectly assigned category values fall off the diagonal.

Error tables summarize the main characteristics of confusion among categories. The diagonal elements contain the test features that are correctly categorized. The diagonal sum is the total number correct. The proportion correct is the total number correct divided by the total number tested. The percent correct can be obtained by multiplying the proportion correct by 100.

Per category accuracy may be extracted from the error table. Two types of accuracy may be calculated, a *user's accuracy* and a *producer's accuracy*. The user relies on the data layer to determine the category for a feature. The user is most often interested in how often a feature is mis-labeled for each category. In effect, the user wants to know how many features that are classified as a category (the row total) are truly from that category (the diagonal element for that row). Thus, the user's accuracy is defined as the number of correctly assigned features (the diagonal element) divided by the row total for the category. The producer, on the other hand, knows the true identity of each feature and is often interested in how often these features are assigned to the correct category. The producer's accuracy is defined as the diagonal element divided by the column total.

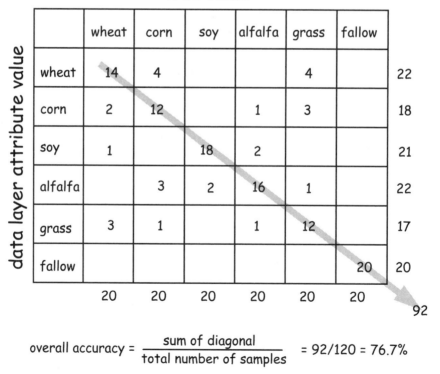

Figure 14-12: An error table succinctly summarizes the attribute accuracy for categorical variables.

Error Propagation in Spatial Analysis

While we have discussed methods for assessing positional and attribute accuracy, we have not described how we determine the effects of input errors on the accuracy of spatial operations. Clearly, input error affects output values in most calculations. A large elevation error in DEM cells will likely cause errors in slope values. If slope is then combined with other features from other data layers, these errors may in turn propagate through the analysis. How do we assess the propagation of errors and their impacts on spatial analysis?

There are currently no widely applied, general methods for assessing the effect of positional errors on spatial models. Research is currently directed at several promising avenues, however the range of variables and conditions involved has confounded the development of general methods for assessing the impacts of purely positional errors on spatial models.

Several approaches have been developed to estimate the impacts of attribute errors on spatial models. One approach involves assessing errors in the final result irrespective of errors in the original data. For example, we may develop a cartographic model to estimate deer density in a suburban environment. The model may depend on the density of housing, forest location, type, and extent, the location of wetlands, and road location and traffic volumes. Each of these data sources may contain positional and attribute errors.

Questions may arise regarding how these errors in our input data affect the model predictions for deer density. Rather than trying to identify how errors in the input propagate through to affect the final model results, we may opt to perform an error assessment of our final output. We would perform a field survey of deer density and compare the values predicted by the model with the values observed in the field. For example, we might subdivide the study area into mutually exclusive census areas. Deer might be counted in each census area and the density calculated. We have replicated values from each census area, so we may calculate a mean and a variance, and the difference between modeled and observed values might be compared relative to the natural variation we observe among different census areas. We could also census an area through time, for example, on successive days, months, or years, and compare the difference between the model and observed values for each sample time.

It may not be possible or desirable to wait to assess accuracy until after completing a spatial analysis. Input data for a specific spatial analysis may be expensive to collect. We may not wish to develop the data and a spatial model if errors in the input preclude a useful output. After model application, we may wish to identify the source of errors in our final predictions. Improvements in one or two data layers may substantially improve the quality of our predictions, for example, better data on forest cover may increase the accuracy of our deer density predictions.

Error propagation in spatial models is often investigated with repeated model runs. We may employ some sort of repeat simulation model that adds error to data layers and records the impacts on model accuracy. These simulation models often employ a standard form known as a *Monte Carlo simulation*. The Monte Carlo method assumes each input spatial value is derived from a population of values. For example, land cover may range over a set of values for each cell. Further, model coefficients may also be altered over a range. In a cartographic model, the weights are allowed to range over a specified interval when layers are combined.

A Monte Carlo simulation controls how these input data or model parameters are allowed to vary. Typically, a random normal distribution is assumed for continuous input values. If all variables save one are held constant, and several model runs performed on different, random selections of the variable, we may get an indication of how a variable affects the model output. We may find that the spatial model is insensitive to large changes in most of our input data values, but sensitive to small changes in a few. For example, predicted deer density may not change much even when land cover varies over a wide range of values, but may depend heavily on housing density. However, we may also find a set of input data, or a range of input data or coefficients, that substantially control model output.

A Monte Carlo or similar simulation is a computationally intensive technique. Thousands of model runs are often required over each of the component units of the spatial domain. The computational burden increases as the models become more involved, and as the number of spatial units increases. However, it is often the only practical way with which to assess the impacts of uncertainties on spatial analyses, uncertainties both in the input data and the parameters and methods in combining them.

Summary

Data standards, data accuracy assessment, and data documentation are among the most important activities in GIS. We cannot effectively use spatial data if we do not know its quality, and the efficient distribution of spatial data depends on a common understanding of data content.

Data may be inaccurate due to several causes. Data may be out of date, collected using improper methods or equipment, or by unskilled or inattentive persons.

Accuracy is a measure of error, a difference between a true and represented value. Inaccuracies may be reported using many methods, including a mean value, a frequency distribution, or a threshold value. An accuracy assessment or measurement applies only to a specific data set and time.

Accuracy should be recognized as distinct from precision. Precision is a measure of the repeatability of a process. Imprecise data collection often leads to poor accuracy.

Standards have been developed for assessing positional accuracy. Accuracy assessment and reporting depend on sampling. A set of features is visited in the field, and the true values collected. These true values are then compared to corresponding values stored in a data layer, and the differences between true and database values quantified. An adequate number of well distributed samples should be collected. Standard worksheets and statistics have been developed.

Data documentation standards have been developed in the United States. These standards, developed by the Federal Geographic Data Committee, are known as the Content Standard for Digital Geospatial Metadata. This standard identifies specific information that is required to fully describe a spatial data set.

Suggested Reading

Arbia, G., Griffith, D., & Haining, R. (1999). Error propagation and modeling in raster GIS: overlay operations. *International Journal of Geographical Information Science*, 12:145-167.

Balazinska, M., Deshpande, A., Franklin, M.J., Gibbons, P.B., Gray, J., Hansen, M, Liebhold, M., Nath, S., Szalay, A., & Tao, V. (2007). Data management in the Worldwide Sensor Web. *Pervasive*, 6:30-40.

Blakemore, M.(1984). Generalization and error in spatial data bases. *Cartographica*, 21:131-139.

Bolstad, P., Gessler, P., & Lillesand, T. (1990). Positional uncertainty in manually digitized map data, *International Journal of Geographical Information Systems*, 4:399-412.

Chong, A.K. (1999). A technique for spatial sampling and error reporting for image map bases. *Photogrammetric Engineering & Remote Sensing*, 65:1195-1198.

Comber, A.J., Fisher, P.F., Harvey, F., Gahegan, M., & Wadsworth, R. (2006). Using metadata to link uncertainty and data quality assessments. *Progress in Spatial Data Handling*, 6:279-292.

DeBiase, D., DeMers, M., Johnson, A., Kemp, K., Luck, A.T., Plewe, B., & Wentz, W. (2006). *Geographic Information Science and Technology Body of Knowledge*. Association of American Geographers.

DeBiase, D., Corbin, T., Fox, T., Francica, J., Green, K., Jackson, J., Jeffress, G., Jones, B., Jones, B., Mennis, J., Schuckman, K., Smith, C., & Van Sickle, J. (2010). The new Geospatial Technology Competency Model: Bringing workforce needs into focus. *URISA Journal*, 22:55-72.

Dunn, R., Harrison, A.R., & White, J.C. (1990). Positional accuracy and measurement error in digital databases on land use: an empirical study. *International Journal of Geographical Information Systems*, 4:385-398.

Fisher, P. (1991). Modelling soil map unit inclusions by Monte Carlo simulation, *International Journal of Geographical Information Systems*, 5:193-208.

Goodchild, M.F. and Gopal, S. (1989). *The Accuracy of Spatial Databases*. London: Taylor and Francis.

Guptill, S.C. & Morrison, J.L.(Eds.) (1995). *Elements of Spatial Data Quality*. New York: Elsevier.

Harmel, R.D., Smith, D.R., King, K.W., & Slade, R.M. (2009). Estimating storm discharge and water quality data uncertainty: A software tool for monitoring and modeling applications. *Environmental Modeling and Software*, 24:832-842.

Heuvelink, G. (1999). *Error Propagation in Environmental Modeling with GIS*. London: Taylor and Francis, London.

Heuvelink, G., Brown, J.D., & van Loon, E.E. (2007). A probabilistic framework for representing and simulating uncertain environmental variables. *International Journal of Geographical Information Science*, 21: 497-513.

Jones, R.R., McCaffrey, K.J.W., Wilson, R.W., & Holdsworth, R.E. (2004). Digital field data acquisition: towards increased quantification of uncertainty during geological mapping. *Geological Society of London Special Publications*, 239:43-56.

Kassenberg, D., & De Jong, K. (2005). Dynamic environmental modeling in GIS: 2. Modeling error propagation. *International Journal of Geographical Information Science*, 19:623-637.

Lodwick, W.A., Monson, W., & Svoboda, L. (1990). Attribute error and sensitivity analysis of map operations in geographical information systems. *International Journal of Geographical Information Systems*, 4:413-427.

Lowell, K., and Jaton, A. (1999). *Spatial Accuracy Assessment: Land Information Uncertainty in Natural Resources*. Chelsea: Ann Arbor Press.

Hunsacker, C.T., Goodchild, M.F., Friedl, M.A., & Case, T.J. (2001). *Spatial Uncertainty in Ecology: Implications for Remote Sensing and GIS Applications*. Springer-Verlag: New York.

Thapa, K. and Bossler, J. (1992). Accuracy of spatial data used in geographic information systems, *Photogrammetric Engineering and Remote Sensing*, 58:841-858.

Walsh, S. J., Lightfoot, D. R., and Butler, D. R. (1987). Recognition and assessment of error in geographic information systems, *Photogrammetric Engineering and Remote Sensing*, 53:1423-1430.

Study Questions

14.1 - Why are standards so important in spatial data?

14.2 - Can you describe processes or activities that are greatly helped by the existence of standards?

14.3 - What are the differences between accuracy and precision?

14.4 - How do mean and frequency thresholds differ in the way they report positional error?

14.5 - What are some of the primary causes of positional error in spatial data?

14.6 - Describe each of these with reference to documenting spatial data accuracy: positional accuracy, attribute accuracy, logical consistency, and completeness.

14.7 - What is the NSSDA, and how does it help us measure positional accuracy?

14.8 - What are the basic steps in applying the NSSDA?

14.9 - What are the constraints on the distribution of sample points under the NSSDA, and why are these constraints specified?

14.10 - What are good candidate sources for test points in assessing the accuracy of a spatial data layer?

14.11 - How are errors in nominal attribute data often reported?

14.12 - What are metadata, and why are they important?

15 New Developments in GIS

Introduction

As every economist knows, predicting the future is fraught with peril. Near-term predictions may be safe; if times are good now, they will probably be good next month. However, the farther he reaches into the future, the more likely he'll be wrong. This chapter describes technologies that may become widespread. It discusses future trends, with the expectation that many of these speculations will prove inaccurate.

Many changes in GIS are based on advances in computers and other electronic hardware. Computers are becoming smaller and less expensive. This is true for both general purpose machines and for specialized computers, such as ruggedized, portable tablet computers. The wizards of semiconductors continue to dream up and then produce impossibly clever devices. Given current trends, we should not be surprised if at some future time a pea-sized device holds all the published works of humankind. Computers may gain personalities, recognize us as individuals, respond entirely to voice commands, routinely conjure three-dimensional images that float in space before our eyes. These and other developments will alter how we manipulate spatial data.

Changes in GIS will also be due to the growing ubiquity of high speed, wireless and wired connections. How we interact with our spatial data changes if it is always available. We can more easily see how things should be in the field, and compare them to how they are, for example, a wiring diagram for a roadside telephone interchange panel, or a site plan vs. stakeout. An agricultural field's fertilization history, the size and type of replacement bulb to order for a damaged streetlight, or a bridge's past inspection records may all be available at any time, streamlining maintenance and management.

Change is also due to increased sophistication in GIS software and users, and increased familiarity and standardization. Change will be driven by new algorithms or methods, for example, improved data compression techniques that speed the retrieval and improve the quality of digital images. Specialized software packages may be crafted that turn a multi-day, technically complicated operation into a few mouse clicks. These new tools will be introduced as GIS technologies continue to evolve and will change the way we gather and analyze spatial data.

GNSS

Three trends will dominate GNSS innovation over the next decade - multi-GNSS receivers, miniaturization, and system integration.

Multi-GNSS receivers will continue to take advantage of distinct satellite constellations. The history of GNSS is marred by difficulties in decimeter or better positioning in difficult environments. Dual GPS/GLONASS systems already exist, and systems that simultaneously support the Galileo and Chinese Compass system will be developed, further increasing availability and reliability. Receivers will commonly have hundreds of channels, and track tens of satellites even when under heavy forest canopies, in canyons, and among tall buildings, bringing real-time precise positioning to everyone.

GNSS receivers will cost less, shrink, and increase accuracy for some time to come, and these improvements will spur even more widespread adoption of this technology (Figure 15-1). Microelectronic miniaturization is helping shape the GNSS market. As GNSS use grows and manufacturing methods improve, single chip GNSS systems have emerged, and these chips are decreasing in size. GNSS chips smaller than a postage stamp are available, including some that may be integrated into common

electronic devices. Many vendors are well on a path to system integration, and it will become more common to embed the antenna, receiver, supporting electronics, power supply, and differential correction radio receivers in a single piece of equipment. Some of these integrated systems are smaller than most GNSS antennas of a decade ago, and systems will continue to shrink. A button-sized GNSS is not far off.

As receivers shrink in size and cost, it becomes practical to collect positional information on smaller individual objects. While GNSS is unlikely to help you find your keys, small GNSS receivers will collect spatial data for smaller objects. For example, a few years ago it was uneconomical to track objects smaller than a cargo ship. Now trucks or containers are routinely followed. In the near future it may be common to track individual packages.

Personal GNSS receivers are rapidly finding applications outdoors. Miniaturization to wrist-size GNSS now allow hikers, runners, and bicyclists to monitor their location, speed, and route, review performance, and log miles travelled (Figure 15-2). Small GPS may be paired with satellite communicators to allow contact from anywhere on the surface of the Earth, allowing both the adventurer and those left behind instant access to positional information.

GNSS miniaturization means we will directly collect much more data in the field than in times past. A city engineer may study traffic patterns by placing special-purpose GNSS receivers into autos. How long does the average commute take? How much of the time is spent sitting at stop signs or lights, and where is the congestion most prevalent? How is traffic affected by weather conditions? Analyses of traffic networks will become substantially easier with small-unit GNSS. Disposable GNSS receivers may be pasted, decal-like, on windshields by the thousands, to transmit their data back to a traffic management center.

Figure 15-1: A miniaturized GNSS unit that may be embedded in a range of electronic devices.

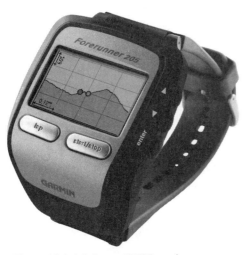

Figure 15-2: Miniature GNSS receivers are available to record speed, distance, and log tracks, that may later be downloaded and analyzed (courtesy Garmin Inc.).

Ubiquitous, inexpensive, or free differential correction signals are substantially improving the accuracy commonly achieved with GNSS. Many states in the U.S., and national governments abroad will establish more complete coverage. Virtual Reference Station (VRS) networks promise to allow real-time, sub-meter and even near centimeter level positioning in real time. Commercial solutions will be further developed and made less expensive. Centimeter-level positional accuracy will soon be available in the field in real-time, in turn supporting many new GNSS applications.

GNSS systems will add functions, including the ability to take photos or videos and attach them to geographic features in a database (Figure 15-3). The old adage "a picture is worth a thousand words" may be modified to, "a picture saves a thousand hours." These systems will greatly aid planning, management, and analysis by more easily providing images in GIS. For exam-

Figure 15-3: GNSS receivers with built-in cameras can be used to record the location (**A**), images (**B**), and attributes (**C**) of objects (courtesy TOPCON).

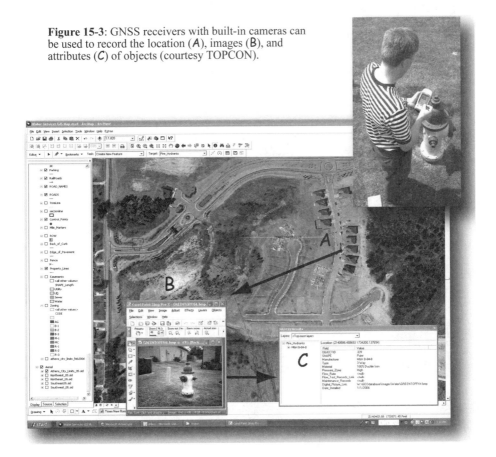

ple, the type, relative location, and condition of public utilities such as fire hydrants may be described with both photos and alphanumeric data collected in a database. If a work order is required to repair a hydrant, a photograph may be taken in the field and tagged to the work order. This photograph may be inspected to verify the type of hydrant, perhaps identify the tools needed for repair, or to recognize that specific parts are required for maintenance.

Fixed and Mobile Three-Dimensional Mapping

GNSS is also being combined with new advances in ground-based laser scanning to increase the scope, accuracy, and efficiency of spatial data collection. Three-dimensional scanning devices have been developed that measure the horizontal and vertical location of features (Figure 15-4). This scanning is necessary because many features are modi-fied over time, for example, buildings are extended, extra supports may be added to towers, or oil refineries may be re-plumbed. Inventories must be updated to record the features as built, rather than as designed or observed during the previous inventory. A three-dimensional scanning laser may be combined with a precise GNSS receiver to measure the x, y, and z coordinates of important features. The GNSS is used to determine the location of the scanning laser. The horizontal and vertical offsets from the scan point are measured by the laser. These measurements are combined with coordinate geometry to calculate the precise positions for all features scanned in the field.

The trend of mutli-technology integration will accelerate for rapid, centimeter-level mobile mapping. Multi-channel GNSS combined with other positioning systems will provide highly accurate locations, and 3-dimensional laser scanners and 360 degree

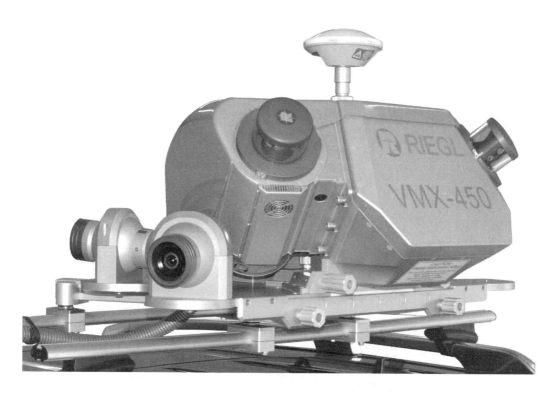

Figure 15-4: Portable, 3-dimensional scanners are in large-scale production, and allow rapid, accurate collection of x, y, and z positions along with image and other data (courtesy Riegl Systems).

Figure 15-5: Mapping systems combined with GNSS, 3-dimensional imaging lasers, optical imaging systems, and other measurement systems to provide integrated 3-d measurements in real time (courtesy Topcon).

image data collection will allow faster collection of X, Y, and Z coordinates (Figure 15-5).

Combined with GNSS systems and mounted on mobile platforms, 3D laser mapping systems will collect highly accurate data accessible by anyone with a traffic-enabled GNSS. Approaching drivers may be forewarned, travel times calculated and new suggested routes identified. One can imagine self-driving automobiles that navigate via a combined GNSS/LiDAR/GIS, using systems to avoid collisions via real-time distance measurements and wireless communications with "nearby" automobiles.

Such mobile systems will help improve the currency and accuracy of digitized transportation networks, and anything visible

from them. Every road can be digitized while driving, as well as every building, light pole, sign, bench, tree, or any other three-dimensional structure visible from them. Efforts will move from a focus on the development of integrated, turn-key data collection systems to software and methods that automate workflow, so that data may travel from the device to the database with as little human intervention as possible.

Highly accurate, current data may spur the adoption of "heads-up" displays (HUD) in automobiles. Current use of HUD concentrated in military aircraft, where images of important data are projected onto the pilot's windscreen. This allows the pilot to monitor vital information, like who's trying to blow up her plane, without looking down. Similar

Figure 15-6: Head's up displays may appear in autos, improving driver safety and trip efficiency under inclement weather (courtesy General Motors).

systems have been under development for automobiles for some time, but have yet to be deployed, in part because sufficient high quality data are not available for road location. This barrier may be surmounted in the not too-distant futures, as centimeter-level infrastructure data become more common.

Automobile HUD systems would rely on multiple technologies. A GNSS would locate the vehicle to within a few 10's of centimeters in real time. Three-dimensional data on road centerline, edges, curbs, adjacent poles, and other important features would be identified for the trajectory ahead. Aids, such as virtual illumination on the windshield, might identify the road edge when visibility is poor (Figure 15-6), flag upcoming hazards or turns, or warn of unexpected conditions. When combined with an on-vehicle laser scanner, the system may identify stationary object within the roadway, or unexpected changes in conditions.

Datum Modernization

New datum realizations will be calculated, based on improved measurements. Datum shifts among ITRF based datums will remain small, but planned datum adjustments for the NAD83 datums over the next decade will likely result in quite large shifts. These shifts will be for the most part due to changes in the way the datums are calculated, and only slightly to improved measurements.

As described in Chapter 3, the continents are moving about the Earth on plates, sometimes at rates exceeding an 2.5 cm (an inch) a year. Over several decades this leads to changes in the relative positions among points on different plates (Figure 15-7). In addition, geodesists must factor the total amount of movement into their development of datums, because measurements have been made over several decades, so the relative positions of monuments depend on both the time and location of the respective measurements. Further, the calculations require we

establish a stationary reference frame against which to measure points. Because of differences in how we account for crustal movement, and other factors, there are large differences between the NAD83 family of datums, used primarily in North America, and the ITRF datums, used by most of the rest of the World.

The ITRF is an Earth-centered system based on measurements of the X, Y, and Z location of points and their velocities. It places the origin of the adopted ellipsoid at the best estimate of center of mass of the Earth at the time of each adjustment. The post 1986 NAD83 datums are also Earth-centered. In contrast to the ITRF system, the NAD83 datums have not adopted the best measurements of the Earth's center, but rather a position compatible with older NAD83 datums. This center assigned a value relative to average crustal velocities on the North American tectonic plate, rather than a global network. There were many

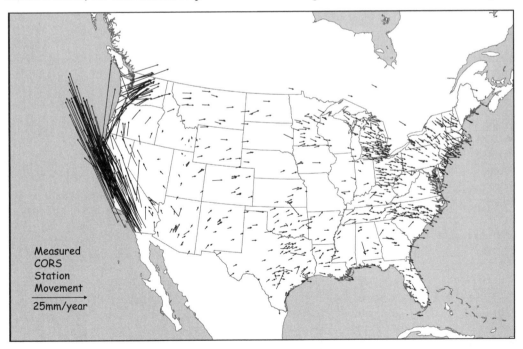

Figure 15-7: Velocity vectors of the Earth's surface, measured at CORS stations in North America. Note the relatively high velocities in western California towards the North, and the general eastward trend of most other stations. These movements must be factored into increasingly accurate datum realizations.

good reasons for maintaining the old origin, primarily because it maintained coordinate compatibility with older datums, could be deployed rapidly, and depended on a relatively dense network of continuously operating stations within North America.

With the development of the ITRF and active participation of the NGS and the integration of CORS stations into this network, there is strong impetus to harmonize datums in North America with international efforts. Improved global measurements will support more accurate horizontal and vertical datum development, and help support precise, rapid, centimeter-level positioning worldwide.

The NGS has published a "National Geodetic Survey Ten-Year Plan" for 2008-2018 which describes a path forward for datum modernization. This plan does not commit to harmonizing the NAD83 and ITRF systems, although it suggests such harmonization will be considered, and a decision taken based on stakeholder and other input. If such a harmonization occurs, datum shifts between one and two meters will be common from current to future North American datums.

The Ten-Year Plan also describes the process for improving the vertical datum for North America. Improvements will be based on gravity measurements across the hemisphere, accounting for changes in gravity fields over time, and also for the rises in mean sea level. Much as with the horizontal datum, it will be based at least in part on an integrated, global set of satellite-based measurements, and tied to ITRF measurements. The new vertical datum will allow the calculation of heights tied to a measurement epoch.

National Adjustment of 2011 (NA2011)

The U.S. National Geodetic Survey (NGS) developed a national adjustment of nearly 80,000 passive control marks using GNSS that best integrates measurements up through 2011. Passive marks are those without continuous measurements and estimates of velocity, and are the vast majority of benchmarks historically used as control of which local surveys are based. The NA2011 attempts to make all passive point coordinates fully consistent with the Continuously Operating Reference Station (CORS). The NA2011 is needed because of changes in CORS positions as a result of a recalculation of their positions, known as the NAD83(2011) Epoch 2010.00.

NA2011 yields ellipsoid heights inconsistent with the GEOID09 model, so a new GEOID12 model is also being estimated. Both will be reflected on new data sheets for control points, and appropriate transformation procedures between older and the new reference coordinates will likely be developed.

Improved Remote Sensing

Spatial data collection will be substantially improved with the continuing advances in remotely sensing. More satellites, higher spatial and temporal resolution, improved digital cameras, and new sensor platforms will all increase the array of available data. We will be able to sense new phenomena, and locate previously measured features with increased precision and accuracy.

Satellite-based systems will continue to increase in resolution and coverage, and in particular the frequency of data collection. The Pleiades system is a salient example of this trend (Figure 15-8).Two satellites, the first launched in late 2011, the second pro-

gramed for late 2012, will provide daily revisits of 0.7 m panchromatic and 2.8 m multispectral imaging. Similar improvements in resolution and coverage are in progress for other satellite image providers, increasing the frequency and types of images available for medium to high-resolution mapping.

Parallel improvements continue in aerial image acquisition. Aerial cameras increase in spatial resolution, meaning increasing availability of detailed images, with higher radiometric sensitivity, yielding a broader range of constraints. Many systems have higher radiometric breadth, leading to routine collection of more than the visible

Figure 15-8: An early image from the first Pleiades satellite, taken over coastal Dubai. This European-based, private-public partnership will produce daily, global, high-resolution images (courtesy CNES).

wavelength spectrums. National aerial acquisition programs are integrating these improvements, with NASS images commonly provided at a one-meter resolution image, up from the common two-meter resolution a few years past. The USGS and other organizations are providing sub-foot resolution images, perhaps nationwide. Individual light poles, curbs, and even parking lot cracks may be observed in these images, rendering them a rich source of spatial data (Figure 15-9).

These higher spatial resolutions will aid many new endeavors. Detailed land surveys often use imagery with effective resolutions of a few centimeters. Property and asset management involves the inventory and assessment of infrastructure and often uses high-resolution photographs. News organizations, business intelligence and strategic planning, and resource management applica-

tions all provide incentives for improvement in spatial and spectral resolution.

Advances in full-sized and miniature aircraft, known as remotely piloted vehicles (RPVs) or unmanned aerial vehicles (UAVs), are leading to increased availability of a broader range of aerial imagery. NASA, other government laboratories, and aerospace firms have been developing pilotless vehicles to collect imagery and data about the Earth's surface and atmosphere. Some experimental UAVs may fly faster and turn tighter than many planes carrying human pilots. Specialized payloads may be carried cheaply on these crafts, for long periods of time, and in more dangerous conditions than in human-piloted aircraft. Military designs are working their way into civilian applications (Figure 15-10).

RPVs may also be small helicopters or airplanes outfitted with cameras and posi-

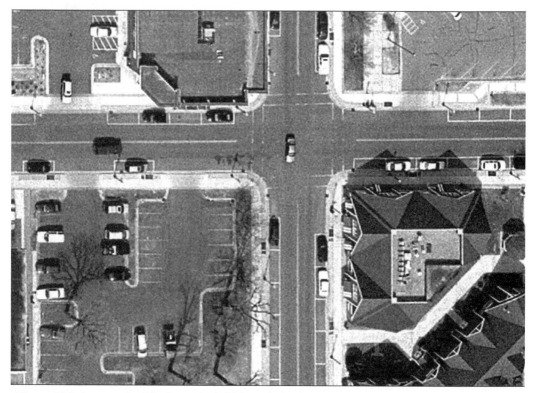

Figure 15-9: An example of the increasingly high resolution image data becoming publicly available. This USGS image is at a six inch (30 cm) resolution and has been orthorectified, and is a suitable source for high accuracy spatial data.

Figure 15-10: A scale RPV fitted with a camera, to collect low- and mid-altitude aerial imagery. (courtesy Aersonde Pty Lmtd)

tioning, control, and telemetry electronics, with pre-programmed flight paths and on-board "intelligence" to modify steering in response to wind, rain, or other in flight conditions. One UAV completed a trans-Atlantic flight with guidance only at take-off and landing, and UAVs are routinely collecting weather data. UAVs will increasingly be used to provide images for coordinate and attributed data collection. Small UAVs may be trucked to a site, where they are flown via remote-control by an operator on the ground. RPVs are used most typically to obtain imagery of a specific set of targets, while self-flying UAVs will increasingly be used

for systematic flight paths, for example, complete photographic coverage of a region at a fixed photo scale.

LiDAR data are another example of improved remote sensing LiDAR systems are improving in quality and declining in cost, with elevation data commonly paired with digital aerial images. Airborne LiDAR systems in the recent past operationally collected a data point ever few square meters, while current systems routinely collects several samples per square meter. Soon, tens or hundreds of samples per square meter will be common, allowing unprecedented spatial definition.

Figure 15-11: A three-dimensional LiDAR data set collected from a low-flying helicopter, recording substantial special detail in this oblique view (courtesy Terra Imaging).

LiDAR data are dropping in cost and county to statewide LiDAR mapping will likely become common. Fusion of LiDAR data with other image and spatial data will continue, and surely create new opportunities and applications.

Three-dimensional GIS will be more widely developed and practiced due to data provided by 3-dimensional, ground or near-ground level lidar (Figure 15-11). LiDAR systems on autos, low-flying helicopters, and high-flying aircraft will provide feature x, y, and z coordinates from various vertical, oblique, and horizontal perspectives, which will be mixed together in forming true three-dimensional characterizations of space. Data development, management, analysis, and visualization is currently taking place across architecture, CAD, surveying, and GIS softwares and disciplines, and substantial devel-

opment and fusing across these disciplines is in the offing.

Cloud-Based GIS

Cloud-based GIS provides data storage, analysis, and display capabilities over the internet, with services usually provided from one or multiple remote location via the internet. Cloud-based computing includes broadly accessible internet mapping applications, but also includes data storage and a full suite of software-supported analytical capabilities, something that has generally followed a local or private network architecture. During the recent past, GIS software primarily resided on a local or closely-networked hard disk, and ran on the central processing unit of a local computer. You downloaded data to a local hard drive, and purchases software to physically install on the local computer, although the software may references licenses or other resources on a (usually) proprietary server computer. Cloud-based computing envisions many of these resources provided from distant sources, with the local computers perhaps serving only as a display and command entry portal - data, software, and processing may all be elsewhere.

Cloud computing has many potential advantages. There may be a lower total cost of ownership, because you may only need to use a set of software occasionally, and can pay as you need it, rather than a fixed price irrespective of total use. Economies of scale in data storage and maintenance or in computing power may be favorable, as well as the centralization of specialized technical support. Additional capacity may be added as needed, as market share grows, or specific project demands increase. Resources may be scaled up or back as needed. New capabilities may be rented or tested more easily, and software functions may better adopt a rental model and pricing structure.

Cloud computing may also provide faster, broader, safer, and more continuous data access. Internet connections are increasing in speed, and solid-state memory in large installations provide faster access yet. Large server facilities may be on continuously, always accessible. Large server arrays may be outfit with proper backup and protection, including data mirroring at distinct locations. Mirroring provides data redundancy, because the same data are stored concurrently at different physical facilities, often miles or even countries apart. If a fire, flood, or other disaster befalls one data server, the mirror image is likely to remain intact.

Internet mapping is perhaps the simplest and most common from of cloud-based computing. Many internet applications allow users to compose maps on a web page. The individual user has some control over the data layers shown, the extent of the mapped area, and the symbols used to render the map. The internet is different from other technologies because it allows a wide range of people to custom-produce maps. Each user may choose her own data and cartographic elements to display. The user is largely free from any data development chores and thus needs to know very little about data entry, editing, or the particulars of map projections, coordinates, or other details required for the production of accurate spatial data. Typically the map itself is the end product, and may be used for illustration, or to support analyses that will be performed entirely within the user's head.

These internet mapping applications are particularly appropriate when a large number of users need to access a limited number of data layers to compose maps. The internet users may select the themes, variables, and symbolization, in contrast to a static map graphic, in which a website cartographer defines the properties of each map.

Because most internet mapping applications are built for users who have little knowledge of spatial data, maps, and analysis, the suite of spatial operations allowed is usually very sparse. Most internet mapping is currently limited to creating simple displays. This is changing, as query, distance functions, and basic tools are provided, albeit in very simple forms.

As noted in a Chapter 7, web mapping services are another step towards a cloud-based GIS model. Data are stored somewhere "on the cloud," and a specific link to them provided. Data are accessible after forging a connection, as if they were from any other disk source, at least as far as the accessing software view. The GIS program doesn't distinguish between local and cloud data once the connection is made.

This brings up one major limitation of the cloud model, its dependence on a fast connection to the cloud of resources. Slow internet speeds become a hindrance, particularly with large data or image sets that characterize many spatial analyses. Each zoom, pan, or layer addition may require a scene to be repainted, involving the movement of billions of pixels through the web connection. While this may be overcome to some extent with local caching, anticipatory downloading, e.g., pulling data in a wider area than that immediately viewed, and other software techniques, but within limits. In many instances there is no substitute for extremely fast internet speeds. Widespread, fast internet should be forthcoming, but access may depend on internet demand as well as supply.

To date, analytical tools delivered over the internet are still quite rudimentary, and may likely remain so for some time. Robust, correct operation of an analytical tool is difficult to provide in many instances, and requires a sizeable investment. Systems for deliver, payment and protection for a broad, interacting suite of geospatial tools will take development, both technically and culturally.

Open GIS

Open Standards for GIS

Open standards in computing seek to reduce barriers to sharing programs, data, and information. Spatial data structures may be very complex, perhaps more than many other kinds of data. Data may be raster or vector, real or binary, or represent point, line or area features. In addition, different software vendors may elect to store their raster imagery using different formats, and data may be delivered on different physical media, or formatted different ways. If a person orders an image in one format, but her computing system does not support the physical media on which the data are written, or does not understand the file structures used to store the image, then she may not be able to use these data. Incompatible systems are generally described as non-interoperable, and open standards seek to remove this non-interoperability.

The development of open standards in computing is driven by the notion that the larger user community benefits when there are no technical barriers that inhibit the free exchange of data and methods. Open standards seek to establish a common framework for representing, manipulating, and sharing data. Open standards also seek to provide methods for vendors and users to certify compliance with the standard. Standards have been developed in a number of endeavors, for example, the ISO 9600 format for CD ROMs is a defined standard allowing any manufacturer, data developer, or user to build, read, write, or share data on CD devices.

Businesses and many other organizations by their nature have a proprietary interest in the spatial data entry, storage, and methods they produce. Many vendors survive by the revenue their GIS products generate, and so have a strong interest in protecting their investments and intellectual property. However, the developers also may spur adoption of their GIS packages and speed up the development of complementary software by making the internal workings of some portions of their GIS packages public knowledge, for example, by publishing the data structures and formats used to store their spatial data. Thus, these vendors also have a strong interest in making parts of their system open to the public.

Open standards for spatial data are the responsibility of the OpenGIS Consortium. The OpenGIS consortium has developed a framework to ensure interoperability. They do this by defining a general, common set of base data models, types, domains, and structures, a set of services needed to share spatial data, and specifications to ease translation among different representations that are compatible with the OpenGIS standards. Data developed by a civil engineer and stored in a raster format on a Unix version of Arc/Info should be readily accessible to a soil scientist using GRASS on an OS-X Apple system.

Open standards in GIS are relatively new. While most of the large software vendors, data developers, and government and educational organizations are members of the OpenGIS consortium, some components of the standard are still under development. In the future, there will be increased emphasis on compliance to the OpenGIS standards.

Open Source GIS

Open source software is different from most other software in that it is distributed free, along with the source code. The open source organization (www.opensource.org) requires that the software is not by design restricted to a specific operating system or other technology, that there can be no royalties, and that there be no explicit discrimination against fields of endeavor, persons, or groups. But the main, defining characteristic of open source software is an open, grassroots network of collaborators developing, documenting, and freely sharing source code.

There are open source software of many types, from operating systems to word processors, and including GIS. Open source GIS software projects are directed at a range of applications, and notable examples

include the development of general-purpose GIS (e.g., GRASS, FMaps) to specific utilities (e.g., MapServer for web-based spatial data display, query, and analysis) or toolkits to support GIS software development (e.g., GDAL, shapelib).

Open source use is a large and growing phenomenon due to many reasons. High software costs are driving many organizations towards open source software. Licenses for some commercial products are tens to hundreds of thousands of dollars annually for some large organizations. If these organizations employ staff programers, open source GIS may meet geoprocessing needs at a reduced cost.

Many organizations use open source GIS because commercial products may not provide the required functions or capabilities. Three-dimensional structural analysis tools may exist that meets the requirements of a mining company, and so they may develop specific applications. This development may be more efficient and less expensive in an open source environment.

Open source use is expanding in many countries because of specific governmental initiatives. China, India, and Brazil have all supported open source software in general, and operating systems in particular, to maintain independence from foreign firms, reduce costs to government and local business, and develop local information technology expertise. Because these nations are home to more than a third of the world's population, their actions alone are substantially increasing the use of open source GIS.

A Hybrid Model

Proprietary software vendors may adopt a hybrid software approach, where they interact with open software and systems. This has taken many guises. Some may simply support standards, and ensure their systems may access and generate industry standard data forms. But a fuller approach provides the code in a mix of open and proprietary parts. Base code may be provided free, with a charge for extensions or some set of additional capabilities.

Alternately, there may be charges for the base code, but enough source code or adherence to open standards that open source extensions can be easily added later. This allows for the development of an "ecosystem" of extension around a base application, both proprietary and open source.

Summary

GIS are a dynamic collection of conceptual models, tools, and methods that use spatial data. As such they will continue to evolve. What becomes standard practice in the future may be quite different from the methods we apply today. However, the fundamental set of knowledge will remain unchanged. We will still gather spatial and attribute data, adopt a spatial data model to conceptualize real world entities, and use map coordinates to define positions in space. The coordinates are likely to remain based on a standard set of map projections, and we will combine the spatial data of various classes of entities to solve spatial problems. This book is an attempt to provide a foundation to effectively use spatial analysis tools. I hope it has provided enough information to get you started, and has sparked your interest in learning more.

Suggested Reading

The worldwide web is the best source for information about new developments and trends in spatial data acquisition, analysis, and output. In contrast to previous chapters, nearly all the suggested readings are websites. We apologize that many links may be short-lived, but the reader is directed to search for similar and additional sites for the most current information.

www. gislounge.com

www.nasa.gov, general NASA entry point

www. gis.com, an ESRI-sponsored website, general information

www.usgs.gov, public domain data from the USGS

eros.usgs.gov, another common USGS entry point

www.epa.gov/geospatial/

www.opengeospatial.org, open GIS consortium description

www.freegis.org

lidar.cr.usgs.gov, LiDAR data and systems description

www.gpsworld.com/

www.digitalglobe.com, QuickBird high resolution satellite data

www.satimagingcorp.com/

www.gisuser.com

www.directionsmag.com

Study Questions

15.1 - Which of the described new technologies is likely to have the largest impact in GIS over the next five years? Why?

15.2 - What are areas of spatial data entry, analysis, output, or storage that are in dire need of innovation or new and better methods? What is a major bottleneck to further advancement of spatial information science and technology?

15.3 - What is Open Source GIS? How will this change spatial computing?

Appendix A: Glossary
Terms used in GIS and Spatial Data Development and Analysis

Accuracy: The nearness of an observation or estimate to the true value.

Active remote sensing system: A system that both emits energy and records the energy returned by target objects.

Adaptive sampling: A method to increase sampling efficiency by increasing the spatial sample frequency in areas with higher spatial variability.

Adjacency: Two area objects that share a bounding line are topologically adjacent.

Affine coordinate transformation: A set of linear equations used to transform from one Cartesian coordinate system to another. The transformation applies a scaling, translation, and rotation.

Almanac: Important system information sent by each GPS satellite, and recorded by a GPS receiver to obtain current satellite health, constellation status, and other information helpful for GPS positioning.

Arc: A line, usually defined by a sequence of coordinate points.

ArcGIS: A GIS software package produced by Environmental Research Systems, Inc., of Redlands, California.

Area feature: A polygon, collection of contiguous raster cells, or other representation of a bounded area. The feature is characterized by a set of attributes and has an inside and an outside.

ASCII: American Standard Code for Information Interchange. A set of numbers associated with a symbol used in information storage and processing. Numbers are between 0 and 255 and may be represented by a single byte of data.

Aspect: The direction of steepest descent on a terrain surface.

Atmospheric distortion: Image displacement due to the bending of light as it passes through the atmosphere.

Attribute: Non-spatial data associated with a spatial feature. Crop type, value, address, or other information describing the characteristics of a spatial feature are recorded by the attributes.

Autocad Geospatial: A suite of GIS software systems produced by Autodesk, Inc., of San Rafael, California.

AVHRR: Advanced Very High Resoluton Radiometer. A satellite system run by the National Oceanographic and Atmospheric Administration to collect visible, thermal, and infrared satellite images of the globe each day. The system has up to a 1.1 km resolution.

Base station: GNSS recording station over a precisely surveyed location, used in differential correction.

Beacon receiver: A GNSS receiver capable of decoding beacon base station signals transmitted by the U.S. Coast Guard beacon stations.

Bearing: A direction, usual specified as a geographic angle measured from some base line, e.g., true north.

Benchmark: A monumented, precisely surveyed location for which coordinates are known to a high degree of accuracy.

Bilinear interpolation: A method for calculating values for a grid location based on a linear combination of nearby grid values.

Binary classification: A classification of spatial objects into two classes, typically denoted by a 0 class and a 1 class.

Binary operation: A spatial operation with two inputs.

Bit: A binary digit. A bit has one of two values, on or off, zero or one. This is the smallest unit of digital information storage and the basic building block from which all other computer data and programs are represented.

Boolean algebra: Conditions used to select features with set algebraic conditions, including and, or, and not conditions.

Buffer: A buffer area is a polygon or collection of cells that are within specified proximities of a set of features. A buffer operation is one that creates buffer areas.

Bundle adjustment: The simultaneous removal of geometric distortion and production of orthophotographs from a number of aerial images.

Byte: A unit of computer storage consisting of 8 binary digits. Each binary digit may hold a zero or a one. A byte may store up to 256 different values.

C/A code GPS: Coarse acquisition code, a GPS signal used to make for rapid, relatively low-accuracy positional estimates. Accuracies without further corrections are typically from a few to tens of meters.

CAD/CAM: Computer Aided Design/Computer Aided Mapping. Software used primarily by design engineers and utilities managers to produce two and three dimensional drawings. Related to GIS in that coordinate information are input, manipulated, and output. These systems typically do not store map-projected coordinates, and do not have sophisticated attribute entry and manipulation capabilities.

Carrier-phase GPS: Relatively slow but accurate signal used to estimate position. Position may be determined to within a few centimeters or better.

Cartesian coordinate system: A right-angle two or three-dimensional coordinate system. Axes intersect at 90 degrees, and the interval along each axis is linear.

Cartographic modeling: The combination of spatial data layers through the application of spatial operations.

Cartographic object: A digital representation of a real-world entity.

Cartometric map: A map produced such that the relative positions of objects depicted are spatially accurate, within the limits of the technology and the map projection used.

Centroid: A central point location for an area feature, often defined as the point with the lowest average distance to all points that define the area boundary.

Classification: A categorization of spatial objects based on their properties.

Clients: Programs that request data from a server.

Clip (overlay): The vertical combination of two data layers, with a clip layer typically designated that defines the extent and location of output areas, and that preserves only the data from the non-clip layer for the clip area.

Cluster sampling: A technique of grouping samples, to reduce travel time among samples while maintaining sample number.

Code-phase GPS: see C/A code.

COGO: Coordinate Geometry, the entry of spatial data via coordinate pairs, usually obtained from field surveying instruments.

Compass: Chinese satellite-based positioning systems

Conformal coordinate transformation: A registration that requires scale changes to be equal in the x and y directions.

Conformal projection: A map projection is conformal when it preserves shape for some portions of the map.

Conic projection: A map projection that uses a cone as the developable surface.

Connectivity: A record or representation of the connectedness of linear features. Two linear features or networks are connected if they may be traversed without leaving the network.

Continuous surface: A variable or phenomena that changes gradually through two-dimensional space, e.g., elevation or temperature.

Contour line: A line of constant value for a mapped variable.

Control points: Point locations for which map projection and database coordinate pairs are known to a high degree of accuracy. Control points are most often used to convert digitized coordinates to standard map projection coordinates.

Convergent circle: A circle used in defining facet for a triangulated irregular network, that passes through three points, and does not contain any other points.

Coordinates: A pair or triplet of numbers used to define a position in space.

Coordinate transformation: The conversion or assignment of coordinates from a non-projected coordinate system to a coordinate system, typically via a system of linear mathematical equations.

Cost surface: A spatial depiction of the cost of traveling among locations in an area.

Cubic convolution: A method of calculating grid values based on a weighted combination of 16 nearby grid cells.

Cylindrical projection: A map projection that uses a cylinder as the developable surface.

Data independence: The ability to make changes in data structure in a database management system that are transparent to users or applications that use data.

Data model: A method of representing spatial and aspatial components of real-world entities on a computer.

Database management system (DBMS): A collection of software tools for the entry, organization, storage, and output of data.

Datum: A set of coordinate locations specifying horizontal positions (for a horizontal datum) or vertical positions (for a vertical datum) on the Earth surface.

Datum adjustment: A re-calculation of a datum based on additional measurements.

Datum shift: The change in horizontal or vertical point location the results from a datum adjustment.

Datum transformation: A method or set of equations that allows the calculation of a point location in a one datum based on coordinates expressed in a different datum.

Declination: The angle between the bearing towards True North and the bearing towards Magnetic North.

DEM: Digital Elevation Model, a raster set of elevations, usually spaced in a uniform horizontal grid.

Developable surface: A geometric shape onto which the Earth sphere is cast during a map projection. The developable surface is typically a cone, cylinder, plane, or other surface that may be mathematically flattened.

Diaphragm: A camera component that functions like the iris of the human eye, to control the amount of light available to fall on the film or CCD recording surface, and to improve focus.

Differential GNSS: GNSS positioning based on two receivers, one at a know location and one at a roving, unknown location. Data from roving receivers are corrected by the difference error computed at the known location.

Digitize: To convert paper or other hardcopy maps to computer-compatible and stored data.

Digitizing table: A device with a flat surface and input pointer used to digitize hardcopy maps.

Dilution of precision (DOP): See position dilution of precision (PDOP).

Dissolve: An operation that removes lines separating adjacent polygons that are considered equal, based on some characteristic or measure. A dissolve operation is typically applied based on equal values of variables that are contained in a table associated with the data layer.

DLG: Digital Line Graph, vector data developed and distributed by the United States Geological Survey.

DOQ: Digital Orthophoto Quadrangle, an orthographic photograph provided in digital formats by the USGS. Most tilt and terrain error have been removed from DOQs.

Domain: The range of values a variable may take.

DRG: Digital Raster Graphics, a digital version of USGS fine- to medium-scale maps.

Dual-frequency GPS receiver: A receiver capable of measuring the L1 and L2 broadcast signals, and using these to estimate highly accurate and precise positions, typically to centimeter levels.

Easting: The axis approximately parallel to lines of equal latitude in UTM and a number of other standard map projections.

Electromagnetic spectrum: A range of energy wavelengths, from X-rays through radar wavelengths. The electromagnetic spectrum is typically observed at wavelengths emitted by the Sun or by objects on Earth, covering wavelengths from the visible to the thermal infrared region.

Ellipsoid: A mathematical model of the shape of the Earth that is approximately the shape of a flattened sphere, formed by rotating an ellipse.

Ellipsoidal height: Height measured from an ellipsoidal surface to a point on the surface of the Earth.

Endlap: The end-to-end overlap in aerial photographs taken in the same flight line.

Entity: A real world item or phenomenon that is represented in a GIS system or database.

Ephemeris: Information on GNSS satellite orbits, required by GNSS receivers to computer satellite position, range distance, and receiver position.

Epsilon band: A band surrounding a linear feature that describes the positional error relative to the feature location.

Equal-area classification: A classification method that assigns classes such that each class corresponds to an equal area.

Equal-interval classification: A classification method that assigns an equally spaced set of classes across the range of a variable.

ERDAS: A GIS and remote sensing image processing software package owned and developed by Leica Geosystems, St. Gallen, Switzerland.

ETM+: Enhanced Thematic Mapper, a scanner carried on board Landsat 7, providing image data with resolutions of 30 meters for visible through mid infrared, 15 meter panchromatic, and 60 meter for thermal wavelengths.

Facet: A triangular face in a TIN.

False northing: A number added to coordinates in a map projection, usually to avoid negative coordinate locations within the area of a map projection.

Feature: An object or phenomenon in the landscape. A digital representation of the feature is often called a cartographic feature.

Feature generalization: The incomplete representation of shape defining coordinates for entities represented in a GIS.

Fiducial marks: Also known as fiducials, precisely scribed marks that are recorded near the edges of aerial images, and used to remove systematic camera distortion and to register images.

FIPS: Federal Information Processing Standards code - a set of numbers for defined political or physical entities in the United States. There are FIPS codes for each state, county, and other features.

Friction surface: A raster surface used in calculating variable travel costs through an area. The friction surface represents the cost per unit distance to travel through a cell.

Flatbed scanner: An electronic device used to record a digital image of a hardcopy map or image.

Flow direction: The direction water will flow from a point, usually an azimuth or bearing angle assigned to a raster cell.

Friction surface: See cost surface.

FTP: File Transfer Protocol, a standard method to transfer files across a computer network.

Functional dependency: Property of a set of items in a database table. If one item is functionally dependent on another, that means knowing the value of one item guarantees we know the corresponding value of the second item.

Galileo: A European-based GNSS system.

Generalization: The simplification of shape or position that inevitably occurs when features are mapped.

Geocentric: A measurement system that uses the center of the Earth as the origin.

Geocoding: The process of assigning a geographic or projection coordinate to a data item that is based on a street address, town, and state or country.

Geodetic datum: A reference system against which horizontal and/or vertical positions are defined. It typically consists of a sphere or ellipsoid and a set of point locations precisely defined with reference to that surface.

Geodesy: The science of measuring the shape of the Earth and locations on or in the Earth.

Geoid: A measurement-based model of the shape of the Earth. The geoid is a gravitational equipotential surface, meaning a standard surface of equal gravitational pull. The geoid is used primarily as a basis for specifying terrain or other heights.

Geographic North: The northern axis of rotation of the Earth. By definition true north lies at $90°$ latitude.

GeoMedia: A GIS software package produced by Intergraph, Inc., of Huntsville, Alabama.

GIS: A geographic information system. A GIS is a computer-based system to aid in the collection, maintenance, storage, analysis, output, and distribution of spatial data and information.

GLONASS: Global Navigation Satellite System. A Russian developed and maintained system for coordinate measurement and positioning.

Global operation: A spatial operation where the output location, area, or extent comes from operations on the entire input area or extent.

GNSS: Global Navigation Satellite System. A constellation of satellites plus a ground control segment that allows precise location on or above the Earth. This includes GPS, GLONASS, and other satellite navigation system.

Gnomonic projection: A map projection with the projection center placed at the center of the spheroid.

GRASS: An open-source GIS software system.

Graticule: Lines of latitude and longitude drawn on a hardcopy map or represented in a digital database.

Gravimeter: An instrument for measuring the strength of the gravitational field.

Great circle distance: The shortest distance between two points on the surface of the Earth. This distance follows a great circle route, defined as the route on the surface defined by a plane that intersects the starting and ending point and the center of the Earth.

Grid North: The direction parallel to the northing axis in a projected, Cartesian coordinate system.

Greenwich meridian: The line of equal longitude passing through the Royal Observatory in Greenwich, England. This line was taken as zero, by convention, for the system of longitude measurements for the world.

GRS80: Geodetic Reference Surface of 1980, an ellipsoid used for map projections in much of North America.

Helmert transformation: A method to transform among horizontal datums.

Hierarchical data model: A method of organizing attribute data that structures values in a tree, typically from general to more specific.

High-pass filter: A raster operation that identifies large or high-frequency differences between cells.

Hydrography: Geographic representation of water features.

Hypsography: Geographic representation of height features.

Ikonos: A high resolution imaging satellite system. Ikonos provides 1-meter panchromatic and 3-meter multispectral image data.

Inner join: A combination of two data tables in a database management system based on a key column. The output table combines rows by matching values in the key column, and saves only rows that have matching key values in both tables.

Interpolation: The estimation of variables at unsampled locations from measurements at sampled locations. Interpolation methods are usually understood to use a formula with all parameters that are pre-determined, meaning that parameter values used in the formula do not depend on the data values.

Intersection (overlay): The vertical combination of two data layers, typically restricted to the extent of one data layer but preserving the data contained in both data layers for that extent.

Interval/ratio scale: A measurement scale that records both order and absolute difference in value for a set of variables.

Items: Variables or attributes in a data table, typically viewed as the columns of the table. These are the types of essential characteristics used to described each feature in the geographic data set, e.g., area, depth, and water quality for a lakes data set.

Idrisi: A GIS system developed by the Graduate School of Geography of Clark University, Worcester, Massachusetts.

IDW: Inverse Distance Weighted interpolation, a method of estimating values at unsampled locations based on the value and distance to sampled locations.

Infrared image: An image that records reflectance in the near infrared wavelengths, typically including 0.7 to 1.1 micrometers.

JPEG: An image compression format.

Kernel: An arrangement of cells and values used as a multiplication template in raster analysis.

Key: An item or variable in a relational table used to uniquely identify each row in the table.

Kriging: An interpolation method based on geostatistics, the measurement of spatial autocorrelation.

Landsat: A NASA project spanning more than three decades and seven satellites that proved the capabilities of space-based remote sensing of land resources.

Latitude: Spherical coordinates of Earth location that vary in a north-south direction.

Law of sines: A trigonometric relationship that allows the calculation of unknown triangle edge lengths from known angles and edge lengths.

Leveling surveys: Surveys used to measure the relative height difference between sets of points.

Lidar: Laser detecting and ranging, the use of pulse laser measurements to identify the height, depth, or other properties of features.

Linear referencing: See geocoding.

LIS: A Land Information System, a name originally applied for GIS systems specifically developed for property ownership and boundary records management.

Local operation: A spatial operation where the output location, area, or extent comes from operations on that same extent.

Logical model: A conceptual view of the objects we portray in a GIS.

Longitude: Spherical coordinates of Earth location that vary in an east-west direction.

Manifold: GIS software package produced by CDA International, of Carson City, Nevada.

Map algebra: The combination of spatial data layers using simple to complex spatial operations.

MapInfo: GIS software package produced by MapInfo, Inc., of Troy, New York.

Map projection: A systematic rendering of features from a spheroid or ellipsoid representing the 3-dimensional Earth to a map surface.

Meridian: A line of constant longitude.

Magnetic North: The point where the northern lines of magnetic attraction enter the Earth. Magnetic North does not occur at the same point as "True" or Geographic North. In the absence of local interference a compass needle points towards magnetic north. The magnetic north pole is currently located in northern Canada.

Metadata: Data about data, that describes the properties of a spatial data set, including the coordinate system, extent, attribute types and values, origin, lineage, accuracy, and other characteristics needed for effective evaluation and use of data.

Metes and bounds survey: A survey method based on distance and sometimes angle measurements from known or monumented points.

Minimum mapping unit (MMU): The smallest area resolved when interpreting an aerial or satellite image, or when mapping area features from a source data set.

Moving window: A usually rectangular arrangement of cells that shifts in position across a raster data set. At each position an operation is applied using the cell values currently encountered by the moving window.

MSS: Multi-spectral Scanner, an early satellite imaging scanner carried by Landsat satellites.

Modifiable areal unit problem: The dependence of aggregate area statistics on the size and shape of the aggregation units.

MODIS: Moderate Resolution Imaging Sensor. A later generation imaging scanner that is part of NASA's Mission to Planet Earth. Provides high spectral resolution, frequent global coverage, and moderate spatial resolution of from 250 to 1000 meters.

Molodenski transformation: A method to transform among geodetic datums.

Multispectral: An image, film, or system that records data collected from multiple wavebands.

Multi-tier architecture: A database management system design where there are multiple levels of clients above a server.

Nadir point: The point directly below the aircraft, usually near the center of an aerial image.

NAD27: North American Datum of 1927, the adjustment of long-baseline surveys to establish a network of standardized horizontal positions in the early 20th century.

NAD83: North American Datum of 1983. The successor to NAD27, using approximately an order of magnitude more measurements and improvements in analytical models and computer power. The current network of standard horizontal positions for North America.

NAVD29: North American Vertical Datum of 1929, an adjustment of vertical measurements to establish a network of heights in the early 20th century.

NAVD88: North American Vertical Datum of 1988, the successor vertical datum to NAVD29.

Neighborhood operation: A spatial operation where the output location, area, or extent comes from operations on an area larger than, and usually adjacent to the input extent.

Network: A connected set of line features, often used to model resource flow or demand through real-world networks such as road or river systems.

Network center: A location on a network the provides or requires resources.

NLCD: National Land Cover Data set, a Landsat Thematic Mapper (TM) based classification of landcover for the United States.

NOAA: National Oceanic and Atmospheric Administration, the U.S. government agency that oversees the development of national datums.

Node: An important point along a line feature, where two lines meet or intersect.

Nominal scale: A measurement scale that indicates the difference between values, but does not reflect rank or absolute differences.

Northing: The axis in the approximately north-south direction in UTM and other standard coordinate systems.

Normal forms: A standard method of structuring relational databases to aid in updates and remove redundancy.

N-tuple: A group of attribute values in a database management system.

NWI: National Wetlands Inventory data compiled by the U.S. Fish and Wildlife Service over most of the United States. These data provide first-pass indications of wetland type and extent.

Object: See cartographic object.

Object-oriented data model: A data model that incorporates encapsulation, inheritance, and other object-oriented programming principles.

Open source software: Computer programs that provide the source code to any user, typically easily accessible through a web portal.

Operation, spatial: The manipulation of coordinate or attribute data.

Optical axis: A ray approximately perpendicular to the film or image plane in a camera and parallel to the center of the lens barrel, that may be thought of as the primary direction of incoming light.

Ordinal scale: A scale the represents the relative order of values but does not record the magnitude of differences between values.

Orthogonal: Intersecting at a 90 degree angle.

Orthographic view: Horizontal placement as would be seen from a vertical viewpoint at infinity. There is no terrain or tilt-perspective distortion in an orthographic view.

Orthographic projection: A map projection with the projection center an infinite distance from the map surface.

Orthometric height: Height measured from the Geoid surface to a point on the surface of the Earth.

Orthophotograph: A vertical photograph with an orthographic view. Orthophotographs are created by using projection geometry and measurements to remove tilt, terrain, and perspective distortion from aerial photographs.

Outer join: A combination of two data tables in a database management system based on a key column. The output table appends those rows in a second table that match values in the key column. Null values are placed in joined-table columns from the second table where there is no match to the first table.

Overlay: The "vertical" combination of two or more spatial data layers.

Overshoot: A digitized line that extends past a connecting line.

Panchromatic: An image, film, or system that records in only one wavelength band, and resulting in gray scale (black and white) images.

Parallax: The relative shift in position of features due to a shift in viewing location.

Passive remote sensing system: A system that does not emit the radiation it records from target objects.

PDOP: Positional Dilution of Precision, a figure of merit used to represent the quality of the satellite geometry when taking GPS readings. PDOPs between 1 and 6 are preferred for most applications, and lower is better.

Perspective convergence: The apparent decrease in inter-object distance as the objects are farther away, for example, the apparent convergence of two railroad rails as the recede into the distance.

Perspective view: A view on a location that includes some relief or perspective distortion, meaning the location of objects may be distorted if their relative distance to the camera varies considerably.

Pixels: Picture elements that make up and image, these are the individual grid cells that record or display a brightness or color in an image.

Plan curvature: Terrain curvature along a contour.

Planar topology: The enforcement of intersection for line and area features in a digital data layer. Each line crossing requires an explicit node and intersection.

Plane surveying: Location surveying methods suitable under the assumption that the surveyed lands form a planar surface, i.e., that distortions due to the Earth's curvature may be ignored.

Platten: The flat back portion of a film camera against which the film rests while an image is collected.

Pointer: An address stored in a data structure pointing to the next or related data elements. Pointers are used to organize data and speed access.

Polygon: A closed, connected set of lines that define an area.

Polygon inclusion: An area different in some characteristic from the recorded attributes of the polygon, but not resolved.

Position dilution of precision (PDOP): An index of the geometric distribution of a set of satellites for the purposes of estimating and controlling position accuracy. PDOPS typically range between 1 and 20, and lower PDOPs on average result in higher positional accuracies.

Precision: The repeatability of a measure or process.

Prime meridian: See Greenwich meridian.

Profile curvature: Terrain curvature in the direction of steepest descent.

Proximity function: See buffer.

Public Land Survey System (PLSS): A land measurement system used in the western United States of America to unambiguously define parcel location.

QGIS: An open-source GIS.

Quad-trees: A raster data structure based on successive, adaptive reductions in cell size within a data layer to reduce storage requirements for thematic area data.

Query: Requests or searches for spatial data, typically applied via a database management system.

Radial lens distortion: The displacement of objects in an image due to small lens imperfections, usually radially inward or outward.

Random sample pattern: A sampling pattern where sample location is determined by a random process.

Range pole: A pole used in surveying to raise a GNSS antenna, survey prism, or other survey instrument above the ground. Range poles are often used in GNSS data collection to raise an antenna and thereby obtain better PDOPs, and improved accuracy.

Raster data model: A regular "grid cell" approach to defining space. Usually square cells are arranged in rows and columns.

Record: A collection of attributes stored for a specific instance of an entity.

Registration: The conversion or assignment of coordinates from a non-projected coordinate system to a coordinate system, typically via an affine transformation.

Relations: See relational table.

Relational algebra: A set of operations on database tables specified by E.F. Codd for the consistent manipulation of data in a database.

Relational table: A data table in a relational database management system.

Relief displacement: Apparent horizontal distortion of features due to height differences relative to the nadir point in a vertical aerial image.

Remote Sensing: Measuring or recording information about an object or phenomena without contacting the object.

Resampling: The recalculation and assignment of cell values when changing cell size and/or orientation of a raster grid.

RMSE: Root Mean Square Error, a statistic that measures the difference between true and predicted data values for coordinate locations.

Rubbersheeting: The use of polynomial or other nonlinear transformations to match feature geometry.

Run-length coding: A compression method used to reduce storage requirements for raster data sets. The value and number of sequential occurrences are stored.

Schema: A compact graphical representation of a database conceptual models, entities, and the relationships among them.

Scope: The spatial extent of input for a spatial operation.

Secant lines: Lines of intersection between a developable surface and a spheroid in a map projection.

Selection operation: The identification of a set of objects based on their properties.

Semi-major axis: The larger of the two radial axes that define an ellipsoid.

Semi-minor axis: The smaller of the two radial axes that define an ellipsoid.

Semivariance: The variance between values sampled at a given lag distance apart.

Server: A computer or a program component that stores data, and provides subsets of data in response to requests.

Set algebra: A method for specifying selection criteria based on comparison operators less than, equal to, greater than, and perhaps others.

Shaded relief map: A depiction of the brightness of terrain reflection with a given sun location.

Sidelap: Edge overlap of photographs taken in flightlines.

Shutter: A system for controlling the time or amount of light reaching a detecting surface.

Skeletonizing: Reducing the width of linear features represented in raster data layers to a single cell.

Sliver: Small, spurious polygons at the margins or boundaries of feature polygons that are an artefact of imprecise digitizing or overlay.

Slope: The change in elevation over a change in location, usually measured over some fixed interval, e.g., the change in height between two points 30 meters apart. Slope is usually reported as a percent slope, or as a degree angle measured from horizontal.

Snap distance: A distance threshold defined in digitizing or other spatial analysis. Point features, nodes, or vertices within the snap are moved to be coincident, to occupy the same location.

Snap tolerance: See snap distance.

Snapping: Automatic line joins during vector digitizing or layer overlay. Nodes or vertices are joined if they are within a specified snap distance.

Spaghetti data model: Vector data model in which lines may cross without intersecting.

Spectrum: see electromagnetic spectrum.

Spherical coordinates: A coordinate system based on a sphere. Location on the sphere surface is defined by two angles of rotation in orthogonal planes. The geographic coordinate system of latitude and longitude is the most common example of a spherical coordinate system.

Spheroid: A mathematical model of the shape of the Earth, based on the equation of a sphere.

Spirit leveling: An early leveling survey technique in which horizontal lines were established between survey stations, and relative height differences determined by measured marks on leveling rods.

Spline: A smoothed line or surface created by joining multiple constrained polynomial functions.

SPOT: Systeme Pour l'Observation de la Terre, a satellite imaging system providing 10 to 20 meter resolution images.

SQL: Structured Query Language, a widely adopted set of commands used to manipulate relational data.

SSURGO: Fine resolution digital soil data corresponding to county level soil surveys in the United States. Produced by the Natural Resource Conservation Service.

Standard parallels: Lines of intersection between a developable surface and a spheroid in a map projection.

STATSGO: Coarse resolution digital soil data distributed on a statewide basis for the United States Most often derived from aggregation and generalization of SSURGO data.

State Plane Coordinates: A standardized coordinate system for the United States of America that is based on the Lambert conformal conic and transverse Mercator projections. State plane zones are defined such that projection distortions are maintained to be less than 1 part in 10,000.

Stereo pairs: Overlapping photos taken from different positions but of substantially the same area, with the goal of using parallax to interpret height differences within the overlap area.

Stereographic projection: A map projection with the projection center is placed at the antipode, the point on the opposite side of the spheroid from the projection intersection point with the spheroid.

Stereophotographs: A pair or more of overlapping photographs that allow the perception of three dimensions due to a perspective shift.

Structured Query Language (SQL): A standard syntax for specifying queries to databases.

Survey station: A position occupied, and from which measurements are made, during a land survey.

Systematic sample: A sampling pattern with a regular sampling framework.

Transaction manager: A component of a database management system that processes requests from clients, and passes them to a server.

Traverse: A series of survey stations spanning a survey. Traverses are closed when the return to the starting point, and open when they do not.

Terrestrial reference frame: The set of measured points and their calculated coordinates the are used to define a geodetic datum.

Thematic layer: Thematically distinct spatial data organized in a single layer, e.g., all roads in a study area placed in one thematic layer, all rivers in a different thematic layer.

TIFF: Tagged Image File Format, a widely-supported image distribution format. The Geo-TIFF variant comes with image registration information embedded.

TIGER: Topologically Integrated Geographic Endcoding and Referencing files, a set of structures used to deliver digital vector data and attributes associated with the U.S. Census.

TIN: Triangulated Irregular Network, a data model most commonly used to represent terrain. Elevation points are connected to form triangles in a network.

TM: Thematic Mapper, a high-resolution scanner carried on board later Landsat satellites. Provides information in the visible, near infrared, mid infrared, and thermal portions of the electromagnetic spectrum.

tntMIPS: An image processing and GIS software package produced by Microimages, Inc., of Lincoln, Nebraska.

Topology: Shape-invariant spatial properties of line or area features such as adjacency, contiguity, and connectedness, often recorded in a set of related tables.

Trigonometric leveling: Measurement of vertical positions or height differences among points by the collection of vertical angles and distance measurements.

Triangulation Survey: Horizontal surveys conducted in a set of interlocking triangles, thereby providing multiple pathways to each survey point. This method provides inherent internal checks on survey measurements.

True North: See Geographic North.

Unary operation: An operation that has only one input:

Undershoot: A digitizing error in which a line end falls short of an intended connection at another line end or segment.

Union: The vertical combination of two spatial data layers, typically over the combined extents of the data layers, and preserving data from both layers.

United States Survey Foot: An official distance used for survey measurements in the United States of America that is slightly different in length from the international definition of a foot.

USGS: United States Geological Survey - the U.S. government agency responsible for most civilian nationwide mapping and spatial data development.

UTM: Universal Transverse Mercator coordinate system, a standard set of map projections developed by the U.S. Military and widely adopted for coordinate specification over regional study areas. A cylindrical projection is specified with a central meridian for each six-degree wide UTM zone.

Variable distance buffer: A buffering variant where the buffer distance depends on some value or level of a feature attribute.

Vector data model: A representation of spatial data based on coordinate location storage for shape-defining points and associated attribute information.

Vertical datum: A reference surface against which vertical heights are measured.

Vertex, vertice: Points used to specify the position and shape of lines.

WAAS: Wide Area Augmentation System, a satellite-based transmission of correction signals to improve GPS positional estimates, largely through the removal of ionsopheric and atmospheric effects.

Wavelength: The distance between peak energy values in an electromagnetic wave.

WGS84: World Geodetic System, a Earth-centered reference ellipsoid used for defining spatial locations in three dimensions. Very similar to GRS80 ellipsoid. Commonly used as a basis for map projections.

Zenith angle: The angle measured between a vertical line upward from a point on the Earth and the line from that point to the Sun.

Appendix B: Sources of Geographic Information

General Information

www.gis.com - software, data, education overview from ESRI Inc.

www.gislounge.com - news, data, information on GIS

spatialnews.geocomm.com - general GIS information, news, and data

www.csiss.org/ - center for spatially integrated social science

Spatial Data

Global, Continental Data

www.fao.org/geonetwork/srv/en/main.home

www.naturalearthdata.com/ - integrated, homogenized spatial data for cartography

webgis.wr.usgs.gov/globalgis/ - global data and atlas layers from USGS and the American Geological Institute

www.esri.com/data/free-data/index.html - map services and downloadable data, some global data, but with a stronger U.S. focus

gcmd.gsfc.nasa.gov/ - NASA's global change master data directory, with links to spatial climate, land use, biological, earth sciences, and other data sets

www.landcover.org - Global landcover facility, largely supported by NASA

www.asian.gu.edu.au - Spatial data for Asia

www.ciesin.org - Center for International Earth Science Information Network

www.eurogi.org - Umbrella organization for spatial data in Europe

www.eurogeographics.org - European mapping agencies cooperative venture

eusoils.jrc.ec.europa.eu/data.html - soils data for Europe

edcsnw4.cr.usgs.gov/adds - Spatial data for Africa

U.S.

geo.data.gov/geoportal/ - U.S. spatial data

www.census.gov/geo/www/tiger/index.html - TIGER data from the U.S. Census Bureau

factfinder2.census.gov/faces/nav/jsf/pages/index.xhtml - Census data in custom downloads

seamless.usgs.gov/ - U.S. National Mapping Information

datagateway.nrcs.usda.gov/ - USDA geospatial data portal

http://www.nass.usda.gov/research/Cropland/SARS1a.htm - U.S. annual cropland digital data www.fws.gov/gis/index.html - U.S. Fish and Wildlife Service data portal

www.fws.gov/wetlands - U.S. National Wetlands Inventory Data

soildatamart.nrcs.usda.gov/ - soils data

water.usgs.gov/maps.html - Spatial data for water resources

eros.usgs.gov - EROS Data Center portal, general U.S. government site for image and map data

landcover.usgs.gov - NLCD and other landcover data for the U.S.

www.esri.com/data/free-data/index.html - map services and downloadable data, some global data, but with a stronger U.S. focus

www1.nga.mil/ProductsServices/Pages/default.aspx - National Imagery and Mapping Agency, formerly the U.S. Defense Mapping Agency

U.S. States

agdc.usgs.gov/data - Alaska Geospatial Data Clearinghouse

www.asgdc.state.ak.us - Spatial data for Alaska

http://agic.az.gov/ - Arizona Geographic Information Council

http://atlas.ca.gov/ - California geodata clearinghouse

www.dnr.state.co.us - Colorado environmental and other spatial data

www.cdphe.state.co.us/gis/ - Colorado public health GIS data

magic.lib.uconn.edu - Connecticut Map and Geographic Information Center

www.stateplanning.delaware.gov/information/gis_data.shtml- Deleware Spatial Data Clearinghouse

www.dep.state.fl.us/gis - Florida Department of Environmental Protection, spatial data

www.gis.state.ga.us - Spatial data from the Georgia state government

www.hawaii.gov/dbedt/gis - Hawaii statewide data and programs

gis.idaho.gov - Idaho GIS and remote sensing data

www.isgs.uiuc.edu/nsdihome - Illinois natural resources spatial data clearinghouse

www.iowagis.org/ - Iowa spatial data clearinghouse

www.kansasgis.org/ - State of Kansas spatial data

kygeonet.ky.gov/ - Kentucky state government GIS data site

lagic.lsu.edu - Louisiana Geographic Information Center

www.maine.gov/megis/ - State of Maine data catalogue

www.state.ma.us/mgis - Massachusetts state GIS data

www.mcgi.state.mi.us/mgdl/ - Michigan geographic information data library

www.mngeo.state.mn.us/chouse/ - Minnesota state government spatial data

msdis.missouri.edu - Missouri spatial data

nris.mt.gov/gis/ - Montana natural resources data

dnr.ne.gov/databank/spat.html - Nebraska spatial data clearinghouse

www.state.nj.us/dep/gis - New Jersey Department of Environmental Protection data

www.gis.ny.gov - New York State spatial data clearinghouse

www.nconemap.com/ - North Carolina spatial data clearinghouse

njgin.state.nj.us/NJ_NJGINExplorer/index.jsp - North Dakota spatial data clearinghouse

gis.oregon.gov/ - Oregon spatial data library

www.pasda.psu.edu - Pennsylvania spatial data

www.edc.uri.edu/rigis - Rhode Island spatial data

www.gis.sc.gov/ - South Carolina Department of Natural Resources

www.sdgs.usd.edu - South Dakota Geological Survey

http://www.tnris.org/ - Texas natural resources spatial data

www.agrc.utah.gov/sgid.html - Utah statewide geographic information database

www.vcgi.org/ - Vermont GIS data waterhouse

gisdata.virginia.gov - Virginia Geospatial and Statistical Data Center

http://wa-node.gis.washington.edu/geoportal/catalog/main/home.page - Washington State spatial data clearinghouse

http://gio.wi.gov/Resources/Data/tabid/253/Default.aspx - Wisconsin land information clearinghouse

http://wygl.wygisc.org/wygeolib/catalog/main/home.page - Wyoming spatial data

Data for non-U.S. Countries

freegisdata.rtwilson.com/ - a compilation of data sources

www.auslig.gov.au - spatial data and information for Australia

www.idee.es - geospatial portal for data, catalogs, and services for Spain

professionnels.ign.fr/24/donnees-et-logiciels-gratuits/telechargement.htm, spatial data for France

www.ordsvy.gov.uk - British National Mapping Agency data

www.geobase.ca/geobase/en/index.html - Spatial data for Canada

geogratis.ca - More spatial data for Canada

www.nrcan.gc.ca/earth-sciences/home - Earth sciences information for Canada

nibis.lbeg.de/cardomap3/ - Data for Germany

sedac.ciesin.columbia.edu/mexico.html - population data for Mexico

http://idechg.chguadalquivir.es/geoportal/en/services/downloads.html - South African data

Images

http://remotesensing.usgs.gov/index.php - U.S. USGS land remote sensing data archive

www.eurimage.com - Satellite image sales for most of Europe

www.rsi.ca - RADARSAT radar imaging data

www.spaceimaging.com - High resolution satellite imagery

www.digitalglobe.com - High resolution satellite imagery

www.spot.com - Pleiades and SPOT satellite imagery

landsat.gsfc.nasa.gov - Landsat 7 gateway

modis-land.gsfc.nasa.gov - MODIS land surface data

photojournal.jpl.nasa.gov - NASA images of the Earth, other planets, and astronomy

http://www.fsa.usda.gov/FSA/apfoapp?area=home&subject=prog&topic=nai - High resolution aerial photographs of the U.S.

Software

usa.autodesk.com - Autodesk Map

grass.fbk.eu/ - GRASS GIS

www.bentley.com/ - Microstation GIS

www.erdas.com - Imagine image processing and raster GIS software

www.esri.com - ArcGIS and related products

www.idrisi.com - IDRISI and Cartalinx spatial data analysis and entry software

www.intergraph.com - MGE and Geomedia GIS software

www.mapinfo.com - MapInfo GIS

www.microimages.com - TNTmips map and image processing system

www.mapwindow.org/ - Mapwindow GIS

www.qgis.org/ - QGIS software

News, Journals, Industry Information

www.asprs.org - American Society for Photogrammetry and Remote Sensing, publishers of Photo-grammetric Engineering and Remote Sensing

www.geoinfosystems.com - Geospatial solutions journal

www.geoplace.com - GIS news

www.gpsworld.com - developments in GPS

www.wkap.nl - Kluwer, publishers of GeoInformatica

www.urisa.org - Urban and Regional Information Systems Organization, publishers of the URISA Journal

Organizations, Standards

www.aag.org - American Association of Geography

www.fgdc.gov - U.S. Federal Geographic Data Committee

www.igu-online.org/site/ - International Geographical Union

www.geog.ubc.ca/cca - Canadian Cartographic Association

www.ngs.noaa.gov - U.S. National Geodetic Survey

www.soc.org.uk - Society of Cartographers

www.ucgis.org/ - university consortium dedicated to furthering geographic information science

www.urisa.org - Urban and Regional Information Systems Organization

Appendix C: Useful Conversions and Information

Length
1 meter = 100 centimeters
1 meter = 1000 millimeters
1 meter = 3.28083989501 International
 feet
1 meter = 3.28083333333 U.S. survey
 feet
1 kilometer = 1000 meters
1 kilometer = 0.62137 miles
1 mile = 5280 feet

Area
1 hectare = 10,000 square meters
1 square kilometer = 100 hectares
1 acre = 43,560 square feet
1 square mile = 640 acres
1 hectare = 2.47 acres
1 square kilometers = 0.3861 square
 miles

Angles
1 degree = 60 minutes of arc
1 minute = 60 seconds of arc
decimal degrees =
 degrees + minutes/60+seconds/3600
180 degrees = 3.14159 radians
1 radian = 57.2956 degrees

Spherical angles on a globe:

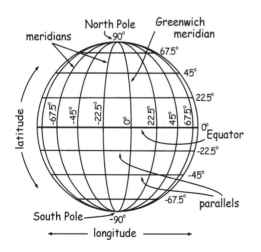

Horizontal angles in a projected coordinate system - Azimuth on a flat
map:

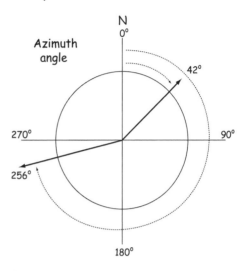

Scale

Scale value	1 centimeter distance on map equals a distance on the ground of:
1:5,000	50 meters
1:10,000	100 meters
1:25,000	250 meters
1:50,000	500 meters
1:100,000	1000 meters

Scale value	1 inch distance on a map equals a distance on the ground of:
1:6,000	500 feet
1:15,840	1,320 feet
1:24,000	2,000 feet
1:62,500	5,208 feet
1:100,000	8,333 feet

State Plane Zones

UTM Zones - USA

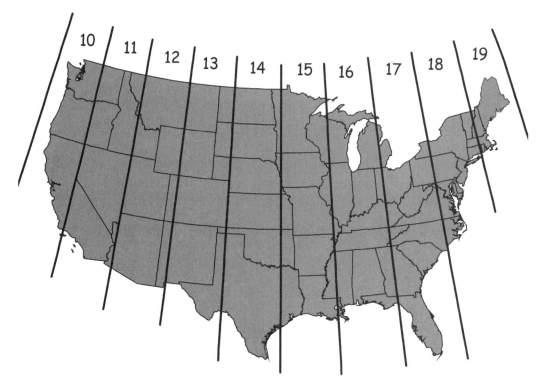

Trigonometric Relationships

sine (a) = A/H
cosine (a) = B/H
tangent (a) = A/B
cotangent (a) = B/A
secant (a) = H/A
cosecant (a) = H/B

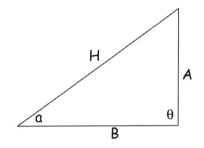

Coordinate Geometry

Coordinate
geometry (COGO)

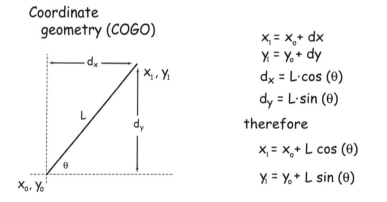

$x_1 = x_0 + dx$

$y_1 = y_0 + dy$

$dx = L \cdot \cos(\theta)$

$dy = L \cdot \sin(\theta)$

therefore

$x_1 = x_0 + L \cos(\theta)$

$y_1 = y_0 + L \sin(\theta)$

If we know the location of a point, x_n, y_n, and have measured the azimuth and distance to another point x_u, y_u. What are the coordinates for the unknown point, x_u, y_u ?

Suppose $x_n = 12$, $y_n = 3$, $D = 6.8$, and azimuth = 242°

From above,

$dy = D \cdot \cos(\theta)$
$dx = D \cdot \sin(\theta)$
$x_u = x_n - dx$
$y_u = y_n - dy$

We can calculate θ from the azimuth,
a = 242°

θ = 242 - 180 (see figure)

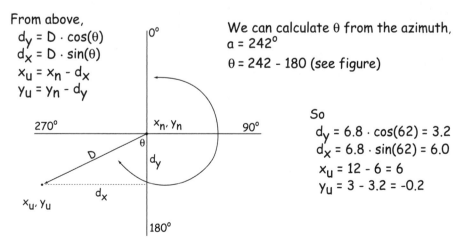

So

$dy = 6.8 \cdot \cos(62) = 3.2$
$dx = 6.8 \cdot \sin(62) = 6.0$
$x_u = 12 - 6 = 6$
$y_u = 3 - 3.2 = -0.2$

Distance between two points, P (x_p, y_p) and W (x_w, y_w)

$$d = \sqrt{(x_p - x_w)^2 + (y_p - y_w)^2}$$

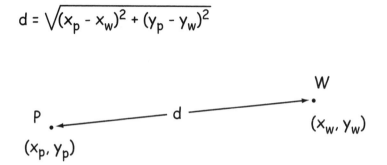

Useful relationships for oblique triangles:

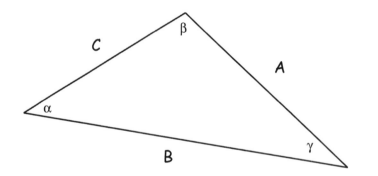

Law of sines

$$\frac{A}{\sin\alpha} = \frac{B}{\sin\beta} = \frac{C}{\sin\gamma}$$

Law of cosines

$$A^2 = B^2 + C^2 + 2BC \cdot \cos\alpha$$

$$B^2 = A^2 + C^2 + 2AC \cdot \cos\beta$$

$$C^2 = A^2 + B^2 + 2AB \cdot \cos\gamma$$

Equation of a circle, with center at $C\,(x_c, y_c)$ and radius r

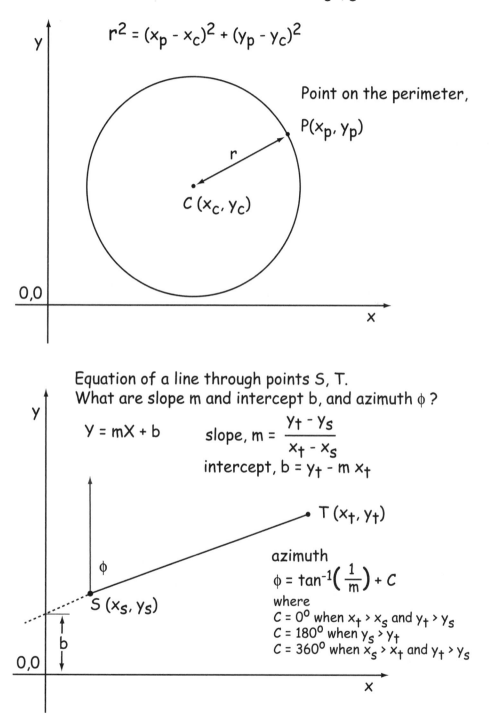

$$r^2 = (x_p - x_c)^2 + (y_p - y_c)^2$$

Point on the perimeter,

$P(x_p, y_p)$

r

$C\,(x_c, y_c)$

0,0

x

Equation of a line through points S, T.
What are slope m and intercept b, and azimuth ϕ ?

$$Y = mX + b$$

slope, $m = \dfrac{y_t - y_s}{x_t - x_s}$

intercept, $b = y_t - m\,x_t$

$T\,(x_t, y_t)$

azimuth

$$\phi = \tan^{-1}\left(\frac{1}{m}\right) + C$$

where
$C = 0^\circ$ when $x_t > x_s$ and $y_t > y_s$
$C = 180^\circ$ when $y_s > y_t$
$C = 360^\circ$ when $x_s > x_t$ and $y_t > y_s$

ϕ

b

$S\,(x_s, y_s)$

0,0

x

Coordinates of a point, when angle and distance to a baseline are known

Suppose d, γ, x_1,y_1 and x_2,y_2 are known

what are x_3,y_3?

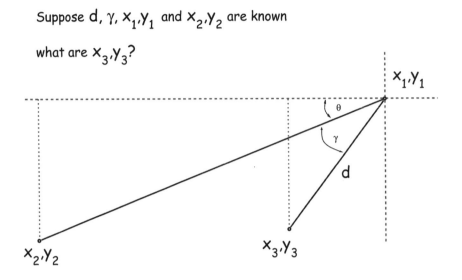

We can calculate θ from x_1,y_1 and x_2,y_2

$$\theta = \tan^{-1}\left(\frac{y_1-y_2}{x_1-x_2}\right)$$

and we can relate the unknown x_3, y_3 to known quantities

$$x_1-x_3 = d\,\cos(\theta+\gamma)$$
$$x_3 = x_1 - d\,\cos(\theta+\gamma)$$

in a similar fashion

$$y_3 = y_1 - d\,\sin(\theta+\gamma)$$

Coordinates of a point, when angle and distance to a baseline are known

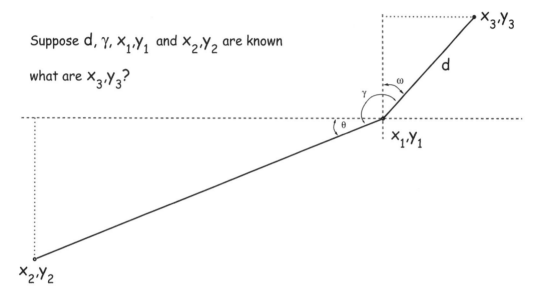

Suppose d, γ, x_1,y_1 and x_2,y_2 are known

what are x_3,y_3?

From the figure, above, $\omega = \gamma - 90° - \theta$

We can calculate θ from x_1,y_1 and x_2,y_2

$$\theta = \tan^{-1}\left(\frac{y_1 - y_2}{x_1 - x_2}\right)$$

and we can relate the unknown x_3, y_3 to known quantities

$$x_3 - x_1 = d \sin(\omega)$$
$$x_3 = x_1 + d \sin(\omega)$$

in a similar fashion

$$y_3 = y_1 + d \cos(\omega)$$

Perpendicular Line at a Known Point

If we know the equation of a line, $y = mx + b$, what is the equation of a perpendicular line that intersects the known line at the known point, x_i, y_i?

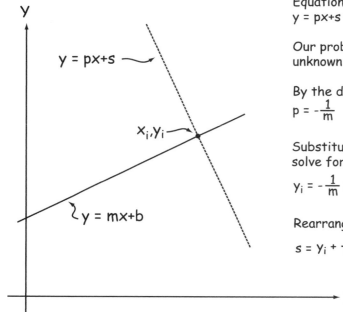

Equation for unknown line
$y = px + s$

Our problem is to find the unknown values for p and s.

By the definition of perpendicular,
$p = -\frac{1}{m}$

Substitution into the equation to solve for s,
$y_i = -\frac{1}{m} x_i + s$

Rearranging,
$s = y_i + \frac{1}{m} x_i$

Intersection of Two Lines

If we know the equation of two lines that intersect, what is the coordinate of their intersection point, x_i, y_i?

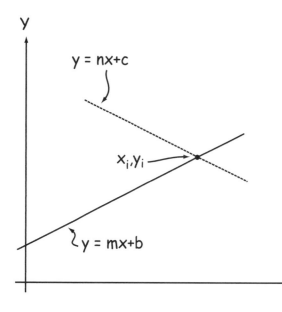

Equation for line 1 is $y = mx + b$
Equation for line 2 is $y = nx + c$

At the intersection,

$nx_i + c = mx_i + b$, and by rearranging

$x_i (n - m) = b - c$, so

$x_i = (b - c) / (n - m)$

Substitution into either equation to solve for y_i,

$yi = \frac{(b - c)}{(n - m)} + c$

Shortest Distance from a Point to a Line

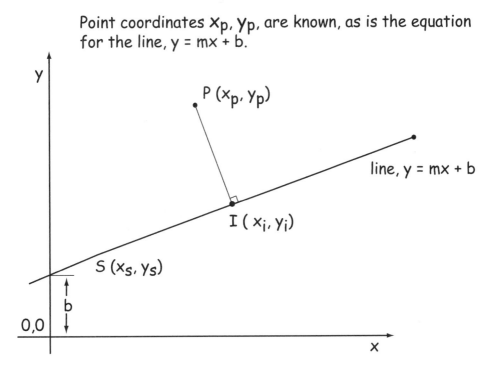

Point coordinates x_p, y_p, are known, as is the equation for the line, $y = mx + b$.

We need to find the coordinates of the point I, xi, yi.
We know the point lies on the line:

$y = mx + b$,   and the perpendicular line $y = -\frac{1}{m}x + s$

We know m, and we can solve for s because the perpendicular line goes through P, so

$$s = y_p + \frac{x_p}{m}$$

Since we now know the equations for both lines, we may apply the formulas on the previous pages for the intersection point of two lines to calculate x_i, y_i, and then use the Pythagorean formula to calculate the distance from P to I.

Intersection of Two Lines of Known Origins and Azimuths

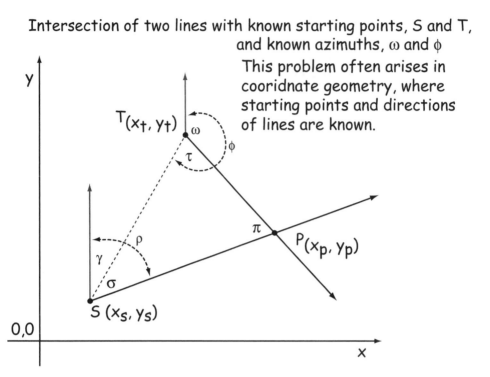

Intersection of two lines with known starting points, S and T, and known azimuths, ω and ϕ

This problem often arises in cooridnate geometry, where starting points and directions of lines are known.

We know the coordinates for S and T, and the azimuth angles ρ and ω. Our goal is to find the coordinates for P.

We see from the figure that the angle $\sigma = \rho - \gamma$, where ρ is the azimuth for the line segment SP, and γ is the azimuth for line segment ST.

We may calculate γ and ϕ from the azimuth formula,

$$\gamma = \tan^{-1}\left(\frac{X_t - X_s}{Y_t - Y_s}\right) + C, \qquad\qquad \phi = \tan^{-1}\left(\frac{X_s - X_t}{Y_s - Y_t}\right) + C,$$

where C values are determined as shown in the description of the azimuth formula on the previous pages.

Then calculate $\sigma = \rho - \gamma$, and $\tau = \phi - \omega$. Then $\pi = 180 - \tau - \sigma$, because the sum of interior angles for a triangle always equals 180.

From the law of sines, the length $SP = ST \cdot \dfrac{\sin(\pi)}{\sin(\tau)}$, where the length

ST is determined from the Pythagorean formula.

Then $X_p = X_s + SP \cdot \sin(\rho)$, and $Y_p = Y_s + SP \cdot \cos(\rho)$

Deflection of a Tangent Line and Circle

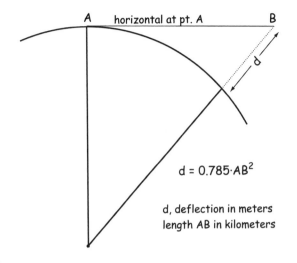

$$d = 0.785 \cdot AB^2$$

d, deflection in meters
length AB in kilometers

Appendix D: Answers to Selected Study Questions

Chapter 1

1.2: I recently used Google Map (http://maps.google.com/maps?tab=wl) to help plan a trip to Granada, Spain. Data collection consisted of a search with the keywords Granada, Hotel, and Spain. The analysis consisted of the hotel quality ranking, location relative to sites I wanted to visit, and cost. Communication involved sending an image and map to friends.

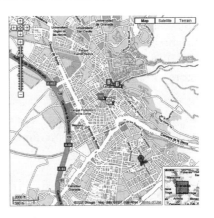

1.4: GIS software differ from other software primarily in tracking geographic coordinate location, tieing these locations to attribute data, and storing and processing large quantities of data. While many softwares are designed to store and analyze large volumes of data (e.g., video editing), and some other softwares focus on coordinates (e.g., computer assisted design programs for three-dimensional objects), GIS records coordinates that are tied to real, physical locations. Coordinates are defined relative to a physical origin, usually some point on the Earth surface, or the near the center of the Earth, and stored in the computer. Sets of points are combined to characterize the location and shape of geographic features, and non-spatial attributes are associated with these features.

1.6: By our definition in this chapter, paper records and maps are not a GIS, because they are not computer based. However, they do serve in our collection, storage, analysis, and output of spatial data and information, so some would argue that they are a GIS, just an extremely low technology version.

Chapter 2

2.2: Our multiple levels of abstraction from the physical, "real" world usually include a data model, data structures, and machine code. The data model describes the real-world objects with a subset of simple objects and relationships. Data models typically encompass our mental image of how the real world entities are connected, shaped, or related. These models may often be illustrated by box and arrows diagrams. Data structures are how these objects are organized in a computer, for example, what parts go in what files, or how the files are linked on to another. Machine code are the 0's and 1's used to store information.

2.4: a) interval/ratio, b) nominal, c) nominal, or ordinal (if read along a brightness gradient), d) ordinal, e) nominal, f) interval/ratio

2.6: Topology is the study of spatial relationships, and it is important in GIS because topological vector data structures have certain positive properties. Topological relationships such as adjacency, connectivity, proximity, and overlap are important in structuring and analyzing data, and are often helpful in ensuring data quality. Many topological characteristics are invariant to warping or bending spatial features. This is important because we often warp spatial data through map projections.

2.8: Mixed cells may be a problem under several conditions, for example, when there are very different values for materials within the cell, or when we are interested primarily in one factor that is a minority presence in a cell but not recorded. Mixed cells may be addressed by decreasing the cell size, by carefully developing the assignment rule for cell values when there are mixed constituents, or by recording multiple attributes for each cell, including the identity and proportion of values in each cell.

2.10: a, b, and f. For c, latitudes vary from N90 to S 90; for d, seconds are always < 60; for e, N,S,E,W denominations are always positive.

2.12: a) 1   b) 10111   c) 100000000   d) 100   e) 1011   f) 1010   g) 11   h) 10100

2.14: a) 14   b) 9   c) 3   d) 133   e) 8   f) 10   g) 145   h) 240

2.16: A TIN is a set of x,y, z points connected in a network of triangles such that each line of every triangle is not crossed by another line, and a circle passing through the three vertices of a triangle does not contain another triangle.

2.18: We compress data when data volumes are too large, particularly for raster data sets. Cells are recorded for each location in a raster area, and gigabytes to terabytes are often stored. Vector data sets typically record shape-defining locations, and only where features of interest occur, for example, a road line. This contrasts with a raster representation which records a set of cells for a road, plus cells for the surrounding area where there is no road.

2.20: An object data model defines "natural" objects, from the point of view of the model designer, that encompasses spatial and attribute properties, as well as functions or operations that may be specific to that object. Rather that breaking data into thematic layers, components from many themes may exist within an object. Objects may relate to other objects through specific or specialized, unique correspondences or connections.

2.22: Run length codes, by row are:
4:c, 5:a
2:a, 2:b, 3:d, 2:a
3:e, 5:f, 1:e
9:a
2:c, 3:a, 1:b, 1:f, 1:d, 1:e

Chapter 3

3.2 a) 6,372,400   b) 6,356,500   c) 6,344,647

3.4: An ellipsoid is a solid shape based on the rotation of an ellipse. An ellipse is a near circular shape, defined by the equation at right, where x and y are the center of the ellipse and r_1 and r_2 specify how large and flattened the ellipse is. An ellipse becomes a circle when $r_1 = r_2$, and a spheroid is solid based on the rotation of a circle.

Ellipse equation:

$$1 = \frac{(x - x_o)^2}{r_1^2} + \frac{(y - y_o)^2}{r_2^2}$$

Flattening factor:

$$f = \frac{r_1 - r_2}{r_1}$$

3.6: Ellipsoids have different radii because they are fit for different purposes, with different data, or using different methods. Some ellipsoids were fit long ago, using primarily surface measurements for a region or continent. Other ellipsoids use a wider array of measurements, and are optimized for specific countries. Some datums seek to fit the entire world, and may be based partly or solely on more recent satellite-based observations. Any times the purpose, data, or methods differ, the ellipsoids may differ.

3.8: Magnetic north is at the point where lines of magnetic attraction converge, and a weightless magnet, if suspended in a frictionless media, would point straight down towards the center of the Earth. Magnetic north is currently located near Greenland.

Geographic north is the northern intersection of the Earth's axis of rotation with the Earth's surface. It is located in the Arctic Ocean.

3.10: Multiple datums exist because we have improved datums through time, and because we develop different datums for different purposes. Datums are required for measurements, and so most governments estimated datums when a sufficient number of points were surveyed. Additional points with improved methods will led to subsequent estimations, or versions, of national datums, in most cases with higher accuracies. Satellite and other measurement capabilities developed in the second half of the 20th century added global datums, increasing the number of available datums for most locations.

3.16: A developable surface is a mathematical, geometric surface onto which a points are projected from a spheroid or ellipsoid. This developable surface may be mathematically "unrolled" to depict a flat map. Planes, cones, and cylinders are the most common developable surfaces.

3.18: The great circle distances are:

DOG to Neah A : 4,314 km

Key West to DOG: 2,630 km

Neah A to Key West: 4,646 km

3.20: The UTM coordinate system defines map projections for all portions of the globe. Areas between 80° S latitude and 84° N latitude are divided into 6o wide zones, each zone running from the equator to the northern or southern limit. Separate transverse Mercator projections are fit to each zone. Negative zone values are avoided by specifying false eastings and northings, coordinate values added to intermediate projection coordinates.

3.22: Chile - TM, because it is narrow with the main territorial axis is north-south;
Nepal, Kyrgyzstan, and The Gambia, LCC, because they are relatively narrow and have main axes oriented east-west.

3.24: The public land survey system (PLSS) is a systematic subdivision of land carried out in the U.S. for the purpose of uniquely identifying property boundaries. Principle meridians and baselines are established, and township and range lines surveyed parallel to these at 6 mile intervals. The township/range grid is further subdivided into 1 mile squares, in turn subdivided into smaller units. The PLSS is not a coordinate system.

Chapter 4

4.2: 1:20,000 is a larger scale, because the number resulting from the division will be larger than 1:1,000,000.

4.4: a - exaggeration; b - simplification; c - omission; d - simplification

---

4.6: A computer screen is now the most common map media; millions of maps are rendered each hour through applications like google map and mapquest. Paper is the second most common media, and is most used when a hardcopy form is required.

4.8: A large scale map typically shows more detail, because each feature is drawn larger, and there is more opportunity to show variation in shape.

4.10: Snapping is the automatic movement or joining of digitized features when they are within a specified snap distance. Points (nodes or vertices) may be moved to nearby points or to lines when they are "close enough". This avoids undershoots, when points do not attach to lines or other points when they are intended to, and overshoots, where ending nodes or points are digitized across a line, leaving a small "dangling arc." Overshoots and undershoots are undesirable because they may cause incorrect topology (gaps or unclosed polygons with undershoots), and redundant storage or lines (overshoots).

4.12: a-overshoot; b-undershoot; c-duplicate digitized line; d-psuedonode; e-dangle; f-dangle; g-missing label.

4.14: A spline is a line smoothly fit through a set of points. Splines are used to increase vertex density without substantially slowing the digitizing process, particularly for smoothly-curving features, such as river meanders or winding roads. Splines fit piecewise polynomial functions while imposing smooth join points.

4.16: Manual digitizing involves fixing a map or displaying a scanned image, and manually positioning a pointing device to indicate the location of each node, vertex, or other shape-defining coordinate. Digitized data are in vector form. Scan digitizing uses a machine to record differences in colors or brightness for a map or image document, usually into a raster grid. Lines, points, and areas are defined by some thresholding technique, and lines or points may be thinned and converted to a vector form, as needed. Manual methods have the advantage of low costs for small maps, inexpensive equipment requirements, feature interpretation by humans when using substandard maps, and relatively little training. Scan digitizing is inexpensive for large numbers of very detailed maps, may be automated, and may be more consistent.

4.18: Map registration fixes a map or image to a ground coordinate system so that the coordinates of any point in the media may be determined easily. The process consists of identifying control points that are visible in both the image/map and on the ground, collecting coordinates of these points in both the image/map system and the projected "ground" coordinate system, fitting a system of transformation equations to the coordinate data sets, and applying these transformation equations to the image to convert it to the projected ground coordinate system.

4.20: An affine transformation uses a system of linear equations to estimate the ground easting (E) and northing (N) values from image x and y values. The equations are of the form:

$$E = a_0 + a_1 \cdot x + a_2 \cdot y \qquad\qquad N = b_0 + b_1 \cdot x + b_2 \cdot y$$

This is a linear transformation because the x and y variables are not multiplied together or raised to a power larger than 1, by definition the equation of a straight line.

4.22: The average positional error is likely to be the same or larger than the RMSE. The RMSE is usually minimized, or closely related to a minimized quantity when statistically fitting the coordinate transformation. If we collected a representative sample, we expect the RMSE to be approximately equal to the average error. However, if our sampling was inadequate or biased, often it is in areas where we have difficulty identifying good control points, and hence our RMSEs tend to be larger in these locations.

4.24: Transformation b is the most likely to have lowest average error at independently measured points. It depends on the distribution and number of control points, but in most cases higher order polynomials overfit, and while exhibiting lower RMSE values, they have larger errors.

4.26: Metadata are the data about data. They describe the extent, type, coordinate system, lineage, attributes, and other important characteristics of a spatial data set. Metadata are required to evaluate the adequacy of a data set for an intended use.

Chapter 5

5.2: GNSS is based on range distance measurements from multiple satellites to "triangulate" a location. Orbiting satellites transmit radio signals along with precise positioning and timing information. The current distance between a satellite and a receiver is a range measurement. A GNSS receiver combines multiple, simultaneous range measurements to estimate location in near-real time.

5.4: Typically 4 satellites are required for a 3-dimensional fix, although a fix may be determined under some assumptions with data collection from 3 satellites over a short period of time.

5.6: GNSS data range in accuracy, from sub-centimeter for the highest accuracy using carrier phase methods, to tens of meters using real-time C/A positioning. Accuracies are highest when using high quality receiving systems in flat terrain, with no buildings, trees, or other structures to block views of the sky. Accuracies also improve when satellites are widely spaced.

5.8: Figure d depicts the lowest PDOP, with the widest distribution of satellites, closely followed by figure b. Figure a has the highest PDOP, with the tightest distribution.

5.10: Differential positioning is based on the simultaneous measurement of GNSS signals at both a known, base location, and at unknown roving stations. The small errors in range measurement may be calculated for each position measurement at the base station. These range errors may be applied in reverse for corresponding rover data, thereby improving the accuracy of position measurements.

5.12: GNSS accuracy typically decreases as terrain becomes more varied, or when canopy or buildings obstruct a portion of the sky. Positional accuracy decreases because sub-optimal constellations of satellites are more likely to be observed. Satellites are in closer proximity, and measurements are less independent, and hence to not reinforce each other to improve accuracy.

5.14: WAAS is the Wide Area Augmentation System, a real-time differential correction system designed to aid navigation in U.S. civil aviation. Correction factors are derived from a nationwide network of control stations and broadcast from a geostationary satellite located over the equator. The system is designed primarily for aviation and related uses in North America.

5.16: COGO stands for coordinate geometry. COGO is the calculation of coordinates based on angle and distance measurements.

5.18:

Starting point P0, X = 10,128.3, Y = 6,096.4

Point ID	Azimuth	Distance	Delta X	Delta Y	X	Y
P1	32.4	122	65.4	103.0	10,193.7	6,199.4
P2	91.7	207	206.9	-6.1	10,400.6	6,193.3
P3	123.3	305	222.0	-209.2	10,622.6	5,984.1
P4	212.5	193	-103.7	-162.8	10,518.9	5,821.4
P5	273.9	206	-205.5	14.0	10,313.3	5,835.4
P6	355.5	145	-1.3	144.9	10,312.1	5,980.4

Chapter 6

6.2: The electromagnetic spectrum is the range of electromagnetic energy frequencies observed. Broadly, this spans from 0 to infinity, however we are most interested in the subset of primary frequencies emitted by the sun. Specifically, we are interested in the ultraviolet through infrared portions of solar radiation, from 0.01 through 1000 μm (1,000,000 μm equals 1 meter). Principal regions of interest are the visible (0.4 to 0.7 μm, approximately equally split in the blue, green, and red portions of the spectrum), the near infrared (0.7 to 1.1 μm), and the mid infrared portions (2.5 to 8 μm). Radar wavelengths are important in remote sensing, most often generated from a device, and range from 0.75 cm to 1 m.

6.4: Film is a layered sandwich of emulsions on a polyester base material. The emulsion is sensitive to light, and reacts to darken in a measure proportional to the amount of light (exposure) the layer receives. Different emulsions are sensitive to different spectral regions. Panchromatic film is typically sensitive to visible wavelengths, from 0.4 to 0.7 μm. Color films typically contain three dye layers, sensitive to the blue, green, and red wavelengths (normal color), or green, red, and infrared wavelengths (color infrared film). Spectral reflectance curves plot the sensitivity versus wavelength.

Digital cameras are similar, except that light is typically split by wavelength and directed to separate receptor electronics, one each for each portion of the spectrum observed. Light generates a voltage or current proportional to the light energy, and in this way an image is formed.

6.6: The most common format is the 9-inch mapping camera, in mid to late stages of transition from primarily film-based to primarily electronic sensors. Film is more familiar with nearly 70 years of use and development, and may be less expensive for existing organizations and small projects because the systems are already in hand, and operational. Digital cameras have the advantage of an inherent digital format, obviating subsequent scanning, and may be sharper due to electronic image motion compensation and other image processing. Digital film systems are perhaps more complicated, but also more flexible, and more easily integrated with GNSS, flight control, and other aviation electronics.

6.8: Distortion magnitude varies with mapping cameras, depending on terrain, tilt, camera characteristics, and scale. For vertical photos, typically defined as those with camera axis tilt of less than 3 degrees, errors are typically between 10 and 70 meters over moderate terrain. Errors may be reduced to a meter or less by applying a full photo orthocorrection, a process that analytically removes most tilt and terrain distortion through the three-dimensional geometric analysis and transformation.

6.10: Stereo photographic coverage is the intentional overlap of sequential photographs in a flight line (end lap) and photos in adjacent flight lines (side lap). Overlap provides views of the same set of objections from two different locations. These "perspective views" take advantage of a phenomenon called parallax to reconstruct three-dimensional positions from two-dimensional images.

6.12: Terrain distortion is removed by applying inverse equations that describe the magnitude of terrain-caused distortion. Three dimensional objects that are projected onto a two-dimensional plane are shifted horizontally when they are at different heights. This shift is also dependent on the angle at which the objects are viewed. We may remove the distortion by measuring the height of each point and knowing the viewing angle from the camera location to each point.

6.14: Photointerpretation is the process of converting images into spatial information, typically by an experienced human analyst, or photointerpreter. The photointerpreter uses size, shape, color, brightness, texture, and location to assign or identify characteristics to features of interest.

6.16: The four systems, from Landsat ETM+ through Quickbird represent a range of resolutions (from 30 m through 0.6 m), spectral ranges (full color through near and mid infrared), per scene coverage (from tens of thousands of square kilometers through a few tens of square kilometers), and more the two week to less than two day repeat times. Finally, costs rise markedly along this gradient. Although near the end of its functional life at the time of this writing, the ETM+ data were available for hundreds of dollars for a full scene, while the higher resolution data were from tens to hundreds of thousands of dollars for an equivalent area.

6.18: Image types are selected if they measure the phenomena of interest to the required level of spatial and attribute accuracy, are within the technical capabilities of the organization, have an acceptable probability of successful data collection, and fit with the available budget for acquisition and processing.

Chapter 7

7.2: Do the data provide the required information, for the required area, at the necessary level of categorical and spatial detail, and at the accuracy needed for the intended use?

7.4: Edge-matching is the process of ensuring consistency in features across the edges of mapping projects, areas, and physical maps. When adjacent areas are mapped at different times, by different methods, or by different people there may be incongruent features on either side of the mapping boundary. Roads may not match in location or type, rivers may end abruptly, or the vegetation or elevation change in an impossible manner. Edge-matching attempts to resolve these differences, and if possible, remove errors across mapping boundaries.

Chapter 8

8.2: Database management systems are computer software tools that aid in the entry, organization, analysis, distribution, and presentation of data.

8.4: A one-to-one relationship among table means that for every row in one table that in some way is matched to a row in another table, there is only one row in the second table that matches. A many-to-one relationship means that one row in a table may match many rows in a second table. Note that by match, we do not mean completely match. Usually we are using a column in each table to match the tables; the rows are considered to match when the match column has the same value in both tables.

8.6: Osel, NumT

8.8: The eight basic operations are illustrated in Figure 8-8 and Figure 8-9. They are restrict, project, product, divide, union, intersect, difference, and join.

8.10: Sets from or conditions will have the same number or more members than the component conditions. Sets from and conditions will have the same number or fewer members than the component conditions.

8.12: a) Australia, Finland, Netherlands, Norway.
b) Australia, France, Japan.
c) Japan, South Africa, Spain, USA.
d) Japan, Spain, USA.
e) Australia, Finland, Netherlands, Norway.
f) Finland, Netherlands, Norway.

8:14:

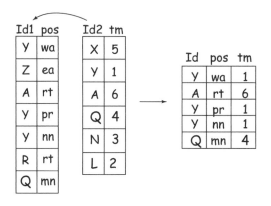

8.16: First normal form has no repeat columns. Second normal form tables are in first normal form, but in addition they have a key, and every other variable is functionally dependent on the key. This means if we know the value of the key variable, we know the value of the functionally dependent variable. Third normal form is in second normal form, with the additional proviso that there are no transitive functional dependencies. This means that no non-key attributes are functionally dependent on other non-key attributes.

8.18: ID-> Size, Shape, Color, Age, Source;
Size -> Color, Age
Color -> Age
Source-> Age, Color, Shape, Size;

Chapter 9

9.2: Selection operations apply criteria to features, and identify features that meet those criteria. The criteria may apply to spatial characteristics, for example, the size, shape, or location of a polygon; they may apply to non-spatial attributes of the features, for example the value or condition of an attribute.

9.4: a) B or C; b) A and B; c) [A and B] and not C; d) [B or C] and not [B and C]

9.6: Large are white, medium light gray, small darker grey.

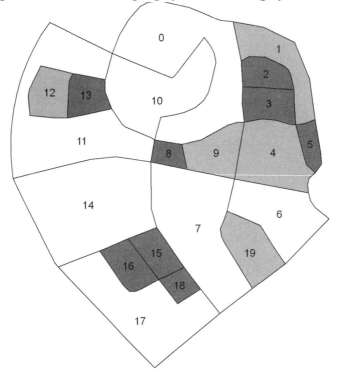

9.8: The modifiable area unit problem arises because statistics for aggregated areas depend on the aggregations. We may combine adjacent areas, and calculate sums, means, medians, and other attributes of the areal units. If we are selective about how we aggregate, we may change these statistics solely by changing the aggregation units.This is the modifiable areal unit problem. The zoning effect is how aggregate statistics change with zone boundaries. The area effect is how statistics change when changing the size of aggregation areas.

9.10:

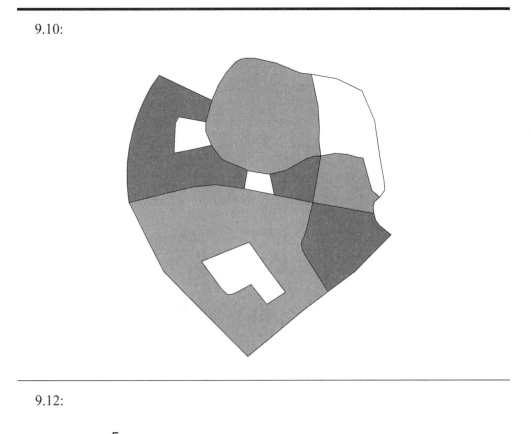

9.12:

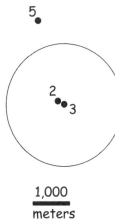

ID	distance
1	250
2	0
3	1500
4	500
5	0
6	500
7	250
8	250

9.14: Features are combined "vertically", in a synthesis of spatial and non-spatial characteristics.

9.16: The minimum dimension (point, line, or polygon) is chosen because to do otherwise courts ambiguity. If two lower dimension features are coincident with higher-dimension features, it is unclear how the attributes should be recorded in the resultant features. For example, if two points fall within a polygon, the polygon attributes may be unambiguously associated with each point. It is unclear or at best cumbersome to assign both sets of point attributes to an output polygon.

9.18: The sliver problem usually arises when two linear features attempt to represent the same boundary. Slight variations in the representations create small gaps and overlaps in coverage, resulting in many small "slivers." While typically less than 0.1% of the total feature area, they can be a substantial proportion of the total number of features, thereby expanding data tables and slowing operations. This problem is resolved by avoiding them, through snapping or fuzzy boundaries, or fixing them after they are identified by their size or shape.

9.20: Network models are connected linear graphs through which resources flow, or to which movement may be constrained. There may be both source and demand features connected to these networks. Networks are different from many other spatial models in that movement or occurrence is limited to the network, and they often track time-varying

Chapter 10

10.2: Compatible cell sizes are often required in raster operations because otherwise the input is often ambiguous. If one input cell is substantially larger or mismatched to another, it may be uncertain which input cell value to choose.

10.4:

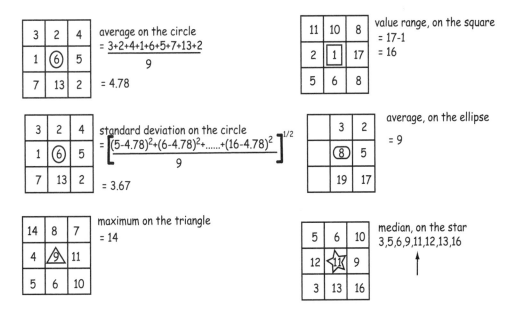

10.6: C1 = 1, C3 = 1, C4 = 1, C10 = 4.

10.8: C9 = 1, C10 = N, C11 = N, C12 = N

10.10: In most systems, a NULL value is returned to logical operations AND, OR, and NOTs when applied to NULL values on either side of the operation.

10.12:

1	1	0	0
0	0	0	1
1	1	0	0
1	0	0	1

and

0	1	0	1
1	0	0	1
1	1	1	0
1	0	1	1

=

0	1	0	0
0	0	0	1
1	1	0	0
1	0	0	1

1	1	0	0
0	0	0	1
1	1	0	0
1	0	0	1

or

0	1	0	1
1	0	0	1
1	1	1	0
1	0	1	1

=

1	1	0	1
1	0	0	1
1	1	1	0
1	0	1	1

10.14: The kernel is an array of cells and values that are applied in a moving window to an input raster. The kernel array elements are coefficients in a multiplied summation across all kernel cells. Kernels may have any size and shape, but are usually square, and with an odd-number of cells, rows, and columns, so an unambiguous central cell may be identified. The figure below, from Chapter 10, shows a 3 by 3 mean kernel applied to an input raster data set.

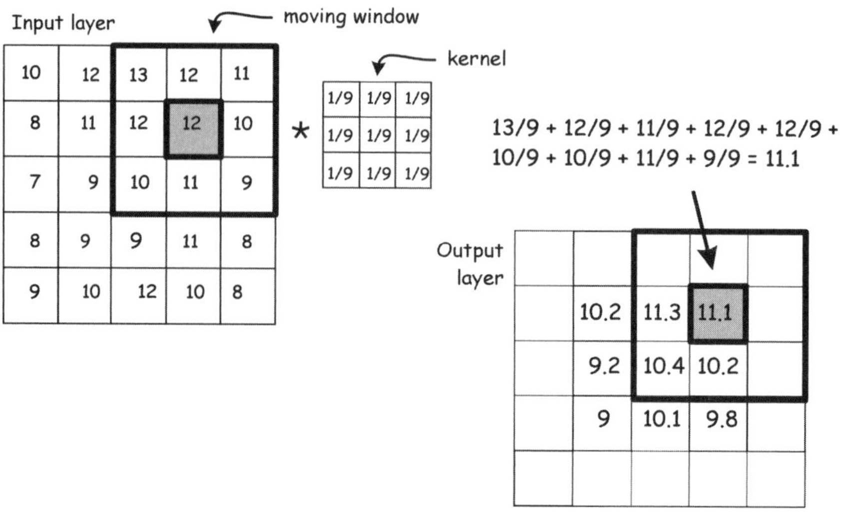

10.16: The kernel is used in calculating slope and aspect, specifically the change in z with a change in y, or dz/dy.

10.18:

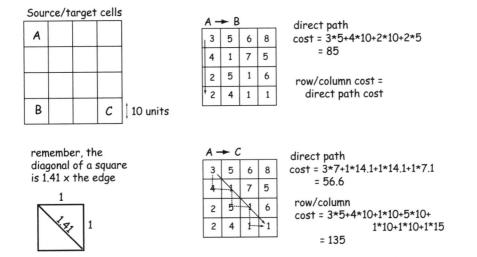

Source/target cells

remember, the
diagonal of a square
is 1.41 x the edge

A → B

3	5	6	8
4	1	7	5
2	5	1	6
2	4	1	1

direct path
cost = 3*5+4*10+2*10+2*5
     = 85

row/column cost =
   direct path cost

A → C

3	5	6	8
4	1	7	5
2	5	1	6
2	4	1	1

direct path
cost = 3*7+1*14.1+1*14.1+1*7.1
     = 56.6

row/column
cost = 3*5+4*10+1*10+5*10+
                1*10+1*10+1*15
     = 135

Chapter 11

11.2: Digital elevation models are created by a variety of methods. Leveling or other ground surveys are used to measure relative height differences across profiles, using optical and electronic instruments to measure distance and vertical and horizontal angles. Photo-based methods rely on parallax, the relative displacement of objects depending on their distance from an observation point. Downward looking images taken from aircraft or satellites may be combined with knowledge of aircraft position and ground surveys to create DEMs. Laser and radar measurements from airborne platforms are a third common method for DEM creation. Return times are recorded for electromagnetic signals sent form the aircraft, and used to calculate terrain height relative to the aircraft. These are combined with precise positioning information for the aircraft and with previous ground surveys to produce accurate DEMs.

11.4:

windows	4-nearest cell	3rd-order finite difference

a)

110	113	118
112	114	119
111	117	121

$\div 10 \div$

$dz/dx = \dfrac{119 - 112}{20} = 0.35$

$dz/dy = \dfrac{113 - 117}{20} = -0.20$

$dz/dx = \dfrac{118 + 2*119 + 121 + -110 - 2*112 - 111}{80} = 0.40$

$dz/dy = \dfrac{118 + 2*113 + 110 + -121 - 2*117 - 111}{80} = -0.15$

b)

67	63	62
65	64	64
70	68	66

$dz/dx = \dfrac{64 - 65}{20} = -0.05$

$dz/dy = \dfrac{63 - 68}{20} = -0.25$

$dz/dx = \dfrac{62 + 2*64 + 66 + -67 - 2*65 - 70}{80} = -0.14$

$dz/dy = \dfrac{62 + 2*63 + 67 + -66 - 2*68 - 70}{80} = -0.21$

c)

18	23	17
21	24	19
20	22	18

$dz/dx = \dfrac{19 - 21}{20} = -0.1$

$dz/dy = \dfrac{23 - 22}{20} = 0.05$

$dz/dx = \dfrac{17 + 2*19 + 18 + -18 - 2*21 - 20}{80} = -0.09$

$dz/dy = \dfrac{17 + 2*23 + 18 + -18 - 2*22 - 20}{80} = -0.01$

11.6

712	709	707	703	704
710	^a706	704	700	696
708	705	705	^c697	700
711	^b709	705	696	694
714	712	708	703	698

a) dz/dx = -0.25, dz/dy = 0.18,
     slope = 16.97°,
     aspect = 124.0°

b) dz/dx = -0.26, dz/dy = -0.29,
     slope = 21.27°,
     aspect = 42.4°

c) dz/dx = -0.36, dz/dy = 0.11,
     slope = 20.78°,
     aspect = 107.24°

11.8

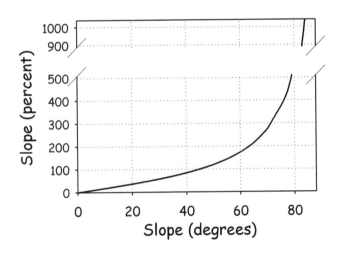

11.10

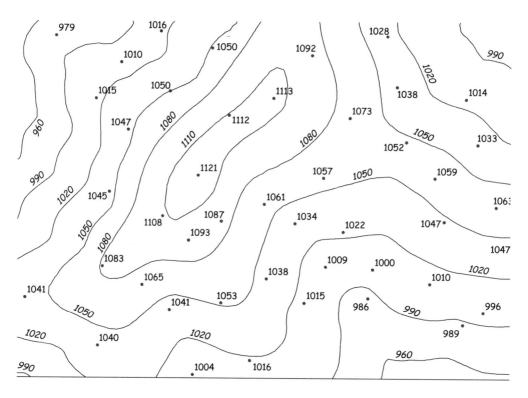

11.12

| 378 | cell and elevation value | — main watershed boundary — potential stream course |

162	108	67	103	56	66	130	214	153	122	70		56	91	165
169	160	101	120	95	115	119	202	212	121	55	43	101	158	261
254	224	182	158	214	142	208	249	225	129	58	121	137	253	344
323	312	204	191	214	228	300	345	195	126	58	105	188	298	381
338	334	267	307	231	194	200	190	176	114	63	141	199	277	278
438	471	405	344	228	242	194	137	103	81	111	103	198	262	195
550	550	387	304	301	330	245	257	175	110	163	204	225	206	144
660	557	502	414	451	378	396	329	180	148	242	349	293	191	148
604	639	490	442	433	425	406	264	169	169	278	401	297	241	167
742	666	536	443	340	294	265	202	221	227	339	342	260	260	245
799	630	509	438	456	414	304	344	337	322	359	377	387	375	308
767	685	608	578	457	426	318	442	371	421	430	330	275	292	226
734	789	721	578	512	421	443	512	506	503	378	315	227	213	173
668	765	826	728	579	558	489	534	513	366	330	244	266	190	170
705	767	784	785	761	675	607	545	440	275	226	202	165	104	55

11.14: A solar incidence angles, illustrated as θ, below, is determined by the terrain factors of local slope and aspect at the point of interest, and the sun location.

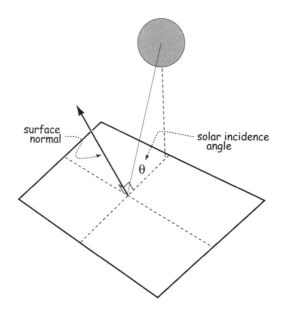

11.16: A shaded relief map depicts terrain reflectance with the Sun in a given position. The surface is shaded relative to the amount of energy the surface would receive and hence reflect with a given Sun position. This incident energy is a function of the Sun's position, and the local slope/aspect. These combine to create an incident angle at each point, as illustrated above. Sites with incident angles closer to 90 degrees are shaded more brightly, while those with incidence angles close to 0 would be show darker shades. Valleys, ridges, and other terrain features may be visualized through this variation in tone.

Chapter 12

12.4:

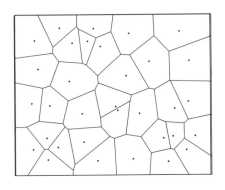

12.6:

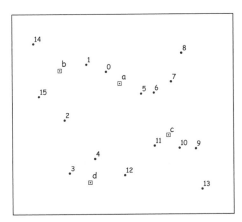

Values, for unknown points:

ID	X	Y	Z
a	155,859.6	4,477,146.0	2097.7
b	147,592.6	4,478,884.2	1871.8
c	162,515.7	4,469,981.2	2104.8
d	151,704.9	4,463,749.2	2184.2

Example calculation, for point a
Closest 3 known points are 0, 5, and 6

ID	X	Y	Z
0	153,951.9	4,478,714.6	2040.6
5	158,827.3	4,475,746.9	2283.2
6	160,607.9	4,475,874.2	1933.5
a	155,859.6	4,477,146.0	2097.7

Distances from a to these points are calculated by the Pythagorean formula:

$$d = \sqrt{(X_1 - X_2)^2 + (Y_1 - Y_2)^2}$$

As an example, for the distance 0 to a,

$$d = \sqrt{(153{,}951.9 - 155{,}859.6)^2 + (4{,}478{,}714.6 - 4{,}477{,}146.0)^2}$$
$$= 2469.9$$

The distances 5 to a is 3280.9, and 6 to a is 4915.6 (check yourself)

The IDW estimate for a is

$$Z_a = \frac{\sum \frac{Z_i}{d_{ia}^n}}{\sum \frac{1}{d_{ia}^n}} \qquad \text{but remember n=1}$$

$$Z_a = \frac{\dfrac{2040.6}{2469.9} + \dfrac{2283.2}{3280.9} + \dfrac{1933.5}{4915.6}}{\dfrac{1}{2469.9} + \dfrac{1}{3280.9} + \dfrac{1}{4915.6}} = 2097.7$$

12.8: A trend surface interpolator estimates coefficients for an equation globally - all data are used to estimate the coefficients for a prediction equation, and apply across the entire sample region. A kriging interpolator uses estimates of global and local variation, specifically spatial autocorrelation, to estimate coefficients for prediction equations. In this way, samples can influence predictions depending on the observed spatial autocorrelation and distribution of samples.

12.10:

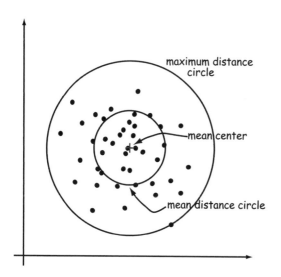

12.12:

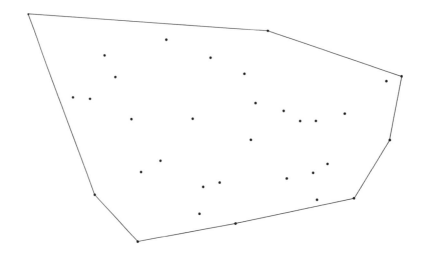

Chapter 13

13.2: Criteria often must be refined because they are expressed in a way that may not be directly applied. Distances or groupings may be ambiguously described, e.g., near, far, large, or small, and often these must be quantified before entry into a cartographic model.

13.4: A discrete weighting has specific, distinct categories. Roads may be passable or impassable for large vehicles, rivers deep or shallow, or forests evergreen or conifer dominated. Weightings may be defined which are specific to these categories, e.g., passable roads receive a weight of 0.5, while impassable roads a weight of zero. In contrast, weightings may also be continuous, in that each road may have a measured width, e.g., 12.4 meters, and the weight may be some continuous function of this width, e.g., width *132.7.

13.6: An agent-based model depicts the actions and behavior in individual elements in a network or space. These elements are typically mobile, somewhat autonomous, and governed by relatively few simple rules of action/reaction. Agent-based models are an attempt to see how simple individual behaviors can result in complex group behavior, and to identify emergent properties from those interactions among large groups.

Chapter 14

14.2: Data transfer is perhaps the process most commonly aided by spatial data standards. Among other things, standards require that spatial data sources, methods, and characteristics be described in a consistent manner. Spatial data standards provide a predictable way of organizing data so that it may be transferred among organizations. Standards allow data to be transferred without loss of information.

14.4: The mean reports the central error, the average error one would expect when sampling from a population. A frequency threshold describes the percentage of errors above or below a value. This gives some general notion of the likelihood of large or small errors.

14.6: Positional accuracy reports how close the represented locations of objects are near the true locations of the objects. Attribute accuracy reflects how often the value of a categorical attribute is correct (discrete) or how close and interval/ratio attribute is to the true value (continuous). Logical consistency does not imply either spatial or attribute accuracy, but just that multiple themes or types of are consistent, e.g., there are no roads in a lake, fires on salt flats, or oil deposits in granitic rocks.

14.8: The steps in applying the NSSDA are 1) identify test points, 2) identify source for "true" points, and extract truth corresponding to test points, 3) Calculate the positional error for each true/test pair, 4) record the error data in a standardized table, which includes calculation of error statistics, 5) create the documentation/metadata describing the accuracy assessment.

14.10: Good candidate points are any features that are well-defined and may be visible on both the data set to be tested and in the source used for truth. This often means constructed features, for example road intersections, curbs, manhole covers, geodetic markers, utility poles, or fire hydrants or other relatively immobile features.

14.12: Metadata are the "data about data." They are important because the describe the characteristics about any data we might wish to use. The allow us to evaluate the data suitability for intended uses, maintain the investment over multiple organizations or changes in personnel, and help us in explaining or describing our data to others.

A

accuracy 497, 565
  assessing interpolation 498
  attribute 564, 567, 577
  calculations 573
  geometric 223
  mean absolute error 497
  NSSDA 571
  positional 564, 567, 568–574
  precision, compared 568
  producer 578
  standard measurements 571
  statistics 570
  test points 571
  user 578
adjacency 356
aerial camera
  digital 230–232
  mechanical 232–233
aerial images 287, 288
  high resolution 288
  NAIP 289
aerial photographs 223–249
  atmospheric distortion 240
  camera formats 232
  compared to satellite imagery 262
  film 232
  geometric correction 243
  infrared 233
  large-format camera 232
  orthographic 234
  panchromatic 233
  parallax 242
  photointerpretation 247
  relief displacement 236
  softcopy workstation 245
  stereo coverage 241
  systematic error 239
  tilt 237
aerial photography
  sources 264
affine 154
All 115
almanac 187
AND queries 322
AND, see Boolean algebra
ArcGIS 14
ASCII 58
aspect 445, 448
  defined 450
  kernel 449
attribute
  nominal 31
attribute table 42
attributes 310, 314
  interval-ratio 32
  nominal 31

  ordinal 32
AUTOCAD 16
autocorrelation 488, 489, 493
AVHRR 273
azimuthal map projection 105

B

bandwidth 506–507
  optimum 507
bandwidth parameter 504
bearings 203
benchmark
  data sheet 95
benchmark, see datum, benchmark
binary 58
  large object 32
  mask 422
Boolean algebra 354–356
bootstrap 498
buffer 371
buffering 370, 371
  calculations 375
  compound 374
  fixed distance 372
  nested 374
  raster 372
  variable distance 375
  vector 372–376

C

camera
  diaphragm 229
  focal plane 229
  formats 232
  large-format 232
  system errors 239
carrier phase 186
cartographic model 525–541
  defined 521–524
  flowchart 534
  importance ranking 532
  rankings 528
  weighting 530
cartographic models
  criteria 527
  weightings 527
cartographic object 25
cartography 164
  labels 167
  legends 168
cartometric 133
cartometric map 161
cell 44
cell assignment 46
cellular automata 548
census 301
characteristic hull 502

classification 352, 359–366
  binary 360
  equal area 363
  equal interval 362
  natural breaks 364
clip 384, 384–385, 420
COGO 202–205
completeness 568
concatenated key 330
conformal 156
conic projection 105
containment 358
Content Standard for Digital
      Geographic Metadata 172
continuously operating reference
      station 88, 93
contour 54, 460
contour lines 54, 478
contours 54
control points 153
  DOQ 161
convex hull 500
coordinate
  Cartesian 28, 82
  COGO 202
  conversion 118
  conversion, degree minute second
    to decimal degrees 31
  definition 26
  DMS 31
  geographic 81
  latitude 93
  longitude 93
  spherical 29
  transformation 153–158
coordinate geometry 202
coordinate geometry, see COGO
core area 474, 499–508
CORS 88, 93
cost surface 434–437
covariance 429
cross-correlation 488
curvature 448–450
cylindrical projection 105

D
data
  binary 58
  climate 300
  discrete 47
  domain 172, 310
  global 273
  items 310
  metadata 171
  national 273
  National Hydrologic Dataset 282
  spatial data transfer standard 171

data compression 61
  pyramids 62
  quad trees 62
  run length codes 61
data model 26, 33
  areas 36
  nodes 35
  object 55
  raster 34, 44
  relational 43
  TIN 34
  topological vector 38
  vector 33
data quality 565–580
data sheet 95
data, EPA river reach 284
database 307, 310
  attribute 310
  attributes 314
  boolean queries 322
  candidate key 333
  concatenated key 330
  data independence 309
  denormalized 338
  first normal form 335
  functional dependency 333
  hybrid 316, 319
  inner join 327
  item 310
  key 314, 329
  logical and physical design 313
  logical design 313
  normal forms 332–338
  NOT queries 322
  OR queries 322
  outer join 327, 328
  physical design 313
  primary key 333
  primary operations 316
  query 321
  records 314
  relational 310, 313
  relational algebra 315
  schema 313
  second normal form 335
  SQL 324
  super key 333
  third normal form 338
  tuples 314
datum 83–93
  adjustment 85
  benchmark 84
  data sheet 95
  geodetic 83
  NAVD88 98
  NGVD29 98
  realization 83

shift 90
  transformation 91, 118–119
  vertical 96
datum transformation 91
DBMS 307
declination 82
degrees 30
DEM 54, 278–282, 443–464
  aspect 445
  creation 278
  hydrology 452
  multiple representations 54
  shaded relief 463
  slope 445
  viewsheds 458
detector
  spot 251
developable surface 105
difference 316
differential correction 196
digital aerial camera 230–232
digital elevation model
  hydrology 551
digital elevation model, see DEM
Digital Line Graph, see DLG
digital orthophoto quad, see DOQ
Digital Raster Graphics, see DRG
digital terrain model 54
digitizing 140–152
  editing 150–152
  error 145
  errors 143
  GNSS 207
  GPS 206
  hardcopy 142
  manual 140–147
  on-screen 140
  overshoots 145
  skeletonizing 149
  snap tolerance, snap distance 147
  snapping 145
  splines 147
  stream mode 145
  undershoots 145
d-infinity flow direction 454
dissolve 368–369
distance
  great circle 103
distortion 117
  atmospheric 240
  camera tilt 237
  map projection 101, 107
  relief 237
  terrain 237
DLG 276
domain 172, 310
DOQ 247, 288

drainage 454
DRG 278
DTM 54
dynamic heights 98

E

Earth
  Erotasthenes 73
  radius 74
easting 114
  false 115
edge detection 428
editing 150
electromagnetic spectrum 225
electronic distance meter 201
ellipsoid 75
  Airy 77
  Austrailian 77
  Bessel 77
  Clarke 1880 75, 77
  Clarke1866 77
  GRS80 77
  radius 77
  semi-major axis 75
  semi-minor axis 75
  specifying 75
ellipsoidal height 79
ellipsoids 77
emulsions 233
entities 25
  area 36
  line 35
  point 34
entity 310
entity-relationship diagram 313
ephemeris 187
epsilon distance 576
ERDAS 17
error 145
  atmospheric distortion 240
  camera 239
  digitizing 145
  GPS 190
  inclusion 47
  propagation 579
  radial lens 239
  relief displacement 237
  RMSE 155
  scale 143
  tilt distortion in photographs 237
error table 577
ETM+ 256
exaggerated, map generalization 138

F

feature inconsistency 152

Federal Geodetic Control Committee
      of the United States FGDC,
      see Federal Geodetic
      Control Committee of the
      United States
Federal Information Processing
      Standard, see FIPS
FGDC, see Federal Geodetic Data
      Committee
FGDC, see Federal Geographic Data
      Committee
file type 63
film 232
   panchromatic 233
FIPS 307
floodplain 298
flow direction 452
   d8 453
   d-infinity 454
flowchart 534
flowcharts 526
foot
   international 112
   U.S. survey 112
format
   file 63
friction surface 435
function
   local 412–423, 424
   logical 413
   moving window 424
   neighborhood 424–432
   nested 417
   nested raster 417
   reclassification 415
   zonal 433
functional dependency 333
fused, map generalization 138

G
Geary's C 491
generalization 138
   boundary 38
   feature 138
   Lang method 148
   line 147
   map 137–138
   raster 46
geocoding 302, 395–396
   census 302
geodesy 71
geodetic accuracy 100
geodetic datum 83
GeoEye 252
geographic north 82
geoid 78
geoidal height 79

geoidal separation 79
GeoMedia 15
GIS
   components 13
   ERDAS 17
   hardware 13
   organizations 19
   QGIS 17
   societal push 6
   software 13–18
   technological pull 7
GIS definition 1
Global Navigation Satellite System 7
GLONASS 185
gnomonic projection 106
GNSS 91, 93, 161
   applications 206
   base station 195
   carrier phase 186
   control points 160
   control segment 184
   differential correction 193, 196
   dilution of precision 192
   DOP 191
   dual frequency 191
   efficiency 211
   multipath 191
   PDOP 191
   range 188
   range pole 210
   rangefinder 213
   real-time differential 197
   satellite segment 184
   terrain obstruction 210
   tracking 213
   user segment 184
GNSS, see Global Navigation Satellite
      System
GNSS. See Global Navigation System
GPS 93, 133, 185
   almanac 187
   atmospheric effects 190
   C/A code 186
   carrier 186
   carrier signal 186
   collars 214
   ephemeris 187
   error 190
   error assessment 565
   field digitizing 206
   ionospheric effects 190
   P code 186
   post-processed differential 196
   WAAS 199
   wildlife tracking 214
GRASS 17
graticule 132

gravimeter 80
great circle distance 102, 103
greater than (>), see set algebra
grid 132

H

hardware 13
HARN 88
height
　dynamic 98
　ellipsoidal 79
　geoidal 79
high accuracy reference network 88
high precision geodetic network 88
hillshade 463
HPGN 88
hydrology
　d8 flow direction 453
　d-infinity flow direction 454
　flow direction 452
　functions 452
　models 549
　pits 455
　sinks 455
　specific catchment area 458
　stream power index 458
　watershed 452, 456
　wetness index 458

I

IDRISI 16
Idrisi 16
IERS 89
IFOV 250
importance rating 532
inclusions 37
infrared film 233
inheritance 56
inner join 327
instantaneous field of view 250
International Earth Rotation
　　　Service 89
international foot 112
International Terrestrial Reference
　　　System 89
interpolation 473–496
　bilinear 162
　cubic convolution 162
　exact 481
　fixed radius 481–483
　IDW 483–486
　kriging 493–496
　lag distance 493
　nearest neighbor 162
　splines 486
　Thiessen polygons 480
　trend surface 491

variograph 494
intersect 384
intersection 316, 384–385
interval attribute 32
Inverse Distance Weigthed
　　　Interpolation, see
　　　interpolation, IDW
is 573
isolines 478
items 310
ITRF 89
ITRS 89

J

join 316
　inner 327, 328
　natural 327
　outer 327, 328

K

karst 455
kernel 427
　edge detection 428
kernel mapping
　bandwidth 504
kernel
　high pass example 429
　mapping 503–508
key 297, 314, 329, 333
　candidate 333
　concatenated 330
　primary 315, 333
kriging, see interpolation, kriging

L

labels 167
lag tolerance 494
Lambert conformal conic, see map
　　　projection, Lambert
　　　conformal conic
land cover
　NLCD 290
landcover
　NASS 293
LANDIS 552–555
Landsat 255–256
laser 279
　rangefinder 213
latitude 29, 81, 93
LCC, see map projection, Lambert
　　　conformal conic
legends 168
less than (>), see set algebra
leveling
　spirit 96
　trigonometric 96
LiDAR 245, 260–262, 278, 280

discrete return 260
waveform 260
lidar 280
line thinning 147, 148
Lineage 568
linear referencing 395
logical consistency 567
logical database design 313
longitude 29, 81, 93

M

magnetic declination 82
magnetic north 82
majority filter 425
Manifold 16
Manual 144
map
    boundaries 139
    cartometric 161
    choropleth 134
    contour 135
    dot-density 135
    feature 134
    graticule 132
    grid 132
    inset 132
    isopleth 135
    media 143
    minimum mapping unit 249
    neatline 132
    north arrow 132
    registration 140
    scale 132, 135–137, 139, 143
    scalebar 132
    shaded relief 463
    transformation 153
map algebra 408–410, 420
map generalization 137
map projection 71
    azimuthal 105
    conic 105
    cylindrical 105
    developable surface 105
    distortion 101, 107
    gnomonic 106
    Lambert conformal conic 107, 111
    orthographic 106
    state plane 110–113
    stereographic 106
    transforming across 163
    transverse Mercator 107
    UTM 113–115
map registration
    affine 154
    conformal 156
    polynomial 156
MapInfo 15

mapping
    core area 499–508
MAUP, see Modifiable Areal Unit
        Problem
mean absolute error 497
mean circle 499
meridian
    Greenwich 81
    prime 81
metadata 171–580
    Australia and New Zealand 174
    definition 171
    profile 174
MicroImages 17
minimum mapping unit 249, 295
minutes 30
MMU 249, 290, 298
MMU, see minimum mapping unit
model
    agent-based 548
    cartographic 527
    cartographic model detailed
        example 533–539
    cell-based 547
    hydrologic 549
    logical 55
    network 393
    simple spatial 542
    spatial 523
    spatial data and error 565
    stochastic 552
    topological vector 38–42
    traffic 394
models
    LANDIS 552
Modifiable Areal Unit Problem 366–
        367
modifiable areal unit problem 474
MODIS 273
Monte Carlo simulation 579
Moran's I 490–491
morphometric features 450
Most 521
moving window 424
MUIR soils data 296
multipath GNSS signal 191
multi-tiered architecture 311

N

NAD27 88, 112
NAD83 88, 112
    versions 88
NAPP, see National Aerial
        Photography Program
NASS 293
National Aerial Photography
        Program 291

national climatic data center 300
National Elevation Dataset 280
National Geodetic Vertical Datum of
        1929
    see NGVD29
National Hydrologic Dataset 282
National Standard for Spatial Data
        Accuracy 571
National Wetland Inventory 294–295
natural join 327
NAVD88 98
NED, see National Elevation Dataset
neighborhood 424
neighborhood operators 351
network
    allocation centers 393
    center capacity 394
    drainage 454
    route selection 391–393
    transit costs 390
network models 390–395, 521
    path analysis 392
    resource allocation 393
    traffic 394
NGVD29 98
NHD, see National Hydrologic
        Dataset
NLCD 290–291
node 144
nodes 35
north
    geographic 82
    magnetic 82
North American Datum of 1927:see
        NAD27
North American Datum of 1983:see
        NAD83
North American Vertical Datum of
        1988,see NAVD88
north arrow 132
northing 114
    false 115
not equal to (), see set algebra
NOT, see Boolean algebra
NSSDA, see accuracy, NSSDA
NSSDA, see National Standard for
        Spatial Data Accuracy
nugget 494, 495
number
    ASCII 59
    binary 58–59
NWI, see National Wetlands
        Inventory

O
object data model 55
omitted, map generalization 138

on-screen query 352
operation
    clip 420
    global 410
    kernel 427
    logical 413
    majority 425
    moving window 424
    neighborhood 410, 424
    nested 417
    overlay 418
    reclassification 415
    spatial 347–351
operator
    buffering 370
operators
    adjacency 356
    buffer 371
    containment 358
    dissolve 368–369
    global 349, 351
    local 349
    neighborhood 349, 351
    selection 352
OR, see Boolean algebra
ordinal attribute 32
orthometric height 79
orthogonal 28
orthographic
    aerial photograph 237
    aerial photographs 234
    view 133
orthographic projection 106
orthophotograph 276
orthophotographs 287, 288
outer join 327, 328
overlay 377–388, 418
    calculation 386
    clip 384
    intersect 384
    raster 378, 420
    slivers 385
    union 384
    vector 379
overshoots 145

P
parallax 242
Parallels 81
parallels
    standard 107
passive systems 226
PDOP 191
perspective
    center 243
    view 234
perspective convergence 237

photogrammetry 228
  softcopy 245
photointerpretation 247
pit
  fill 456
  sink 456
pits 455
pixel 231
plan curvature 448, 450
plane surveying 201
PLSS, see Public Land Survey System
plumb bob 80
plumb line 201
point thinning 147
pointer 59
pointers 59
polygon
  inclusions 37
polygons 36
precision 568
primary key 315
profile curvature 448
profile path 460
profile plot 460
project 316
projection
  azimuthal 105
  conic 105
  cylindrical 105
  gnomonic 106
  Lambert conformal conic 107, 111
  map 71, 101
  UTM 113, 163–164
  v.s. transformation 163
  vs. transformation 163
proximity functions 370
Public Land Survey System 120–122
pyramids 62

Q

QGIS 17
quad trees 62
query 316, 321
  AND 322
  AND, OR, NOT 322
  NOT 322
  on-screen 352
  OR 322
QuickBird 252

R

RADAR 227, 262
range 188
range measurement in GNSS 188
range pole 210
raster 34, 44–51
  advantages 50

attribute table 48
cell assignment 47
cell dimension 44
cell size 44
compared to vector 48–51
comparison to vector 51
conversion to vector 51
coordinates 44
definition 34
pyramids 62
resampling 162
reclassification 359–366, 415
  binary 360
recoding 359
reference frame 83
  terrestrial 83
registration 140, 153
  control points 153
relational algebra 315
relational data model 43
relief displacement 236
remote sensing 223
  sources 264, 287
resampling 162
  nearest neighbor 162
resource allocation 393
restrict 316
river reach data 284
RMSE 155, 157
rmse 497
RMSE, see root mean square error
root mean square error 573–574, 575
root mean squared error 497
route selection 391–393
run length codes 61
RUSLE 542

S

sampling 475
  adaptive 477
  cluster 477
  for accuracy assessment 573
  random 477
  systematic 475
satellite imagery 250–264
  classification 262
  compared to aerial
    photographs 262
  ETM+ 256
  GEOEYE-2 253
  high resolution 252
  Ikonos 252
  Landsat 255
  QuickBird 252
  Quickbird 252
  RADAR 262
  sources 264

SPOT 254
TM 256
WorldView-2 253
scale 132, 135–137, 139
   aerial photographs 229
   error 143
   non-constant in aerial
      photograph 237
scalebar 132
scanner 149
scope, spatial 349
SDTS 171
seconds 30
selection 352, 356
   adjacency 356
semi-major axis 75
semi-minor axis 75
semi-variance 494
set algebra 352
shaded relief 463
shapefiles 59
Shuttle Radar Topography
      Mission 281
sill 495
similar triangles
   photographs 244
simple spatial model 542
simplified, map generalization 138
sinks 455
skeletonizing 149
slivers 385
slope 445
   3rd-order finite difference 448
   4 cell 447
   calculation 446
   kernel 449
smoothing
   line 147
   raster to vector conversion 52
smoothing, line 148
snapping 145
   distance 146
   line 146
   node 146
softcopy 245
softcopy photogrammetry 245
software 13
   ArcGIS 14
   AUTOCAD 16
   ERDAS 17
   geomedia 15
   GRASS 17
   Idrisi 16
   Manifold 16
   MapInfo 15
   MicroImages 17
   QGIS 17

terrain analysis 464
spatial autocorrelation 488
spatial covariance 429
spatial data analysis 347
spatial data transfer standard 276
spatial fields 488
spatial interpolation 474
spatial model 523
spatial operation 347–351
spatial regression 491
spatial scope 349
spatio-temporal model
   hydrology 549
spatio-temporal models 521–524,
      545–555
   stochasic 552
specific catchment area 458
spectral range 223
spectral reflectance 226
spherical coordinate 29
spirit leveling 96
spline 147
splines 147, 486
SPOT 254
SQL 324
SRTM, see Shuttle Radar Topography
      Mission
SSURGO 295
standard
   media 564
standards
   analysis 561
   certification 561
   documentation 564
   format 564
   media 564
   professional 561
   spatial data 564
state plane coordinate system, see map
      projection, state plane
STATSGO 295
stereo photographs 241
stereographic 106
stereomodel 241
stream power index 458
survey
   bearings 203
   station 203
surveying 159, 201–205
   closed traverse 202
   COGO 204
   control points 159
   geodetic 84
   leveling 96
   open traverse 202
   optical 201
   plane 201

station 202
traverse 202
triangulation 85
symbol size 165

T

table 42
    inner join 327
    join 326
    key 314, 329
    outer join 327, 328
    query 316, 321
    raster 48
    relate 326
    relational 310
tables 307
terrain analysis 443–445
terrain feature 450
thematic layer 27
Thematic Mapper, see TM
theodolite 201
Thiessen polygons 480
TIGER 301
tilt
    camera angles 238
time geographic density
            estimation 509
TIN 34
TIN, see triangulated irregular network
topology 38–42, 357
    adjacency 40
    line 40
    planar 39
    point 40
    polygon 41
    tables 40, 41
transformation 153–158
    coordinate 153
    datum 118–119
    Helmert 92
    Molodenski 92
    vs. projection 163
transit 201
transit costs 390
traverse 202
triangulated irregular network 53, 54
Triangulation survey 85
transformation
    v.s. projection 163
tuple 310
tuples 314
type 310

U

U.S. survey foot 112
undershoots 145
union 316, 384, 384–385

universal soil loss equation 542
universal transverse Mercator, see
            UTM
USLE 542
UTM 113–115

V

variogram 494
vector
    advantages 51
    areas 36
    attribute table 42
    compared to raster 48–51
    conversion to raster 51
    data model 34–42
    definition 33
    smoothing 52
vertex 144
viewsheds 458
visibility 458

W

WAAS 199
watershed 452, 456
web mapping service
weightings 527
wetness index 458
WGS84 88
    versions 88
WMS. See web mapping services
World Geodetic System of 1984:see
            WGS84

Z

zenith angle 74
zonal function 433